TORIAL GUIDE TO FOSSILS

ard Ramon Case

g for fossil specimens is rewarding —
ly if you know what you're looking
d what you've found. Until now,
g a reference that identifies and classi-
ssils to be found around the world has
arder than uncovering the finds them-
   Such extensive publications have
ither extremely expensive, or out of
or years.

orial Guide to Fossils solves this prob-
y covering everything you'd want to
about fossils in a reasonably priced
volume. It will help you quickly de-
a familiarity with the broad fossil
um of both extinct and extant life
   Recently discovered and described
ens rarely presented outside scientific
s are included, complete with typical
" fossils.   This "find" is profusely
ted with *more than 1300 photographs
ne drawings*.   Numerous specimens
een photographed from several per-
ves to give you a multi-dimensional
f the fossils.

nprehensive faunal listing for each fos-
ylum further helps you classify and
gue fossils, and figure captions provide
nore details. The handy *Guide* spans
hing from the lowest forms of life —
rotozoa — to the highest forms of
nals.

torial Guide to Fossils is the most up-
e and complete compendium of facts
ple about each of the varied fossil flora
auna within the rock strata of our
. If you're a researcher, teacher, an
enced amateur, weekend fossil collec-
student, this book will excite you and
ain you, while helping you better
your glimpses into the past.

A
Pictorial
Guide to
Fossils

Books by

Gerard R. Case

Fossil Shark and Fish Remains of North America
Fossils Illustrated
Handbook of Fossil Collecting
Fossil Sharks: A Pictorial Review

A
Pictorial
Guide to
Fossils

Gerard R. Case

VNR VAN NOSTRAND REINHOLD
_____ New York

Van Nostrand Reinhold
115 Fifth Avenue
New York, New York 10003

Van Nostrand Reinhold International Company Limited
11 New Fetter Lane
London EC4P 4EE, England

Van Nostrand Reinhold
480 La Trobe Street
Melbourne, Victoria 3000, Australia

Nelson Canada
1120 Birchmount Road
Scarborough, Ontario M1K 5G4, Canada

16 15 14 13 12 11 10 9 8 7 6 5 4 3

Library of Congress Cataloging in Publication Data

Case, Gerard Ramon.
  A pictorial guide to fossils.

  Includes index.
  1. Paleontology—Pictorial Works. I. Title.
QE714.C32   560'.22'2   81-10504
ISBN 0-442-22651-9   AACR2

FOREWORD

Fossils occupy a unique niche in the world of nature. As relic animal and plant remains, fossils serve as insights into nature's past.

The collecting of fossils can be an exciting and pleasurable experience, whether as a hobby or a profession. Finding a specimen and realizing its age, provides us with a knowledge and appreciation of the progression of life throughout the earth's history.

Consequently, as more people become interested in fossil collecting, they will want to identify, classify, and understand the treasures that they have found. Locating a book or guide that encompasses the multitude of fossils, not only those of one's own region, but those that are found around the world, has been difficult. Such publications have been either expensive, out of print, or lacking the breadth of information needed.

There are many books ranging from very general texts to those containing a complexity of terms and illustrations.

Probably, the most useful to all is a guidebook that contains a simplification of terms and a complement of well-illustrated fossils, and provides the reader with the information desired.

A Pictorial Guide to Fossils by Gerard R. Case is designed to aid the collector in the identification of a broad spectrum of fossils. Like the past publications of Mr. Case (*Fossil Shark and Fish Remains of North America,* 1967; *Fossils Illustrated,* 1968 and 1971; *Handbook of Fossil Collecting,* 1972; and *Fossil Sharks: A Pictorial Review,* 1973) it employs the pictorial aspect of diversity and ages of fossils—worldwide! In conjunction with illustrating the aesthetics of fossils, this present book can serve as a stepping stone to more advanced studies in the myriad wonders of paleontology.

Richard D. Hamell
Rochester, New York

Note: The scientific community is currently shifting from the use of Lower/Middle/Upper to Early/Middle/Late in formation level designations. Either set of terms is acceptable usage.

Synonyms or obsolete generic names are placed in parentheses after the current genus name.

PREFACE

This work is an illustrated presentation of past life forms that inhabited the earth millions of years ago. Whether extinct or extant, their fossils provide a permanent record of their life traces over millions of years.

The illustrations in and of themselves are quite aesthetic. The reader can readily appreciate the beauty in nature's design of the various animals pictured within.

There are many typical forms of fossils presented, as well as "index" or guide fossils, and newly discovered and described specimens rarely presented outside of scientific journals.

Many specimens have been photographed from several perspectives to give the reader a multidimensional view of the fossils. The author has taken special care to secure the finest drawings and photographic illustrations for the book.

A Pictorial Guide to Fossils will be useful to beginners as well as to professional scientists, teachers, and students.

A comprehensive faunal listing for each fossil phylum is included in its respective chapter to aid the reader in classifying and cataloging fossils.

Sufficient text is included as well as figure captions to give more detailed information to the reader.

The book begins with the lowest form of life—the Protozoa (one-celled animal), and continues to the highest form of life—the mammal.

Much cooperation from both professional scientists and amateur collectors has enabled the author to produce this work. The book is dedicated to these people and to those with an inquiring mind.

Gerard R. Case
Jersey City

CONTENTS

A Pictorial Guide to Fossils

TIME SCALE

ERA	PERIOD			EPOCH	MILLIONS OF YEARS AGO	IMPORTANT	EVENTS
CENOZOIC	QUATERNARY			HOLOCENE			
		NEOGENE		PLEISTOCENE	0.01	MODERN MAN	
	TERTIARY			PLIOCENE	2.0	ICE AGE MAMMOTHS MASTODONS HORSES PRIMITIVE ELEPHANTS	
				MIOCENE	6.0	TRUE WHALES PRIMITIVE HOMINIDS	
		PALEOGENE		OLIGOCENE	26	GRASSES GIANT SHARKS PRIMITIVE WHALES	
				EOCENE	36	PRIMITIVE HORSES OREODONTS	
				PALEOCENE	58		
MESOZOIC	CRETACEOUS				65	MAMMALS ABUNDANT DINOSAURS BECAME EXTINCT FLOWERING PLANTS	
	JURASSIC				135	DINOSAURS CONIFERS	
	TRIASSIC				190	CYCADS DINOSAURS PRIMITIVE MAMMALS	
PALEOZOIC	PERMIAN				225	MAMMAL LIKE REPTILES	
	CARBONIFEROUS	PENNSYLVANIAN			280	REPTILES PRIMITIVE REPTILES COAL SWAMPS	
		MISSISSIPPIAN			310	INSECTS AMPHIBIANS CRINOIDS	
	DEVONIAN				345	SHARK-LIKE FISHES ARMORED FISHES	
	SILURIAN				400	PRIMITIVE LAND PLANTS EURYPTERIDS	
	ORDOVICIAN				430	PRIMITIVE FISH	
	CAMBRIAN				500	CORALS TRILOBITES	
PRECAMBRIAN					570	PRIMITIVE MARINE ANIMALS	
					1500	GREEN ALGAE	
					3000	BLUE GREEN ALGAE	
					4500	AGE OF THE EARTH	

R.D.HAMELL   1981

A Pictorial Guide to Fossils

1 **PROTOZOA** (Foraminifera and Fusilinids)

Protozoans are one-celled aquatic organisms, in most cases, microscopic in size. In life, they are composed of a jellylike (protoplasmic) substance. Some protozoans, e.g., foraminifera, secrete a test. Upon death their tests remain as minute shells—with or without surface ornamentation or detail. The common *Amoeba* is one of these sarcodinid protozoa. Fossilized protozoans may represent either animal or plant—it is difficult to determine this for many of the early or ancient species. There are four distinct subphyla of protozoa: Mastigophora (Late Cretaceous to Recent); Sarcodina (Cambrian to Recent); Sporozoa; and Infusoria. Of these, only Mastigophora and Sarcodina are usually represented in the fossil record.

In the subphylum Sarcodina we have the order Foraminifera. These are by far the most common of the fossils of protozoa found in rocks. Essentially marine animals, a few have been found in brackish or fresh waters. The fossil tests range in size from microscopic (0.02 mm) to several inches (110 mm). The Tertiary forms are especially large (e.g.,*Nummulites,* see Fig. 1-6). Most foraminifera contain internal gas chambers which may aid in floating or locomotion of the animal. Upon breaking open a fossil test of a given foram, one can observe these chambers.

Some formations, such as sands of Eocene age, in particular, the Vincentown Formation of central New Jersey, contain an abundance of these tests. At this locality, there are various types of foraminifera found in a bryozoan matrix. A typical species from this formation is *Nodosaria* sp. (see Fig. 1-2) which is usually preserved with at least five or six segments (chambers) intact. Due to the washing techniques in sieves, the originally complete elongate forms can easily be broken. The foraminifera have a diversity of forms and ornamentation, e.g., the leaf-shaped and dorsoventrally flattened *Neoflabellina* sp. (see Fig. 1-1), and the three-dimensional globular forms of *Globigerina* (see Fig. 1-3). Others are coin-shaped and flattened dorsoventrally, such as *Nummulites (Lepidocyclina)* (see Fig. 1-6) and are so abundant as to be the major constituent of some rocks, such as the limestones that comprise the blocks which make up the great pyramids of Egypt.

Another interesting group of Sarcodina which are regularly found as fossils, are the fusulinids. These large rice-grained shaped chambered animals comprise large portions of the Carboniferous limestone in the midwestern portion of the United States of America. These rocks are commonly called "rice-rocks" (see Figs. 1-4 and 1-5).

Fusulinids are most abundant in the middle Pennsylvanian (Desmoinesian) strata of Missouri and eastern Kansas, especially near Topeka. They are also known in smaller numbers in such states as: Ohio, Iowa, Illinois, Oklahoma, Texas, Utah, Colorado, and Wyoming.

1-1. Late Cretaceous. *Neoflabellina* sp. A quite distinctive leaf-shaped foraminifera test. Marlbrook Marl, Taylor Group, Howard County, Arkansas. ×40. (SEM).

1

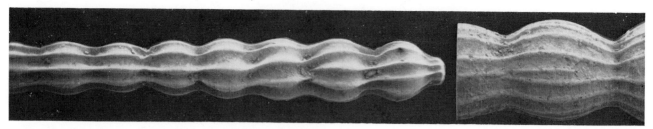

1-2. Late Cretaceous. *Nodosaria* sp. Elongate, ribbed and chambered foraminifera common to the Late Cretaceous to Early Middle Eocene. Navarroan Group, Hunt County, Texas. Closeup specimen from the Vincentown Formation (Early Middle Eocene) of Central New Jersey—for comparison. Both specimens approx. ✕ 20. (SEM).

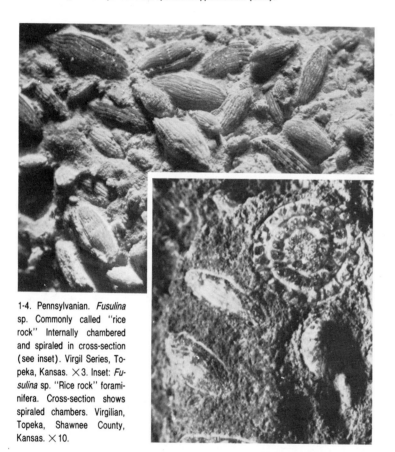

1-4. Pennsylvanian. *Fusulina* sp. Commonly called "rice rock" Internally chambered and spiraled in cross-section (see inset). Virgil Series, Topeka, Kansas. ✕ 3. Inset: *Fusulina* sp. "Rice rock" foraminifera. Cross-section shows spiraled chambers. Virgilian, Topeka, Shawnee County, Kansas. ✕ 10.

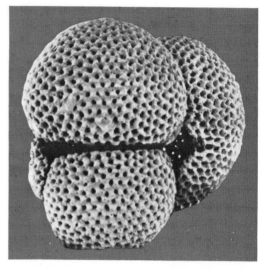

1-3. Late Cretaceous. *Globigerina cretacea* Orbigny. Globular, chambered tests cemented together, with pores, and, on some species, spines extending from the tests. Navesink Formation, Atlantic Highlands, Monmouth County, New Jersey. ✕ 60. (SEM).

1-6. Eocene. *Nummulites ghizensis*. Flattened test with faint tracings of radial ridges and internally chambered. Hundreds upon hundreds of thousands of these camarid forams make up the building stones of the great pyramids. From a limestone block of the Pyramid Chephren, (Ghizeh), Cairo, Egypt. Approx. 3 cm.

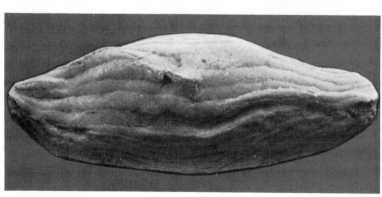

1-5. Pennsylvanian. *Fusulina* sp. An isolated foram called "Rice rock" from the Middle Pennsylvanian (Desmoinesian), Henry County, Missouri. ✕ 17. (SEM).

2 PORIFERA (Sponges)

The porifera or parazoa (sponges) represent one of the earliest stages of coordination between individual cells, and made their first appearance in our seas over one billion years ago.

They are aquatic animals, mostly marine, funnel or vase shaped, and in life have a covering supporting a calcareous or siliceous skeleton. The skeleton and its associated spicules are composed of a series of penetrating canali (for water circulation). The water escapes through openings on the upper or outer surface of the animal. The porifera are bottom dwelling (benthonic) and attached by their bases (sessile) to the sea floor. Calcareous varieties of porifera are found at the bottom of shallow (neritic) seas, while siliceous forms tend to reside in deeper waters, often to abyssal depths (below 610 meters). The spicules of the porifera solidify and become an integral part of the main skeleton. They form a fenestration or trellis (framework) to the skeleton. Sponges vary in size from almost microscopic (some approximating the head of a pin) to over 1 meter in width and/or height. In general, fossil sponges are found in a rather poor state of preservation.

Fossil sponges have been found as far back as the Lower Cambrian, while isolated spicules of sponges have been recovered from Precambrian rocks. Siliceous sponge fossils are far more common to the Paleozoic, while fossil calcareous sponges are only traced as far back as the Ordovician. The latter types are far more abundant in the Cretaceous and Tertiary. No combination of siliceous and calcareous spicules in sponge skeletons has ever been found in association with the same individual animal, and it is unlikely that they ever will. It would seem that the porifera possess either a calcareous or a siliceous skeletal structure.

The fossil listing for the silicispongiae (siliceous sponges) is as follows (genera): CAMBRIAN: *Chancelloria, Choia, Eiffelia, Leptomitus, Multivasculatus, Protospongia,* and *Vauxia.* ORDOVICIAN: *Archaeoscyphia, Astylospongia, Branchiospongia, Dystactospongia, Eospongia, Hindia, Hudsonospongia, Kiwetinokia, Lissocoelia, Nevadocoelia, Petrosites,* and *Zittella.* SILURIAN: *Astylospongia.* DEVONIAN: *Dictyospongia, Hydno-*

ceras, and *Prismodictya.* MISSISSIPPIAN: *Hindia* and *Titusvillia.* PENNSYLVANIAN: *Rhakistella.* TRIASSIC: *Stellispongia.* JURASSIC: *Pachyteichisma* and *Stellispongia.* CRETACEOUS: *Coeloptychium, Coscinopora,* and *Ventriculites.* Isolated siliceous spicules: PENNSYLVANIAN: *Asteratinella, Geodites, Hyalostellia,* and *Reniera.*

Listing for the calcispongiae (calcareous sponges): ORDOVICIAN: *Camarocladia* and *Nidulites.* PENNSYLVANIAN: *Amblysiphonella, Cotyliscus, Cystauletes, Girtycoelia, Heterocoelia* (see Fig. 2-3), *Maenadrostia, Talpaspongia,* and *Wewokella.* TRIASSIC: *Corynella.* CRETACEOUS: *Barroisia* and *Corynella.*

In Incertae Sedis (spongelike fossils which may or may not be related to the porifera): *Receptaculites* (see Fig. 2-2) and *Ischadites,* are noteworthy.

The sponge borer (shell borer) *Cliona* (see Figs. 2-4 and 2-5) is a most curious fossil. The form genus of *Cliona cretacica* Fenton and Fenton appears to have left tubes extending into the shells of host mollusca. These sponge-like colonies were indeed carnivorous, and they have attached themselves to, and replaced entire shells of molluscans (see Fig. 10-24 in the Chapter Mollusca-Pelecypoda, and Fig. 2-4 of this chapter, where the sponge borer *Cliona* has taken the place of the internal aragonitic shield of a belemnoid cephalopod). Now on the other hand, the boring pelecypod *Pholas* (see Fig. 10-20) drills round even holes into belemnoid shields and shell surfaces (see Fig. 2-6).

Present-living sponges are in demand mostly due to their high absorption of water, and a large industry has arisen in the gathering and preparation of these modern porifera in the Mediterranean sea as well as in the shallower port areas near the coast of Florida.

Note: The taxonomic position of the calcareous colonial animals known as stromatoporoids has been debated as to their close affinity to the corals (coelenterata) and the porifera. Recent work by Hartman and Goreau (1970) and Stearn (1972) suggests that stromatoporoids are more closely related to the sclerosponges. Stromatoporoid specimens are illustrated in this book on pages 212 and 213, in the chapter on Plants.

3

2-1. Ordovician. Siliceous sponges (species indeterminate). Weathered out of marble. From Sam Creek, near Upper Takaha, New Zealand. Approx. 4 cm. Courtesy: Geology Museum, Victoria University, Wellington, New Zealand.

2-2. Devonian. *Receptaculites australis.* An entire colony of extinct colonial animals related to the sponges. From the Baton River, West of Nelson, New Zealand. Approx. 12 cm. Courtesy: Geology Museum, Victoria University, Wellington, New Zealand.

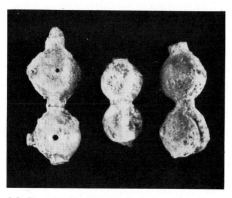

2-3. Pennsylvanian. *Heterocoelia beedei* Girty. Silicified sponges from the Graford Formation, Bridgeport, Wise County, Texas. Largest specimen 1.5 cm.

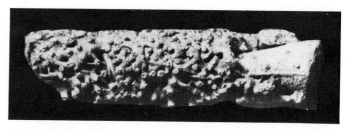

2-4. Late Cretaceous. *Cliona retiformis* Stephenson. A sponge borer. It attached itself to the shield of the cephalopod *Belemnitella* and replaced the aragonite shield. Note Phragmacone ''plug'' at the right end of the specimen. Navesink Formation (Middle Maestrichtian), Atlantic Highlands, Monmouth County, New Jersey, 5 cm in length.

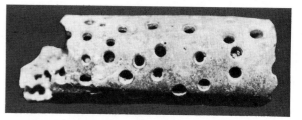

2-6. Late Cretaceous. *Cliona* sp.? (Probably the borer pelecypod: *Pholas,* see Fig. 10-20.) caused these evenly drilled holes into the aragonite shield of the cephalopod *Belemnitella americana* (Morton), not replacing it as is the case with *Cliona retiformis* Stephenson (see Fig. 2-4 for comparison). Navesink Formation (Middle Maestrichtian), Monmouth Group, Hornerstown, Monmouth County, New Jersey. Approx. 3.5 cm in length (fragment).

2-5. Late Cretaceous. *Cliona cretacica* Fenton and Fenton. Boring clionoid sponge commonly attaching itself to the shells of *Exogyra* and *Pyncnodonte* (see Figs. 10-22, 10-26, and 10-27). Navesink Formation (Middle Maestrichtian), Holmdel, Monmouth County, New Jersey. ×1.5.

3 COELENTERATA (Corals, Conularids, Jellyfishes and Graptolites)

The largest part of the coelenterate group is, of course, the corals. These ubiquitous aquatic animals form colonies and, in death, their skeletons build up massive lime deposits (limestone bioherms). The designs of corals are as diversified as the number of species.

The coelenterates are not just restricted to the corals alone. This large group also includes the medusae ("comb-jellies" or "jellyfishes"). The conularids (popularly called "cone-in-cone" fossils) are in the subphylum Cnidaria.

The Graptozoa (graptolites) are, at present, assigned to the phylum Hemichordata, but they are included in this chapter as a temporary convenience.

Class Anthozoa (corals): The class Anthozoa (corals) is divided into essentially three subclasses: Ceriantipatharia, Octocorallia, and Zoantharia. Ceriantipatharia and Octocorallia are virtually unknown in the fossil record. We will be concerned here with the Zoantharia and will list the rugose, tabulate and scleractinid corals. The Zoantharia is made up of simple and/or compound corals with septa (radial walls of the coral) that are imperforated. The rugose corals (also known as "horn corals") have epitheca ("wrinkled surfaces"). Their range is Ordovician to Permian. The faunal listing for the rugose corals is as follows (genera): ORDOVICIAN: *Favistella (Columnaria), Lambeophyllum,* and *Streptelasma.* SILURIAN: *Acanthophyllum (Strombodes), Amplexus, Arachnophyllum, Chonophyllum, Cystiphyllum (Conophyllum), Diplophyllum, Disphyllum, Duncanella, Entelophyllum, Enterolasma, Fletcheria, Porpites, Ptychophyllum,* and *Rhabdocyclus.* DEVONIAN: *Acrophyllum, Amplexiphyllum, Amplexus, Aulacophyllum, Billingsastrea, Blothrophyllum, Calceola, Charactophyllum, Cyathophyllum* (see Fig. 3-1), *Cylindrophyllum, Disphyllum, Diversophyllum, Enterolasma, Eridophyllum* (see Fig. 3-2), *Grypophyllum, Hadrophyllum, Hallia, Heliophyllum* (see Figs. 3-5 and 3-8), *Hexagonaria, Homalophyllum, Meseophyllum (Cystiphylloides), Microcyclus, Neozaphrentis, Pachyphyllum, Siphonophrentis, Stereolasma, Synaptophyllum, Tabulophyllum* (see Fig. 3-14), *Tortophyllum, Triplophyllum (Heterophrentis),* and *Zaphrentis.* MISSISSIP-PIAN: *Baryphyllum, Carcinophyllum (Axophyllum), Dibunophyllum, Dipterophyllum, Hapsiphyllum, Hexagonaria, Lithostrotion, Lophophyllidium, Lophophyllum, Neozaphrentis* (see Fig. 3-17), and *Palaeosmilia.* PENNSYLVANIAN: *Amplexus, Axophyllum, Canina, Lithostrotion, Lithostrotionella, Lophamplexus, Lophophyllum* (see Fig. 3-18), and *Neozaphrentis.* PERMIAN: *Duplophyllum, Heritschia, Leonardophyllum, Lophamplexus, Lophophyllum* (see Fig. 3-20), *Malonophyllum,* and *Waagenophyllum.*

Tabulate corals have long and slender corallites with imperforate walls and a number of horizontally arching tabulae. The corallites divide and give rise to new corallites. Their range is Ordovician to Mesozoic. The artificial subdivision of the Anthozoa is Alcyonaria—tabulate corals that have calcareous corallum with rough spicules.—Tubular outgrowths of the zooids (animal living chambers) with "budding" taking place. Range: Ordovician to Recent. Examples of the alcyonarians are: *Favosites* (see Figs. 3-3, 3-4, and 3-6), *Halysites* (see Fig. 3-7), and *Syringopora* (see Fig. 3-10). The faunal listing for the tabulate corals is as follows (genera): ORDOVICIAN: *Billingsaria, Calapoecia, Chaetetes, Favosites, Halysites, Palaeofavosites, Propora,* and *Tetradium.* SILURIAN: *Alveolites, Coenites, Halysites, Palaeofavosites, Plasmopora, Propora, Romingeria, Striatopora, Syringopora, Thamnopora,* and *Thecia.* DEVONIAN: *Alveolites* (see Fig. 3-12), *Aulocaulis, Aulocystis (Ceratopora), Aulopora, Chonostegites, Cladochonius, Coenites, Emmonsia, Favosites* (see Figs. 3-3, 3-4, and 3-6), *Halysites* (see Fig. 3-7), *Heliolites, Plasmopora, Pleurodictyum* (see Fig. 3-9), *Romingeria, Striatopora, Syringopora* (see Fig. 3-10), *Thamnopora* (see Fig. 3-11), and *Trachypora.* MISSISSIPPIAN: *Cladochonius, Thamnopora, Striatopora,* and *Syringopora.* PENNSYLVANIAN: *Aulopora* (see Fig. 3-19), *Cladochonius, Striatopora, Syringopora,* and *Thamnopora.* PERMIAN: *Cladochonius, Striatopora,* and *Thamnopora.*

The scleractinids are simple-composite corals with their septa arranged in six cycles. Their geological range is

5

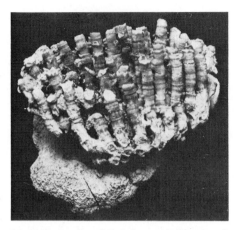

3-1. Middle Devonian. *Cyathophyllum robustum* Hall. Onondaga Shale, Oran Valley, New York. ×2.

3-2. Middle Devonian. *Eridophyllum seriale* Edwards and Haime. Onondaga Limestone (glacial drift), Geneseo, New York. Silicified specimen. 7.5 x 15 cm.

3-3. Middle Devonian. *Favosites* sp. Onondaga Shale, Oran Valley, New York. ×3. An example of "honey-comb" coral.

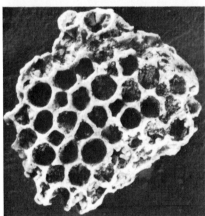

3-4. Middle Devonian. *Favosites* sp. Onondaga Limestone (glacial drift). Collected in Pleistocene gravels, Holmdel, New Jersey. ×1.5.

3-5. Middle Devonian. *Heliophyllum confluens* Hall. Hamilton Shale, East Avon, New York. Each corallite approximately 5 cm.

3-6. Middle Devonian. *Favosites radiatus.* Ludlowville Formation, Hamilton Group, East Bethany, New York. 6 cm in width.

Triassic to Recent. Faunal listing is as follows (genera): TRIASSIC: *Complexastrea (Confusastrea), Elysastraea, Montlivaltia, Palastraea, Thamnasteria,* and *Thecosmilia.* JURASSIC: *Astrocoenia, Complexastrea (Confusastrea), Dungulia, Elysastraea, Montastrea, Montlivaltia, Thamnasteria,* and *Thecosmilia.* CRETACEOUS: *Astrocoenia, Axosmilia, Blothrocyathus, Cladophyllia, Diploastrea, Dungulia, Flabellum, Goniopora, Lophosmilia, Micrabachia, Montastrea, Montlivaltia, Parasmilia, Pleurocora, Thamnasteria, Thecosmilia, Tiarasmilia,* and *Trochocy-* *athus.* PALEOCENE: *Flabellum* (see Fig. 3-20), and *Trochocyathus.* EOCENE: *Acrohelia, Balanophyllia* (see Fig. 3-25), *Discotrochus, Endopachys, Flabellum, Haimesiastraea, Platytrochus, Trochocyathus,* and *Turbinolia.* OLIGOCENE: *Acrohelia, Antiguastrea, Astreopora, Astrocoenia, Diploastrea,* and *Montastrea.* MIOCENE: *Astrhelia* (see Figs. 3-23 and 3-24), and *Septastrea.* PLIOCENE: *Septastrea.*

Notes on corals: The total Upper Devonian coelenterate faunal list from the "B" Member of the Hay River For-

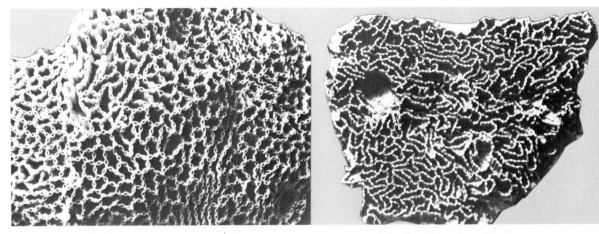

3-7. Middle Devonian. *Halysites catenularia* (Linnaeus). Onondaga Limestone, Falls of the Ohio, Indiana. Both specimens approximately 15 cm.

3-8. Middle Devonian. *Heliophyllum halli* Edwards and Haime. Hamilton Shale, Arkona, Ontario, Canada. ✕2.

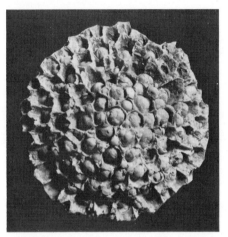

3-9. Middle Devonian. *Pleurodictyum convexum* (Orbigny). Hamilton Shale, Pt. Colbourne, Ontario, Canada. Approx. 8 cm. in diam.

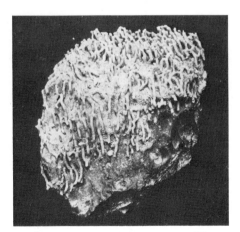

3-10. Middle Devonian. *Syringopora perelegans* Billings. Onondaga Limestone (glacial drift), Geneseo, New York. Specimen silicified. 8 x cm.

mation, near Enterprise, Northwest Territories, Canada, is as follows (from Lasmanis, 1971): *Alveolites multiperferatus* Lecompte, *Aulopora* sp., *Charactophyllum* sp., *Macgeea proteus* Smith, *Phillipsastrea macouni* Smith, *Stromatopora* sp., *Tabulophyllum mcconnelli* (Whiteaves), *Thamnopora cervicornis* Blainville, and *Thamnopora* cf. *polyforata* (Schlotheim).

The Upper Devonian coelenterate *Plumalina* (see Fig. 3-16) may not be part of the Anthozoa (corals). Recent work by Sass and Rock (1975) indicates the possible affinity of *Plumalina* with the hydroids.

Hexagonaria, a rugose coral from the Devonian/Mississippian is an interesting form. Agatized specimens are known as "Petoskey stones" from the shores of Lake Michigan at Petoskey, Emmet County, Michigan.

The alcyonarian tabulate corals *Favosites* (see Figs. 3-3, 3-4, and 3-6) and *Halysites* (see Fig. 3-7) are commonly known as "Honey-comb corals" and "Chain corals," respectively.

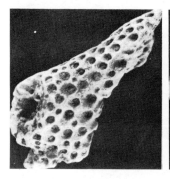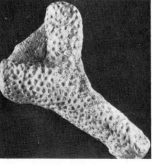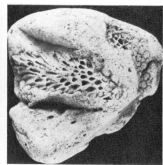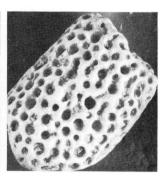

3-11. Middle Devonian. *Thamnopora* sp. Onondaga Limestone (glacial drift). Collected in Pleistocene gravels, Holmdel, Monmouth County, New Jersey. All specimens approx. ×1.

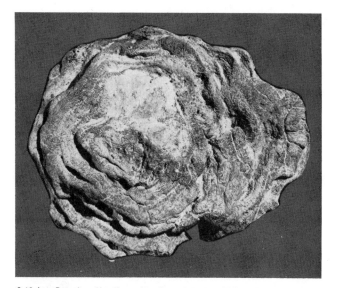

3-12. Late Devonian. *Alveolites multiperforatus* Lecompte. "B" Member, Hay River Formation, Enterprise, Northwest Territories, Canada. 5 cm in width.

3-13. Late Devonian. *Phillipsastrea macouni* Smith. "B" Member, Hay River Formation, Enterprise, Northwest Territories, Canada. ×1.5.

3-14. Late Devonian. *Tabulophyllum mcconnelli* (Whiteaves). "B" Member, Hay River Formation, Enterprise, Northwest Territories, Canada. ×1.

3-15. Late Devonian. *Macgeea proteus* Smith. "B" Member, Hay River Formation, Enterprise, Northwest Territories, Canada. ×1.5.

3-16. Late Devonian. *Plumalina plumaria* Hall. "Feather coral." Genesee Group, Ithaca, New York. ×1.25.

3-17. Mississippian. *Neozaphrentis tenella* (Miller). Oolite Limestone Beds, South of Bloomington, Indiana. 3.5 cm.

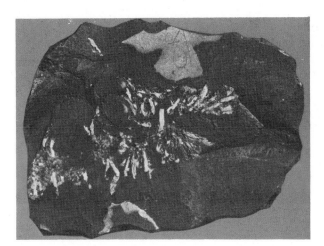

3-19. Middle Pennsylvanian. *Aulopora* sp. on the plant *Calamites*. Stark Shale Member, Dennis Limestone Formation, Bronson Group, Missourian Series, Crescent, Pottawatamie County, Iowa. 15 cm in length.

3-18. Pennsylvanian. *Lophopyllum profundum* Edwards and Haime. Graham Formation, Wayland Shale, Coleman County, Texas. 5.5 cm.

3-21. Late Cretaceous. *Cyclolites ellipticus* Lamarck. Anthozoan—large coral head. Santonian Stage, Montsech de Meyá, Lerida, Spain. ×1.5.

3-20. Paleocene. *Flabellum atlanticum*. Left: top view, right: side view of a coral. Hornerstown Formation (Danian), Cream Ridge, Monmouth County, New Jersey. ×2.

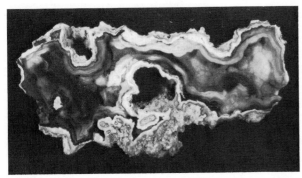

3-22. Miocene. Tampa Bay Coral (replaced by banded agate and drusy quartz in the vugs or openings). Ballast Point, Tampa Bay, Hillsborough County, Florida. 16.5 cm.

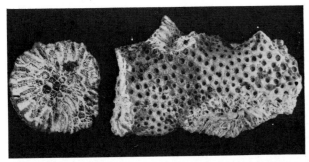

3-23. Late Miocene. *Astrhelia palmata* (Goldfuss). Branch of colonial coral. Yorktown Formation (Sahelian), Beaufort County, North Carolina. 4 cm in diameter, and 8 cm in length.

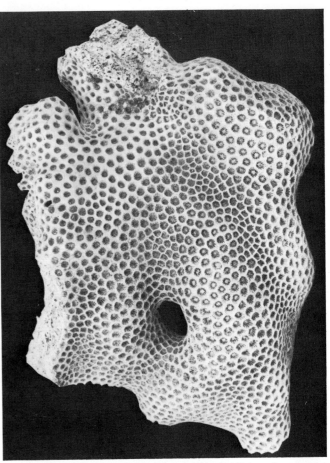

3-24. Late Miocene. *Astrhelia palmata* (Goldfuss). Colonial coral. Yorktown Formation (Sahelian), Beaufort County, North Carolina. Approx. 14 cm in length.

3-25. Eocene. *Balanophyllia desmophyllum* Edwards and Haime. Anthozoan corallum. Paspotansa Member of the Piscataway, Maryland. 3.5.

Class Scyphozoa (medusae/jellyfishes) and class Hydrozoa (siphonophores/hydrozoans): The most curious of the coelenterates ("hollow-gut" animals), are the medusa and the hydra (the jellyfish and hydrozoan). Such soft-bodied animals are hard to imagine being found in a fossil state, preserved in the rocks of the ages. But, nevertheless, it is true, some of these fragile creatures do preserve as witnessed by the recent discoveries of medusae in the ubiquitous siderite concretions found in the northern Illinois coal-mining regions. A good example of a fossil "jellyfish" is *Octomedusa* (see Fig. 3-26). This is a soft-bodied animal with eight arms. Newly discovered and perfectly preserved hooded medusae of the genus *Essexella* (see Fig. 3-27) are turning up in the tailings of Pit 11 in Will County, Illinois by the hundreds. They are there to see: beautifully preserved soft-bodied creatures. Hydrozoans are even rarer as fossils. They have, nevertheless, been turning up also at Pit 11 in the siderite concretions.

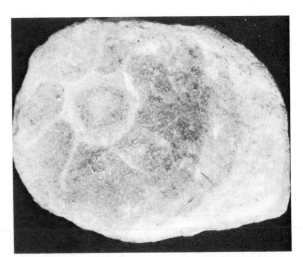

3-26. Pennsylvanian. *Octomedusa pieckorum* Johnson and Richardson. "Jellyfish" (medusae) in an ironstone (siderite) concretion. Essex (marine) fauna, Mazon Creek, Grundy County, Illinois. ×2. Courtesy: Field Museum of Natural History, Chicago.

3-27. Pennsylvanian. *Essexella asherae* Foster. A "hooded" medusa ("jellyfish") preserved in a half of a siderite concretion. Essex (marine) fauna, Mazon Creek, Will County, Illinois. Slightly less than actual size.

The hydra lives generally in fresh water (ponds, streams, and lakes) where it is attached by its basal disk, but not permanently anchored to the bottom, thus giving it the mobility it needs to search out food. The hydra lives on insect larvae and microscopic crustaceans. Some species of hydra are indiginous to a marine environment, but the fossils of marine forms are poorly preserved—if known at all.

Conversely, the medusae are "free agents" that manage to raise and lower themselves by means of gas bladders. These gas bladders also enable the creature to anchor itself at the surface of the water where it awaits its prey. The medusae sting with needle-like cells (nematocysts) attached to their tentacles or tendrils, which hang suspended from their bodies, thus "stunning" their potential food sources. As is the case with the hydrozoans, the medusae are hermaphroditic.

Conularia or "cone-in-cone": The problematical conularids are temporarily classed with the Scyphozoa—the true medusae or jellyfishes—in the subphylum Cnidaria of the phylum Coelenterata. They are curious animals, pyramidal-shaped (usually four-sided), with finely distributed transverse striations or markings (see Figs. 3-28 and 3-29).

In 1821, Sowerby described the first genus name *Conularia* which gave rise to the family name. General scientific opinion has it that the conularids are not sponges or echinoderms. They are, however, found in direct association in many marine deposits with: sponges, bryozoans, mollusks, corals, echinoderms, brachiopods, and arthropods (trilobites).

Some scientists believe that the Conulariida should be classed in a group of their own, while others believe that

3-28. Pennsylvanian. *Conularia* sp. "Cone-in-cone." Graham Formation, Wayland Shale, Wise County, Texas. 2 cm in length.

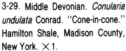

3-29. Middle Devonian. *Conularia undulata* Conrad. "Cone-in-cone." Hamilton Shale, Madison County, New York. ×1.

3-30. Middle Silurian. *Mongraptus clintonensis* (Hall). Rhabdosomes with only a single row of "recurved" thecae (bent points!). Williamson Shales, Rochester, New York. ×5.

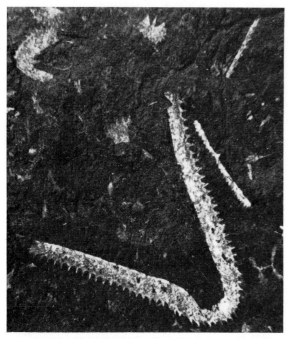

3-31. Ordovician. *Isograptus* sp. Marine graptolites preserved in mud (black) shale, fragments of rhabdosomes with pronounced theca. Aorangi Mine, Northwest Nelson, New Zealand. 3.5 cm. Courtesy: Geology Department, Victoria University, Wellington, New Zealand.

3-32. Ordovician. *Loganograptus* sp. Remains of a graptolite colony showing radiating rhabdosomes with pronounced theca. Aorangi Mine, Northwest Nelson, New Zealand. 4 cm across whole specimen. Courtesy: Geology Department, Victoria University, Wellington, New Zealand.

they are an extinct phylum related to the Annelida (worms).

They have been found in the fossil state in all types of rock matrices (limestones, shales, and sandstones). There are presently 125 described species. The range of the Conulariida is from the Early Cambrian to the Triassic. No living form seems to be even distantly related to this group of organisms.

Graptolites: The Graptozoa or "graptolites," as they are more commonly known, are colonies formed by the unison of individual rhabdosomes with extending branches housing the zooids (animals) called thecae. The colony is asexual and the animals reproduce by the "budding" process. The sicula is the central conical cusp from which the colony extends its radiation.

Most graptolites were free-floating (vagile or planktonic), while certain species were sessile or benthonic (attached to the sea floor). Graptolites are essentially found in all types of sedimentary rocks, especially in black shales (where they are best preserved). Their stratigraphic range was short-lived (Cambrian to Mississippian). Their geographic range, however, was world-wide (cosmopolitan). Their remains are most abundant in the Ordovician and common during the Silurian period, and are important in correlation.

There are two major orders for the graptolites—the Dendroidea and the Graptoloidea. There are 21 known

genera for the sessile forms, while the planktonic or epi-planktonic types have 108 known genera. See Figures 3-30 through 3-33 for examples of the graptolites.

3-33. Ordovician. *Climacograptus bicornis* (Hall). Isolated rhabdosomes. Kings Falls Limestone, Trenton Group, St. Johnsville, Montgomery County, New York. 2.5 cm.

4 ECHINODERMATA (Cystoids, Blastoids, Edrioasteroids, Crinoids, and Starfishes)

The phylum Echinodermata is divided into four groups (subphyla): the Homalozoa, Crinozoa, Asterozoa, and Echinozoa. We are primarily interested here only in the subphyla Crinozoa and Echinozoa. The subphylum Crinozoa comprises the following classes: Cystoidea (the cystoids), Edrioasteroidea (the edrioasteroids), Blastoidea (the blastoids), and Crinoidea (the crinoids). The subphylum Echinozoa is composed of the classes: Stelleroidea (starfishes)—with its three subclasses: Asteroidea, Auluroidea, and Ophiuroidea; Echinoidea (the sea-urchins and sand-dollars)—which will be discussed separately in the following chapter; and the Holothuroidea (the sea-cucumbers). Of the seven classes and three subclasses of the phylum Echinodermata, the following are extinct: classes, Cystoidea, Edrioasteroidea, and Blastoidea; and the subclass Auluroidea.

Class Cystoidea: The cystoids are marine organisms with attached stems and a calyx (like the crinoids), but without perfect arms. These forms existed only during the Paleozoic era and were more commonly distributed during the Ordovician (when they reached their peak) and Silurian periods.

The calyx of the cystoid has sutured polygonal plates (varying in some species up to several hundreds and as few as ten plates in others—e.g., *Calix* sp. has about 2000 small plates) comprising the whole head, while there is a regular series of basal plates attached to the column (stem). The mouth of the animal is located near the center of the ventral side of the calyx and has (on some specimens) 2 or more branching ambulacra (food grooves) and small plates near the aperture and the ambulacra. Vestigial unbranched arms (if at all present in some species) are connected to the distal end of the ambulacra. These arms are composed of either uniserial (single) or biserial (double) row or rows of plates with a ventral groove covered by small plates. There exists an irregularly placed anal opening just below the mouth area. The calyx has small pores, lozenge-shaped, which appear on one half of each of two or more adjoining plates-forming a "rhomb." The pore-rhombs of *Caryocrinites* (see Fig. 4-1) are quite

distinctive and show a transverse striation. Other typical cystoid genera (all from the Lower to Upper Ordovician) are *Echinosphaerites, Heliocrinites, Paleosphaeronites,* and *Sphaeronites.*

Cystoids are probably the rarest of the preserved fossils in the phylum Echinodermata. Their stratigraphic range is from the Lower Ordovician to the Upper Devonian.

Class Edrioasteroidea (Lower Cambrian to Mississippian): Edrioasteroids have a large theca (sack-like body) with a "starfish" shaped ambulacra on the aboral (dorsal) part of the theca. The mouth of the animal is located in the center of the ambulacral grooves of the ambulacra. The anal area is located below the mouth area on the adoral (ventral) surface.

A typical edrioasteroid is *Cooperidiscus* (Upper Devonian, see Fig. 4-7a and 7b), which shows the ambulacra in a circular form. The particular specimens shown do not have enough clear detail for the reader to see the mouth or the anal areas. The specimens are covered by thin layers of matrix. When properly prepared, these anatomical details can be uncovered. This group of invertebrates was restricted to the Paleozoic era.

Class Blastoidea (from the Greek: Blastus = bud) (Silurian to Permian): Blastoids have an ovoid calyx attached to a stem or column (although the stems or columnals are rarely preserved attached to the calyxes in specimens). The calyx has no radiating arms coming from the mouth area, but rather, in some species, a set of thread-like branching arms that extend out from either side of the midline (waist) of the calyx. These arms are attached to arm bases on the radial plates. The entire calyx consists of the uppermost plates—the distals (or deltoids) including the ambulacra containing the centrally located mouth area; the radials which are approximately at the waistline; and the lower and larger plates, the basals, with their attachments (by stem ossicles) to the stems or columns. The design of the ambulacra varies from a linear pattern or form (see Fig. 4-3) to the more common "petal-like" (petaloid) form of *Pentremites* (see Fig. 4-5). Rarely well preserved in the fossil record on specimens of blastoids are

4-1. Middle Silurian. *Caryocrinites ornatus* Say. Ornate cystoid head. Rochester Shale, Niagara Gorge, New York. ×1.25.

4-2. Silurian. *Troosticrinus reinwardti* (Troost). Blastoid crown head. Brownsport Formation, Niagaran Series, Decatur County, Tennessee. 17 mm.

4-3. Devonian. *Nucleocrinus bondi* Thomas. Blastoid crown head. Sheffield Formation, Cedar Valley Limestone, Cedar Rapids, Johnson County, Iowa. 2.5 cm.

4-4. Mississippian. *Orophocrinus fusiformis*. An isolated blastoid crown head with branchioles. Kinderhook Series, Le Grand, Iowa, 4 cm. Courtesy: Geological Enterprises, Ardmore, Oklahoma.

4-5. Mississippian. *Pentremites* sp. Three views of an isolated blastoid crown head—clockwise from left to right: Bottom view, top view, and profile view. Oolitic Limestone Beds, Bedford, Lawrence County, Indiana. 1.5 cm.

4-6. Mississippian. *Pentremites godoni* (DeFrance). An isolated blastoid crown head with nearly complete branchioles. Illinois. 4.5 cm. Courtesy: Geological Enterprises, Ardmore, Oklahoma.

the branchioles, composed of tiny rodlets extending upward and outward from the margins of the ambulacra area (see Figs. 4-4 and 4-6). Other typical blastoid genera: DEVONIAN: *Cordyloblastus, Devonoblastus,* and *Nucleocrinus* (see Fig. 4-3); MISSISSIPPIAN to PENNSYLVANIAN: *Pentremites*; and LOWER CARBONIFEROUS: *Nymphaeoblastus* and *Phaenoschisma.*

Class Crinoidea (Lower Ordovician to Recent): The crinoids are the largest diversified group of the phylum Echinodermata, and by far some of the most popularly collected fossils next to cephalopods and trilobites. The crinoids, or as they are popularly called "sealilies," are strictly marine animals. Most have calyxes attached to stems (columns), while others are free-swimming forms such as *Anteodon* (see Fig. 4-52) and *Saccocoma* (Upper Jurassic to Lower Cretaceous, see Fig. 4-53). We will discuss first the attached forms of which there are many fossil as well as modern types. The crinoid has both soft and hard parts to its anatomical makeup. For the purposes of this book, which deals only with fossil forms, we will discuss the hard parts or those parts which are generally found preserved in fossil specimens.

The calyx (cup) is that part of the crinoid theca which is located between the stem and the origin of the free arms. The calyx is made up of many crystalized (calcite) individual plates which support the body of the animal. The entire area comprising the calyx and pinnules (which come in uniserial and biserial arms), etc., is called the head or the crown of the crinoid. The lowermost part of the crinoid's calyx is called the basal cup. The arms of the crinoid are attached to the top of the basal cup. The basal cup in turn is attached to the stem or column. The column has branches or smaller columnals called cirri, as well as several branches at the base of the column where they act as holdfasts to anchor the entire animal to the ocean floor (see Fig. 4-20). In cross-section, the ossicles or discs have either elliptical, stelliform, or pentagonal designs; while in center design the majority are circular, and a few are

4-7. Late Devonian. *Cooperidiscus alleganius* (Clarke). Edrioasteroids. Conowango Group, Schoharie, Schoharie County, New York. 8 and 6 cms.

quadrangular or crescentic. The designs vary with the differing species. For an example of the stelliform type see Fig. 4-51.

Crinoids (Comatulida, the featherstars) are still present in our modern seas—there are estimates that several hundred species exist today. (In the fossil record, the figure is 5500 species as represented by over 1000 genera.) Recent crinoids are not considered to be typical since they are "vagile," stalkless, and lacking a complete armory of tegmen and ambulacra. Most modern crinoids have been found in very shallow waters and in depths of from 4,000 to 9,000 meters. They are found in tropical and temperate zones, and in the frigid waters of both the arctic and antarctic. The earliest forms of fossil crinoids were discovered in rocks of Ordovician age.

Faunal listing for crinoids (genera): ORDOVICIAN: *Alisocrinus, Anomalocrinus, Archaeocrinus, Canistrocrinus (Tanaocrinus), Carabocrinus, Composocrinus, Cremacrinus, Cupulocrinus, Daedalocrinus, Dendrocrinus, Diabolocrinus, Drymocrinus, Dystactocrinus, Ectenocrinus, Glyptocrinus, Heterocrinus, Hybocrinus, Hybocys-*

tites, Iocrinus, Isotomocrinus, Merocrinus, Ottawacrinus, Palaeocrinus, Periglyptocrinus, Porocrinus, Protaxocrinus, Ptychocrinus, Reteocrinus, Rhaphanocrinus, Sygcaulocrinus, and *Xenocrinus.* SILURIAN: *Alisocrinus, Allocrinus, Ampheristocrinus, Anisocrinus, Asaphocrinus, Botryocrinus, Calliocrinus, Carpocrinus, Clidochirus, Clonocrinus, Cremacrinus, Calceocrinus, Crotalocrinites, Cyathocrinites, Cyliocrinus, Cytocrinus, Dendrocrinus, Dimerocrinites, Emperocrinus, Eucheirocrinus (Calceocrinus), Eutaxocrinus, Gazacrinus, Gissocrinus, Glyptocrinus, Gnorimocrinus, Homocrinus, Hormocrinus, Ichthyocrinus, Lampterocrinus, Laurelocrinus, Lecanocrinus, Lyonicrinus, Lyriocrinus, Macrostylocrinus, Marsupiocrinus, Myelodactylus, Mysticocrinus, Nyctocrinus, Paragazacrinus, Parisocrinus, Patelliocrinus, Paulocrinus, Periechocrinus, Petalocrinus, Pisocrinus, Protaxocrinus, Pycnosaccus, Sagenocrinites, Siphonocrinus, Thalamocrinus, Wilsonicrinus,* and *Zophocrinus.* DEVONIAN: *Acanthocrinus, Allagecrinus, Anamesocrinus, Ancyrocrinus* (see Fig. 4-11), *Aorocrinus, Arachnocrinus, Arthroacantha, Bactrocrinites* (see Fig. 4-10), *Botryocrinus, Clar-*

keocrinus, *Comanthocrinus (Stereocrinus)*, *Cordylocrinus*, *Corematocrinus*, *Corocrinus*, *Cradeocrinus*, *Ctenocrinus*, *Culicocrinus*, *Cyttarocrinus*, *Dactylocrinus*, *Dedadocrinus*, *Deltacrinus*, *Dimerocrinites*, *Dolatocrinus* (see Figs. 4-15, 4-16, and 4-17), *Edriocrinus*, *Eucalyptocrinites*, *Euryocrinus*, *Eutaxocrinus* (see Fig. 4-9), *Gennaeocrinus*, *Gibertsocrinus*, *Gissocrinus*, *Hallocrinus*, *Halysiocrinus*, *Haplocrinites*, *Haplocrinus*, *Hexacrinites*, *Hybochilocrinus (Allagecrinus)*, *Ichthyocrinus*, *Lasiocrinus*, *Logocrinus*, *Maragnicrinus*, *Megistocrinus* (see Figs. 4-19 and 4-20), *Melocrinites*, *Myelodactylus*, *Myrtillocrinus*, *Pagecrinus*, *Protaxocrinus*, *Pterinocrinus*, *Pyxidocrinus* (see Fig. 4-18), *Schultzicrinus*, *Scytalocrinus*, *Sphaerotocrinus*, *Synaptocrinus*, *Synbathocrinus*, *Taxocrinus*, *Technocrinus*, *Thamnocrinus*, *Thylacocrinus*, *Triacrinus* (see Fig. 4-8), and *Vasocrinus*. MISSISSIPPIAN: *Abrotocrinus* (see Fig. 4-31), *Acrocrinus*, *Actinocrinites* (see Fig. 4-35), *Adinocrinus*, *Agaricocrinites* (see Figs. 4-23 and 4-35), *Agassizocrinus*, *Allagecrinus*, *Allocatillocrinus*, *Alloprosallocrinus*, *Amphicrinus*, *Amphoracrinus*, *Anartiocrinus*, *Aorocrinus*, *Ascetocrinus* (see Fig. 4-25), *Aulocrinus*, *Barycrinus*, *Batocrinus* (see Fig. 4-27), *Blothrocrinus*, *Brahmacrinus*, *Cactocrinus* (see Figs. 4-24, 4-26, 4-28, 4-29, and 4-37), *Camptocrinus*, *Catillocrinus*, *Cercidocrinus*, *Coeliocrinus*, *Cosmetocrinus*, *Culmicrinus* (see Fig. 4-37), *Cyathocrinites* (see Figs. 4-30 and 4-35), *Cydrocrinus*, *Dasciocrinus*, *Dedadocrinus*, *Dichocrinus* (see Figs. 4-24, 4-26 and 4-32), *Dichostreblocrinus*, *Dinotocrinus*, *Dizygocrinus*, *Eratocrinus*, *Eretmocrinus* (see Fig. 4-33), *Erisocrinus*, *Eucatillocrinus*, *Eucladocrinus*, *Eupachycrinus*, *Euryocrinus*, *Eutrochocrinus*, *Forbesiocrinus*, *Gilbertsocri-*

nus, *Gilmocrinus* (see Fig. 4-26), *Goniocrinus*, *Graphiocrinus*, *Halysiocrinus*, *Histocrinus*, *Hybochilocrinus (Allagecrinus)*, *Hylodecrinus*, *Hypselocrinus*, *Kallimorphocrinus*, *Linocrinus*, *Macrocrinus* (see Figs. 4-34 and 4-35), *Megistocrinus*, *Mespilocrinus*, *Metichthyocrinus*, *Nipterocrinus*, *Onychocrinus* (see Figs. 4-22, 4-23 and 4-36), *Pachylocrinus* (see Figs. 4-24, 4-26, 4-35 and 4-41), *Paradichocrinus*, *Parichthyocrinus*, *Parisocrinus*, *Pentaramicrinus*, *Phanocrinus* (see Fig. 4-42), *Physetocrinus*, *Platycrinites* (see Fig. 4-44), *Poteriocrinites*, *Pterotocrinus*, *Rhodocrinites*, *Sarocrinus* (see Fig. 4-35), *Scytalocrinus* (see Figs. 4-35, 4-36 and 4-38), *Springericrinus*, *Steganocrinus*, *Stinocrinus*, *Strotocrinus*, *Synbathocrinus* (see Fig. 4-39), *Talarocrinus*, *Taxocrinus*, *Teleiocrinus*, *Tholocrinus*, *Trophocrinus*, *Ulrichicrinus*, *Uperocrinus*, *Wachsmuthicrinus*, and *Zeacrinites* (see Fig. 4-43). PENNSYLVANIAN: *Aatocrinus*, *Aesiocrinus* (see Fig. 4-45), *Alcimocrinus*, *Allocatillocrinus*, *Amphicrinus*, *Apographiocrinus*, *Athlocrinus*, *Cibolocrinus*, *Cromyocrinus*, *Delocrinus*, *Dichostreblocrinus*, *Elibatocrinus*, *Endelocrinus*, *Erisocrinus*, *Ethelocrinus* (see Fig. 4-49), *Galateacrinus*, *Graphiocrinus*, *Isoallagecrinus* (see Fig. 4-46), *Kallimorphocrinus*, *Lasanocrinus*, *Laudonocrinus*, *Lecythiocrinus*, *Neozeacrinus*, *Oklahomacrinus*, *Paradelocrinus*, *Paragassizocrinus*, *Parulocrinus*, *Perimestocrinus*, *Phacelocrinus*, *Pirasocrinus*, *Plaxocrinus*, *Plummericrinus*, *Protencrinus* (see Fig. 4-47), *Schistocrinus*, *Sciadiocrinus*, *Stellarocrinus*, *Synerocrinus*, *Talanterocrinus (Aexitrophocrinus)*, *Ulocrinus*, *Utharocrinus*, *Woodocrinus* (see Fig. 4-40), and *Zeacrinites*. PERMIAN: *Apographiocrinus*, *Cromyocrinus*, *Delocrinus*, *Endelocrinus*, *Eri-*

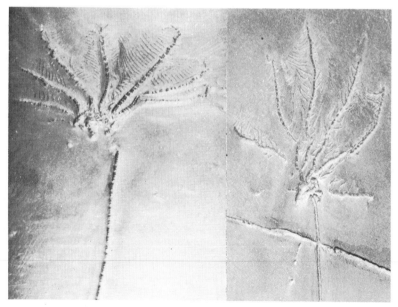

4-8. Early Devonian. *Triacrinus elongatum* Follman. Two crinoid specimens in black slate. Hunsrück Shales, Bundenbach, West Germany. Approx. 8.5 cm each.

4-9. Early Devonian. *Eutaxocrinus prognatus* Schmidt. An ornate crinoid calyx. Hunsrück Shale, Bundenbach, West Germany. 6.5 cm. Courtesy: Geological Enterprises, Ardmore, Oklahoma.

4-10. Early Devonian. *Bactrocrinus nanus* (Roemer). Three examples of a most common Lower Devonian crinoid. Note the lengthy anal sac in the calyx area. Hunsrück Shale, Bundenbach, West Germany. 7.5 cm.

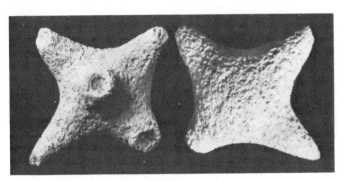

4-11. Early Devonian. *Ancyrocrinus spinosus* Hall. Two views of the spiny base (root) of a crinoid. Sellersburg Limestone, Indiana. 5 cm dia.

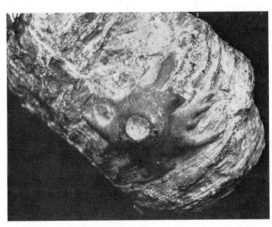

4-12. Middle Devonian. Crinoid anchor on the coral *Heliophyllum halli* Edwards and Haime. Hamilton Shale, Geneseo, Livingston County, New York. 2 cm dia.

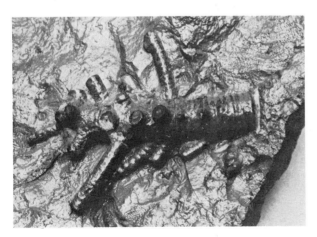

4-13. Middle Devonian. Crinoid root system (species indeterminate). Hamilton Shale, Geneseo, Livingston County, New York. 9 cm.

4-14. Middle Devonian. Crinoid stem on glacial polished section. Onondaga Limestone, Central New York State. 15 cm length.

socrinus, Graphiocrinus, Nebraskacrinus, Neozeacrinus, Perimestocrinus, Plummericrinus, Spaniocrinus, Stuartwellercrinus, and *Ulocrinus.* TRIASSIC: *Balanocrinus, Encrinus* (see Fig. 4-50), and *Erisocrinus.* JURASSIC: *Apiocrinites, Balanocrinus,* and *Pentacrinites* (see Fig. 4-51). CRETACEOUS: *Balanocrinus, Bourgueticrinus, Marsupites, Rhizocrinus,* and *Uintacrinus.*

Notes on crinoids: The four basic types of crinoids are: inadunata; flexibilia; camerata; and articulata. We will only discuss here the camerates and flexibles. The Camerata are crinoids with rigid or fixed plates making up the calyx. *Platycrinites* (see Fig. 4-44) is a good example of a camerate crinoid. The differing genera of the camerate crinoids are distinguished by the size and shape of the dorsal cups. There is a great variability in the dorsal cup design, and the many genera can show an evolutionary trend depending upon the design of the cup, i.e., plate pattern, plate flatness, indentations (concavities), etc.

The dicyclic crinoids (Class Flexibilia—the flexibles Middle Ordovician to Late Permian) are perhaps not as numerous as the camerates, but nevertheless, are quite well represented in the fossil record. A good example of a flexible crinoid is *Ichthyocrinus* from the Silurian. Flexible crinoids differ from the camerates in that their lower branchials are part of the dorsal cup—but without rigidity in their attachment to the cup.

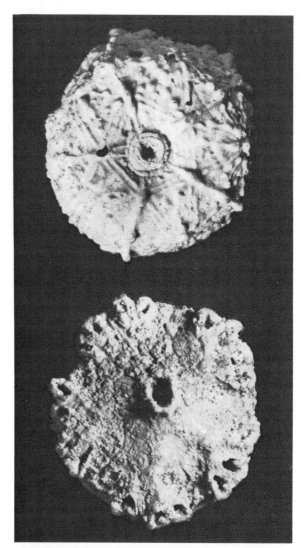

4-15. Middle Devonian. *Dolatocrinus bellulus* Miller and Gurley. Crinoid crown, top and bottom views. Hamilton Group, Indiana. 3 cm dia.

4-16. Early Devonian. *Dolatocrinus lyoni* Miller and Gurley. Crinoid crown, top and bottom views. Helderberg Group, Indiana. 4 cm.

Class Stelleroidea (starfishes): The stelleroids or "starfishes," as they are more popularly known, are echinoderms which have a pentagonal (5-rayed) body design, made up of five arms radiating from a central disklike structure or body. This body, including its "armlike" appendages (see Fig. 4-65), is made up of small plates or ossicles which are not at all rigid, but are loosely joined to one another, allowing flexibility and mobility of the animal. These ossicles scatter when the animal dies, and, for this reason, complete or nearly diagnostic skeletons are scarce in the fossil record.

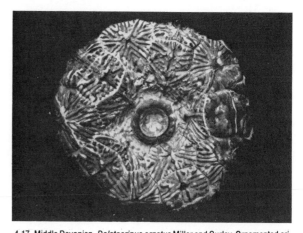

4-17. Middle Devonian. *Dolatocrinus ornatus* Miller and Gurley. Ornamented crinoid crown. Hamilton Shale, East Avon, Livingston County, New York. 3.5 cm dia.

4-18. Devonian. *Pyxidocrinus collensi*. Crinoid crown head. Italy 5.5 cm dia.

4-19. Devonian. *Megistocrinus reeftonensis*. Crinoid calyx (flattened) with pinnules. Reefton, New Zealand. 13 cm. Courtesy: Geology Museum, Victoria University, Wellington, New Zealand.

4-21. Mississippian. Crinoidal limestone block showing fragmented columnals in various views. Arkansas. 7.5 x 8 cm.

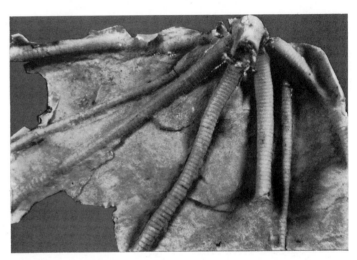

4-20. Devonian. *Megistocrinus reeftonensis*. The "holdfast" of the root of the same specimen (Fig. 4-19). Reefton, New Zealand. 14 cm. Courtesy: Geology Museum, Victoria University, Wellington, New Zealand.

4-22. Mississippian. *Onychocrinus distensus*. A complete crown and columnal (stem.) Keokuk Limestone, Morning Sun, Louisa County, Iowa. Crown 5 cm. Courtesy: Geological Enterprises, Ardmore, Oklahoma.

The "arms" or "feet" (whichever you prefer!) of the ophiuroid in particular, can move about in a sinuous or "snake-like" motion. They can latch onto objects, as well as encircle prey, and travel easily over the sandy bottoms and rocks of the ophiuroid's environment.

Modern starfishes are rather abundant, whereas the fossilized forms are quite rare—the fact being that so few complete specimens have been found preserved. The reason for this has already been discussed above. The range of the fossil forms is from the Early Ordovician to Recent.

Stelleroidea is divided into three subclasses. They are: Somasteroidea, Asteroidea, and Ophiuroidea. The subclass Somasteroidea (somasteroids) are the more primitive starfish types, dating back to the rocks of Early Ordovician age. A particularly good example of a somasteroid is *Villebrunaster* from the Early Ordovician of southern France.

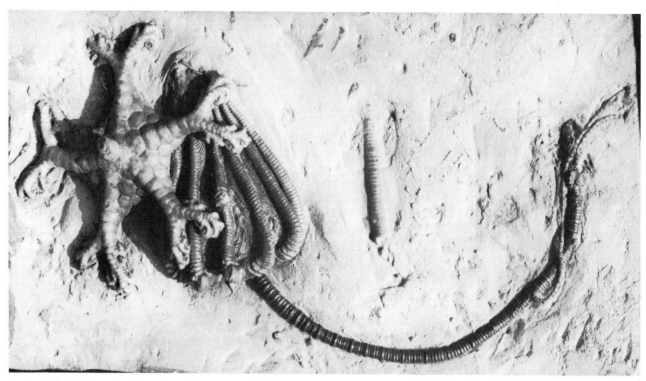

4-23. Mississippian. *Onychocrinus ulrichi* Miller and Gurley on *Agariocrinites splendens*. Calyxes and columnal *(Agariocrinus)*. Osagian Group, Kinderhook Series, Crawfordsville, Montgomery County, Indiana. 8 x 16 cm. Courtesy: Siber and Siber Ltd., Aathal, Switzerland.

4-24. Mississippian. The late B. H. Beane of Le Grand, Iowa with a large limetone slab containing his magnificient crinoid finds at Le Grand, including: *Cactocrinus imperator* Laudon, *Pachylocrinus,* and *Dichocrinus.* Courtesy: Des Moines Register and Tribune.

4-25. Mississippian. *Ascetocrinus rusticellus* (White). Crinoid calyx. Burlington Limestone, Chester Series, Augusta, Iowa. 5.5 cm.

4-26. Mississippian. A. *Dichocrinus bozemanensis*. B. *Cactocrinus imperator* Laudon. C. *Gilmocrinus iowensis*. D. *Pachylocrinus* sp. E. *Dichocrinus multiplex*. All specimens in a large limestone block measuring approximately 35 x 40 cm. Gilmore City Formation, Kinderhook Series, Gilmore City, Pocahontas County, Iowa.

The species of Somasteroidea differ from the species of Asteroidea in having virgalia (rods) present on the ambulacral surface of the arms. These ambulacral plates radiate outward from the mouth area (which is centrally located on the body of the somasteroid) with the virgalia (which are not free pinnules, but are skin-connected to the arm surfaces and thus form the oral (lower) wall of the animal) anchored to the outer plates of the ambulacra and directed towards the peripheral (outer) edges of the arms. The somasteroids have sparsely distributed spicules as their only kind of defensive armor. The somasteroids were most likely the "root" stock for the asteroids and the ophiuroids. The range of the somasteroids is Early Ordovician to Recent.

The subclass Asteroidea (as well as Ophiuroidea) was probably an offshoot from Somasteroidea. The earliest presence of the asteroids (and the ophiuroids) in the fossil record is in the Lower Ordovician Period. The asteroids differ from the somasteroids by the lack of virgalia on the arms. The asteroids have ambulacral structures on the oral surface (particularly evident on *Xenaster* from the Lower Devonian of Germany) with distinctive and regularly ar-

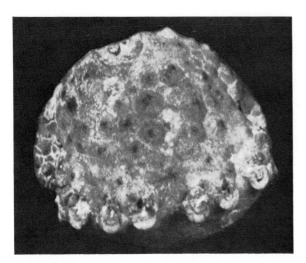

4-27. Mississippian. *Batocrinus wachsmuthi* (White). Crinoid cup (crown). Keokuk Limestone, Warsaw, Kosciusko County, Indiana. 22 mm dia.

4-28. Mississippian. *Cactocrinus imperator* Laudon. Complete calyx and partial columnal, in matrix. Gilmore City Formation, Kinderhookian, Gilmore City, Pocahontas County, Iowa. × 1.5.

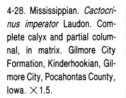

4-29. Mississippian. *Cactocrinus imperator* Laudon. Complete calyx with cup (crown) and pinnules. Gilmore City Formation, Kinderhook Series, Gilmore City, Pocahontas County, Iowa. 5 cm in height.

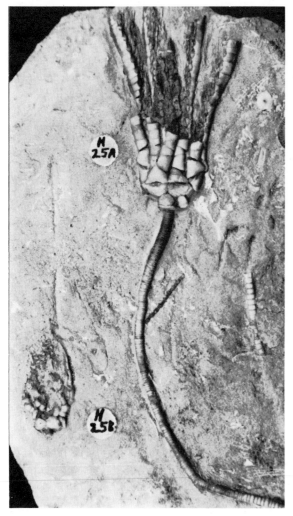

4-30. Mississippian. *Cyathocrinites multibranchiatus* (Lyon and Casseday) (N25A specimen), and *Scytalocrinus validus* Wachsmuth and Springer (lower left–N25B). Complete crinoid calyx and columnal, and a partial calyx. Edwardsville Formation, Crawfordsville, Montgomery County, Indiana. × 1.5.

4-31. Mississippian. *Abrotocrinus unicus* (Hall). Crinoid calyx with partial columnal. Keokuk Limestone, Northeastern Missouri. × 1. Courtesy, Geological Enterprises, Ardmore, Okla.

4-32. Mississippian. *Dichocrinus inornatus* Wachsmuth and Springer. Complete calyx. Hampton Formation, Kinderhook Series, Le Grand, Iowa. 3.2 cm.

4-33. Mississippian. *Eretmocrinus* sp. Calyx and columnal. Gilmore City Formation, Kinderhook Series, Gilmore City, Pocahontas County, Iowa. ×1.5.

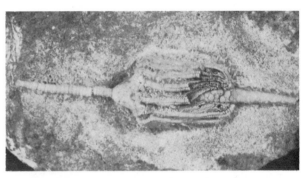

4-34. Mississippian. *Macrocrinus verneuilianus* Shumard. Complete calyx and partial columnal. Burlington Limestone Formation, Morning Sun, Louisa County, Iowa. ×1.

4-35. Mississippian. *Cyathocrinites* gorbyi, *Macrocrinus mundulus*, *Agaricocrinites splendens*, *Sarocrinus granilineus*, *Scytalocrinus*, *Pachylocrinus aequalis*, and *Actinocrinites gibsoni*. Keokuk Limestone, Crawfordsville, Montgomery County, Indiana. 20 x 25 cm. Courtesy: Geological Enterprises, Ardmore, Okla.

4-36. Mississippian. *Onychocrinus ramulosus* and *Scytalocrinus robustus*. Complete crinoid calyxes and partial columnals. Keokuk Limestone Formation, Crawfordsville, Montgomery County, Indiana. ×2. Courtesy: Geological Enterprises, Ardmore, Okla.

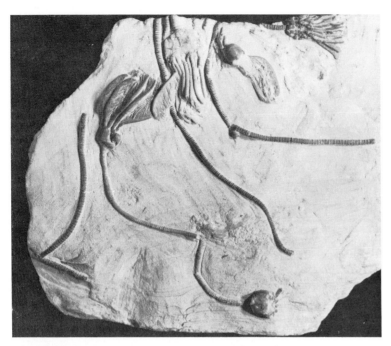

4-37. Mississippian. *Cactocrinus imperator* Laudon, *Rhodocrinites douglassi serpens* (Laudon), and *Culmicrinus* sp. Gilmore City Pocahontas County, Iowa. ×3.

4-38. Mississippian. *Scytalocrinus validus* Wachsmuth and Springer. Complete calyx and partial columnal. Borden Group, Osagian, Crawfordsville, Montgomery County, Indiana. 4.5 cm.

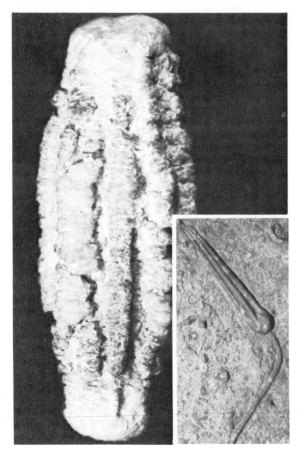

4-39. Mississippian. *Synbathocrinus robustus* Shumard. Worn, but complete calyx. St. Louis Group, Indiana. 5 cm. Inset figure: *Synbathocrinus swallovi* for comparison of calyxes. Calyx: 7.5 cm. Inset Courtesy: Geological Enterprises, Ardmore, Oklahoma.

4-40. Mississippian. *Woodocrinus macrodactylus* (de Koninck). Calyx with partial columnal. Carboniferous Limestone, Richmond, Yorkshire, England. ×3.

4-41. Mississippian. *Pachylocrinus lacunosus* (Miller and Gurley). Complete calyx. Gilmore City Formation, Kinderhook Series, Gilmore City, Pocahontas County, Iowa. ×2.

4-42. Mississippian. *Phanocrinus formosus* (Worthen). Spread-out calyx resting upon the lace bryozoan *Archimedes*. Golconda Formation, Chester Series, Anna, Illinois. ×2.

4-43. Mississippian. *Zeacrinites wortheni* (Hall). Isolated crinoidal calyx and stem. Gilmore City Formation, Kinderhookian, Waterloo, Black Hawk County, Iowa. ×2. Courtesy: Geological Enterprises, Ardmore, Oklahoma.

4-44. Mississippian. *Platycrinites hemisphericus* (Meek and Worthen). A. Complete calyx with a "twisted" columnal. B. Complete isolated calyx. Both specimens from the Edwardsville Formation, Montgomery County, Indiana. (A-X2). Courtesy: Black Hills Institute of Geological Research, Hill City, South Dakota. (B-X2). Courtesy: Geological Enterprises, Ardmore, Okla.

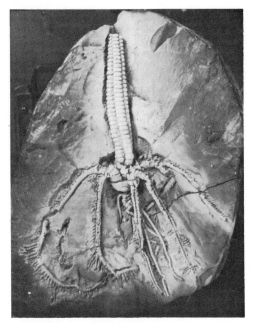

4-45. Pennsylvanian. *Aesiocrinus magnificus* Miller and Gurley. Stem, calyx with crown, and long anal sac. Lane Shale, Kansas City Group, Kansas City, Clay County, Missouri. ✕1.

4-46. Pennsylvanian. *Isoallagecrinus* sp. Posterior view of a calyx. La Salle Formation, Pontiac, Oakland County, Michigan. ✕1.5.

4-47. Pennsylvanian. *Protencrinus atoka* Strimple. Complete calyx. Savanna Formation, Desmoinesian, Coal County, Oklahoma. ✕1.5.

4-49. Late Pennsylvanian. *Ethelocrinus magister* (Miller and Gurley). Complete calyx with partial stem. ✕2.5. Virgil Series, Doniphan Shale, Lecompton Formation, Ace Hill Quarry, Plattsmouth, Sarpy County, Nebraska.

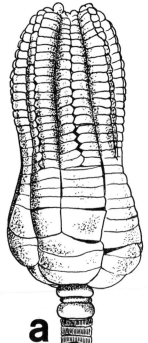

a   **b**

4-48. Pennsylvanian. Crinoid columnal showing broken-off "branchlets" of columnals. Lake Bridgeport, Wise County, Texas. ✕1.

4-50. Triassic. *Encrinus liliiformis* Lamarck. a. Drawing of a complete calyx with a partial columnal. b. Two views of another specimen. West Germany. a. ✕2., b. ✕1.

4-51. Late Jurassic. *Pentacrinites asteriscus* Meek and Hayden. Isolated columnals shaped like starfish. Sundance Formation, Wymore, Wyoming. Entire matrix approx. 3 cm square.

4-52. Late Jurassic. *Anteodon pennatus* (Schlotheim). Free-swimming sea lilly (nonattached crinoid). Solnhofener Plattenkalk, Malm Zeta (Liassic), Solnhofen, Bavaria, Germany, 8.5 cm.

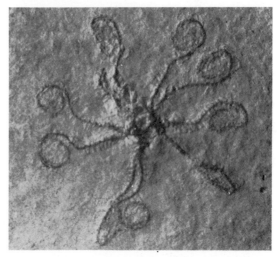

4-53. Late Jurassic. *Saccocoma pectinata* (Goldfuss.) Free "floating" crinoid. Solnhofener Plattenkalk, Lias, Eichstätt, Bavaria, West Germany. 3.5 cm. Courtesy: Siber and Siber Ltd., Aathal, Switzerland.

ranged "armor" plates distributed towards the periphery of the arms (in the places where the virgalia would be present on a somasteroid). The oral face, as its name implies, contains the mouth area. The asteroids have prehensile "tube" feet (with sucker disks) which provide them with locomotion slow as it may be! The aboral side of the asteroid is armored with regular distributed and patterned thick "plates" which act as defensive armor. The anal opening is located on the aboral face as well. *Hudsonaster* and *Devonaster* (see Figs. 4-60 and 4-62) are good examples of strongly armored asteroid stelleroid echinoderms. The range for asteroids is Early Ordovician to Recent.

The subclass Ophiuroidea (ophiuroids) or "brittle-stars," as they are commonly known, lack the bulky and thick body and arms of the asteroids (and somasteroids). The term "brittle-star" refers to the unique ability of a particular group of the ophiuroids which can shed an arm when agitated or in danger. The ophiuroids have sinuosity (as discussed earlier) and the animal makes snake-like motions to propel itself across the ocean bottom. The ophiuroids are also known as "wrigglers" (as opposed to the term "crawlers"—which is used for the asteroids). A

good example of a typical non-shedding ophiuroid is *Ophioderma* (see Figs. 4-69 and 4-70).

The ophiuroids have long and slender arms extending out from a circular or disk-shaped body (see Figs. 4-64, 4-66, 4-67, 4-68, 4-69 and 4-70), and they are lightly armored. The aboral face of the animal does not contain an anal opening, rather there exists on the opposite side in the intermedial section of the test, a small slit or opening, inconspicuous as it is, called the madreporite—this is the anal area of the animal. The oral side does indeed contain a mouth, centrally located and pentagonally shaped, as well as the arms, which radiate outwardly from the periphery of the oral cavity (mouth). Therefore, the arms are

4-54. Early Devonian. *Medusaster* sp. Starfish on black slate. Bundenbacher Schiefer, Bundenbach, Rheinland, Germany. Half size.

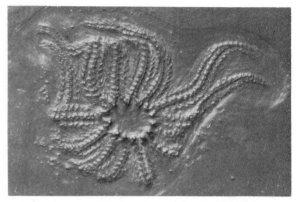

4-55. Early Devonian. *Medusaster rhenanus*. Starfish on slate. Bundenbacher Scheifer, Bundenbach, Rheinland, Germany. Half size.

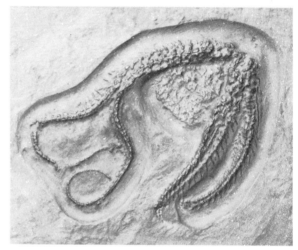

4-56. Early Devonian. *Medusaster rhenanus*. Starfish on slate Bundenbacher Schiefer, Bundenbach, Rheinland, Germany. × 1.

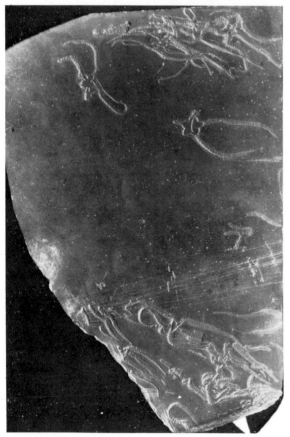

4-57. Early Devonian. *Furcaster palaeozoicus* Stürtz. A slab of slate with 11 complete and partial specimens. Hunsrück Shale, Bundenbach, Germany. Slab is 19 x 26 cm. Courtesy: Geological Enterprises, Ardmore, Oklahoma.

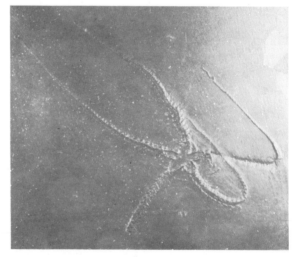

4-58. Early Devonian. *Furcaster palaeozoicus* Stürtz. Starfish in ''blackboard'' slate. Bundenbacher Schiefer, Bundenbach, Rheinland, Germany. × 1.

4-60. Middle Devonian. *Devonaster (Palaester)* sp. Starfish. Hamilton Shale, Hamilton, Madison County, New York. 7 cm dia.

4-59. Early Devonian. *Eospondylus primigenius* (Steurtz). Starfish in black slate. Bundenbacher Schiefer, Bundenbach, Rheinland, Germany. 12 cm overall. From: Müller and Zimmermann "Aus Jahrmillionen," Veb Gustav Fischer Verlag, Jena 1962.

4-61. Pennsylvanian. Orphiuroid "brittle star" (species indeterminate) from the Estuary Shales, Bluefield, Mercer County, Virginia. ×4.

4-62. Middle Devonian. *Devonaster (Palaester) eucharis* Hall. Starfishes in limestone. Hamilton Shale, Hamilton, Madison County, New York. 7.5 cm.

a distinctive part of the disk's skeleton, and as such, in most cases are usually found associated with the fossil specimen. As discussed earlier, the ophiuroids were doubtlessly derived from the more primitive somasteroids. The range of the ophiuroids is also from the Early Ordovician to the present time.

Finally, the following groups (not previously mentioned in this chapter) are worthy of noting: the Paracrinoidea, Edrioblastoidea, Parablastoidea, the Camptostromatoidea, Stylophora, Homostelea, and Homoiostelea.

4-63. Late Jurassic. *Pentasteria longispina* Hess. Starfish. Lower part of the Upper Jurassic. Wiessenstein, Kanton Solothurn, Switzerland. 14 cm. Courtesy: Siber and Siber Ltd., Aathal, Switzerland.

4-65. Jurassic. *Tylasteria berthandi* (large starfish) and *Dermaster boehmi* (smaller starfishes). Schinznach, Switzerland. Overall matrix size: 25 x 35 cm. Courtesy: Siber and Siber Ltd., Aathal, Switzerland.

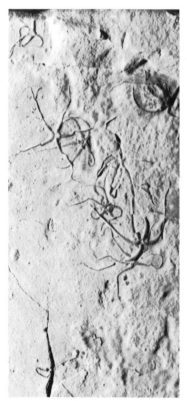

4-64. Middle Jurassic. *Ophiopinna elegans* (Heller). Starfishes. Callovian, La Voulte, Ardeche, France. 5 x 6 cm. Courtesy: Geological Enterprises, Ardmore, Oklahoma.

4-66. Jurassic. *Geocoma planata*. Starfish. Solnhofen Limestone, Kelheim, Bavaria, West Germany. 7 cm. Courtesy: Siber and Siber Ltd., Aathal, Switzerland.

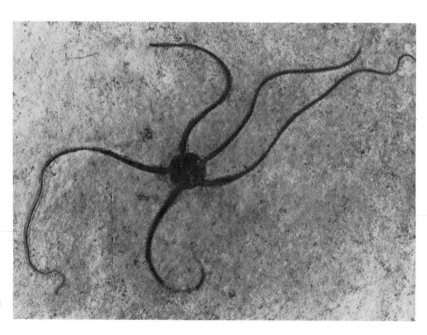

4-67. Jurassic. *Geocoma carinata* (Münster). Starfish. Malm Zeta (Liassic), Solnhofen Limestone, Solnhofen, Bavaria, Germany. 4 x 2.5 cm.

4-68. Jurassic. *Geocoma carinata* (Münster). Brittle starfish. Solnhofen Limestone, Zandt, Bavaria, Germany. 4 cm. Courtesy: Siber and Siber Ltd., Aathal, Switzerland.

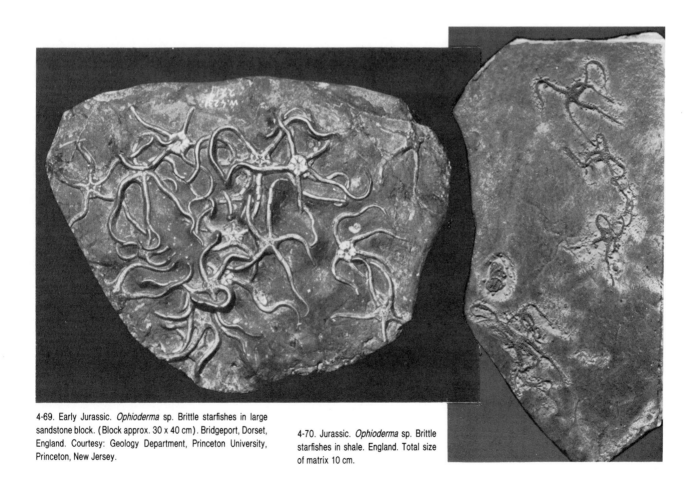

4-69. Early Jurassic. *Ophioderma* sp. Brittle starfishes in large sandstone block. (Block approx. 30 x 40 cm). Bridgeport, Dorset, England. Courtesy: Geology Department, Princeton University, Princeton, New Jersey.

4-70. Jurassic. *Ophioderma* sp. Brittle starfishes in shale. England. Total size of matrix 10 cm.

5 ECHINOIDEA (Sea-urchins, Sand-dollars, and Sea-cucumbers)

The holothurians (sea-cucumbers) are poorly known in the fossil record. It is only under special conditions that these soft bodied organisms are preserved at all. The class Holothuroidea dates back to the Early Devonian. Their recorded occurrence in the Ordovician is presently in doubt. In the Pennsylvanian Mazon Creek siderite nodules an exceptionally well preserved fossil sea-cucumber has been found: *Achistrum* sp. (see Fig. 5-1).

The echinoids are part of the phylum Echinodermata, but they are purposely set aside for this chapter and not included in with the echinoderm groups or classes of the previous chapter.

Echinoids (the "sea-urchins") are either globular in shape—the cidarids being a good example (see Fig. 5-4)—or have flattened tests such as the order Clypeaster-oidea ("sand-dollars") —examples being *Scutella* (see Fig. 5-18) and *Dendraster* (see Fig. 5-22).

The regular echinoids: The globular shaped tests (and in some cases, heart shaped ones (see Fig. 5-7 for example) are the typical "spiny" sea-urchins that sluggishly move about the bottom of the ocean. Their tests (body shells or skeletons) are made up of numerous plates of varying design as well as ambulacral plates which are compound. Spines emanate from the interambulacral plates. The designs of the spines vary from the lengthy, slender or needle-like types of many of the modern species, to the rotund and "Indian club"-like spines of *Balanocidaris* (see Fig. 5-3) of the Triassic. As is obvious, the spines are primarily a defensive weapon, but they also assist in the locomotion of the animal. The compound ambulacral plates of Mesozoic and Cenozoic echinoids are of two types— and these types are indicative of distinctive species. The two basic types or styles of plates are: 1. Diadematoid, and 2. Echinoid. The diadematoid compound plates show one large plate centered between two or more smaller ones, while the echinoid compound plate shows the larger plate at the bottom or even lower than the smaller plates. While many Paleozoic echinoids have rows of interambulacral plates, quite a few have more than one row in these inter-ambulacral areas and, as stated before, it is in these areas

that the spines protrude from the test.

The tests of echinoids are either rigid (post-Paleozoic species) or composed of overlapping, thin "shingle-like" plates (Paleozoic species, see Fig. 5-2 for a good example). The majority of post-Paleozoic echinoids seem to have rigid tests made of complex plates fitting up against each other like bricks in a wall of a house. The rigid tests of echinoids are more widely distributed from the Triassic to the Present. The rigid tests plates can best be seen in the specimens of the two views of *Echinocorys* (see Fig. 5-12) from the Late Cretaceous of England. The mouth of the echinoid is on the ventral (lower or oral) side of the animal, while the anus in most species is situated somewhere in or near the posterior (rear) interambulacral area (see Fig. 5-9, left specimen and Fig. 5-14, top specimen).

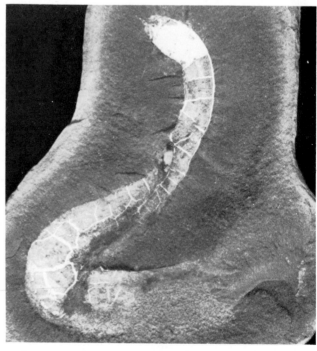

5-1. Pennsylvanian. *Achistrum* sp. A holothuroid animal ("sea-cucumber") in a siderite concretion. Francis Creek Shale (Mazon Creek), Pit 11, Kankakee County, Illinois. ×1.

5-2. Late Devonian. Fragment of the test of a cidarid echinoid. Species indeterminate. "B" Member, Hay River Formation, Enterprise, Northwest Territories, Canada. ×2.

5-3. Late Triassic. *Balanocidaris glandifera* Goldfuss. Isolated large spine from a cidarid echinoid. Hoselkus Formation, California. 32 mm.

5-4. Late Jurassic. *Plegiocidaris coronata* (Goldfuss). Complete sea-urchin test. Badener Schichten, Mellikon, Switzerland. 5 cm dia. Courtesy: Siber and Siber Ltd., Aathal, Switzerland.

5-5. Early Cretaceous. *Cyphosoma texanum* (Roemer). Dorsal (top) and ventral (bottom) views of a complete echinoid test. Comanche Peak Formation, Fredericksburg Group, Comanche Series, San Saba River Valley, Fredericksburg, Texas. 5 cm. dia.

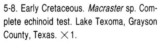

5-6. Early Cretaceous. *Cyphosoma texanum* (Roemer). Complete sea urchin test. Comanche Series, Coryell County, Texas. ×1.5.

5-7. Early Cretaceous. *Heteraster (Enallaster) texanus* (Roemer). Complete echinoid. Comanche Series, Coryell County, Texas. ×1.5.

5-8. Early Cretaceous. *Macraster* sp. Complete echinoid test. Lake Texoma, Grayson County, Texas. ×1.

A note about the spine distribution on the tests of cidaroid echinoids: The cidaroid tests (see Figs. 5-4 and 5-15) have a series of tubercles in the ambulacral regions. The large "club-like" spines (*Balanocidaris*, see for example, Fig. 5-3) fit onto the heads of these tubercles.

The irregular echinoids: Spatangoid echinoids as well as the holectypoids and cassiduloids are probably the most numerous types in the fossil record. These include the heart-shaped *Heteraster (Enallaster)* (see Fig. 5-7) and *Macraster* (see Fig. 5-8) among others. A typical spatangoid echinoid, and one which by now most every collector is familiar with is *Eupatagus* (see Fig. 5-16) from the Eocene of Florida.

5-9. Late Cretaceous. *Hemipneustis striatordiatus* (Leske). Large echinoid test. Maestrichtian Chalk Formation, Maastricht, The Netherlands. 10 cm. 3 views of the same test.

5-11. Late Cretaceous. *Galerbis vulgaris.* Complete urchin test. Senonian, Geschiebe, Ostsee Coast, Ahrenshoop, Germany. 2 cm.

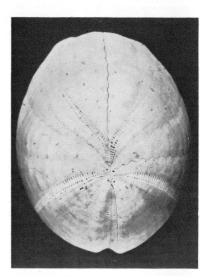

5-10. Late Cretaceous. *Hemipneustis striatordiatus* (Leske). Dorsal (top) view of a complete echinoid test. Maestrichtian Chalk Formation, Maastricht, The Netherlands. 8.5 cm.

5-12. Late Cretaceous. *Echinocorys scutata* Leske. Two views of a complete echinoid (urchin) test. Chalk Formation, Andover, Hampshire, England. 5 cm dia.

Common representatives of the cassiduloids are *Cassidulus* and *Hardouinia* (see Figs. 5-13 and 14 for the latter), as well as *Louenia* (see Fig. 5-20). The large echinoid *Hemipneustis* (see Figs. 5-9 and 10) from Holland, is another good example of a cassiduloid type echinoid.

There are many varieties of clypeastroids (sand-dollars), but only a few are illustrated here. *Encope* (see Fig. 5-17), *Scutella (Albertella)* (see Figs. 5-18 and 5-21), and

5-13. Late Cretaceous. *Hardouinia subquadrata* Conrad. Complete echinoid test. Ripley Formation, Union County, Mississippi. 33 mm.

5-14. Late Cretaceous. *Hardouinia subquadrata* Conrad, Dorsal and ventral views of a complete test. Ripley Formation, Union County, Mississippi. 4.5 cm.

5-15. Late Cretaceous. a. *Stereocidaris sceptrifera.* b. *Tylocidaris clavigera* (König). Two examples of cidarid echinoids from the Upper Chalk, Coraguinum Zone (Turonian), Micheldever, Hants, England. Approx. 4 cm. Courtesy: Geological Enterprises, Ardmore, Oklahoma.

5-16. Eocene. *Eupatagus floridanus* Clark. Dorsal and ventral views of a complete echinoid test. Ocala Limestone, Ocala, Marion County, Florida. 6.5 cm.

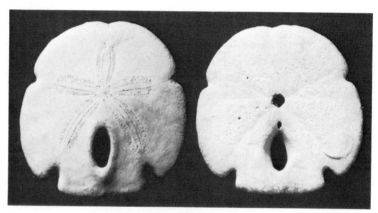

5-17. Miocene. *Encope tamiamiensis* Mansfield. Dorsal and ventral views of a "keyhole" sand dollar test. Tamiami Formation, Ochopee, Florida. 4.5 cm dia.

5-18. Miocene. *Scutella (Albertella) alberti* Conrad. Complete sand dollar test. Choptank Formation, Solomons, Calvert County, Maryland. 10 cm.

5-19. Miocene. *Astrodapsis (Astrodaspis) cuyamenus* (Kew). Complete sand dollar test. Neroly Formation, Cuyama Valley, California. 6.5 cm.

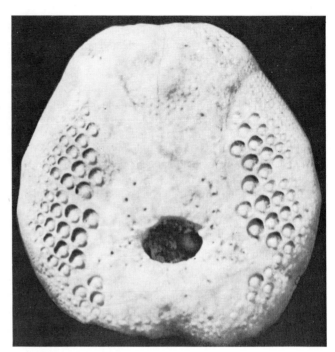

5-20. Miocene. *Louenia forbesi*. Ventral view of a complete echinoid test. Mannum Formation, Southern Australia. 3 cm.

5-21. Miocene. *Scutella (Albertella) alberti* Conrad. Internal mold of a sand dollar test. Hawthorne Formation, Ft. Meade Phosphate area, Polk County, Florida. Approx. half size.

5-22. Pliocene. *Dendraster gibbsi* (Rémond). Complete sand dollar test. San Joaquin Formation, Kettleman Hills, Kern County, California. Approx. half size.

Dendraster (see Fig. 5-22) are all typical examples of the clypeastroid types. *Encope* is commonly called the "keyhole" sand-dollar, for an obvious reason. This large notch or hole on the test of *Encope* and related species is called the lunule.

Clypeastroids have numerous short spines which cover the entire test and give the animal a "furry" appearance.

Notes on echinoids in general: Echinoids, like the crinoids and other echinoderms, are necessarily confined to a marine environment, where they have been found in great depths as well as in shallow areas. As a rule, the sexes are separated, although some cases of hermaphrodism are known to exist. While certain species of echinoids have prehensile tube feet with sucker discs attached for locomotion, the majority of echinoids move about on their spines as if on stilts. The large club-like spines of the cidaroids were used to balance the animal as it lay in wave washed areas. Some species of echinoids, such as the spatangoids, live in existing burrows in deep waters.

The clypeastroids live in colonies on the mud or sandy bottoms of the oceans and just recline supinely, digesting bits of food particles in the sediments that they lie upon. The flattened tests of the clypeastroids are first found in the Tertiary (Paleocene/Eocene) where they successfully competed with the ubiquitous spatangoid types.

For a faunal listing of the echinoids (sand-dollars and sea-urchins) see the addenda on page 495.

6 ANNELIDA (Worms)

The phylum Annelida or Annulata (worms) is a most elusive group of invertebrates, in terms of the fossil record. Because of their soft, segmented and sometimes cuticle covered bodies, they are quite rarely found in a whole state in rocks. Extraordinarily preserved specimens such as the "accordion worm" *Lecathylus* (see Fig. 6-2) and the long, tapering "garden-worm" type (see Fig. 6-3) are exceptions to the rule.

For the most part, all that remains of most fossil worms are either sediment filled-in trails (see Figs. 6-4 and 6-5) or the calcareous tubes (see Fig. 6-6), some of which are preserved grotesquely twisted around on top of one another (see Fig. 6-7).

The worms possess very few hard parts that lend themselves to fossilization; but on rare occasions, the jaw sections containing "teeth" (called scolecodonts) have been found in shales or heavy mudstones (see Fig. 6-1). These scolecodonts (worm jaws) are composed of calcium carbonate material and are found throughout the geologic ages. They are especially abundant in the Carboniferous rocks (Ordovician, Devonian, Mississippian, and Pennsylvanian). See Fig. 6-1 for examples of these worm jaws. On occasion, the teeth of scolecodonts have been mistaken for conodonts, an entirely different phylum, although there are some superficial similarities between scolecodont teeth and the "tooth-like" conodonts. See the following chapter for a comparison.

Faunal listing for the Annelida (genera): CAMBRIAN: *Arenicolites, Canadia, Ottoia, Skolithos (Scolithus)* and *Worthenella.* ORDOVICIAN: *Arabellites* (see Fig. 6-1), *Cornulites, Eunicites, Leodicites, Lumbriconereites* (see Fig. 6-1), *Nereidavus, Oenonites, Protarabellites, Protoscolex,* and *Staurocephalites.* SILURIAN: *Arabellites, Cornulites, Eunicites, Ildraites* (see Fig. 6-1), *Lecathylus* (see Fig. 6-2), *Lumbriconereites, Nereidavus, Oenonites* (see Fig. 6-1), *Skolithos (Scolithus)* and *Spirorbis.* DEVONIAN: *Arabellites, Cornulites, Eunicites, Ildraites, Lumbriconereites, Nereidavus, Oenonites, Palaeochaeta, Paulinites,* and *Spirorbis.* MISSISSIPPIAN: *Scalarituba* and *Spirorbis.* PENNSYLVANIAN: *Spirorbis.* JURAS-

SIC: *Lumbricaria* (see Fig. 6-4). CRETACEOUS: *Hamulus* (see Figs. 6-5 and 6-7), *Serpula* (see Fig. 6-6), and *Rotularia (Tubulostium).*

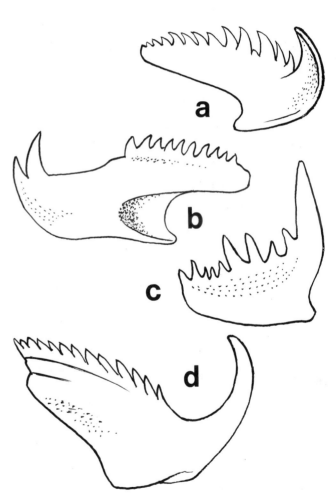

6-1. Typical Ordovician-Silurian scolecodonts (worm jaws). a. *Lumbriconereites webbi* Stauffer, Ordovician of Minnesota, ×72. b. *Ildraites duplex* Eller, Silurian of New Yrok, ×50. c. *Oenonites acinaceas* Eller, Silurian of New York, ×40. d. *Arabellites giganteus* Stauffer, Ordovician of Minnesota, ×60. Redrawn from Shimer and Schrock. 1944.

6-2. Silurian. *Lecathylus gregarius.* Bodies of two "accordion worms." Racine Dolomite, Illinois. × 1.5.

6-4. Jurassic. *Lumbricaria.* Possible worm "trails" or "burrows" in silicified mud (turned to lithographic limestone). Also thought to be fish coprolites. Upper Malm, Lithographischer Schiefer, Eichstät, Bavaria, Germany. 8 x 5.5 cm.

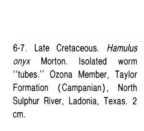

6-5. Late Cretaceous. *Hamulus* sp. Impression of a fragment of a worm burrow in limonitic sandstone. Navesink Formation (derived), Middle Maestrichtian, Holmdel, Monmouth County, New Jersey. × 1.

6-6. Early Cretaceous. *Serpula* sp. Worm "tubes" in matrix. Paluxy River, Somervell County, Glen Rose, Texas. 5 cm.

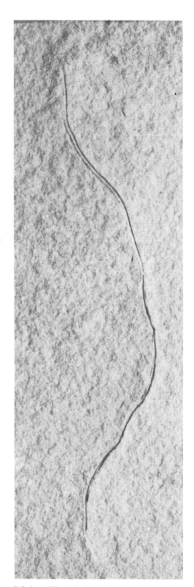

6-3. Late Mississippian. Worm body impression in lithographic limestone. Species indeterminate. Bear Gulch Limestone, West of Lewistown, Montana. × 1.

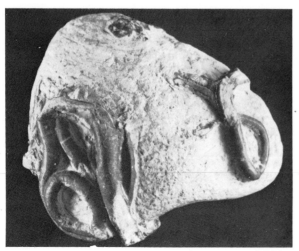

6-7. Late Cretaceous. *Hamulus onyx* Morton. Isolated worm "tubes." Ozona Member, Taylor Formation (Campanian), North Sulphur River, Ladonia, Texas. 2 cm.

7 CONODONTS

Conodonts are common microscopic remnants belonging to a presently unknown fossil life form. Their geologic range is from Middle Cambrian to the Late Triassic.

The "toothlike" fossils are composed of calcium phosphate, a mineral/chemical composition that is also present in the teeth of vertebrates—particularly fishes. By comparison, the worm jaws called scolecodonts (see preceding chapter), which are not related to the conodonts, have a chemical composition of calcium carbonate with a thin outer layering high in carbon, low in calcium, and containing no sulphur.

Some paleobiologists believe that the conodonts are parts of fishes—perhaps the gill denticles or for that matter, possibly "throat" teeth or denticles; but, clearly they are *not* the teeth from the oral cavity, as most Paleozoic fishes to whom the conodonts are contemporaneous have identifiable tooth designs, and their teeth are not so minute as the conodonts. Others have attributed these "toothlike" fossils to any of the following animal groups: worms, crustaceans, or gastropods. One recent theory proposes that the teeth of conodonts may in fact belong to a previously unknown animal, similar in habit, and even superficially in design, to *Amphioxus*. The "teeth" it seems, correspond to known conodont types, i.e., the bar, blade and platform styles, and are paired in opposition to each other, and centrally located in the body cavity of the recently discovered "conodont" animal. It appears that this is where the conodontodemus is located. The animals are no bigger than 7 cm in length by 2 cm in height. These new conodont animals come from the Late Mississippian Bear Gulch lithographic limestones of central Montana.

Isolated conodont *teeth* have been known since 1856. However, it was not until over 110 years later that the "possible" animals which may or may not be the owners of these ubiquitous "teeth" had been discovered. Nevertheless, many students of conodont fossils, particularly those concerned with index fossils, including oil company personnel who study these "teeth" in sediments to determine the presence of hydrocarbons to aid in the discovery of new oil deposits, do not accept this new animal as the host-owner of the conodont assemblages. In this respect, and in deference to those scientists who do not believe that the conodont animal has at last been found, we will go on to discuss the isolated "teeth" that are called conodonts.

Conodonts come in either cone, bar, blade, or platform styles. Figure 7-2 is a good representative of a bar conodont, while a platform conodont is shown in Figs. 7-1 and 7-5. The blade type is shown in Figs. 7-3 and 7-4. A cone conodont is shown in Fig. 7-6.

Conodonts have been used as index fossils for quite some time now, since it seems that distinct form/species types are represented throughout correlated formations in geological history. That is to say that these conodonts can be used to determine stratigraphic correlation and distribution in the fossil record.

It appears that most common assemblages of these conodont "teeth" are found in direct association with fish fossils, especially in carbonaceous black fissile shales of the Pennsylvanian Period.

Some typical conodonts in the fossil record: CAMBRIAN: *Furnishina, Hertzina,* and *Muellerina.* ORDOVICIAN: *Acodus, Amorphognathus, Belodus, Cardiodella, Chirognathus, Curtognathus, Dichognathus, Drepanodus* (see Fig. 7-6), *Erismodus, Loxodus, Multioistodus, Oulodus, Paltodus, Polycaulodus, Scolopodus, Trichognathus,* and *Ulrichodina.* SILURIAN: *Plectospathodus, Polygnathoides,* and *Spathognathus.* DEVONIAN: *Ancyrodella* (see Fig. 7-1), *Angulodus, Apatognathus, Bryantodus, Diplododella, Hibbardella, Hindeodella, Lonchodina, Nothognathella, Palmatodella, Polygnathus,* and *Prioniodina.* MISSISSIPPIAN: *Bactrognathus, Doliognathus, Gnathodus, Hindeodella, Pinacognathus, Prioniodina, Scaliognathus, Spathognathodus,* and *Trichognathus.* PENNSYLVANIAN: *Gondolella, Hindeodella* (see Fig. 7-5), *Idiognathodus, Metalonchodina* (see Fig. 7-2), *Ozarkodina* (see Fig. 7-3), *Neognathodus, Polygnathodella, Spathognathodus,* and *Streptognathodus.* PERMIAN: *Idiognathodus, Neogondodella,* and *Sweetognathus.* TRIASSIC: *Furnishius, Neospathodus,* and *Platyvillosus.*

As was the case with descriptions of early shark's teeth, the various generic and species descriptions of conodonts from a given formation or horizon are from the same individual type of animal. The gnathal or dental elements (paired as they are) of conodonts probably make up one individual type's dentition. Have we now already discovered the conodont animal—or are we still searching for it?

7-1. Late Devonian. *Ancyrodella* sp. Dorsal (oral) (top) and three-fourths (bottom) views of a distinctive platform conodont. Independence Formation, eastern Iowa. ×55.

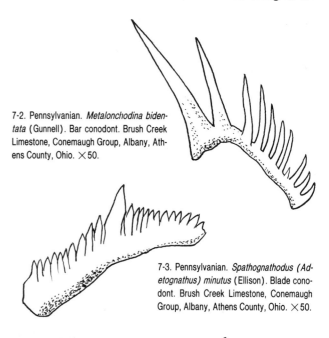

7-2. Pennsylvanian. *Metalonchodina bidentata* (Gunnell). Bar conodont. Brush Creek Limestone, Conemaugh Group, Albany, Athens County, Ohio. ×50.

7-3. Pennsylvanian. *Spathognathodus (Adetognathus) minutus* (Ellison). Blade conodont. Brush Creek Limestone, Conemaugh Group, Albany, Athens County, Ohio. ×50.

7-4. Pennsylvanian. *Ozarkodina delicatula* (Stauffer and Plummer). Blade conodont. Brush Creek Limestone, Conemaugh Group, Albany, Athens County, Ohio. ×50.

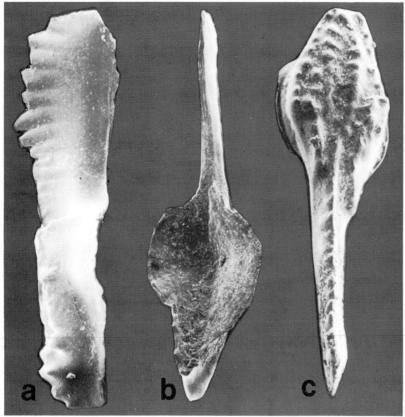

7-6. Early Ordovician. *Drepanodus arcuatus* Branson and Mehl. A typical cone conodont. Jefferson City Formation, Missouri. ×55.

7-5. Pennsylvanian. *Hindeodella* sp. A bar conodont. 3 views: a. Profile. b. Ventral (aboral). c. Dorsal (oral). Hertha Limestone, St. Charles, Madison County, Iowa. ×55.

8 BRYOZOA

Bryozoans or "moss animals," as they are commonly called, are primarily marine invertebrates, although a very few species have been known to inhabit fresh water. Bryozoans are colonial animals. They are similar to the corals (see Chapter 3), occurring as mats of many individual, microscopic animals (zooids). Generally under 1 mm in size, they organize colonies that grow to 2.5 cm or more.

Mass burials involving scores of bryozoan colonies are a large zoological constituent of many Carboniferous rocks representing fossil reefs. Many limestones of Mississippian and especially Pennsylvanian age contain large spiraled axes (sections) of the "screwlike" or "drill-like" central support of *Archimedes* (see Figs. 8-1 through 8-3), as well as the "lacy" or fenestrated "body fans" of the cyclostome bryozoan *Fistulipora* (see Fig. 8-6).

The bryozoans are broken down into two subphyla: the entoprocta, and the ectoprocta, which includes freshwater and marine varieties. This latter subphyla is represented by a ubiquitous fossil record.

The bryozoans are further divided into five distinctive orders. They are: Ctenostomata (extant—Ordovician to Recent—with very few fossils), Cyclostomata (extant—Ordovician to Recent—with a large fossil record), Trepostomata (extinct—Ordovician to Permian—with numerous species during the Paleozoic), Cryptostomata (extinct—Ordovician to Permian—also with numerous species during the Paleozoic), and finally Cheilostomata (extant—Cretaceous to Recent—with numerous species).

Faunal listing of ctenostomes as follows (genera): ORDOVICIAN: *Ropalonaria (Rhopalonaria)* and *Vinella*. DEVONIAN: *Allonema, Asodictyon,* and *Ropalonaria (Rhopalonaria)*.

Faunal listing of cyclostomes as follows (genera): ORDOVICIAN: *Anolotichia, Berenicea, Ceramophylla, Ceramoporella, Coeloclema, Corynotrypa, Crepipora, Diastoporina, Favositella, Mitoclema,* and *Proboscina*. SILURIAN: *Ceramopora, Cheilotrypa (Chilotrypa), Fistulipora,* and *Phacelopora*. DEVONIAN: *Botryllopora, Buskopora, Hederella, Hernodia, Phacelopora, Prismopora, Reptaria,* and *Scalaripora*. MISSISSIPPIAN:

Cheilotrypa (Chilotrypa), Coscinotrypa (Coscinium), Evactinopora, Glyptopora, and *Meekopora*. PENNSYLVANIAN: *Cyclotrypa* and *Fistulipora* (see Fig. 8-6). PERMIAN: *Dybowskiella, Hexagonella,* and *Meekopora*. JURASSIC: *Idmonea*. CRETACEOUS: *Cardioecia, Entalophora, Filifascigera, Lichenopora, Proboscina, Spiropora,* and *Stomatopora*. EOCENE: *Erkosonea, Hornera, Leiosoecia, Partretocycloecia,* and *Pleuronea*. OLIGOCENE: *Crisisina*. MIOCENE: *Tretocycloecia*. PLEISTOCENE: *Crisia* and *Psilosolen*.

Faunal listing of trepostomes as follows (genera): ORDOVICIAN: *Amplexopora, Aspidora, Atactoporella, Batostoma, Bythopora, Constellaria, Dekayella, Dekayia, Eridotrypa, Hallopora, Hemiphragma, Homotrypa, Leioclema, Mesotrypa, Monotrypa, Monticulipora (Monticuliporella), Nicholsonella, Phylloporina, Prasopora, Rhombotrypa, Stigmatella, Stromatotrypa,* and *Subretepora*. SILURIAN: *Batostomella, Hallopora, Idiotrypa, Leioclema, Pseudohonera, Subretepora,* and *Trematopora*. DEVONIAN: *Atactoechus, Eridocampylus, Eridotrypella, Hallopora, Leioclema,* and *Petalotrypa*. MISSISSIPPIAN: *Leioclema*. PENNSYLVANIAN: *Leioclema* and *Tabulipora (Stenopora)* (see Fig. 8-7). PERMIAN: *Leioclema*.

Faunal listing of cryptostomes as follows (genera): ORDOVICIAN: *Arthroclema, Arthropora, Arthrostylus, Escharopora, Helopora, Nematopora, Pachydictya, Phyllodictya, Polypora, Rhinidictya,* and *Stictoporella*. SILURIAN: *Clathropora, Diamesopora, Helopora, Hemitrypa, Lichenalia, Pachydictya, Phaenopora, Polypora, Semicoscinium,* and *Stictotrypa*. DEVONIAN: *Acanthoclema, Acrogenia, Anastomopora, Ceramella, Fenestrapora, Hemitrypa, Intrapora, Loculipora, Paleschara, Penniretepora, Polypora, Ptilodictya, Ptylopora (Ptilopora), Semicoscinium, Streblotrypa, Sulcoretepora (Cystodictya), Taeniopora,* and *Unitrypa*. MISSISSIPPIAN: *Actinotrypa, Archimedes* (see Figs. 8-1 through 8-3), *Bactropora, Coeloconus, Dichoporaria, Fenestralia, Fenestrellina, Hemitrypa, Lyropora, Penniretepora, Ptilopora, Rhombopora, Streblotrypa, Thamiscus,* and *Wor-*

thenopora. PENNSYLVANIAN: *Fenestrellina* (see Fig. 8-4), *Polypora* (see Fig. 8-5), and *Sulcoretepora.* PERMIAN: *Minilya, Polypora, Rhabdomeson, Stenopora,* and *Streblotrypa.*

Faunal listing of cheilostomes as follows (genera): CRETACEOUS: *Alderina, Aplousina, Callopora, Crassimarginatella, Diacanthopora, Euritina, Floridinella, Lunulites (Reptolunulites), Membraniporidra, Otionella,* *Perigastrella, Smittipora (Velumella),* and *Stichocados.* EOCENE: *Acanthodesia, Alderina, Aplousia, Callopora, Cellaria, Conopeum, Coscinopleura, Cribrilina, Diplotresis, Euritina, Hincksina, Mamillopora, Metracolposa, Ochetosella, Periporosella, Porella, Schizorthosecos, Steganoporella, Tremotoichos, Trigonopora,* and *Tubucellaria.* MIOCENE: *Adeona, Cyclocolposa, Discoporella, Floridina, Holoporella, Perigastrella,* and *Stylopoma.*

8-1. Middle Mississippian. *Archimedes wortheni* (Hall). Lace bryozoan with spiraled fronds. Warsaw, Illinois. 3.5 cm.

8-2. Mississippian. *Archimedes* sp. Fragment of a lace bryozoan with spiraled fronds. Princeton, Kentucky. ×1.

8-3. Mississippian. *Archimedes* sp. Isolated spiral fronds. Oolitic Limestone Beds, South of Bloomington, Indiana. Each spiral approximately 3 cm.

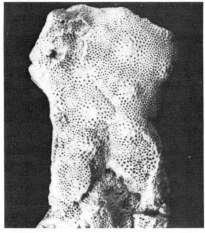

8-4. Pennsylvanian. *Rhombopora lepidodendroides* Meek. Large colony with numerous "autopore" apertures. Belknapp Limestone Member, Harpersville Formation, Stephens County, Texas. 6 cm.

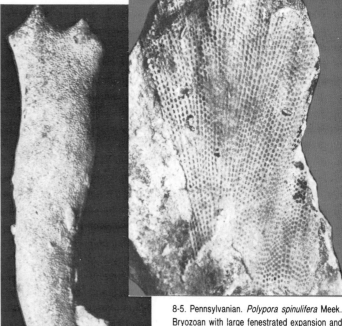

8-5. Pennsylvanian. *Polypora spinulifera* Meek. Bryozoan with large fenestrated expansion and numerous autopore apertures on individual branchlets. Belknapp Limestone Member, Harpersville Formation, Stephens County, Texas. 6 cm.

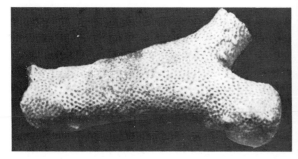

8-6. Pennsylvanian. *Fistulipora carbonaria* Ulrich. Branch fragment of a moss animal (bryozoan). Graham Formation, Bunger Limestone, Stephens County, Texas. 33 mm in length.

8-7. Permian. *Tabulipora (Stenopora) cava* Moore. Large bryozoan colony branchlet. Choaza Formation, West Texas. 53 mm in length.

9 BRACHIOPODA

Brachiopods are marine bivalves with unequal valves, but bilateral symmetry.

There are two basic classes of brachiopods—the inarticulates and the articulates. The shells of inarticulates are composed of a chitinous-phosphatic material, while those of the articulate brachiopods are composed of calcite. The pediculate brachiopods are held in anchorage by a fleshy stalk extending from the pedicle-beak area (a notch or hole at the posterior of the pedicle valve) and attached on the other end to a shell or to the sandy bottom itself.

The pedicle valve is normally the ventral valve and the brachial valve is the dorsal valve. The articulation of the brachial with the pedicle valve is called the hinge.

The inarticulate brachiopod has a pedicle, but in general it lacks an articulated hinge. A good example of an inarticulate brachiopod is *Leptolobus* (see Fig. 9-1). The Early Paleozoic was the dominant time for inarticulates.

The articulates are the largest group of the phylum Brachiopoda and they are quite abundant in the fossil record, especially in the Paleozoic era.

The valves are articulated by means of a tooth and socket arrangement on the posterior margin of the valves. The oldest brachiopods have been found in rocks of Early Cambrian age. The articulate brachiopods are numerous in their distinctive group designs. Among these are the orthids, the terebratulids, the pentamerids, the rhynchonellids, the strophomenids, and the spiriferids among others.

The average size for fossil brachiopods is 50 mm, although some species have attained sizes up to 8 cm. The smallest sized brachiopod specimens found in the fossil record are less than 5 mm (for an example of the latter see Fig. 9-23).

Most brachiopods have ornamented shells, while a few species are smooth or naked. The ornamentation can be caused by foldings, plications, serrations, striations (costae) and concentric or radial lines or folds. The concentric lines usually indicate growth lines for the shells.

There are ten orders of brachiopods, four of which are in the Inarticulata, while the remaining six are in the Articulata. They are, in the class Inarticulata, orders: Lingulida (Early Cambrian to Recent), Acrotretida (Early Cambrian to Recent), Paterinida (Early Cambrian to Middle Ordovician), and Obolellida (Early Cambrian to Middle Cambrian). In the class Articulata, orders: Orthida (Early Cambrian to Late Permian), Strophomenida (Early Ordovician to Early Jurassic), Pentamerida (Middle Cambrian to Late Devonian), Rhynchonellida (Middle Ordovician to Recent), Spiririfirida (Middle Ordovician to Jurassic), and Terebratulida (Early Devonian to Recent).

Faunal listing of atremates in the Inarticulata (genera): CAMBRIAN: *Dicellomus, Lingulepis,* and *Lingulella.* ORDOVICIAN: *Elkania, Leptolobus* (see Fig. 9-1), and *Lingulasma.* SILURIAN: *Dinolobus, Lingula, Monomerella,* and *Trimerella.* DEVONIAN: *Barroisella* and *Lingula.* MISSISSIPPIAN: *Lingula.* PENNSYLVANIAN: *Trigonoglossa.* CRETACEOUS: *Lingula.*

Faunal listing of neotremates in the Inarticulata (genera): CAMBRIAN: *Acrothela, Acrotreta, Conodiscus, Dictyonina, Linnarssonella, Micromitra, Obolella, Paterina, Prototreta (Homotreta), Schizambon,* and *Yorkia.* ORDOVICIAN: *Conotreta, Craniops (Pholidops), Petrocrania, Philhedra, Schizocrania,* and *Trematus.* SILURIAN: *Craniops (Pholidops).* DEVONIAN: *Craniops (Pholidops), Lindstroemella, Orbiculoidea, Petrocrania, Philhedra, Roemerella,* and *Schizobolus.* PENNSYLVANIAN: *Orbiculoidea* and *Petrocrania.* PERMIAN: *Orbiculoidea.* JURASSIC: *Discinisca.* MIOCENE: *Discinisca.* PLIOCENE: *Discinisca.*

Faunal listing of paleotremates in the Articulata (genera): CAMBRIAN: *Kutorgina* and *Rustella.*

Faunal listing of impunctate Articulata (genera): CAMBRIAN: *Apheoorthis, Billingsella, Diraphora, Eoorthis, Finkelnburgia, Huenella, Nisusia, Ocneororthis, Palaeostrophia, Plectotrophia,* and *Wimanella.* ORDOVICIAN: *Ancistrorhynchus, Anomalorthis, Apheoorthis, Archaeorthis, Austinella, Camerella, Campylorthus, Catazyga, Clarkella, Cyclospira, Desmorthis, Diaphelasma, Diparelasma, Doleroides, Finkelnburgia, Glyptorthis, Hebertella* (see Figs. 9-3 and 9-4), *Hesperornomia, Hesperorthis, Mcewanella, Mimella, Multicostella, Nanorthis,*

9-1. Middle Ordovician. *Leptolobus* sp. An inarticulate brachiopod in matrix. Trenton Limestone, New York State. ×3.5.

9-3. Late Ordovician. *Hebertella sinuata* (Hall). Internal view of the brachial valve of an articulate (orthid) brachiopod in matrix. Leipers Limestone, Nashville, Davidson County, Tennessee. ×1.5

9-2. Middle Ordovician. *Pionodema subaequata* (Conrad). Brachial valve of an orthid (articulate) brachiopod. Decorah Shale, Fountain, Minnesota. 2 cm.

9-4. Late Ordovician. *Hebertella* sp. External view of the brachial valve of an articulate (orthid) brachiopod. Richmond Group, Cincinatti Series, Clarksville, Clinton County, Ohio. 2 cm.

Orthambonites, Orthorhynchula, Oxoplecia, Parastrophina, Plaesiomys (Dinorthis) (see Fig. 9-5), *Platystrophia* (see Figs. 9-6 and 9-7), *Plectorthus, Polytoechia, Pomatotrema, Protozyga, Retrorsirostra, Rhynchotrema, Rostricellula, Skenidioides, Syntrophia, Syntrophina, Syntrophopsis, Tetralobula, Triplesia, Tritoechia, Valcourea, Vellamo,* and *Zygospira.* SILURIAN: *Anastrophia, Atrypa, Atrypella, Atrypina, Brooksina, Clorindella (Barrandella), Coelospira, Conchidium, Crispella, Cryptothyrella, Cyrtia, Delthyris, Dictyonella, Dolcrorthis, Eospirifer, Gypidula, Hesperorthis, Hyattidina, Merista, Meristina, Nucleospira, Pentameroides, Pentamerus, Platystrophia, Plectatrypa, Rhipidium, Rhynchotreta, Sieberella, Stegerhynchus, Stricklandia, Trigonirhynchia (Uncinulina), Triplesia, Virgiana,* and *Whitfieldella.* DEVONIAN: *Acrospirifer, Ambocoelia, Anastrophia, Antispirifer, Athyris* (see Fig. 9-17), *Atrypa* (see Fig. 9-15a), *Atrypina, Brachyspirifer, Brevispirifer, Callipleura (Cyclorhina), Camarospira, Camarotoechia* (see Fig. 9-26), *Coelospira, Costellirostra, Costispirifer, Cyrtospirifer* (see Fig. 9-31) *Delthyris* (see Fig. 9-13), *Eatonia, Echinocoelia, Elytha, Emanuella, Eospirifer, Fimbrispirifer, Gypidula, Hypothyridina, Leiorhynchus, Leptocoelia, Martiniopsis, Meristella, Metaplasia, Mucrospirifer* (see Figs. 9-15b, 9-18 through 9-20), *Nucleospira, Orthostrophia,*

9-5. Late Ordovician. *Plaesiomys (Dinorthis) subquadrata* (Hall). Brachial valve of an articulate brachiopod. Richmond Group, Cincinatti Series, Clarksville, Clinton County, Ohio. 28 mm.

9-6. Late Ordovician. *Platystrophia moritura.* Brachial valve of an articulate brachiopod showing the foramen of the pedicle valve. Corryville Member, McMillan Formation, Cincinnati Series, Upper Stone Lick, Ohio. 28 mm.

A

B

9-7. Late Ordovician. *Platystrophia ponderosa* (Fuerste). Two views of an articulate brachiopod: A. Dorsal view of the brachial valve with part of the pedicle view showing (at top). B. Top (posterior) view showing the beak ridge, interarea, and the cardinal extremity. Richmond Member, Maysville Formation, Cincinnati Series, Indiana. 32 mm.

Paraspirifer (see Figs. 9-15d, 9-21 and 9-22), *Paurorhyncha, Pentagonia, Pentamerella, Plethorhynchus, Prosserella, Pugnoides, Pustulia (Pustulina), Skenidium, Spinocyrta (Platyrachella)* (see Figs. 9-24 and 9-25), *Syringospira, Tenticospirifer, Theodossia, Trigonirhynchia, Tylothyris, Uncinulina,* and *Uncinulus.* MISSISSIPPIAN: *Allorhynchus, Athyris, Brachythyris, Cleiothyridina, Composita, Leiorhynchus, Moorefieldella, Paryphorhynchus, Pugnoides, Schumardella, Spirifer, Stenoscisma (Camerophoria), Torynifer,* and *Tylothyris.* PENNSYLVANIAN: *Cleiothyridina, Composita* (see Figs. 9-32d and 9-35), *Crurithyris, Neospirifer, Phricodothyris, Spirifer,* and *Wellerella.* PERMIAN: *Neospirifer, Stenoscisma (Camerophoria), Torynechus (Uncinuloides),* and *Wellerella.* CRETACEOUS: *Peregrinella.*

Faunal listing of pseudopunctate Articulata (genera): ORDOVICIAN: *Bimuria, Christiania, Dactylogonia, Holtedahlina, Leptaena, Leptellina, Ptychoglyptus, Rafesquina* (see Figs. 9-8 and 9-9), *Sowerbyella, Sowerbyites,* and *Strophomena.* SILURIAN: *Brachyprion (Stropheodonta), Fardenia, Liljevallia, Plectodonta,* and *Protome-*

9-8. Late Ordovician. *Rafesquina* sp. Two specimens of brachial valves. The specimen on the reader's right shows a smaller brachiopod (see arrow) adhered to the shell's edge. Cincinnati Series, Cincinnati, Hamilton County, Ohio. ×2.

gastrophia. DEVONIAN: *Cymostrophia, Douvillina, Douvillinella, Hipparionyx, Leptaena, Leptaenisca, Leptostrophia (Rhytistrophia), Megastrophia* (see Fig. 9-16), *Nervostrophia (Sulcatostrophia), Pholidostrophia, Protoleptostrophia, Schuchertella, Stropheodonta* (see Fig. 9-15c, 9-27 and 9-28), *Strophonella,* and *Strophonelloides.* MISSISSIPPIAN: *Orthotetes* and *Schuchertella.* PENNSYLVANIAN: *Derbyia* (see Fig. 9-37), and *Meekella.* PERMIAN: *Derbyia, Leptodus,* and *Meekella.*

Faunal listing of the superfamily Productacea (the productid brachiopods, genera): DEVONIAN: *Anoplia, Buxtonia, Chonostrophia, Eodevonaria, Leptalosia, Lon-*

gispina, Neochonetes, and *Productella.* MISSISSIPPIAN: *Avonia, Buxtonia, Chonopectus, Diaphragmus, Dictyoclostus, Echinoconchus, Leptalosia, Linoproductus, Marginicinctus, Marginirugus, Neochonetes,* and *Setigerites.* PENNSYLVANIAN: *Chonetina, Dictyoclostus* (see Fig. 32a), *Heteralosia, Juresania* (see Figs. 9-32c and 9-33), *Linoproductus, Lissochonetes* (see Fig. 9-34), *Marginifera, Mesolobus* (see Fig. 9-32b), and *Teguliferina.* PERMIAN: *Aulosteges, Institella, Marginifera, Prorichthofenia, Strophalosiina,* and *Waagenoconcha.*

Faunal listing of punctate Articulata (genera): ORDOVICIAN: *Fascifera, Heterorthis, Mendacella, Pauror-*

9-9. Late Ordovician. *Rafesquina alternata* (Emmons). External view of a large pedicle valve. Richmond Member, Waynesville Formation, Weisburg, Indiana. 48 mm.

9-10. Silurian. *Resserella meeki* (Miller). External view of the brachial valve. Niagaran, South of Indianapolis, Marion County, Indiana. 18 mm.

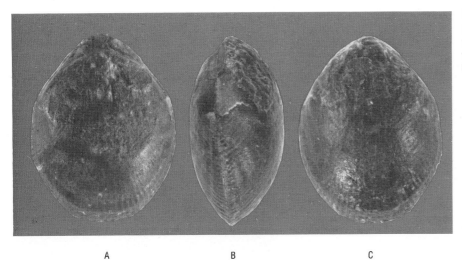

A   B   C

9-11. Early Devonian. *Rensselaerina medioplicata* Dunbar. Three views: A. Dorsal-brachial valve. B. Profile-commissure. C. Ventral-pedicle valve. Helderbergian, Montague, Sussex County, New Jersey. 2 cm.

9-12. Devonian. *Gluteus minimus* Davis and Semken. A fossil of uncertain affinity placed here temporarily with the brachiopods. It was originally thought to be a fish tooth *(Synthetodus)* in the literature. Maple Mill Formation, Kalona, Washington County, Iowa. ×7.5.

9-13. Early Devonian. *Delthyris perlamellosa* (Hall). Brachial view of a spiriferoid brachiopod. Helderbergian, New York State. 32 mm.

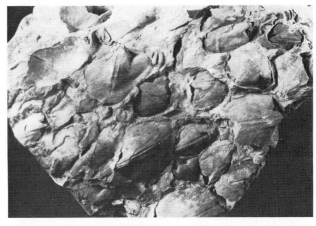

9-14. Devonian. *Reeftonia marwicki* Allan. A mass burial (most in pedicle view) of internal molds of brachiopods. No original shell material present. Reefton, New Zealand. 4 cm across. Courtesy: Geology Museum, Victoria University, Wellington, New Zealand.

9-15. Middle Devonian. a. Brachial view: *Atrypa reticularis* (Linné). b. Pedicle view: *Mucrospirifer mucronatus* var. *prolificum* (Stewart). c. Brachial view: *Stropheodonta demissa* (Conrad). d. Pedicle view: *Paraspirifer bownockeri* (Stewart). Silica Shale Formation, Sylvania, Lucas County, Ohio. Half size.

9-16. Middle Devonian. *Megastrophia concava* (Hall). A large brachial valve. Silica Shale Formation, Medusa Limestone Quarry, Sylvania, Lucas County, Ohio. 38 mm.

9-17. Middle Devonian. *Athyris spiriferoides* (Eaton). Pedicle and brachial views. Moscow Formation, Hamilton Group, Geneseo, Livingston County, New York. 3.5 cm.

9-18. Middle Devonian. *Mucrospirifer mucronatus* (Conrad). Brachial valve (top specimen), pedicle valve (bottom specimen). Silica Shale Formation, Sylvania, Lucas County, Ohio. 4 cm across.

9-19. Middle Devonian. *Mucrospirifer* sp. Brachial view of a spiriferoid brachiopod in matrix. Hamilton Shale, Clockville, New York. ×1.

9-20. Middle Devonian. *Mucrospirifer prolificus*. Brachial view of a spiriferoid brachiopod in limestone matrix. Marcellus Shale, Cortland, Cortland County, New York. ×1.

9-21. Middle Devonian. *Paraspirifer bownockeri* (Stewart). Pedicle valve of a large spiriferoid brachipod. Specimen is pyritized. Silica Shale Formation, Sylvania, Lucas County, Ohio. ×2.

9-22. Middle Devonian. *Paraspirifer bownockeri* (Stewart). Left: pedicle valve. Right: top view showing the fold of the brachial valve (top of specimen), the beak ridge, the interarea, and the cardinal extremity. Specimens are pyritized. Silica Shale Formation, Medusa Limestone Quarries, Sylvania, Lucas County, Ohio. ×1.

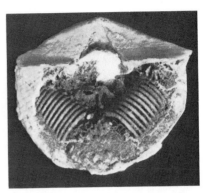

9-23. Early Devonian. *Dicoelosia (Bilobites)* sp. Pedicle and brachial views of a very small ''bilobite'' brachiopod. Helderbergian, New York State. ×14. (Photographs were prepared with a Scanning Electron Microscope).

9-24. Middle Devonian. *Spinocyrtia (Platyrachella) oweni* (Hall). Internal view of the pedicle valve showing the remains of the brachidium. This specimen has its spiralia incomplete and the jugum is missing. Internal paired appendages are rarely preserved in the fossil record. Sellersburg Limestone of Indiana. 33 mm.

9-25. Middle Devonian. *Spinocyrtia (Platyrachella) oweni* (Hall). A complete brachial valve covering the internal part of the pedicle valve where the brachia (brachidium) is located (see Fig. 9-24). Sellersburg Limestone, Indiana. 33 mm.

9-26. Middle Devonian. *Camarotoechia sappho*. Pedicle (left) and brachial (right) valves. Internal molds free from matrix. Hamilton Shale, Hubbardsville, Madison County, New York. 1.5 cm.

9-27. Middle Devonian. *Stropheodonta demissa* (Conrad). Pedicle (left) and brachial (right) valves. Specimen pyritized. Silica Shale Formation, Medusa Limestone Quarries, Sylvania, Lucas County, Ohio. 3.5 cm.

9-28. Middle Devonian. *Stropheodonta demissa* (Conrad). Dorsal view of the pedicle valve. Pyritized specimen. Silica Shale Formation, Medusa Limestone Quarry, Sylvania, Lucas County, Ohio. 3 cm.

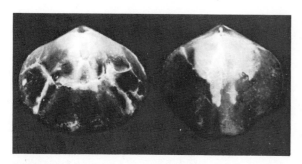

9-29. Late Devonian. *Calvinaria variabilis* (Whiteaves). Brachial views of two specimens. "B" Member, Hay River Formation, Enterprise, Northwest Territories, Canada. 22 mm.

9-30. Late Devonian. *Spinatrypa artica* Warren. Pedicle (left) and brachial (right) valve views. "B" Member, Hay River Formation, Enterprise, Northwest Territories, Canada. 23 mm.

9-31. Late Devonian. *Cyrtospirifer thalattodoxa* Crickmay. Brachial (left) and pedicle (right) valve views. "B" Member, Hay River Formation, Enterprise, Northwest Territories, Canada. 5 cm.

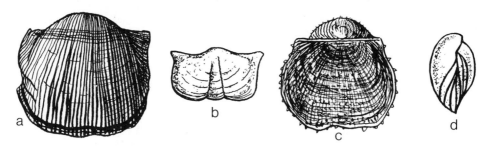

9-32. Pennsylvanian. a.*Dictyoclostus* sp. pedicle view. b. *Mesolobus* sp. pedicle view. c. *Juresania* sp. brachial view. d. *Composita* sp. profile view. Typical Pennsylvanian brachiopods. Upper Brush Creek Limestone, Conemaugh Group, Albany, Athens County, Ohio. ×1.

this, *Pionodema* (see Fig. 9-2), and *Reuschella*. SILURIAN: *Dalmanella, Homeospira, Isorthis, Parmorthis, Resserella* (see Fig. 9-10), and *Rhipidomella*. DEVONIAN: *Amphigenia, Aulacella, Beachia, Cariniferella, Centronella, Cranaena, Cryptonella, Cyrtina, Dicoelosia (Bilobites)* (see Fig. 9-23), *Etymothyris, Isorthis, Levenea, Parazyga, Platyorthis, Prionothyris, Rensselaeria, Rensselandia, Rhipidomella, Rhynchospirina, Schizophoria,* *Stringocephalus, Subrensselandia, Trematospira,* and *Tropidoleptus*. MISSISSIPPIAN: *Cranaena, Cyrtina, Dielasma, Dielasmella, Dimegelasma, Eumetria, Girtyella, Perditocardinia, Reticulariina, Rhipidomella, Schizophoria,* and *Syringothyris*. PENNSYLVANIAN: *Dielasma, Enteletes, Hustedia, Parenteletes, Punctospirifer* (see Fig. 9-36), *Rhipidomella,* and *Rhynchopora*. PERMIAN: *Dielasma, Heterelasma, Hustedia, Punctospirifer,*

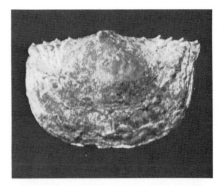

9-33. Pennsylvanian. *Juresania* sp. Pedicle (left) and brachial (right) valve views. Garner Formation, Lone Camp Group, Strawn Series, Wise/Jack County line, Texas. 3 cm.

9-34. Middle Pennsylvanian. *Lissochonetes geinitzianus* (Waagen). Ventral view (pedicle valve). Dickerson Formation, Millsap Lake Group, Strawn Series, Bridgeport, Wise County, Texas. 16 mm.

9-35. Late Pennsylvanian. *Composita subtilita* (Shepard). Brachial (left) and pedicle (right) valve views. Breckenridge Formation, Trinity Group, Cisco Series, Newcastle, Young County, Texas. 2 cm.

9-36. Late Pennsylvanian. *Punctospirifer kentuckiensis* (Schumard). Dorsal view (pedicle valve). Trickham Shale Member, Wayland Formation, Graham Group, Coleman County, Texas. 1.5 cm.

9-37. Late Pennsylvanian. *Derbyia crassa* (Meek and Hayden). Dorsal view (pedicle valve). Breckenridge Formation, Trinity Group, Cisco Series, Newcastle, Young County, Texas. 33 mm.

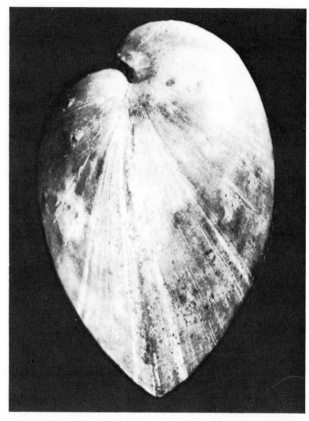

9-38. Late Jurassic. *Microthyridina lagenalis* (Schlotheim). Side view (see Fig. 9-39 for pedicle and brachial valves). Callovian, Upper Cornbrash, Rushden, Northants, England. 32 mm.

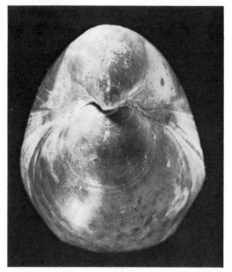

9-39. Late Jurassic. *Microthyridina lagenalis* (Schlotheim). Top view showing both valves (see Fig. 9-38 for side view). Callovian, Upper Cornbrash, Rushden, Northants, England. 22 mm.

9-40. Early Cretaceous. *Rhynchonella latissima* Sowerby. Brachial valve showing the prominent foramen of the pedicle valve. Upper Aptian, Lower Greensand, Faringdon, Berkshire, England. ✕1.5.

9-41. Early Cretaceous. *Kingena wacoensis* (Roemer). Brachial valve view (top) and pedicle valve view (bottom). Main Street Limestone, Crowley, Tarrant County, Texas 1.5 cm.

9-42. Late Cretaceous. *Choristothyris plicata* Say. Brachial valve view (left), side view showing commissure (center), and pedicle valve view (right). Navesink Formation (Middle Maestrichtian), Crosswicks Creek, Hornerstown, Monmouth County, New Jersey. ✕2.

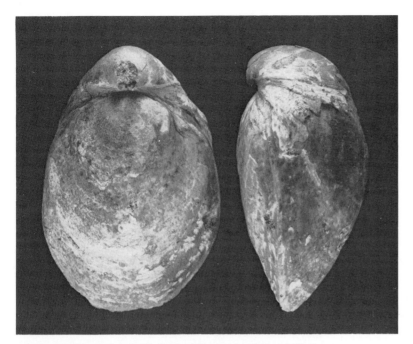

9-43. Eocene. *Oleneothyris harlani* Morton. Brachial valve view (left) and side view (right) with commissure. Hornerstown Formation, New Egypt, Ocean County, New Jersey. ×1.

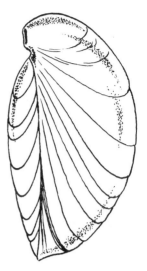

9-44. Eocene. *Oleneothyris harlani* Morton. Side view of a "lampshell" brachiopod (see Fig. 9-43). Hornerstown Formation, New Egypt, Ocean County, New Jersey. ×1.

and *Rhynchopora*. TRIASSIC: *Plectoconcha, Psioidea,* and *Spondylospira*. JURASSIC: *Terebratulina*. CRETACEOUS: *Choristothyris* (see Fig. 9-42), and *Kingena* (see Fig. 9-41). EOCENE: *Argyrotheca, Oleneothyris* (see Figs. 9-43 and 9-44), and *Terebratulina*. MIOCENE: *Dallinella, Terebratalia,* and *Terebratulina*.

The collecting of brachiopods: Brachiopods are probably among the most common invertebrate fossils found in Paleozoic rocks throughout the world, along with crinoid stems, and microfossils (foraminifera). In some formations in the United States, especially in Pennsylvanian age fossil beds, the brachiopods constitute over 50% of the collectible invertebrate fauna at some fossil sites. The Middle to Late Pennsylvanian fossil deposits of western Iowa and eastern Nebraska are good examples of highly productive collecting areas in the midcontinent of the United States. Common brachiopods in these deposits are *Composita subtilita* (see Figs. 9-32d and 9-35) and the productid *Mesolobus* (see Fig. 9-32b).

The productid brachiopods are especially abundant in areas such as the Permian "Glass Mountains" of Texas, and in the Pennsylvanian Upper Brush Creek Limestone beds of the Conemaugh Group in southeastern Ohio. In these Pennsylvanian limestone exposures, we find an abundance of these species (unfortunately, without the spines!): *Neochonetes, Mesolobus* (see Fig. 9-32b), *Dictyoclostus* (see Fig. 9-32a), and *Juresania* (see Figs. 9-32c and 9-33).

Neochonetes and *Mesolobus* were the smaller of these brachiopods, with numerous spines, especially near the

hinge line. The spines helped balance the brachiopod, but at times they caused some difficulties for the animal, as they would tangle themselves up in the neighboring bryozoan and sponge colonies. The spines would catch on the spicules of the sponges and the fenestration of the bryozoa.

Dictyoclostus was a free productid brachiopod with ear and flank spines, some of which attained the phenomenal length of up to 23 centimeters. Several long-spined specimens were found entangled in sponge colonies in the Permian Glass Mountain fossil beds of Texas. *Juresania* was an attached genus with both long and short spines. It pretty much stayed close to the bottom of the ocean floor.

The common brachiopod species found in the Late Cretaceous Navesink Formation sediments in central and southern New Jersey is *Choristothyris plicata* (see Fig. 9-42). This brachiopod is found associated with: the oysters, *Ostrea* (see Figs. 10-34 through 10-39), *Pyncnodonte* (see Figs. 10-22 through 10-25), and *Exogyra* (see Figs. 10-18 through 10-20, 10-27 and 10-28); the clams, *Inoceramus* (see Figs. 10-29 through 10-33), *Dianchora* (see Fig. 10-26), and *Cardium* (see Fig. 10-14), among others; and the belemnite (cephalopod), *Belemnitella americana* (see Fig. 12-75).

In the Eocene of New Jersey, the brachiopod *Oleneothyris harlani* (see Figs. 9-43 and 9-44) is most abundant. This species is so "packed" together in fossil beds of the Hornerstown Formation, that it is difficult to obtain a perfect specimen without breaking a dozen others in the trying.

10 MOLLUSCA (Pelecypoda)

The class Pelecypoda (Bivalvia—also known as the lamellibranchs) are bivalved shelled molluscs living in marine or fresh-water environments. This chapter deals primarily with the fossilized remains of those animals that lived in ancient oceans or bays.

Turning the two attached shells sideways (dorsal view—see Figs. 10-2 (left), 10-17 (top two specimens), and Fig. 10-56 (both specimens)—the hinge line of the shells show the symmetrical orientation), the symmetry of the shells in lateral view is the difference between a pelecypod (mollusc) and a brachiopod (see previous chapter).

The two shells can be ornamented in various ways, e.g., with spines projecting from the ventral and dorsal surfaces, with rugose or fine ribbings—a pattern following the shell's surface limitations, or even devoid of any ornamentation at all (smooth). In the case of the oysters (suborder Ostreina), the ornamentation can be rather eccentric or irregular. For example, see Figures 10-18, 10-19, 10-22, 10-25, 10-27, 10-28, and 10-34 through 10-40.

The valves of pelecypods are articulated by a series of interlocking "teeth and sockets" (see Fig. 10-37, center illustration, and 10-51 and 10-53; note the band of incised grooves near the hinge line of the specimen in Fig. 10-48, these grooves are the "teeth" of the animals). Both the teeth and the sockets are located along the dorsal hinge line. The crenulations ("teeth") pictured in 10-37 and 10-38, are a "gaping" morphology to control the intake of small particulate matter.

The geological range for the pelecypods is from the Middle Cambrian to Recent.

The earliest species were the paleoconchs, a most primitive and degenerate form. Examples of the paleoconchs (Palaeoheterodonta/lamellodonta of the Treatise) are: ORDOVICIAN: ? *Cymatonata (Psiloconcha)*, *Cuneamya*, *Modiolopsis (Orthodesma)*, and *Saffordia*. SILURIAN: *Buchiola* and *Cuneamya*. DEVONIAN: *Buchiola*, *Clinopistha*, *Edmondia*, *Glossites*, *Grammysia* (see Figs. 10-1 and 10-2), *Opisthocoelus (Ontaria)*, *Orthonota*, *Palaenatina*, *Praecardium (Paracardium)*, *Phthonia*, *Prorhynchus*, *Prothyris* (see Fig. 10-4), *Solemya* (Devonian to Recent),

and *Tellinopsis*. MISSISSIPPIAN: *Cardiola (Cardiopsis)*, and *Grammysia*. PENNSYLVANIAN: *Cardiola (Cardiopsis)*, *Chaenomya*, *Clinopistha*, *Edmondia*, and *Prothyris*. PERMIAN: *Chaenomya*, and *Edmondia (Allorisma)* (see Fig. 10-12).

The nuculaceans are part of the taxodont pelecypods and are distinguished by their smooth shell surfaces or finite concentric lines or growth rings on the valve surface. This group is called subclass Palaeotaxodonta in the Treatise. Some examples of this group are: ORDOVICIAN: *Ctenodonta* and *Nuculites (Cleidophorus)*. SILURIAN: *Nuculites*. DEVONIAN: *Nuculites* and *Palaeoneilo*. MISSISSIPPIAN: *Palaeoneilo*. TRIASSIC: *Nuculana* (Triassic to Recent). CRETACEOUS: *Acila*, *Nucula*, *Nuculana*, and *Yoldia*. EOCENE: *Acila*, *Nucula*, and *Nuculana*. MIOCENE: *Acila* and *Nuculana*.

The next superfamily of the Pelecypoda is Arcacea. The arcaceans have elongate valves, some subquadrate; the cardinal line has an angular anterior end, and the valves are rounded at the bottom. The shells have a lengthy and straight hinge line. Thick or thin concentric rings (or striae) are the ornamentation. Examples of this group are: ORDOVICIAN: *Cyrtodonta*, *Ischyrodonta*, *Vanuxemia*, and *Whitella*. SILURIAN: *Megalomoidea (Megalomus)*. DEVONIAN: *Megambonia* and *Paralleledon*. JURASSIC: *Cucullaea (Idonearca)*. CRETACEOUS: *Arca*, *Cucullaea (Idonearca)* (see Fig. 10-16), *Nemodon* (see Fig. 10-45), *Cucullaea* (see Fig. 10-46), and *Glycymeris*. MIOCENE: *Arca* and *Glycymeris* (see Fig. 10-51).

The next group or superfamily in Pelecypoda is Pteriacea of the schizodonts. The valves of these forms are unequal, and one valve may have a convexity (in some species). Surface ornamentation is grooved longitudinally. These forms are sometimes called "winged" clams due to the extended "ears" at the hinge line of their shells. For example: ORDOVICIAN: *Ambonychia (Byssonychia)*, *Cleionychia*, and *Pterinea*. SILURIAN: *Pterinea* and *Ptychopteria (Cornellites)*. DEVONIAN: *Actinodesma (Glyptodesma)*, *Conocardium*, *Dozierella*, *Gosseletia*, *Kochia (Loxopteria)*, *Leptodesma (Leiopteria)*, *Limoptera*

10-1. Middle Devonian. *Grammysia bisulcata* (Conrad). Complete shell. Right valve view in figure. Hamilton Shale, Hubbardsville, Madison County, New York. ×1.

10-2. Middle Devonian. *Grammysia arcuata* (Conrad). Left: specimen in side view showing both valves. Right: right valve. Hamilton Shale, Hubbardsville, Madison County, New York. 5 cm.

10-4. Middle Devonian. *Prothyris* sp. Elongate clam shell. Hamilton Shale, Hubbardsville, Madison County, New York. 5 cm length.

10-3. Middle Devonian. *Ptychopteria* (*Cornellites*) *flabella* (Conrad). An ornate "winged" clam. Hamilton Shale, Hubbardsville, Madison County, New York. 3.5 cm.

10-5. Middle Devonian. *Limoptera* sp. A "winged" clam in matrix, in association with the brachiopod *Camarotoechia*. Hamilton Shale, Hubbardsville, Madison County, New York. ×1.

10-6. Middle Devonian. *Modiomorpha* sp. Left valve of a clam. Hamilton Shale, Hubbardsville, Madison County, New York. 3 cm.

10-7. Middle Devonian. *Paracyclas elliptica* var. *occidentalis* (Hall). Hamilton Shale, Hubbardsville, Madison County, New York. 3.5 cm.

10-8. Middle Pennsylvanian. *Pteria* sp. A small "winged" clam. Labette Black Shale, Labette Formation, Marmaton Group, Des Moines Series, Westphalian D, Madrid, Boone County, Iowa. ×1.

10-9. Late Pennsylvanian. *Astartella concentrica* Conrad. Left valve. Wayland Shale Formation, Graham Group, Cisco Series, Stephens County, Texas. 28 mm.

10-10. Middle Pennsylvanian. *Dunbarella* sp. A primitive pecten. Labette Black Shale, Labette Formation, Marmaton Group, Des Moines Series, Westphalian D, Preston Branch, Des Moines River, Madrid, Boone County, Iowa. ×1.5.

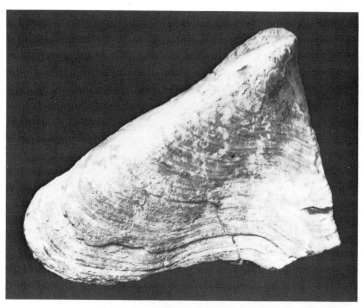

10-11. Early Permian. *Myalina copei* Whitfield. A myalid dysodont clam. Horse Creek Limestone, Moran Formation, Wichita Group, Shackelford County, Texas. 8.5 cm in length.

10-12. Middle Permian. *Edmondia* (*Allorisma*) *terminale* (Hall). Elongate clam. Blaine Formation, Pease River Group, Guadalupian Series, Northcentral Texas. 8 cm in length.

10-13. Early Jurassic. *Chlamys* sp. A mass burial of primitive pectens, all pyritized. (Sinemurian), Lias, Pödinghausen, Westphalia, Germany, Each pecten approx. 6.5 cm.

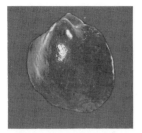

10-14. Late Cretaceous. *Cardium tenuistriatum* Whitfield. Internal cast of a clamshell. Navesink Formation (Middle Maestrichtian), Hornerstown, Monmouth County, New Jersey. ×1.

10-15. Late Cretaceous. *Crassatella (Crassatellites)* sp. Internal cast showing muscle scar attachments. Navesink Formation (Middle Maestrichtian), Hornerstown, Monmouth County, New Jersey. ×1.

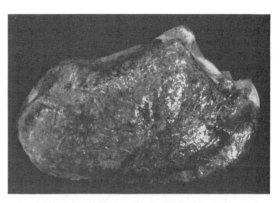

10-16. Late Cretaceous. *Cucullaea (Idonearca) neglecta* (Morton). Internal cast showing muscle scars. Navesink Formation (Middle Maestrichtian), Marlboro, Monmouth County, New Jersey. ×1.5.

10-17. Late Cretaceous. *Glossus (Isocardia)* sp. Four views of the same specimen. Top row, left to right: back and front lateral views showing lunule and hinge line. Bottom row: two views of a right hand valve. Cody Sandstone, 10 miles west of Tensleep, Washakie County, Wyoming. ×1.5.

10-18. Early Cretaceous. *Exogyra texana* Roemer. A primitive oyster. Walnut Clay, Fredericksburg Group (Albian), Decatur, Texas. 7.5 cm in length.

10-19. Early Cretaceous. *Exogyra arietina* Roemer. An ornate oyster. Two views: left valve (left) and side (right) showing curled-up beak. Upper Washita Formation, Pecos, Texas. 2.5 cm.

10-20. Late Cretaceous. *Pholas* sp. A boring or drilling pelecypod that bored holes into the shields of belemnites (see Fig. 2-6) as well as into the shells of *Cucullaea (Idonearca)* and *Inoceramus*. Navesink Formation (Middle Maestrichtian), Crosswicks Creek, Hornerstown, Monmouth County, New Jersey. 1.5 cm.

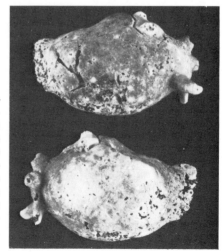

10-21. Late Cretaceous. *Cymella undata* Meek and Hayden. Mass burial of a clam community. Bearpaw Shale Formation (Late Campanian), Maple Creek, Saskatchewan, Canada. 20 cm across.

10-22. Early Cretaceous. *Gryphaea arcuata* (Lamarck). Oyster shell commonly called "devil's toenail." Grayson Marl (Albian), Tarrant County, Texas. ×1.5.

10-23. Late Cretaceous. *Pyncnodonte mutabilis* (Morton) (formerly called *Gryphaea convexa*). Large oyster shell. Navesink Formation (Middle Maestrichtian), Hornerstown, Monmouth County, New Jersey. Approx. half size.

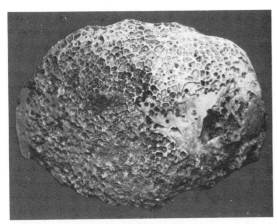

10-24. Late Cretaceous. *Pyncnodonte mutabilis* (Morton). Shell of this large pelecypod almost entirely eaten into by the boring sponge; *Cliona cretacica*. Navesink Formation (Middle Maestrichtian), Crosswicks Creek, Hornerstown, Monmouth County, New Jersey. Approx. half size.

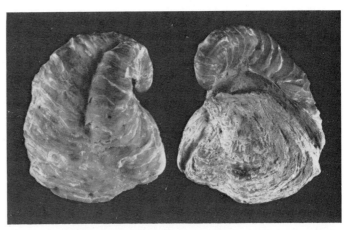

10-25. Early Cretaceous. *Gryphaea graysonana.* Large oyster shell in dorsal and ventral views. Grayson Marl (Albian), Lake Arlington, Tarrant County, Texas. 7 cm.

10-26. Late Cretaceous. *Dianchora echinata* (Morton). A fine example of a ''spiny'' clam shell. Dorsal view (left), side view (center), and internal view of the left or upper valve. The right or lower valve is missing. Species restricted to the Navesink Formation (Middle Maestrichtian), Crosswicks Creek, Hornerstown, Monmouth County, New Jersey × 1.5.

10-27. Late Cretaceous. *Exogyra costata* Say. A common oyster shell of the Atlantic and Gulf Coastal Plains. Navesink Formation (Middle Maestrichtian), Hornerstown, Monmouth County, New Jersey. Approx. half size.

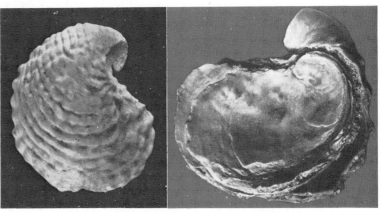

10-28. Late Cretaceous. *Exogyra cancellata* Stephenson. The exterior and interior views of a common large oyster shell. Note the rugose ornamentation on the shell surface (left) and the right valve (right) in place (see picture). Navesink Formation (Middle Maestrichtian), Crosswicks Creek, Hornerstown, Mounmouth County, New Jersey. Approx. half size.

10-29. Late Cretaceous. *Inoceramus* sp. Specimen is squashed flat in a limestone matrix. Cenomanian, Misberg bei Hannover, West Germany. Approx. half size.

10-30. Late Cretaceous. *Inoceramus fragilis* Hall and Meek. Pyritized specimen. Eagle Ford Shale, Cooks Branch Creek, Farmers Branch, Dallas County, Texas. 2.5 cm.

(see Fig. 10-5), *Lunulacardium (Honeoyea), Modiella, Mytilarca (Plethomytilus), Palaeopinna, Pterochaenia,* and *Ptychopteria (Cornellites)* (see Fig. 10-3). MISSISSIPPIAN: *Caneyella, Myalina, Pseudomonotis,* and *Pteronites.* PENNSYLVANIAN: *Bakewellia, Monopteria, Myalina, Pseudomonotis, Pteronites (Aviculopinna),* and *Pteria* (see Fig. 10-8). PERMIAN: *Myalina* (see Fig. 10-11). TRIASSIC: *Daonella, Halobia,* and *Monotis.* JURASSIC: *Buchia* and *Gervillia.* CRETACEOUS: *Buchia, Gervillia (Gervilliopsis), Inoceramus* (see Figs. 10-29 through 10-33), *Pinna,* and *Pteria.* EOCENE: *Pteria.*

PLIOCENE: *Pinna* (see Fig. 10-55).

Our next superfamily is Ostracea or the "oysters" belonging to the suborder Ostreina (of the Treatise). Their geological range is Late Triassic to Recent. These pelecypods are distinguished by their distorted shells (a result of their adhering to rocks or other shells in the ocean bed).

10-31. Late Cretaceous. *Inoceramus* cf. *salisburgensis* Fugger and Kastner. Nacreous shells in a large concretion. Spray Formation (Maestrichtian), Collishaw Point, Hornby Island, British Columbia, Canada. 16.5 cm across matrix.

10-32. Late Cretaceous. *Inoceramus balticus* (Boehm) ssp. *barabini* Morton. Shell in concretion. Spray Formation (Maestrichtian), Collishaw Point, Hornby Island, British Columbia, Canada. 11 cm.

10-33. Late Cretaceous. *Inoceramus simpsoni* Meek. A specimen with nacreous coating ("mother of pearl"). Pierre Shale, Glendive, Dawson County, Montana. Approx. half size.

10-34. Early Cretaceous. *Ostrea travisana*. The interior and exterior views of a large oyster shell. Comanchean Series, Northcentral Texas. 10 cm.

The larger valve is known as the left valve and is convex. The smaller is the right valve (or operculum)—in most species. The shells have a lamellate structure, unequal valves (as stated above), and commonly have rugose or radiating costae (ribs). The genera in the fossil record are as follows: CRETACEOUS: *Exogyra* (see Figs. 10-18, 10-19, 10-27, and 10-28), *Pyncnodonte* (see Figs. 10-22 through 10-25), and *Ostrea* (see Figs. 10-34 through 10-39). EOCENE: *Ostrea* (see Fig. 10-47).

The following are representative genera of 21 superfamilies. They are all listed together to expedite the faunal listing of the remaining pelecypods: ORDOVICIAN: *Aristerella, Colpomya, Cymatonota (Endodesma), Lyrodesma,* *Modiolodon, Modiolopsis (Modiodesma), Rhytimya* and *Whiteavesia.* SILURIAN: *Eurymyella, Goniophora,* and *Newsomella.* DEVONIAN: *Archanodon (Amnigenia), Cimitaria, Crenipecten, Cypricardella, Goniophora, Lyriopecten, Modiomorpha* (see Fig. 10-6), *Nyassa, Paracyclas* (see Fig. 10-7), *Pholadella, Pterinopecten (Vertumnia), Sanguinolites (Sphenotus),* and *Spathella.* MISSISSIPPIAN: *Aviculopecten* and *Cypricardella.* PENNSYLVANIAN: *Anthraconauta (Anthracomya), Astartella* (see Fig. 10-9), *Aviculopecten, Edmondia (Allorisma), Euchondria, Naiadites, Permophorus (Pleurophorus),* and *Schizodus.* PERMIAN: *Acanthopecten, Aviculopecten, Edmondia (Allorisma)* (see Fig. 10-12), *Permophorus (Pleurophorus),* and *Schizodus.* TRIASSIC: *Myophoria* and *Unio.* JURASSIC: *Lima, Pleuromya, Tancredia, Trigonia,* and *Unio.* CRETACEOUS: *Aenona, Anatimya, Anomia, Arctica, Atracta (Diploschiza), Caprina, Cardium* (see Fig. 10-14), *Clavagella, Coralliochamia, Corbicula, Corbula, Crassatella (Crassatellites)* (see Fig. 10-15), *Crenella, Cuspidaria, Cymbophora, Cymella* (see Fig. 10-21), *Cyprimeria, Estea, Glossus (Isocardia)* (see Fig. 10-17), *Legumen, Leptosolen, Lima, Linearia, Liopistha, Lucina, Modiolus (Volsella), Monopleura, Neithea* (see Fig. 10-42), *Panopea (Panope), Pecten* (see Figs. 10-41 through 10-43), *Pholadomya, Plicatula* (see Fig. 10-40), *Protocardia, Scambula, Sphaerium, Tancredia, Tellina, Tellinimera, Tenea, Toucasia, Trigonia, Trigonocallista, Unio,* and *Veniella.* EOCENE: *Dosiniopsis, Dreissena, Lucina, Modiolus (Volsella), Pitar, Tellina,* and *Venericardia.* MIOCENE: *Astarte* (see Fig. 10-52), *Clementia, Cumingia, Diplodonta (Taras), Ensis, Hiatella (Saxicava), Lithophaga, Lyropecten* (see Figs. 10-49 and 10-50), *Macoma, Mactra, Mya, Mytilus, Semele, Tellina, Thracia, Venus,* and *Chione* (see Fig. 10-53). PLIOCENE: *Clementia,* and *Glossus (Isocardia)* (see Fig. 10-56).

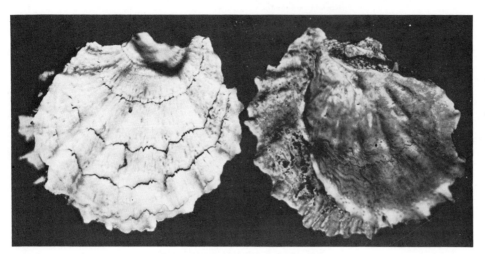

10-35. Late Cretaceous. *Ostrea panda* (Morton). Exterior and interior views of the same valve. Navesink Formation (Middle Maestrichtian), Crosswicks Creek, Hornerstown, Monmouth County, New Jersey. ×2.

10-36. Late Cretaceous. *Ostrea falcata* Morton. Cast with some original shell material. Merchantville Formation (Upper Santonian), St. Georges, New Castle County, Delaware. ×1.5.

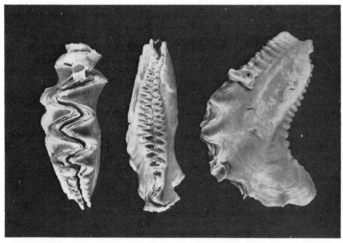

10-37. Late Cretaceous. *Ostrea falcata* Morton. Three views of the same specimen. Left: commissure opening (ventral view), Center: hinge teeth (dorsal view), Right: right valve with attachment scar (to another shell). Navesink Formation (Middle Maestrichtian), Hornerstown, Monmouth County, New Jersey. ×1.

10-38. Late Cretaceous. *Ostrea mesenterica* Morton. Two oyster shells. Navesink Formation (Middle Maestrichtian), Poricy Brook, Middletown, Monmouth County, New Jersey. ×1.5.

10-39. Late Cretaceous. *Ostrea mesenterica* Morton. Oyster shell. Navesink Formation (Middle Maestrichtian), Crosswicks Creek, Hornerstown, Monmouth County, New Jersey. ×1.5.

10-40. Late Cretaceous. *Plicatula urticosa* (Morton). Ornate clam shell. Navesink Formation (Middle Maestrichtian), Hornerstown, Monmouth County, New Jersey. ×1.5.

10-41. Early Cretaceous. *Pecten* sp. Grayson Marl (Albian), Lake Arlington, Tarrant County, Texas. 4 cm.

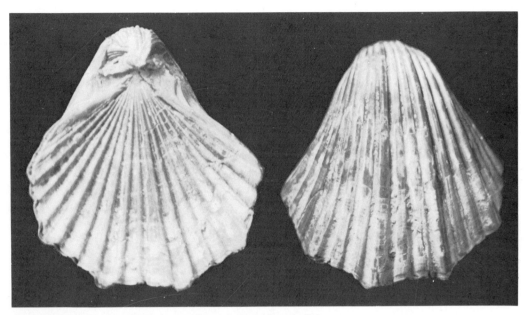

10-42. Early Cretaceous. *Pecten (Neithea) texanus* Roemer. Left and right valves. Fort Worth Limestone, Washita Group, Comanchean Series, Kingston, Marshall County, Oklahoma. 4 cm.

10-43. Late Cretaceous. *Pecten venustus* Morton. A rare and fragile shell. Navesink Formation (Middle Maestrichtian), Crosswicks Creek, Hornerstown, Monmouth County, New Jersey. ×5.

10-44. Late Cretaceous. *Spondylus gregalis* Morton inside the shell of *Pyncnodonte mutabilis* (Morton). Navesink Formation (Middle Maestrichtian), Crosswicks Creek, Hornerstown, Monmouth County, New Jersey. ×1.

10-45. Late Cretaceous. *Nemodon vancouverensis* Meek. Complete bivalve clam. Spray Formation (Maestichtian), Collishaw Point, Hornby Island, British Columbia, Canada. 7.8 cm.

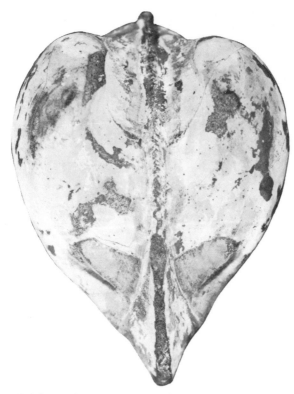

10-46. Eocene. *Cucullaea (Arca) gigantea* Conrad. "Beefsteak" clam. Internal cast of both valves. Aquia Formation (Wilcoxian), Potomac River, Marlborough Point, Prince Georges County, Virginia. Half size.

10-47. Eocene. *Ostrea compressirostra*. Giant oyster shell. Aquia Formation Wilcoxian), Belvedere Beach along the Potomac River, Osso, Prince Georges County, Virginia. Half size.

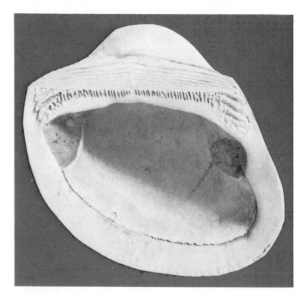

10-48. Eocene. *Arca* sp. Internal aspect of the right valve showing teeth and muscle scars. Aquia Formation (Wilcoxian), Fairview Beach, Potomac River, Prince Georges County, Virginia. ✕1.

10-49. Middle Miocene. *Pecten (Lyropecten) jeffersonius* Say. Left valve. Pungo River Marl Formation (Helvetian), Aurora, Beaufort County, North Carolina. ✕1.

10-50. Middle Miocene. *Pecten (Lyropecten) madisonius* Say. Right valve (exterior and interior views). Pungo River Marl Formation (Helvetian), Aurora, Beaufort County, North Carolina. Approx. half size.

10-51. Late Miocene. *Glycymeris parilis* (Conrad). Interior view of a right valve showing teeth and muscle scars. Yorktown Formation (Sahelian), Beaufort County, North Carolina. Half size.

10-52. Late Miocene. *Astarte cuneiformis* Conrad. Interior and exterior views of a right valve. The interior view (left) shows the teeth and muscle scars. Yorktown Formation (Sahelian), Beaufort County, North Carolina. ×1.5.

10-53. Late Miocene. *Venus (Chione) mercenaria* Linnaeus. Exterior and interior views of the right valve. St. Marys Formation, St. Marys River, Calvert County, Maryland. ×1.

10-54. Pliocene. *Arca* (?) *wagneriana* (Dall). Right valve-exterior view. Caloosahatchie Formation, Dade County, Florida. 10 cm.

10-55. Pliocene. *Pinna pectinata* Linnaeus. Razor clam shells in burial repose. Lillo Formation, Oorderen Sands, Kallo Harbour Works, Antwerpen, Belgium. 15 x 20 cm.

10-56. Pliocene. Top: *Glossus (Isocardia) humanus* Linné. Bottom: *Glossus (Isocardia) lunulata* (Nyst). Sands of Edegem, Belgium. 6 x 5 cm.

11 MOLLUSCA (Gastropoda)

The class Gastropoda is a principal division of the phylum Mollusca. They are a most varied group, both marine and fresh-water as well as land-dwelling. These diversified molluscs carried their homes along with them on land or in the sea.

The hard calcareous shell is made up of secreted lime ($Ca\ CO_3$—calcite or aragonite) from an area inside of the animal's mantle. The animal itself is soft-bodied and needs the shell to dwell in, as well as for protection from its enemies.

There are from one to two pairs of tentacles protruding from the head area. These are sensory in nature. The mouth of the gastropod has tiny bands or rows called radula. These are in fact a series of horn-like teeth with which the animal chews up its food. These horny teeth are rarely preserved in the fossil record, being so minute, and upon the death of the soft-bodied animal, bacterial degradation displaces (with the help of the currents) any trace of these teeth. The teeth are similar in appearance to conodonts and/or scolecodonts (see chapters 6 and 7). The name gastropod means "stomach-foot," and is quite descriptive of the animal's curious habit of thrusting out a lengthy muscle or "foot" with which it moves about.

Gastropods (snails, for example) have a diversity of shell design and structure. Some species live in coiled shells, while others live in non-coiled shells. The names for the various types of shells are almost as numerous as the different species. A few descriptive types are: dextral or right-handed (see Fig. 11-14 for example), advolute (see Fig. 11-34), turbinate (see Fig. 11-1), turriculate (see Fig. 11-21), biconical (see Fig. 11-38), and sinistral or left-handed (see Fig. 11-28)—just to name a few!

Faunal listing of Gastropoda (genera): CAMBRIAN: *Dirhachopea, Helcionella, Hypseloconus, Matherella, Owenella, Palaeacmaea, Pelagiella, Proplina, Scaevogyra, Scenella,* and *Sinuopea.* ORDOVICIAN: *Archinacella, Bucanella, Bucania (Loxobucania), Bucanopsis, Carinaropsis (Phragmostoma), Ceratopea, Clathrospira (Palaeoschisma), Clisospira, Cyclonema (Pleurotomaria), Cyrtolites (Microceras), Ecculiomphalus (Eccylopterus), Ectomaria (Solenospira), Eotomaria (Siroraphe), Euconia,* *Gasconadia, Helicotoma (Palaeomphalus), Holopea (Litiopsis), Lecanospira, Liospira, Lophospira, Loxoplocus, Lytospira, Macluritella, Maclurites (Maclurea), Meekospira (Eulima), Murchisonia (Hormotoma), Naticonema, Omospira, Omphiletina, Ophileta (Ozarkispira), Orospira, Ozarkina, Palaeozygopleura (Loxonema), Palliseria (Mitrospira), Phragmolites (Conradella), Plethospira, Proplina, Pterotheca (Clioderma), Raphistoma, Rhaphistomina (Rotellomphalus), Scalpingostoma, Schizopea (Roubidouxia), Sinuites, Sinuopea, Straparollina, Subulites (Polyphemopsis), Tetranota, Trochonema (Eunema), Trochonemella, Tropidodiscus (Tropidiscus), Pilina (Tryblidium),* and *Ulrichospira.* SILURIAN: *Auriptygma, Bellerophon (Waagenia), Bucanopsis, Coelocaulus (Vetotuba), Cyclonemina, Cyclonema (Pleurotomaria), Eotomaria (Spiroraphe), Euomphalopterus (Bathmopterus), Mourlonia (Ptychomphalina), Naticonema, Palaeozygopleura (Loxonema), Phanerotrema, Pilina (Tryblidium), Poleumita (Polytropis), Pycnomphalus, Rhaphistomina (Rotellomphalus), Sinuites (Sinutropis), Sinuspira, Subulites (Cyrtospira), Tremanotus (Gyrotrema), Trochonema (Eunema)* (see Fig. 11-1), and *Ulrichospira.* DEVONIAN: *Bellerophon (Waagenia), Bembexia, Bucanella, Cyrtonella, Elasmonema (Callonema), Euryzone, Ianthinopsis (Strobeus), Isonema, Mourlonia (Ptychomphalina), Murchisonia (Hormotoma), Naticonema, Palaeozygopleura (Loxonema)* (see Fig. 11-4), *Phanerotrema, Platyceras (Platyostoma)* (see Figs. 11-5 and 11-6), *Pleuronotus, Ptomatis, Ptychospirina (Ptychospira), Scalitina, Serpulospira, Straparolus (Philoxene), Strophostylus (Helicostylus), Trepospira, Tropidodiscus (Tropidiscus),* and *Turbonopsis (Turbo).* MISSISSIPPIAN: *Anematina (Holopea), Baylea, Bellerophon (Waagenia), Bulimorpha (Bulimella), Caliendrum (Foordella), Ceraunocochlis, Eotrochus, Euphemites (Euphemus), Girtyspira (Bulimella), Gosseletina (Gosseletia), Ianthinopsis (Strobeus), Lepetopsis (Patella), Meekospira (Eulima), Mourlonia (Ptychomphalina), Murchisonia (Hormotoma), Naticopsis (Naticodon), Palaeostylus (Pseudozygopleura), Platyceras (Orthonychia), Porcellia, Portlockiella (Portlockia), Rhineoderma, Stegocoelia (Hypegonia), Straparolus (Euom-*

11-1. Silurian. *Trochonema (Eunemia) nitidium* Ulrich and Scofield. Isolated snail shell. Niagara Group, Indiana. 18 mm.

11-2. Silurian. *Omphalotrochus discus.* Primitive marine snail in matrix. Dudley, England. Courtesy: Geology Museum, Victoria University, Wellington, New Zealand.

11-3. Early Devonian. *Tentaculites gyracanthus* (Eaton). A group of cast-off shells of a most primitive snail. Helderberg Series, Manlius Formation, near Albany, New York. × 1.

11-4. Middle Devonian. *Palaeozygopleura (Loxonema) hamiltoniae* (Hall). Internal cast of a turritella-like snail shell. Hamilton Shale, Hubbardsville, Madison County, New York. 2.5 cm.

phalus) (see Fig. 11-7), *Trepospira (Angyomphalus), Worthenia,* and *Yunnania.* PENNSYLVANIAN: *Anematina (Holopea), Anomphalus (Antirotella), Araeonema (Turbina), Baylea (Yvania), Bellerophon (Waagenia), Borestus, Bucanella (Plectonotus), Bucanopsis (Sphaerocyclus), Ceraunocochlis, Cyclozyga, Donaldina, Eoptychia, Eucochlis, Euconospira (Trechmannia), Euphemites (Euphemus), Girtyspira (Bulimella), Glabrocingulum, Goniasma (Goniospira), Helminthozyga, Hemizyga (Hyphantozyga), Hypselentoma, Knightella, Knightites (Cymatospira), Lepetopsis, Leptoptygma, Ianthinopsis (Strobeus), Meekospira, Microdoma, Microptychia, Naticopsis, Orthonema, Phymatopleura, Palaeostylus (Leptozyga), Platyceras, Shansiella* (see Fig. 11-10), *Soleniscus, Stegocoelia, Stephanozyga, Straparolus (Euomphalus)* (see Fig. 11-9), *Streptacis, Trachydomia, Trepospira* (see Fig. 11-8), *Worthenia* (see Fig. 11-8), and *Yunnania.* PERMIAN: *Anthracopupa, Baylea, Cylindritopsis, Lepetopsis, Murchisonia (Hormotoma), Naticopsis, Omphalotrochus, Plocostoma, Procerithiopsis, Straparolus, Trachydomia,* and *Warthia.* TRIASSIC: *Amberleya (Eucyclus), Amphitrochus, Purpuroidea, Talantodiscus,* and *Zygopleura.* JU-

RASSIC: *Amberleya (Eucyclus), Amphitrochus, Aptyxiella, Eucyclomphalus, Globularia, Natica, Nerinea, Nerinella, Planorbis, Pleurotomaria, Pseudonerinea, Ptychomphalus, Purpuroidea, Tylostoma, Valvata,* and *Zygopleura.* CRETACEOUS: *Acteon, Acteonella, Amauropsis* (see Fig. 11-11), *Amberleya, Ampullina, Anchura, Anisomyon, Anomalfusus* (see Fig. 11-12), *Aporrhais, Architectonica, Athleta, Atira, Avellana, Bellifusus, Belliscala, Beretra, Boltenella, Buccinofusus, Bulla, Busycon, Calliostoma, Calyptraphorus, Campeloma, Caricella, Cassiope, Cerithium, Cerithiopsis, Clio, Creonella, Cylichna, Cypraedia, Diodora, Drepanochilus, Drilluta, Ecphora, Ellipsoscapha, Eoancilla, Epitonium, Eucycluscala (Urceolabrum), Fasciolaria, Fulgerca, Fusimitra, Globularia, Goniokasis, Graphidula, Gyrodes, Helix, Hercorhynchus, Hydrotribulus, Lirosoma, Liopeplum, Littorina, Lunatia* (see Fig. 11-13), *Margarites, Mataxa, Mazzalina, Medionapus, Melania, Mesalia, Monodonta, Morea, Napulus, Natica* (see Fig. 11-14), *Nerinea, Nerinella, Odontofusus, Odostomia, Oligoptycha, Ornopsis, Paladmete, Perissolax, Physa, Piestochilus, Planorbis, Polinices, Pseudoliva, Pseudonerinea, Pterocerella, Pugnellus, Purpuro-*

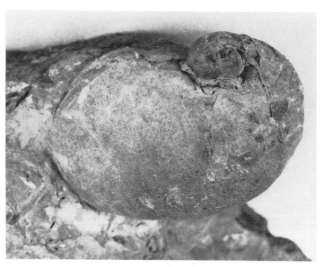

11-5. Devonian. *Platyceras* sp. Internal casts of a typical "coiled" snail shell. Note the small turritella-like *Palaeozygopleura (Loxonema)* shell impression below the right-hand specimen. Grassy Creek Limestone, Conowango Stage, Missouri. ×1.5.

11-6. Devonian. *Platyceras (Platyostoma) ventricosum* Conrad. Internal cast of a common snail shell of the Early Carboniferous Period. Rueben Hill locality, Montague, Sussex County, New Jersey. ×1.

11-7. Middle Mississippian. *Straparolus (Euomphalus) latus* (Hall). Flat, coiled snail with original shell material. St. Genevieve Limestone, Meromecian Series, Rolla, Missouri. 7.5 cm. dia.

11-8. Pennsylvanian. A. *Worthenia* sp. B. *Trepospira* sp. Common snail shells found throughout the Carboniferous Period. Desmoinesian Series, Northeastern Kansas. ×1.

11-9. Pennsylvanian. *Straparolus (Euomphalus) plummeri* Knight. Cast of a coiled snail shell with some original shell material adhering. Caseyville Limestone, Missouri Series, Illinois. 18 mm.

11-10. Pennsylvanian. *Shansiella carbonaria* (Norwood and Pratten). A common pyritized (actually marcasite) replacement of a snail shell found in coal shales. Carbondale Group, Farmington, Illinois. ×2.

11-11. Early Cretaceous. *Amauropsis bulbiformis* (Sowerby). Internal cast of a large marine snail shell. Duck Creek Formation, Lake Benbrook, Tarrant County, Texas. 5 cm in length.

11-12. Late Cretaceous. *Anomalofusus* sp. Posterior and anterior of an ornate marine snail shell. Not an internal cast, but the actual shell material preserved. Ripley Formation, Coon Creek, McNairy County, Tennessee. ×2.

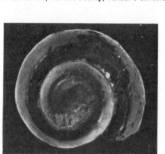

11-13. Late Cretaceous. *Lunatia halli* Gabb. Top view of an internal cast of a coiled marine snail shell. Navesink Formation (Middle Maestrichtian), Crosswicks Creek, Hornerstown, Monmouth County, New Jersey. ×1.5.

idea, Pyrgulifera, Pyropsis, Remera, Ringicula, Sargana, Schizobasis, Scobinella, Seminola, Surcula, Tessarolax, Tornatellaea, Trochactaeon, Trochifusus, Tryoniella, Tulotoma, Turbonilla, Turritella (see Figs. 11-15, 11-16, 11-17, and 11-18), *Tylostoma, Vaginella, Valvata, Vanikoropsis, Viviparus* (see Fig. 11-19), *Volutoderma, Volutomorpha* (see Fig. 11-20), and *Xenophora*. PALEOCENE: *Bittium, Coronia, Ficus, Fulgurofusus, Lapparia, Levifusus, Littorina, Mesalina, Natica, Orthosurcula, Priscoficus, Pyramidella, Tornatellaea,* and *Volutocorbis*.

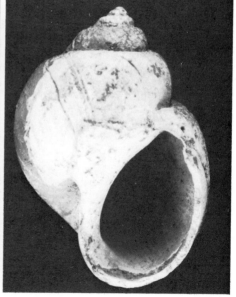

11-14. Late Cretaceous. *Natica* sp. Posterior and anterior views of a marine snail shell with actual shell material preserved. Ripley Formation, Coon Creek, McNairy County, Tennessee. ×2.

11-15. Late Cretaceous. *Turritella encrinoides* Morton. An internal cast of a spiraled snail shell. Navesink Formation, Hornerstown, Monmouth County, New Jersey. ×1.5.

11-16. Late Cretaceous. *Turritella vertebroides* Morton. Internal cast. Merchantville Formation, St. Georges, New Castle County, Delaware. ×1.5.

11-18. Early Cretaceous. *Turritella* sp. Internal cast. Goodland Limestone, Lake Worth, Tarrant County, Texas. 3 cm.

11-19. Early Cretaceous. *Vivaparus* sp. An internal cast. Goodland Limestone, Lake Worth, Tarrant County, Texas. 7.5 cm length.

11-17. Late Cretaceous. *Turritella vertebroides* Morton. Internal cast of a spiraled marine snail shell. Navesink Formation (Middle Maestrichtian), Crosswicks Creek, Hornerstown, Monmouth County, New Jersey. ×1.5.

11-20. Late Cretaceous. *Volutomorpha ponderosa* Whitfield. An internal cast of a large marine snail shell. Navesink Formation (Middle Maestrichtian), Atlantic Highlands, Monmouth County, New Jersey. 11 cm.

EOCENE: *Amaurellina, Ampullina, Ancillopsis, Architechtonica, Bolis, Bruclarkia, Buccitriton, Calicantharus, Calyptraea, Calyptraphorus, Campeloma, Cancellaria, Cantharus, Caricella, Cassidaria, Cerithium (Campanile)* (see Fig. 11-21), *Cirsotrema, Clathrodrillia, Clavilithes, Clio, Cochlespira, Conomitra, Conus, Coronia, Crepidula, Drillia, Ectinochilus, Ficopsis, Ficus, Fossarus, Fusimitra, Galeodea, Harpa, Helix, Lacinia, Lapparia, Latirus, Levifusus, Mangilia, Marginella, Melanella, Mesalia, Murex, Natica, Neptunea, Neverita, Oliva, Olivella, Olivula, Orthaulux, Perissolax, Planorbis, Pleurolimnaea, Priscoficus, Ringicula, Scobinella, Sinum, Siphonalia, Strepsidura, Sulcocypraea, Surculites, Terebra, Tornatellaea, Tritiaria, Tritonalia, Tritonoatractus, Turris, Turritella* (see Figs. 11-22 through 11-27), *Typhis, Urosalpinx, Vaginella, Valvata, Viviparus, Volutocorbis,* and *Volvula.*
MIOCENE: *Acteonina, Alabina, Aphera, Architechtonica, Aurinia* (see Fig. 11-37), *Bittium, Bruclarkia, Buccinum, Bulla, Bulliopsis, Busycon* (see Fig. 11-28), *Caecum, Calliostoma, Calyptraea, Cancellaria* (see Fig. 11-31), *Cavolina, Clathrodrillia, Conus* (see Fig. 11-38), *Crassispira, Crepidula, Crucibulum, Cymatosyrinx, Cypraea, Diodora, Ecphora* (see Figs. 11-32 and 11-33), *Epitonium, Fasciolaria* (see Fig. 11-36), *Forreria, Haliotis, Hastula, Heilprinia, Latirus, Lemintina, Limacina, Liro-*

11-21. Eocene. *Cerithium (Campanile) giganteum* Lamarck. Huge turritella-like marine snail shell. Calcaire sableux à *C. giganteum,* (Middle Lutetian), Damery, Department Marne, France. 54 cm in overall length.

11-22. Eocene. *Turritella mortoni* Conrad. An internal cast of a spiraled snail shell. (See Fig. 11-23). Aquia Formation (Ypresian), Belvedere Beach, Prince Georges County, Virginia. ×1.

11-23. Eocene. *Turritella mortoni* Conrad. Complete spiraled snail shell with original shell material preserved. Aquia Formation (Ypresian), Belvedere Beach, Prince Georges County, Virginia. ×1.

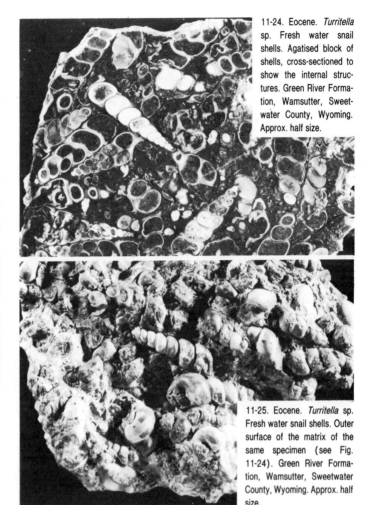

11-24. Eocene. *Turritella* sp. Fresh water snail shells. Agatised block of shells, cross-sectioned to show the internal structures. Green River Formation, Wamsutter, Sweetwater County, Wyoming. Approx. half size.

11-25. Eocene. *Turritella* sp. Fresh water snail shells. Outer surface of the matrix of the same specimen (see Fig. 11-24). Green River Formation, Wamsutter, Sweetwater County, Wyoming. Approx. half size.

11-26. Late Eocene. *Turritella imbricataria* Lamarck. Spiraled marine snail shell. Barton Clay (Bartonian), Barton-on-sea, Hants, England 31 mm in length.

soma, *Lunatia* (see Fig. 11-35), *Lyria, Mangelia, Marginella, Mitra, Mitrella, Molopophorus, Nassa, Nassarius, Natica, Neptunea, Neverita, Nucella, Odostomia, Oliva* (see Fig. 11-30), *Olivella, Opalia, Orthaulax, Pterygia, Ptychosalpinx, Scalaspira, Sinum, Strombus, Sycotypus, Terebra (Subula)* (see Fig. 11-29), *Trigonalia, Trigonostoma, Trophosycon, Turbonilla, Turcica, Typhis, Vermicularia, Vexillum, Xanchus,* and *Xenophora*. PLIOCENE: *Acteonina, Amphissa, Aporrhais* (see Fig. 11-42),

Barbarofusus, Bittium, Busycon, Calyptraea, Cancellaria, Cantharus, Cavolina, Clathrodrillia, Conus, Crepidula, Cymatosyrinx, Cypraea, Epitonium, Forreria, Goniokasis (see Fig. 11-39), *Heilprinia, Littorina, Mangella, Mitrella, Nassarius, Neptunea, Oliva, Olivella, Opalia, Physa, Planorbis, Scalez, Sinum, Strombus, Sycotypus, Terebra, Trigonostoma, Turritella* (see Fig. 11-40), *Vermicularia,* and *Voluta.* PLEISTOCENE: Species essentially the same as the preceding age.

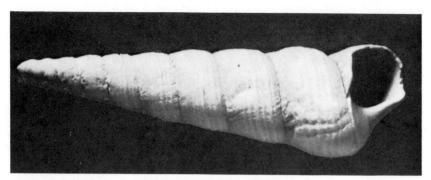

11-27. Late Eocene. *Turritella alveata* Conrad. Spiraled marine snail shell. Moodys Branch Formation (Claibornian), Jackson, Hinds County, Mississippi. 42 mm in length.

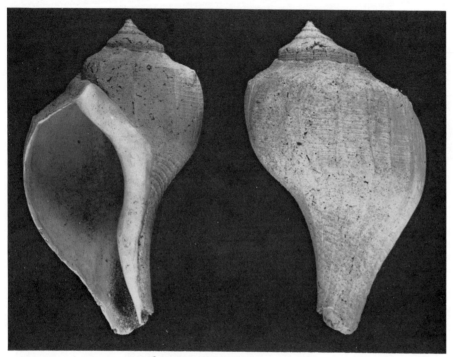

11-28. Late Miocene. *Busycon coronatum* (Conrad). Anterior and posterior views of a large conch shell. Yorktown Formation (Tortonian), Aurora, Beaufort County, North Carolina. 16.5 cm.

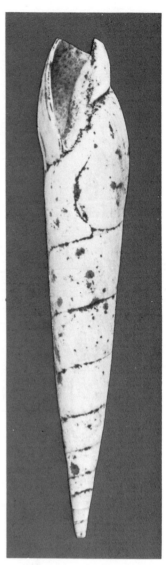

11-29. Miocene. *Terebra (Subula) plicaria* Bast. High spired tropical snail shell. Saché, Department Touraine, France. ×1.5.

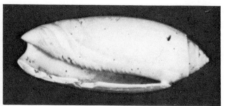

11-30. Late Miocene. *Oliva litterata* Lamarck. Olive shell. Yorktown Formation (Tortonian), Aurora, Beaufort County, North Carolina. ×1.

11-31. Late Miocene. *Cancellaria alternata* Conrad. Posterior and anterior views of an ornate snail shell. Yorktown Formation (Tortonian), Aurora, Beaufort County, North Carolina. ×2.

11-32. Middle Miocene. *Ecphora quadricostata* (Say). Ornate marine snail shell. Pungo River Marl Formation (Helvetian), Aurora, Beaufort County, North Carolina. ×1.

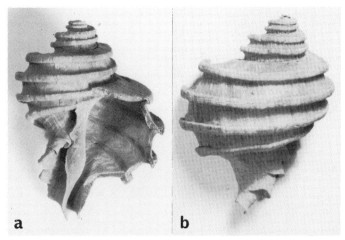

11-33. Middle Miocene, *Ecphora quadricostata* (Say). a. Anterior. b. Posterior. Views of an ornate marine snail shell. (See Fig. 11-32). Pungo River Marl Formation (Helvetian), Aurora, Beaufort County, North Carolina. ×1.

11-34. Late Miocene. *Busycon* sp. 3 views of a large conch shell. Yorktown Formation (Tortonian), Aurora, Beaufort County, North Carolina. 11 cm.

11-35. Late Miocene. *Lunatia heros* (Say). Coiled marine snail shell. Yorktown Formation (Tortonian), Aurora, Beaufort County, North Carolina. ×1.5.

11-36. Late Miocene. *Fasciolaria rhomboidea* Rogers. Large bulbous, spiraled marine snail shell. Anterior and posterior views. Yorktown Formation (Tortonian), Aurora, Beaufort County, North Carolina. 8 cm.

11-37. Late Miocene. *Aurinia mutabilis* (Conrad). Large bulbous marine snail shell. Posterior and anterior views. Yorktown Formation (Tortonian), Aurora, Beaufort County, North Carolna. 8.5 cm.

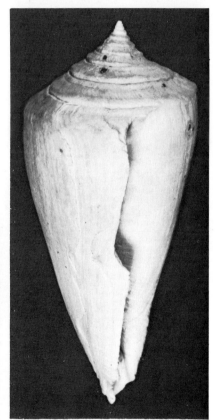

11-38. Late Miocene. *Conus adversarius* Conrad. Complete shell of the "poisonous" snail. Yorktown Formation (Tortonian), Aurora, Beaufort County, North Carolina. ×1.5.

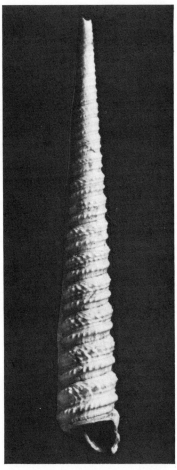

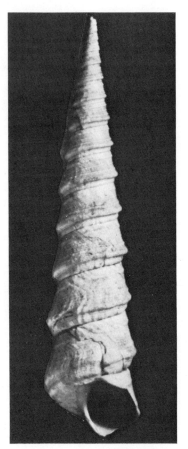

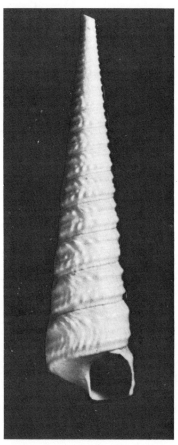

11-39. Pliocene. *Goniokasis (Turritella) kettle-manensis* Arnold. High spired snail shell. San Joaquin Formation, Kettleman Hills, Kern County, California. 2 cm in length.

11-40. Pliocene. *Turritella perattenuata* Heilprin. Spiraled marine snail shell. Caloosahatchie Marl, Moore Haven, Glades County, Florida. 3.5 cm in length.

11-41. Pleistocene. *Turritella communis* Risso. Spiraled marine snail shell. Eemien, Bosch Plaat, Terschelling, The Netherlands (Holland). 3 cm in length.

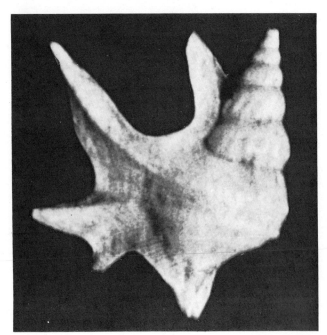

Fossilized gastropod shells are probably some of the most common fossils found in shell beds—especially in some Miocene deposits such as the Calvert Cliffs (now called "Scientist's Cliffs") along the Chesapeake Bay in Maryland. They are almost as commonly distributed in some formations as the varieties of coelenterates (corals) and the pelecypods (clams, oysters and pectens).

At the present time, the genus *Tentaculites* (see Fig. 11-3) is of uncertain origin, but is placed here tentatively with the Gastropoda. Tentaculites are listed as "Mollusca Incertae Sedis" in Shimer and Shrock, 1944, p. 526; and until the Treatise decides their exact taxonomic affinities, this author chooses to recognize their tentative placement in with the class Gastropoda.

Scaphapods and amphineurans are separate classes in the phylum Mollusca and are listed at the end of the Gastropoda section in Shimer and Shrock.

11-42. Pliocene. *Aporrhais scaldensis.* Ornate marine snail shell. Katendijk Formation (Regteren altena), 5th Harbour works, Kallo, Antwerpen, Belgium. 2.4 cm.

12 MOLLUSCA (Cephalopoda)

The cephalopods are a marine group (no representatives of this class are known from fresh water deposits—fossil or recent) of fast-moving invertebrates with a long geological history—from the Late Cambrian to Recent. The following six subclasses comprise the cephalopod group: Nautiloidea (nautiloids), Bactritoidea, Endoceratoidea, Actinoceratoidea, Ammonoidea (ammonoids), and Coleoidea (belemnoids/belemnites, squids, octopi, etc.). The subclasses Nautiloidea, Ammonoidea, and Coleoidea will be discussed in this chapter.

Cephalopods are predatory creatures with eyes and other well-developed sensory organs, and with tentacles with sucker discs or cups. They have the ability to change color to match their background, as well as possessing an "ink sac" used for squirting a dark fluid to cover their escape from an enemy. Their mouths contain a "toothlike" beak with which they tear apart their prey.

The cephalopod class contains the extinct ammonoids (ammonites or "Ammon's stones"), and the nautiloids, squids and octopi. The squids and octopi are the best known and most abundant groups of present day cephalopods, while the nautiloids still exist, descended from the fossil record, although not in any great numbers. They are represented by *Nautilus* and the argonauts.

We are mainly concerned here with the fossil forms of the class Cephalopoda, such as the extinct ammonoids, as well as the older (earlier) straight cephalopods which may have given rise to the coiled forms (ammonoids and nautiloids). We are also interested in this chapter in the primitive or early nautiloids, and the squid-like belemnoids, such as *Passoloteuthis* (see Fig. 12-24), and *Belemnitella* (see Fig. 12-75). The cephalopods evolved during the Late Cambrian Period. They were predominantly elongated with conical shells (see Figs. 12-1 and 12-2 for examples). A group of these evolved into a semi-coiled form (see Fig. 12-3), and eventually gave rise to some of the coiled forms of the ammonoids and nautiloids. Another branch of the straight-shelled forms continued on into the Devonian period, and some species even persisted up into the Carboniferous (particularly, the Pennsylvanian) period.

The "coiled" forms which would become the ammonoids and nautiloids, were most prolific and diversified. In the case of the nautiloids, their evolution was fairly steady from their first appearance in the Late Cambrian with the species *Palaeoceras,* and in the Silurian period with such forms as *Ophioceras (Euophioceras)* and *Lechritochoceras,* to their domination in the Devonian with representatives of *Centroceras* and *Goldringia.* Some Carboniferous nautiloids were *Cooperoceras* (Permian) *Domatoceras,* and *Liroceras.* The representative nautiloids of the Mesozoic were *Cosmonautilus, Eutrephoceras* (see Figs. 12-40 and 12-41), and *Proclydonautilus.* Tertiary forms include *Aturia, Cimomia,* and *Hercoglossa.* The ammonoids (ammonites) probably evolved from earlier (Devonian) nautiloids, possibly from the genus *Bactrites* (in the subclass Bactritoidea) which was a straight cephalopod with a marginal siphuncle arrangement, a feature which evolved in nautiloids. The ammonoids were predominant in the Jurassic and Cretaceous and became extinct at the end of the Mesozoic period.

Faunal listing of the ammonoids (genera): DEVONIAN: *Acanthoclymenia, Agoniatites* (see Figs. 12-4 and 12-5), *Imitoceras, Lobobactrites, Mantioceras* (see Fig. 12-6), *Parawocklumeria, Probeloceras, Soliclymeria, Sporadoceras, Timanites,* and *Tornoceras.* MISSISSIPPIAN: *Cravenoceras, Dimorphoceras, Eumorphoceras, Girtyoceras, Imitoceras, Lyrogoniatites, Muensteroceras, Neoglyphioceras, Prodromites, Prolecanites,* and *Protocanites.* PENNSYLVANIAN: *Agathiceras, Anthracoceras, Bisatoceras, Boesites, Dimorphoceras, Eoasianites, Eothalassoceras, Gastrioceras, Gonioloboceras, Imitoceras, Neodimorphoceras, Neoschumardites, Owenoceras, Paralegoceras, Peritrochia, Pronorites, Prouddenites, Pseudoparalegoceras, Schistoceras, Schumardites, Uddenites,* and *Wellerites.* PERMIAN: *Adrianites, Agathiceras, Artinskia, Cyclolobus, Daraelites, Eoasianites, Imitoceras, Medlicottia, Metalegoceras, Neoschumardites, Paraceltites, Peritrochia, Perrinites, Popanoceras, Properrinites, Propinacoceras, Pseudogastrioceras, Stacheoceras, Timorites, Waagenoceras, Xenaspis,* and *Xenodiscites.* TRIAS-

SIC: *Acrochordiceras, Anasibirites, Arniotites, Arcestes, Ceratites, Clypeoceras, Cochloceras, Columbites, Cuccoceras, Discotropites, Hungarites, Juvavites, Lecanites, Leconteiceras, Meekoceras, Ophioceras, Pseudosageceras, Rhacophyllites, Sagenites, Tirolites, Trachyceras, Tropites, Ussuria, Xenaspis,* and *Xenodiscus.* JURASSIC: *Amaltheus, Arnioceras* (see Fig. 12-22), *Asteroceras* (see Figs. 12-8, 12-9, and 12-12), *Aulocosphinctes, Cardioceras* (see Fig. 12-10), *Cenoceras* (see Fig. 12-17), *Chondroceras (Defonticeras), Dactylioceras* (see Fig. 12-11), *Durangites, Fanninoceras, Garantiana (Garantia), Glochiceras, Haploceras, Harpoceras, Hildoceras* (see Fig. 12-15), *Idoceras, Kosmoceras, Kossmatia, Liparoceras, Lytoceras* (see Fig. 12-14), *Macrocephalites* (see Fig. 12-7), *Mazapilites, Microderoceras* (see Fig. 12-18), *Nebrodites, Normannites, Ochetoceras, Oecoptychius, Oppelia, Oxy-*

noticeras (see Fig. 12-13), *Parechioceras* (see Fig. 12-21), *Parkinsonia* (see Fig. 12-16), *Perisphinctes, Phlycticeras, Phylloceras, Promicroceras, Prorasenia, Reineckeia, Scarburgiceras, Seymourites, Schlotheimia, Streblites, Teloceras* (see Fig. 12-19), *Vermiceras,* and *Zemistephanus.* CRETACEOUS: *Acanthoceras, Acanthodiscus, Acanthoscaphites* (see Figs. 12-28 through 12-30), *Ancyloceras, Baculites* (see Figs. 12-72 through 12-74), *Beudanticeras, Cirroceras (Didymoceras)* (see Fig. 12-36), *Collingnoniceras (Prionotropis), Crioceratites (Crioceras), Desmoceras* (see Fig. 12-35), *Desmophyllites (Schluteria)* (see Fig. 12-67), *Diploceras, Discoscaphites* (see Figs. 12-31 through 12-34, and 12-44), *Distoloceras, Douvilleiceras, Dufrenoyia, Emperoceras, Engonoceras, Euhoplites* (see Fig. 12-37), *Exiteloceras, Gastroplites, Gaudryceras* (see Fig. 12-43), *Hamites, Hamulina, Ho-*

12-1. Sea beach 500,000,000 years ago during the Ordovician Period. Painting by Charles R. Knight, in the Ernest R. Graham Hall of the Field Museum in Chicago. Courtesy: Field Museum of Natural History, Chicago.

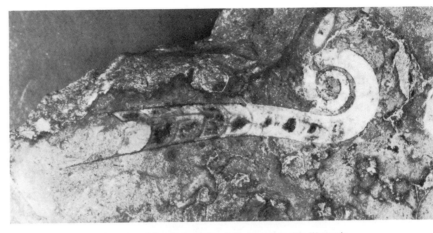

12-3. Ordovician. *Lituites lituus.* A partially coiled cephalopod—the beginnings of the "Ammon's stones" (ammonites). Tjusby, Öland, Sweden. 13 cm in length. Courtesy: Siber and Siber Ltd., Aathal, Switzerland.

12-2. Ordovician. *Michelinoceras (Orthoceras) sociale* Hall. A "straight" cephalopod. Maquoketa Shales, Saint Genevieve, St. Genevieve County, Missouri. 10 cm in length.

12-4. Middle Devonian. *Agoniatites vanuxemi* (Hall). An ammonoid. Cherry Valley Limestone, Nedrow, Onondaga County, New York. 38 mm dia.

12-5. Middle Devonian. *Agoniatites* sp. A primitive ammonoid, sliced to show its internal chambers. Matrix is marble. Erfoud, Morocco. 15 cm dia. Courtesy: Siber and Siber Ltd., Aathal, Switzerland.

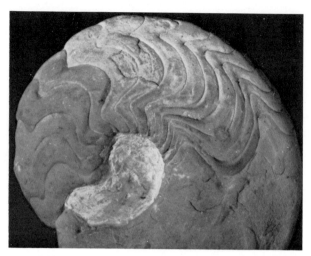

12-6. Late Devonian. *Mantioceras sinuosum* (Hall). An early ammonoid. "B" Member, Hay River Formation, Enterprise, North West Territory, Canada. 9.5 cm across specimen.

12-8. Early Jurassic. *Asteroceras obtusum* (Sowerby). Three ammonoid specimens in matrix. Lower Lias (Sinemurian), Obtusum Zone, Lyme Regis, Dorset, England. 14 cm dia. (the larger ammonoid). Courtesy: Brooklyn Childrens Museum, Brooklyn, New York.

12-7. Middle Jurassic. *Macrocephalites* sp. An ammonoid with crystalized internal chambers (drusy quartz crystals). Madagascar. 5.5 cm. Courtesy: Siber and Siber Ltd., Aathal, Switzerland.

12-9. Early Jurassic. *Asteroceras obtusum* (Sowerby). A fairly large ammonoid. (See Fig. 12-8). Lower Lias (Sinemurian), Obtusum Zone, Lyme Regis, Dorset, England. 18 cm. Courtesy: Siber and Siber Ltd., Aathal, Switzerland.

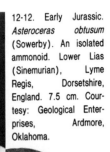

12-12. Early Jurassic. *Asteroceras obtusum* (Sowerby). An isolated ammonoid. Lower Lias (Sinemurian), Lyme Regis, Dorsetshire, England. 7.5 cm. Courtesy: Geological Enterprises, Ardmore, Oklahoma.

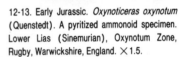

12.10. Late Jurassic. *Cardioceras subexcavatum*. An isolated ammonoid. Upper Oxford Clay (Lower Oxfordian), Warboys, Huntingdonshire, England. 5 cm.

12-11. Early Jurassic. *Dactylioceras commune* (Sowerby). An isolated ammonoid. Upper Lias (Toarcian), Bifrons Zone, Whitby, Yorkshire, England. ✕1.5.

12-13. Early Jurassic. *Oxynoticeras oxynotum* (Quenstedt). A pyritized ammonoid specimen. Lower Lias (Sinemurian), Oxynotum Zone, Rugby, Warwickshire, England. ✕1.5.

12-14. Late Jurassic. *Lytoceras* sp. Newly discovered giant ammonoid. Excavated and reconstructed by the staff of the New Zealand Geological Survey. 1.45 meters dia. Heterian Stage, Whakapirau Road, Taharoa, S. of Kawhia Harbour, New Zealand. Courtesy: New Zealand Geological Survey, Lower Hutt, New Zealand, and with permission of Dr. G. R. Stevens. (Dr. Stevens' children appear in the picture alongside the ammonoid).

12-15. Early Jurassic. *Hildoceras bifrons* (Bruguière). An isolated ammonoid totally replaced by pyrite. Upper Lias (Toarcian), Agen de Aveyron, Department Aveyron, France. 7 cm dia. Courtesy: Siber and Siber Ltd., Aathal, Switzerland.

12-16. Middle Jurassic. *Parkinsonia parkinsoni* (Sowerby). A large isolated ammonoid specimen. Upper Inferior Oolite (Bajocian), Parkinsoni Zone, Normandie, France, 10 cm dia. Courtesy: Siber and Siber Ltd., Aathal, Switzerland.

12-17. Middle Jurassic. *Cenoceras inornatum* (Orbigny). A nautiloid. Left: internal view showing typical "swirled" or recurved nautiloid sutures. Lower Inferior Oolite (Iron-Shot Oolite - Bajocian), Dundry Hill, Bristol, England. ×1.5.

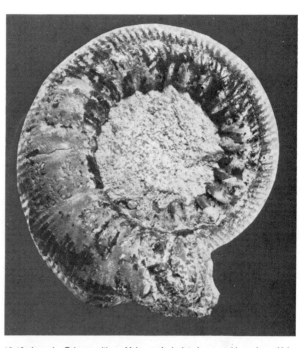

12-19. Jurassic. *Teloceras itinsae* McLearn. An isolated ammonoid specimen. Yakoun Formation, Skiddegate Inlet, Queen Charlotte Islands, British Columbia, Canada. 8 cm dia.

12-18. Early Jurassic. *Microderoceras birchi* Sowerby. An isolated ammonoid specimen. Lower Lias (Sinemurian), Honeybourne, Worcestershire, England. ✕ 2.5.

12-21. Early Jurassic. *Parechioceras* sp. An isolated ammonoid. Lower Lias (Sinemurian), Lyme Regis, Dorsetshire, England. ✕ 1.5.

12-20. Late Jurassic. *Acanthoteuthis speciosa* Münster. Isolated tentacles from a "Tintenfische" (large belemnoid squid). Solnhofen Limestone (Malm Zeta—Liassic), Eichstätt, Bavaria, West Germany. 9 cm in length. Courtesy: Siber and Siber Ltd., Aathal, Switzerland.

12-22. Late Jurassic. *Arnioceras cuneiforme*. An isolated ammonoid (pyritized). Robin Hood's Bay, Yorkshire, England. ✕ 1. Courtesy: Geological Enterprises, Ardmore, Oklahoma.

  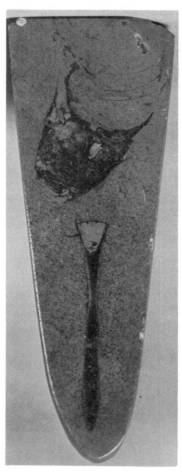 

12-23. Late Jurassic. *Plesioteuthis prisca* (Rüppell). A large "tintenfische" (ink-squirting squid). Note pen in center of body. Lithographischer Schiefer (Upper Malm—Liassic), Schernfeld, Bavaria, West Germany. 28 cm in length.

12-24. Late Jurassic. *Passaloteuthis* sp. Cross-section of a belemnoid in limestone matrix, showing the phragmacone as well as the complete shield. Langenaltheim, Bavaria, West Germany. 16 cm in length.

12-25. Middle Jurassic. *Hibolites semisulcatus* (Münster). A polished cross-section of a belemnoid (belemnite) shield, showing a disarticulated phragmacone in the matrix. Bavaria, West Germany. 14 cm in length.

12-26. Jurassic. *Belemnites vulgaris* Young and Bird. A complete belemnoid shield. Note the apical line (slit) at the tip of the shield (compare with Fig. 12-75). Lias Delta, Dotterhausen, West Germany. 14 cm in length.

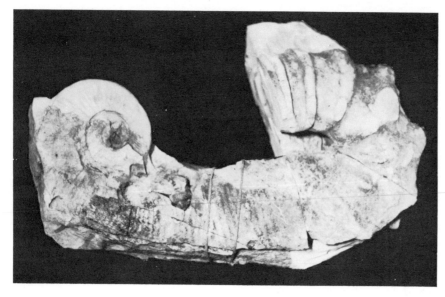

12-27. Early Cretaceous. *Simancyloceras* sp. A huge "scaphites" type of ammonoid (an internal cast). Aptian, Saint André-les-Alpes, Department Hautes Alpes, France. 58 cm in length.

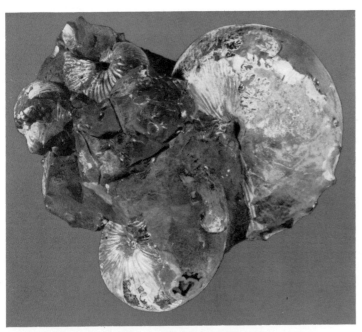

12-29. Late Cretaceous. *Acanthoscaphites nodosus* (Owen). An isolated specimen showing a nacreous coating. Pierre Shale, Glendive, Dawson County, Montana. 5 cm.

12-28. Late Cretaceous. *Acanthoscaphites nodosus quadrangularis* (Owen, non Lopuski). A group of ammonoids in matrix. Pierre Shale, Glendive, Dawson County, Montana. 15.5 cm across matrix. Largest specimen: 9.5 cm.

12-31. Late Cretaceous. *Discoscaphites cheyennensis*. An isolated ammonoid specimen. Fox Hills Sandstone (Maestrichtian), Edgemont, Fall River County, South Dakota. 4.5 cm.

12-30. Late Cretaceous. *Acanthoscaphites nodosus* (Owen). An isolated ammonoid specimen on a piece of matrix. Pierre Shale, Meade County, South Dakota. 7.5 cm. Courtesy: Black Hills Institute of Geological Research, Hill City, South Dakota.

ploscaphites (see Fig. 12-44), *Hypophylloceras (Neophylloceras* (see Figs. 12-48 and 12-49), *Leopoldia, Lytoceras, Metengonoceras, Metoicoceras, Mortoniceras* (see Figs. 12-46 and 12-47), *Neocomites, Neolobites, Nipponites, Nostoceras* (see Fig. 12-50), *Oxytropidoceras, Pachydiscus* (see Figs. 12-51 through 12-55), *Parapuzosia* (see Fig.

12-32. Late Cretaceous. *Discoscaphites nebrascensis*. An isolated ammonoid in matrix. Fox Hills Sandstone (Maestrichtian), Edgemont, Fall River County, South Dakota. 8.5 cm overall (specimen only). Courtesy: Black Hills Institute of Geological Research, Hill City, South Dakota.

12-33. Late Cretaceous. *Discoscaphites conradi* (Morton). An isolated ammonoid. Fox Hills Sandstone (Maestrichtian), Edgemont, Fall River County, South Dakota. 4.5 cm.

12-34. Late Cretaceous. *Discoscaphites nicolletti* (Morton). An isolated ammonoid. Fox Hills Sandstone (Maestrichtian), Edgemont, Fall River County, South Dakota. 5.5 cm.

12-35. Early Cretaceous. *Desmoceras dawsoni* (Whiteaves). An isolated ammonoid. Haida Formation (Albian), Cumshewa Inlet, Queen Charlotte Islands, British Columbia, Canada. 7.5 cm.

12-37. Early Cretaceous. *Euhoplites truncatus* Spath. An isolated ammonoid (pyritized). Lower Gault Clay (Middle Albian), Folkestone, Kent, England. ×1.

12-36. Late Cretaceous. *Cirroceras (Didymoceras)* sp. A coiled or "coprolitic" ammonoid specimen in matrix. Pierre Shale, Fall River County, South Dakota. 20 cm. Courtesy: Black Hills Institute of Geological Research, Hill City, South Dakota.

12-38. Early Cretaceous. *Aegocrioceras* (*Crioceratites*) *krishnaeae*. A loosely coiled ammonoid in matrix. (Hauterivian), France. Approx. ×1. Courtesy: Geological Enterprises, Ardmore, Oklahoma.

12-39. Late Cretaceous. *Acanthoceras stephensoni* Adkins. A cross-section of an ammonoid. Eagle Ford Shale (Turonian), Dallas County, Texas. Approx. half size. Courtesy: Geological Enterprises, Ardmore, Oklahoma.

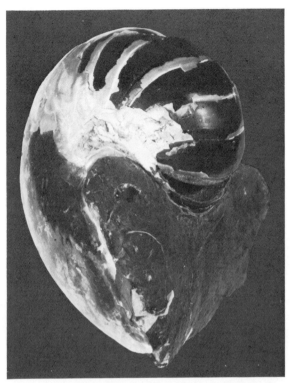

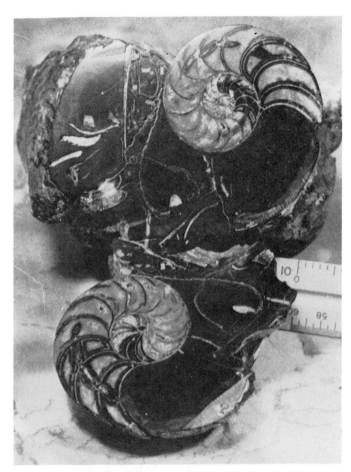

12-40. Late Cretaceous. *Eutrephoceras dekayi* (Morton). An isolated nautiloid specimen. Pierre Shale, Cheyenne River, Bridger, Meade County, South Dakota. Approx. 9 cm.

12-41. Late Cretaceous. *Eutrephoceras dekayi* (Morton). A "chambered" nautiloid cross-sectioned to show the internal chambers. Bear Paw Shales (Late Campanian), Treasure County, Montana. Approx. 6 cm. Courtesy: Black Hills Institute of Geological Research, Hill City, South Dakota.

12-42. Early Cretaceous. *Eutrephoceras* (*Cymatoceras?*) sp. a nautiloid. Lower Greensand (Aptian), Cambridge, England. Approx. half size.

12-43. Late Cretaceous. *Gaudryceras denmanense* Whiteaves. An ammonoid in matrix (the remains of a concretion). Spray Formation (Maestrichtian), Collishaw Point, Hornby Island, British Columbia, Canada. 5.5 cm dia.

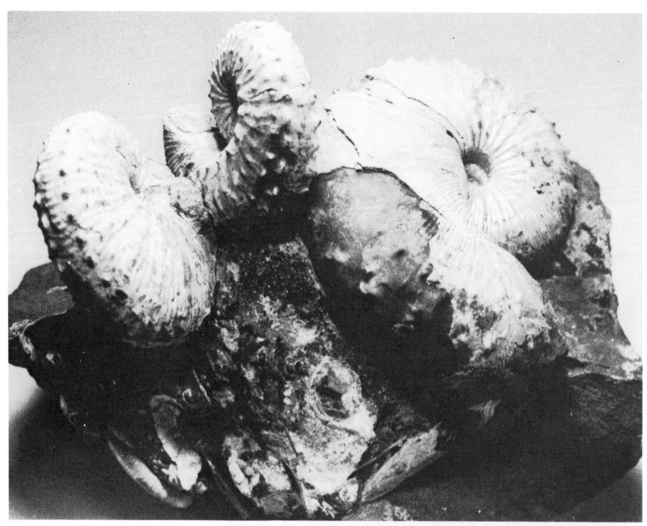

12-44. Late Cretaceous. *Hoploscaphites nebrascensis* and *Discoscaphites* (*Hoploscaphites*) *conradi* (Morton). Two varieties of ammonoids in the remnants of a concretion (matrix). Fox Hills Sandstone (Maestrichtian), Edgemont, Fall River County, South Dakota. Approx. 17.5 cm across. Courtesy: Black Hills Institute of Geological Research , Hill City, South Dakota.

12-45. Early Cretaceous. *Protohoplites* sp. An ornamented ammonoid with an iridescent (nacreous) shell. (Albian), Normandie, France. 7 cm. Courtesy: Siber and Siber Ltd., Aathal, Switzerland.

12-46. Late Cretaceous. *Mortoniceras soutoni* (Baily). An internal cast of a large ammonoid. Umzamba Beds (Senonian), Pondoland, Republic of South Africa. Approx. 65 cm. Courtesy: South African Museum, Cape Town, Republic of South Africa.

12-47. Late Cretaceous. *Mortoniceras soutoni* (Baily). Cross-section of a large ammonoid specimen. (See Fig. 12-46). Umzamba Beds (Senonian), Pondoland, Republic of South Africa. Approx. 26 cm dia. Courtesy: South African Museum, Cape Town, Republic of South Africa.

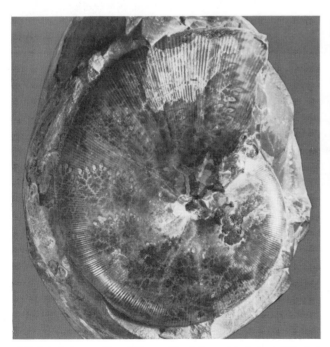

12-48. Late Cretaceous. *Hypophylloceras (Neophylloceras) lambertense* (Usher). A complete ammonoid in matrix (remnants of a concretion). Spray Formation (Maestrichtian), Collishaw Point, Hornby Island, British Columbia, Canada. Approx. 8 cm.

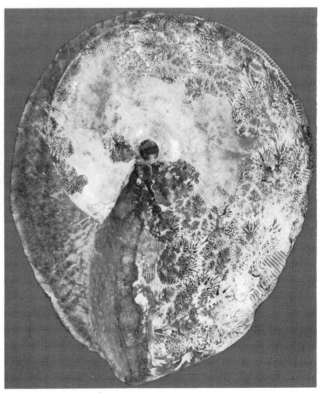

12-49. Late Cretaceous. *Hypophylloceras (Neophylloceras) ramosum* (Meek). A complete ammonoid specimen in a concretion (Matrix). Spray Formation (Maestrichtian), Collishaw Point, Hornby Island, British Columbia, Canada. Approx. 8 cm.

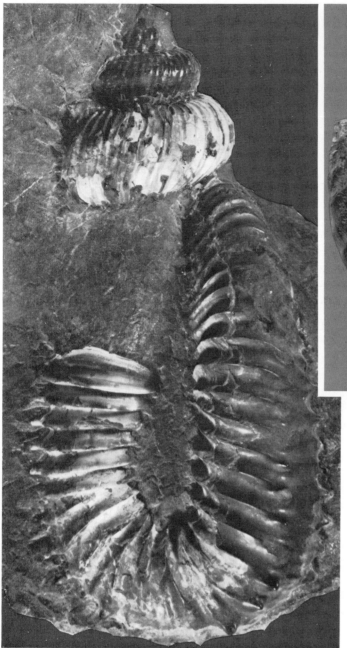

12-51. Late Cretaceous. *Pachydiscus haradai* Jimbo. Internal cast of an ammonoid. Some shell material covering the living chamber. Haslam Formation (Campanian), Englishman's River, Parksville, Vancouver Island, British Columbia, Canada. Approx. 9 cm.

12-50. Late Cretaceous. *Nostoceras hornbyense* (Whiteaves). A coiled or "coprolitic" ammonoid. A bottom-dweller, its shell grew into this incongruous shape. Spray Formation (Maestrichtian), Collishaw Point, Hornby Island, British Columbia, Canada. 13 cm in total length.

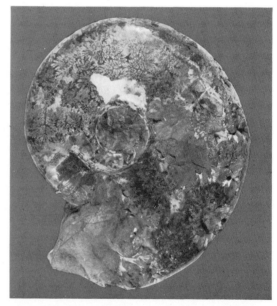

12-52. Late Cretaceous. *Pachydiscus neevesi* (Whiteaves). A complete isolated ammonoid. Spray Formation (Maestrichtian), Collishaw Point, Hornby Island, British Columbia, Canada. Approx. 12 cm.

12-56), *Pictetia, Phlycticrioceras, Phylloceras, Phyllopachyceras* (see Fig. 12-57), *Placenticeras* (see Figs. 12-58 through 12-62), *Prionocyclus* (see Fig. 12-63), *Protohoplites* (see Fig. 12-45), *Pseudophyllites* (see Fig. 12-64), *Ptychoceras, Scaphites* (see Figs. 12-65 and 12-66), *Schloenbachia, Sharpeiceras, Simancyloceras* (see Fig. 12-27), *Sphenodiscus* (see Fig. 12-69), *Spiticeras, Substeuroceras, Texanites, Tissotia,* and *Turrilites.*

The collecting and care of cephalopods: The collecting of ammonite specimens is a challenge in itself, as not too

12-53. Late Cretaceous. *Pachydiscus ootacodensis* (Stoliczka). A complete isolated ammonoid specimen. Spray Formation (Maestrichtian), Collishaw Point, Hornby Island, British Columbia, Canada. Approx. 5 cm.

12-54. Late Cretaceous. *Pachydiscus suciaensis* (Meek). A complete ammonoid specimen. Spray Formation (Maestrichtian), Collishaw Point, Hornby Island, British Columbia, Canada. 12.5 cm across specimen.

12-55. Late Cretaceous. *Pachydiscus haradai* Jimbo (left), and *Pachydiscus (Canadoceras* yokoyami (Jimbo) (right). Two varieties of ammonoids in matrix (remnants of a sandstone concretion). Haslam Formation (Campanian), Benson Creek near Nanaimo, British Columbia, Canada. 14 cm across both specimens.

12-56. Late Cretaceous. *Parapuzosia bradyi.* A "fragment" of a huge ammonoid. The living chamber is lost in this particular specimen. An internal cast with no original shell material preserved. Found in an even larger concretion in the Cody Sandstone outcroppings near Greybull, Big Horn County, Wyoming. Height: approximately 1.5 meters. Courtesy: Greybull Museum, Greybull, Wyoming. (This specimen is on permanent exhibition at the above museum).

12-57. Late Cretaceous. *Phyllopachyceras forbesianum* (Orbigny). A complete isolated ammonoid specimen. Spray Formation (Maestrichtian), Collishaw Point, Hornby Island, British Columbia, Canada. Approx. 8 cm.

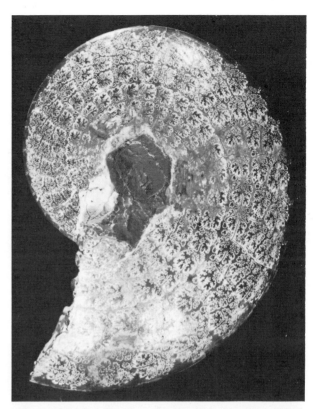

12-58. Late Cretaceous. *Placenticeras intercalare* Meek. A complete ammonoid specimen without the nacreous shell material, showing the complex suture patterns. Pierre Shale, Cheyenne River, Bridger, Meade County, South Dakota. 14 x 18 cm.

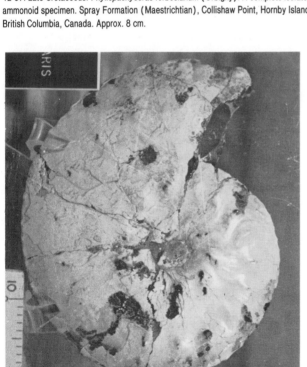

12-59. Late Cretaceous. *Placenticeras intercalare* Meek. A complete ammonoid specimen with nacreous shell material covering the complex suture patterns. (See Fig. 12-58). Pierre Shale, Cheyenne River, Bridger, Meade County, South Dakota. 3.5 x 4 cm. Courtesy: Black Hills Institute of Geological Research, Hill City, South Dakota.

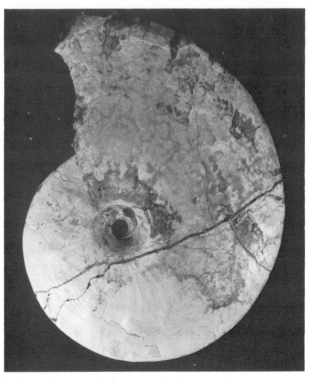

12-60. Late Cretaceous. *Placenticeras meeki* Boehm. A complete ammonoid with nacreous shell coating ("mother-of-pearl"). Bearpaw Formation (Late Campanian), Maple Creek, Saskatchewan, Canada. 11.5 x 14.5 cm.

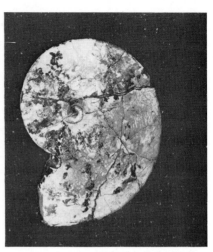

12-62. Late Cretaceous. *Placenticeras whitfieldi* Hyatt. A complete ammonoid specimen. Bearpaw Formation (Late Campanian), Blaine County, Montana. 18 cm.

12-61. Late Cretaceous. *Placenticeras meeki* Boehm. A complete ammonoid specimen in matrix. Pierre Shale, Cheyenne River, Bridger, Meade County, South Dakota. Approx. 17.5 cm. Courtesy: Black Hills Institute of Geological Research, Hill City, South Dakota.

12-63. Late Cretaceous. *Prionocyclus wyomingensis* Meek. Several ammonoid specimens in matrix. Carlile Shale (Turonian), Hot Springs, Fall River County, South Dakota. 17.5 cm across the matrix. Courtesy: Black Hills Institute of Geological Research, Hill City, South Dakota.

many completely preserved internal casts and shells are found lying loose in sands or marls. One must look for larger concretions, for it is within these concretionary "tombs" that most ammonoids are to be found.

Just recently, one of the world's record "giant" ammonites was found in such a concretion. This was a large am-

12-64. Late Cretaceous. *Pseudophyllites indra* (Forbes). A complete ammonoid specimen in matrix (concretion). Spray Formation (Maestrichtian), Collishaw Point, Hornby Island, British Columbia, Canada. Approx. 9 cm.

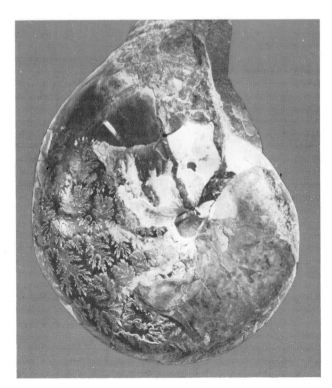

12-67. Late Cretaceous. *Desmophyllites (Schluteria) selwyniana* (Whiteaves). A complete ammonoid specimen showing suture pattern. Cedar District Formation (Campanian), Sucia Island, San Juan County, Washington. Approx. 5 cm.

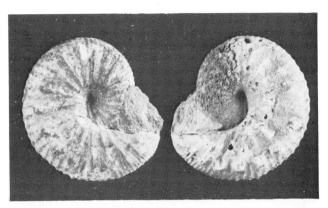

12-65. Late Cretaceous. *Scaphites* sp. Anterior and posterior views of the same ammonoid specimen. Cody Shale, West of Tensleep, Washakie County, Wyoming. 4.5 cm dia.

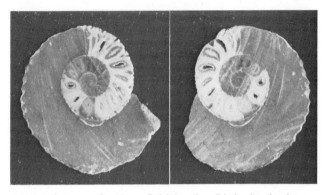

12-66. Late Cretaceous. *Scaphites* sp. Both halves of a polished and sectioned ammonoid specimen (see Fig. 12-65), showing the internal structure of the chambers. Note the large living chamber. Cody Shale, West of Tensleep, Washakie County, Wyoming. Approx. 4.5 cm.

12-68. Late Cretaceous. Various ammonoids (species indeterminate) mixed in with assorted molluscs in a limestone block. Zululand, Province of Natal, Republic of South Africa. Courtesy: South African Museum, Cape Town, Republic of South Africa. Approx. 55 cm across slab.

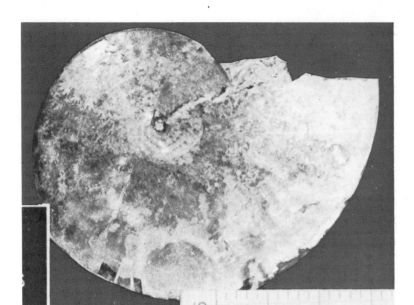

12-69. Late Cretaceous. *Sphenodiscus lenticularis.* A complete ammonoid specimen. Fox Hills Formation (Maestrichtian), Edgemont, Fall River County, South Dakota. 8.3 cm. Courtesy: Black Hills Institute of Geological Research, Hill City, South Dakota.

12-71. Recent. A female "paper" nautilus (argonaut) which had been washed ashore at Betty's Bay, Cape Province, Republic of South Africa. Modern specimens of a fauna are useful in studying (comparison) and understanding fossil material. Approx. ×1.

12-70. Late Cretaceous. Possible aptychius—aperture for the entrance to the living chamber of an ammonoid. Höheres Campan (Upper Campanian), Misburg bei Hannover, West Germany. ×2.

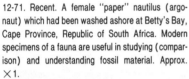

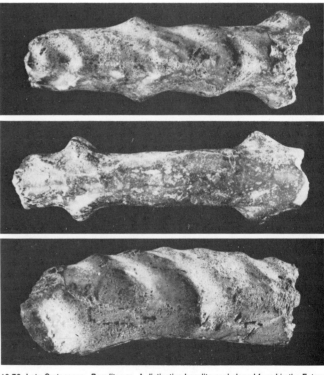

12-72. Late Cretaceous. *Baculites* sp. A distinctive baculite cephalopod found in the Eutaw Formation (Santonian). Note the "lumps" or "bumps" on the sides of the shell. 3 views of the same specimen. Columbus Lock and Dam. Columbus, Lowndes County, Mississippi. ×1.5.

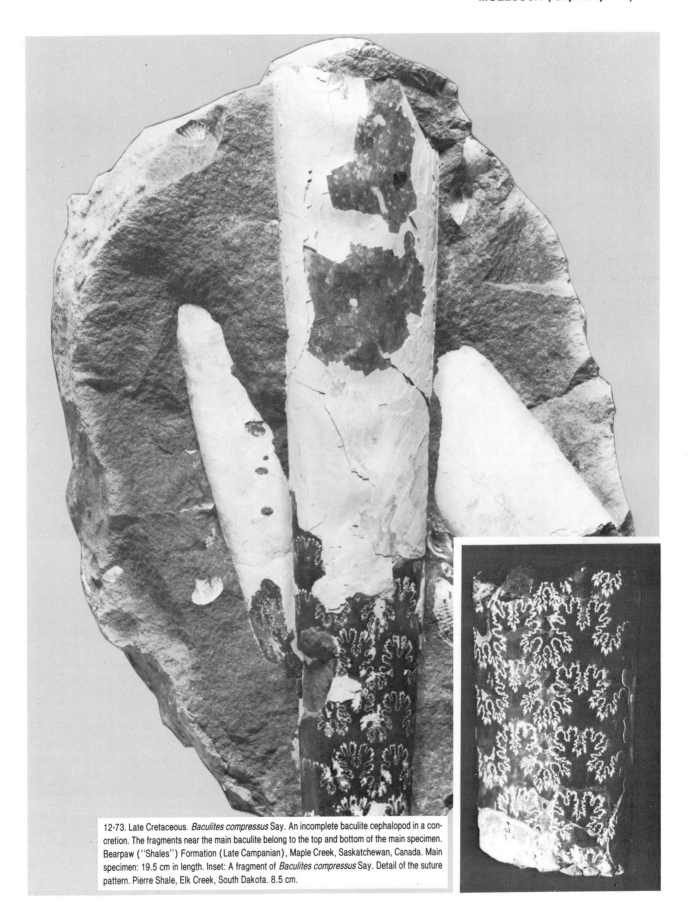

12-73. Late Cretaceous. *Baculites compressus* Say. An incomplete baculite cephalopod in a concretion. The fragments near the main baculite belong to the top and bottom of the main specimen.
Bearpaw ("Shales") Formation (Late Campanian), Maple Creek, Saskatchewan, Canada. Main
specimen: 19.5 cm in length. Inset: A fragment of *Baculites compressus* Say. Detail of the suture
pattern. Pierre Shale, Elk Creek, South Dakota. 8.5 cm.

12-74. Late Cretaceous. *Baculites compressus* Say. A lengthy shell showing the outside shell (nacreous) material still attached. Pierre Shale, Cheyenne River, Bridger, Meade County, South Dakota. ×1. Courtesy: Black Hills Institute of Geological Research, Hill City, South Dakota.

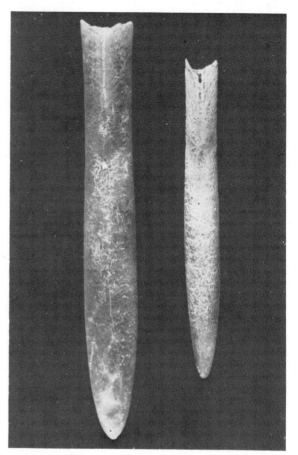

12-75. Late Cretaceous. *Belemnitella americana* (Morton). Two examples of belemnoid shields. The shields are made up of an aragonitic material. Note the apical line (slit) at the bottom (back) of the shield—compare with Fig. 12-26. Navesink Formation (Middle Maestrichtian), Crosswicks Creek, Hornerstown, Monmouth County, New Jersey. ×1.

monite found in New Zealand (see Fig. 12-14). Other such giants have been found before and some were even a great deal larger, such as *Parapuzosia* (see Fig. 12-56), which is currently on exhibit at the Greybull Museum, Greybull, Wyoming.

Most ammonoids (ammonites), however, are rather small and some are quite inconspicuous, ranging in size from ¼ of an inch on up to an average of 8 inches in diameter. These are the majority of the finds that are made by collectors and scientists. The ammonoid usually has to be "prepared" or carefully removed from its concretionary matrix jacket. This requires special care, as a slip of the cold chisel or a bad blow with the hammer could destroy a vital part of the specimen, such as the living chamber area or even the aptychius.

Strangely enough, one of the rarest parts of an ammonoid or nautiloid is the aperture cover or aptychius. These calcareous plates are seldom found associated with the fossilized remains of an ammonoid or a nautiloid. Figure 12-70 may well represent an isolated apertural covering.

13 ARTHROPODA (Trilobita)

The trilobites are an extinct form of marine organisms belonging to the phylum Arthropoda ("joint-footed" animals). Arthropoda is composed of the many varieties of insects, arachnids (which include the scorpions, ticks, mites, and spiders), centipedes, and crustaceans (ostracodes, barnacles, crabs, lobsters, crayfishes, and shrimps).

The trilobites (three-lobes) are represented in the fossil record by numerous species (over 1500 genera representing some 10,000 species) ranging in geologic age from the Lower Cambrian period up to the Middle Permian. Thus, their entire evolution was during the Paleozoic era.

The trilobites have a chitinous shield or body armor which is separated into three distinct parts: a) the cephalon or head region, b) the thorax or body portion, and c) the pygidium or tail region.

The cephalon or head area is composed of the glabella, the central portion or lobe of the head, which has on either side of it, a "cheek." In some species, the cheek has posteriorly (to the rear) directed genal spines (see Figs. 13-11, 13-23, among others). The cheeks (free cheeks) sometimes break away along their suture lines—especially on molted (shed) shields. (See Figs. 13-2, 13-7, 13-8, and 13-9, among others). The compound eyes of most trilobites are situated at the anterior (forward) edges of the cranidium on the cheek lobes. The thorax shield is composed of a number of segments in articulation with each other to give the animal's body flexibility, especially when the animal is threatened. Thus, the trilobite can "roll" itself up into a ball, much like the present day Isopod, the "pill" or "sow" bug. The thoracic segments are further divided (longitudinally) into three "furrows" or lobes. The central one being the axial, while on either side of this are the pleura, or pleural lobes.

The pygidium or abdomen of the shield is a single fused plate divided by a central axial lobe and the lateral lobes (one on either side of the axial). Some species of trilobites have a "telson" or spiked tail appendage called a posterior spine (see Figs. 13-14 and 13-57 through 13-60, among others). While other species have a decorative apron on the posterior edge (and sides) of the pygidium (see Figs.

13-15, 13-22, and 13-48, among others), which are called marginal spines.

Before going too far along with the trilobites, it would be a good idea to briefly discuss a small allied (sister) group of the trilobites, the agnostids or "blind trilobites." These small, extinct creatures (see Fig. 13-19) had a relatively short existence in the Paleozoic seas. Their geological range was from the Lower Cambrian to the Upper Ordovician period. The agnostids in appearance differed radically from the typical trilobites, by having an almost equal sized cephalon and pygidial shield, separated by two or three thoracic segments. The cephalon of most species had a glabella attached to a medial furrow with a prominent frontal lobe as well. There were no eyes on the cephalon shield, indicating that the agnostids like *Peronopsis* in particular spent most if not all of their time burrowing through the bottom sands and muds searching out food. In this case, eyes would not seem to be a necessity. The thoracic area as discussed above, was comprised of only two or three segments separating the cephalon from the pygidium. It is quite obvious that the agnostids could not "roll" themselves up too readily. The pygidium or abdomen shield of the agnostids in most cases (species) had a relatively short axis.

Back again to the typical trilobites! The fossilized remains of most species of trilobites do not show certain features which were no doubt present in the animal when it was alive and roaming the seas, especially the surfaces of the sandy bottoms of the seas or over the nooks and crannies of the coral reefs. These features which are rarely preserved in most sediments (rocks), are the appendages such as legs, antennae and the soft tissue "skirts" or aprons used to protect the legs and the soft ventral (underside) tissues of the animals.

In the case of recently discovered pyritized specimens (see Figs. 13-49, 13-69, and 13-74), from the Devonian "slate" shales of Bundenbach, Western Germany, such appendages can be seen through the process of radiography (X-rays). It is now possible to determine more information regarding the eye structure, soft tissue areas such

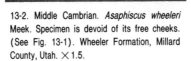

13-2. Middle Cambrian. *Asaphiscus wheeleri* Meek. Specimen is devoid of its free cheeks. (See Fig. 13-1). Wheeler Formation, Millard County, Utah. ×1.5.

13-3. Cambrian. *Conocoryphe sulzeri* (Schlotheim). A complete trilobite specimen. Jinetz Shale, Jince, Czechoslovakia. ×1.

13-1. Middle Cambrian. *Asaphiscus wheeleri* Meek. A complete trilobite. (See Fig. 13-2). Wheeler Formation, Millard County, Utah. ×2.5.

13-5. Middle Cambrian. *Elrathia kingi* (Meek). Several specimens of this most common trilobite in a soft slatey shale. Wheeler Amphitheatre Formation, House Range Mountains, Millard County, Utah. ×1.5.

13-4. Middle Cambrian. *Elrathia kingi* (Meek). A complete trilobite specimen. Note the similarities to *Asaphiscus*. Wheeler Amphitheatre Formation, House Range Mountains, Millard County, Utah. ×3.

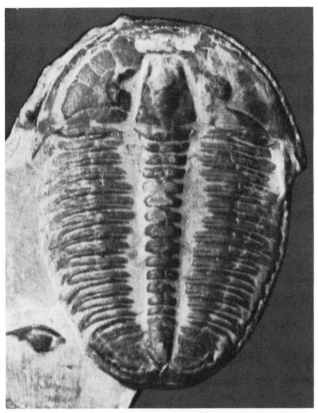

13-7. Middle Cambrian. *Ellipsocephalus hoffi* (Schlotheim). A group of complete trilobite specimens. Stage C, Jinetz Shales, Jince, Bohemia, Czechoslovakia. ×1. Courtesy: Geological Enterprises, Ardmore, Oklahoma.

13-6. Middle Cambrian. *Elrathia kingi* (Meek). Another example to show the facial structure, especially the sutures for the free cheeks. (See Fig. 13-4 for comparison). Wheeler Amphitheatre Shale Formation, House Range Mountains, Millard County, Utah. ×3.

13-8. Middle Cambrian. *Paradoxides* (*Hydrocephalus*) *carens* (Barrande). An almost complete trilobite specimen. The free cheeks of the cephalon are missing. Kambrium Shale, Jince, Bohemia, Czechoslovakia. 10 cm. Courtesy: Siber and Siber, Ltd., Aathal, Switzerland.

13-10. Early Cambrian. *Olenellus vermontanus* (Hall). Cephalon and partial thorax of a trilobite specimen. Parker Formation, Vermont. ✕ 2. Courtesy: Geology Department, Princeton University, Princeton, New Jersey. (PU 42011).

13-9. Middle Cambrian. *Ogygopsis klotzi* (Rominger). An almost complete trilobite specimen. The free cheeks of the specimen are missing. Burgess Shale Formation, Burgess Pass, Field, British Columbia, Canada. ✕ 2.

13-11. Early Cambrian. *Olenellus* sp. An isolated cephalon with slight distortion. Center, Alabama. ✕ 2.

13-12. Early Cambrian. *Olenellus* sp. Several isolated cephalons in dark shale matrix. Parker Formation, Vermont. ✕ 1.

as the heart, kidneys and the digestive system, as well as the modes and means of locomotion of certain species of trilobites.

Much important work is currently being done in the radiography of specimens prior to using mechanical means of preparation of specimens in rock (which inevitably destroys important details only seen through radiography). A leading proponent of this technique of radiography is Professor Dr. Wilhelm Stürmer of Erlangen, West Germany. Dr. Stürmer has installed radiographic equipment in a van and he and his assistants can take this portable X-ray unit to the very quarry floor to determine "in situ" where important specimens of trilobites (as well as the associated fauna of crinoids, sea-urchins, and other invertebrates, and also vertebrates) are positioned in their ecological niche within the now hardened sediments of an ancient sea floor. Dr. Stürmer's work with the radiography of fossils is well known throughout the world; and this author is proud to be able to present, with Dr. Stürmer's permission, just a few of his marvelous trilobite specimens prepared by radiography (see Figs. 13-49, 13-69, 13-74, and 13-76).

AUTHOR'S NOTE: The following lists of genera are culled from Shimer and Shrock (1944); Moore, Lalicker and Fischer (1952); and the Treatise of Invertebrate Paleontology (Trilobita, 1959). It is useful to have an alphabetical listing of generic names in a popular book on fossils.

LOWER AND MIDDLE CAMBRIAN: *Acrocephalops, Agraulos, Albertella, Alokistocare, Andrarina, Anoria, Antagmus, Asaphiscus (Eteraspis), Bailiaspis, Bailiella, Bathyuriscidella, Bathyuriscus, Blainia, Bolaspidella (Deisella), Bolaspis, Bonnia, Bonniella, Burlingia, Callavia, Centropleura, Chancia, Clavaspidella, Conocoryphe, Ctenocephalus, Dawsonia, Dolichometopsis, Ehmania, Ehmaniella (Clappaspis), Elrathia, Elrathiella, Elrathina, Eodiscus, Glossopleura, Glyphaspis, Holmia (Esmeraldina), Hypagnostus, Kochaspis, Kochiella, Kochina, Kootenia, Marjumia, Mexicaspis, Modocia (Armonia), Ogygopsis (Taxioura), Olenellus, Olenoides, Orria, Oryctocara, Oryctocephalites, Oryctocephalus, Pachyaspis, Paedeumias, Pagetia, Paradoxides, Parehmania (Rowia and Thompsonaspis), Peachella, Peronopsis, Po-*

13-13. Early Cambrian. *Olenellus thompsoni* (Hall). An almost complete trilobite specimen. A small portion of the right side of the cephalon is lost. Specimen is in a sandstone matrix. Kinzer Formation, York, York County, Pennsylvania. 16 mm (entire specimen). Inset: *Paedeumias clarki*, from the Lower Cambrian of California. This specimen is shown here as comparison to *Olenellus*. 8 cm. Courtesy: Geological Enterprises, Ardmore, Okla.

13-14. Early Cambrian. *Olenellus vermontanus* (Hall). A complete trilobite specimen and an associated isolated cephalon of another individual. Parker Formation, Appalachian Belt, Vermont. ×1. Courtesy: Geology Department, Princeton University, Princeton, New Jersey. (PU 52235).

13-15. Middle Cambrian. *Olenoides expansus.* An almost complete pygidium with spines. Several of these spines are missing from the right side of the specimen. Marjum Formation, Millard County, Utah. ×4.

13-16. Middle Cambrian. *Paradoxides spinosus* Geinitz. An almost complete trilobite specimen. Specimen is missing its free cheeks on the cephalon. Kambrium Shale, Jince, Bohemia, Czechoslovakia. ×1. Courtesy: Geology Department, Marietta College, Marietta, Ohio.

liella, Poulsenia, Proliostracus, Protolenus, Protypus (Bicaspis), Prozacanthoides, Ptarmigania, Ptychoparella, Ptychoparia, Redlichia, Solenopleurella, Strenuella, Tonkinella, Vanuxemella, Wanneria, Weymouthia, and *Zacanthoides.* UPPER CAMBRIAN: *Acheilops, Aphelaspis, Arapahoia, Blountia, Brassicephalus, Briscoia, Burnetiella (Burnetia), Bynumia, Calvinella, Camaraspis, Cedaria, Chariocephalus, Cheilocephalus (Pseudolisania), Conaspis, Coosella, Coosia, Crepicephalus, Ctenopyge, Dikelocephalus, Dokimocephalus, Dresbachia, Drumaspis, Dunderburgia, Elvinia, Entomaspis, Euptychaspis, Eurekia, Genevievella, Holteria, Housia, Idahoia, Iddingsia, Illaenurus, Irvingella, Kingstonia, Lonchocephalus, Maladia, Maryvillia, Menomonia, Meteoraspis, Modocia, Monocheilus, Norwoodella, Norwoodia, Orygmaspis, Osceolia, Parabolina, Peltura, Plethometopus, Plethopeltis, Prosaukia, Pterocephalia, Ptychaspis, Ptychopleurites, Rasettia (Platycolpus), Saukia, Saukiella, Sphaerophthalmus, Stenopilus, Taenicephalus, Talbotina, Tellerina, Terranovella, Theodenisia (Acheilus), Triarthropsis, Triacrepicephalus, Uncaspis, Welleraspis,* and *Wilbernia.* ORDOVICIAN: *Amphilichas (Acrolichas* and *Tetralichas),*

13-17. Middle Cambrian. *Paradoxides gracilis* Geinitz. An almost complete trilobite specimen. Specimen is missing its free cheeks. Kambrium Shale, Jince, Bohemia, Czechoslovakia. 11 cm. Courtesy: Siber and Siber, Ltd., Aathal, Switzerland.

13-18. Middle Cambrian. *Paradoxides bohemicus* Geinitz. A complete trilobite specimen (with the exception of the spiked tail). Kambrium Shale, Jince, Bohemia, Czechoslovakia. ✕1.5. Courtesy: Geology Department, Marietta College, Marietta, Ohio.

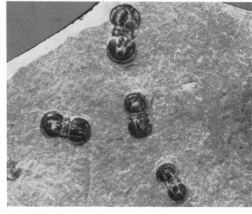

B

13-19. Middle Cambrian. *Peronopsis* (*Agnostus*) *interstricta* (White). A. One complete trilobite specimen (black) and several impressions of the ventral parts of others. B. Four individuals in a travelling pose. Wheeler Shale Formation, Wheeler Amphitheatre, House Range, Millard County, Utah. A: ✕2. B: ✕1.5.

A

13-20. Middle Cambrian. *Zacanthoides romingeri* Resser. A partial thorax and a complete pygidium with all spikes intact. Note an isolated free cheek near the upper left corner of the picture. Stephen Shale Formation, British Columbia, Canada. ✕2.

13-21. Ordovician. *Acidaspis* (*Odontopleura*) *buchii*. A complete trilobite specimen. Jinetz Shale, Jince, Bohemia, Czechoslovakia. ✕1. Courtesy: Marietta College, Marietta, Ohio.

13-22. Middle Ordovician. *Amphilichas (Acrolichas) ottawensis* (Foerste). A complete trilobite pygidium, ornate in characteristics. Trenton Limestone Formation, Ontario, Canada. ✕2.5.

13-23. Late Ordovician. *Amphyxina bellatula* (Savage). Several examples of this gregarious trilobite. No actual chitinous "shell" material, rather molds in limestone. Note impressions of the genal spines extending down from the cephalon. Maquoketa Formation, Cape Girardeau, Cape Girardeau County, Missouri. ✕2.

13-24. Ordovician. *Bathyurus (Asaphus) nobilis* var. *bituberculata* (Samp.). The cephalon and partial thorax of a large trilobite specimen. Limonitic replacement cast. Church Hill Formation, Oretaña Mountain Range, Northern Spain. 17.5 cm.

Apatokephalus (Diplapatokephalus), Basilicus (Basiliella), Bathyurellus, Bathyurus, Bellefontia (Xenostegium), Brachyaspis, Bronteopsis, Bumastus, Calyptaulax (Calliops), Ceratocephala, Ceraurinus, Ceraurus, Cryptolithus, Cybeloides, Encrinurus, Eoharpes, Eomonorachus, Flexicalymene, Homotelus, Hystricurus, Illaenus, Isoteloides, Isotelus, Lichas, Lloydia (Leiostegium), Lonchodomas, Nileus, Otarion (Cyphaspis), Pliomerops, Proetus, Pterygometopus (Achatella), Remopleurides, Seleneceme (Alsataspis), Shumardia, Sphaerexochus, Telephina (Telephus), Thaleops, Triarthrus, and *Vogdesia.* SILURIAN: *Acaste, Arctinurus, Bumastus, Calymene, Ceratocephala, Cheirurus, Dalmanites, Dalmanitina, Encrinurus, Eophacops, Lichas, Odontopleura, Otarion (Cyphaspis), Phacops, Proetus, Scutellum, Sphaerexochus, Trimerus,* and *Trochurus.* DEVONIAN: *Anchiopella (Anchiopsis), Brachymetopus, Calymene, Ceratocephala, Ceratolichas, Cordania, Coronura, Dalmanites, Echinolichas, Gaspelichas, Greenops, Odontocephalus, Odontochile, Otarion (Cyphaspis), Phacops, Proetus, Reedops, Scutellum, Synphoria, Terataspis,* and *Trimerus (Dipleura).* MISSISSIPPIAN: *Brachymetopus, Exochops, Griffithides, Kaskia, Paladin,* and *Phillipsia.* PENNSYLVANIAN: *Ameura, Ditomopyge, Paladin,* and *Sevillia.* PERMIAN: *Ameura, Anisopyge, Delaria, Ditomopyge,* and *Permoproetus.*

Note: Generic names in parenthesis are synonyms of the preceeding generic name.

13-25. Silurian. *Bumastus niagarensis* (Whitfield). A group of trilobite specimens on a limestone matrix. Niagaran of New York. ×1.

13-26. Ordovician. *Cryptolithus ornatus.* An ornately designed trilobite specimen. The pointed tail is missing on this specimen. Bohemia, Czechoslovakia. ×3. Courtesy: Geology Department, Marietta College, Marietta, Ohio.

13-27. Ordovicia. *Cryptolithus tesselatus* Green. A complete trilobite specimen in limestone matrix. Trenton Formation, Martinsburg Group, Swatara Gap, Lebanon County, Pennsylvania. 2.5 cm.

13-28. Ordovician. *Cryptolithus tesselatus* Green. An almost complete trilobite specimen. The left side of the cephalon is missing. Eden Shale Member, Trenton Formation, Martinsburg Group, Swatara Gap, Lebanon County, Pennsylvania. 2.2 cm. Courtesy: Geology Department, Monroe Community College, Rochester, New York.

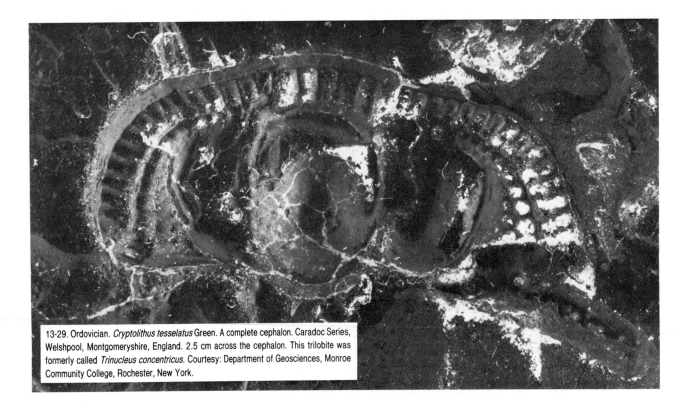

13-29. Ordovician. *Cryptolithus tesselatus* Green. A complete cephalon. Caradoc Series, Welshpool, Montgomeryshire, England. 2.5 cm across the cephalon. This trilobite was formerly called *Trinucleus concentricus*. Courtesy: Department of Geosciences, Monroe Community College, Rochester, New York.

13-30. Middle Ordovician. *Eoharpes ottawensis* (Billings). A complete trilobite specimen with an unusually broad brim. Trenton Limestone Formation, Ontario, Canada. ×1.

13-31. Late Ordovician. *Flexicalymene meeki* (Foerste). A flattened-out specimen. (See Fig. 13-32 for cephalon view). Arnheim Formation, Clarksville, Clinton County, Ohio. ×2.

13-32. Late Ordovician. *Flexicalymene meeki* (Foerste). The frontal view of the cephalon of the flattened specimen pictured in Fig. 13-31. Arnheim Formation, Clarksville, Clinton County, Ohio. 3 cm.

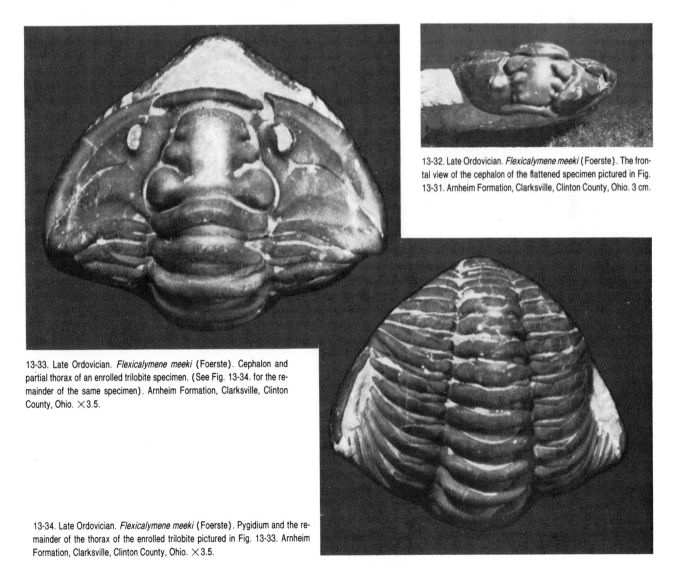

13-33. Late Ordovician. *Flexicalymene meeki* (Foerste). Cephalon and partial thorax of an enrolled trilobite specimen. (See Fig. 13-34. for the remainder of the same specimen). Arnheim Formation, Clarksville, Clinton County, Ohio. ×3.5.

13-34. Late Ordovician. *Flexicalymene meeki* (Foerste). Pygidium and the remainder of the thorax of the enrolled trilobite pictured in Fig. 13-33. Arnheim Formation, Clarksville, Clinton County, Ohio. ×3.5.

13-35. Ordovician. *Homotelus bromidensis* Esker. Dorsal view of a large group of trilobites. Possibly molted shields. Poolville Member, Bromide Formation, Criner Hills, Carter County, Oklahoma. Each specimen approx. 5 cm. Courtesy: Geological Enterprises, Ardmore, Oklahoma.

13-36. Ordovician. *Homotelus bromidensis* Esker. The ventral view of a group of trilobite specimens. Probably "moltings" of the shells of several individuals washed into one area. Poolville Member, Bromide Formation, Criner Hills, Carter County, Oklahoma. 5 cm (each specimen). Courtesy: Geology Department, Princeton University, Princeton, New Jersey.

13-37. Ordovician. *Homotelus bromidensis* Esker. The dorsal view of three trilobites. See Fig. 13-36 for typical ventral view. Poolville Member, Bromide Formation, Criner Hills, Carter County, Oklahoma. Each specimen approx. 5 cm.

13-38. Ordovician. *Homotelus bromidensis* Esker. A large group of gregarious trilobites. Shell cast-offs (or a "mass burial?"). Poolville Member, Bromide Formation, Criner Hills, Arbuckle Mountains, Carter County, Oklahoma. Overall dimensions: approx. 10 x 15 cm. (Each individual approx. 5 cm).

13-39. Ordovician. *Illaenus hispanicus*. The pygidium and partial thorax of a large enrolled trilobite specimen. The specimen is a limonite cast weathered from sandstone. Church Hill Formation, Oretaña Mountain Range, Northern Spain. 5.5 cm across.

13-40. Devonian. *Metacryphaeus venustus*. A large ophisthoparian trilobite in a concretion matrix. These are newly discovered trilobites from the Paleozoic rock of Bolivia. 10 cm. Courtesy: Brooklyn Children's Museum, Brooklyn, New York.

The collecting of trilobites: The trilobites are the most fascinating group of fossils among the invertebrates. They are quite popular with collectors of fossils, and the most common species are in almost every private as well as museum collection. Species such as *Phacops rana milleri* (Stewart) from the Devonian Silica Shales of southern Michigan and northwestern Ohio (see Figs. 13-78 through 13-83), and *Calymene celebra* Raymond from the Silurian beds near Grafton, Illinois (see Fig. 13-51), are in just about everyone's collection.

The trilobites' unusual appearance makes them the most desirous of specimens for study as well as for a collection. It is quite difficult to determine by a fossil specimen itself whether it is in fact a molting of the animal's shield or the remnants of its mortality. But, this does not stop the average collector from seeking out these remarkable fossils for his collection. Collectors will even collect the fragmented remains of these animals, such as the isolated cephalons or the thorax and pygidium. Most collectors prefer complete "flattened-out" specimens, but will have enrolled examples in their collections as well.

There seem to be very few specialists in trilobites among amateur collectors, and among professional researchers or scientists. For this reason, as well as the scarcity of most species, it is most difficult for the collector to try and obtain one or more examples of each known type of trilobite for his or her collection. The professional researcher or scientist, on the other hand, may specialize in only one particular group or even family within the phylum.

The most sought after trilobites by amateur collectors are: the pyritized *Phacops rana milleri* (Stewart), preferably the flattened-out specimens, of the Devonian; the larger varieties such as *Isotelas gigas* DeKay (from the Ordovician of New York State, see Fig. 13-42); and *Homotelus bromidensis* Esker from the Ordovician Arbuckle Mountains of Oklahoma (see Figs. 13-36 through 13-38). On the eastern seaboard of the United States, prized specimens such as *Cryptolithus tesselatus* Green are sought after. This small trilobite (it averages about 1.5 cm in total length) with its distinctive ornamented head shield (a pattern on the rim of small tubercles) is found in the Ordovician rocks of Pennsylvania (see Figs. 13-27 through 13-29). The "shovel-nosed" *Trimerus (Dipleura) dekayi* (Green) of the Devonian of New York State (along with its earlier cousin, *Trimerus delphinocephalus* Green from the Silurian age—see Figs. 13-61, 13-63 through 13-68), are also among the most sought after trilobites by eastern collectors.

Many fine European trilobite collecting localities exist, such as the Devonian Hunsrück shales of Bundenbach, West Germany. *Phacops ferdinandi* (see Fig. 13-75) is a good example of the most common of the trilobites found in the Hunsrückshiefer.

By far, the best collecting areas in Europe for trilobites are in the Democratic Peoples Republic of Czechoslovakia

(CSSR). In this eastern European country, some of the finest examples of trilobites have been discovered over the years. Such species as: *Conocoryphe sulzeri* (Schlotheim) (Fig. 13-3), *Paradoxides (Hydrocephalus) carens* (Barrande) (Fig. 13-8), *Paradoxides spinosus* Geinitz, *P. gracilis* Geinitz, and *P. bohemicus* Geinitz (see Figs. 13-16 through 13-18), from the Cambrian of Czechoslovakia, to the Czechoslovakian Ordovician species of *Acidaspis buchii* (Fig. 13-21) and *Cryptolithus ornatus* (Fig. 13-26), have found their way into collections, both private and museum, throughout the world.

Great Britain is another country where some very fine trilobites have been found over the years. More recently, Spain has turned up some fine specimens in its countryside, especially in the Oretaña Mountains north of Madrid (Ordovician, see Figs. 13-24 and 13-39). *Bathyurus (Asaphus) nobilis bituberculata* (Samp.) and *Illeanus hispanicus,* are just a few of the many newly discovered species at the new "Church Hill" locality in northern Spain.

Even though Europe and North America seem to have a preponderance of fossil localities for trilobites, New Zealand, Canada, and Bolivia have many outcrops from which these fossils can be collected.

All trilobites are found in limestones, shales, and sandstones. With the exception of "weathered-out" specimens, most trilobites have to be freed from their hardrock graves by a geologist's hammer and cold chisels. In the Hamilton (Devonian) beds of New York State, the trilobites are in a hard sandstone, while in the younger limestones, they are in friable black shales. At Hubbardsville in Madison County, New York, the large trilobite *Trimerus (Dipleura) dekayi* (Green) (see Figs. 13-63 through 13-68), is scattered throughout the upper member of the Hamilton Formation in a three foot thick bedding plane which con-

13-41. Middle to Late Ordovician. *Isotelus* sp. A large almost complete opisthoparian trilobite specimen in matrix. A partial loss of the posterior edge (genal spine) of the right side of the cephalon. Cincinnati Group, Wayne County, Ohio. Approx. 13 cm. Courtesy: Black Hills Institute of Geological Research, Hill City, South Dakota.

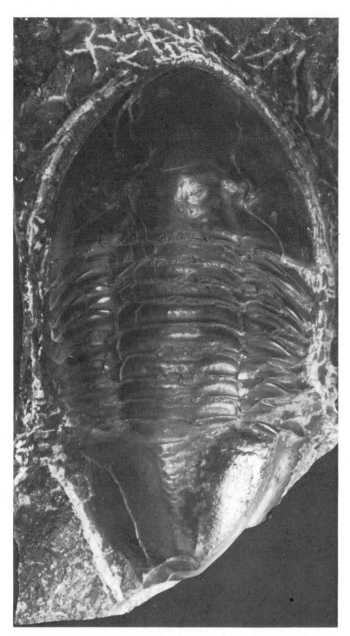

13-42. Middle to Late Ordovician. *Isotelus gigas* Dekay. A large complete opisthoparian trilobite specimen in shale matrix. Trenton Group, Trenton Falls, Oneida County, New York, ×10.

13-43. Early Ordovician. *Kanoshia kanoshensis* (Hintze). An isolated pygidium in limestone. The pygidium of this species appears to have similarities with the pygidium of *Greenops* (see Fig. 13-70) of the Devonian. Utah. ×3. Courtesy: Geology Department, Marietta College, Marietta, Ohio.

13-44. Middle Ordovician. *Triarthrus* sp. An almost complete trilobite specimen. The posterior portion (the pygidium) is missing. Lorraine Formation, Trenton Group, Dolgeville, Herkimer County, New York. ×3.5.

13-45. Middle Ordovician. *Triarthrus* sp. "Trilobite hash"—scraps or parts of various specimens: glabella, genal spines, and fragments of thoraxes. Lorraine Formation, Trenton Group, Dolgeville, Herkimer County, New York. ×3.

13-46. Middle Ordovician. *Triarthrus eatoni* (Hall). A complete specimen in black shale matrix. Compare with specimen in Fig. 13-47. Lorraine Formation, Trenton Group, St. Johnsville, Herkimer County, New York. 2 cm. Courtesy: Geological Enterprises, Ardmore, Oklahoma.

13-47. Middle Ordovician. *Triarthrus eatoni* (Hall). A complete trilobite specimen in black shale matrix. Utica Black Shale, Eastern New York State. 13 mm in length.

tains the associated fauna of pelecypods such as: *Grammysia* (Fig. 10-1 and 10-2); *Cornellites* and *Limoptera,* the winged clams (see Figs. 10-3 and 10-5); and *Modiomorpha* and *Prothyris,* the bottom bed clams (see Figs. 10-6 and 10-4). *Trimerus (Dipleura) dekayi* is also found associated with several varieties of spiriferoid brachiopods, and with the cephalopod *Orthoceras. Trimerus (Dipleura) dekayi* (Green) is rarely found complete at Hubbardsville, rather, isolated cephalons, pygidiums and scattered thoracic segments are the usual findings. On occasion, a reasonably complete molting of the entire shield is found (see Fig. 13-64).

The trilobite parts are difficult to separate from the surrounding matrix during field collecting. Many specimens have to be cleaned and prepared at home or in the laboratory, in a workshop area where excess matrix and partial matrix covering the specimen can be removed with smaller cold chisels and tools such as needles and dental picks. The specimen should be held securely in a vice while preparation is proceeding.

The difficulties of removing excess matrices from specimens is not the case with trilobites from the House Range Mountains in Millard County, Utah, where the Middle Cambrian Wheeler Formation outcrops. The specimens at this locality are found in a fairly soft clay-like shale. The trilobites which are commonly recovered at this locality are the agnostid *Peronopsis* (see Fig. 13-19), and *Elrathia kingi* and its cousin *Asaphiscus wheeleri* (see Figs. 13-1, 13-2, and 13-4 through 13-6).

These species as well as others are easily removed from the matrix by gentle tapping of small cold chisels with a ball peen hammer. It is desirable, though, for the collector to leave the specimen cleaned and prepared on a matrix

13-48. Middle Ordovician. *Triarthrus* sp. Another example of trilobite hash, although these specimens are in far better condition (thoraxes and cephalons) than those in fig. 13-45. Utica Black Shale, Eastern New York State. ×2.5.

13-49. Middle Ordovician. *Triarthrus eatoni* (Hall). A radiograph ("X-ray") positive print of the ventral (bottom) of a trilobite showing antennae and endopodites (walking legs). Specimen was partially pyritised, thereby allowing the success of the radiography. Utica Shale, Rome, Oneida, New York. × 10. (WS 1026). Courtesy: Prof. Dr. W. Stuermer, Erlangen, West Germany.

13-50. Ordovician. *Pseudogyties canadensis* (Chapman). A complete trilobite specimen in shale matrix. Specimen has pronounced genal spines. Trenton Group, Craigleith, Ontario, Canada. ✕3.5.

13-51. Silurian. *Calymene celebra* Raymond. A complete trilobite specimen replaced by dolomite. Edgewood Formation (Niagaran), Grafton, Jersey County, Illinois. ✕2.

13-52. Silurian. *Acaste downingiae* (Murchison). A complete trilobite specimen in matrix. Wenlock Series, Wrens Nest, Dudley, Worcestershire, England. ✕1.5.

13-53. Late Silurian. *Arctinurus* (*Lichas*) *boltoni* (Bigsby). A complete trilobite specimen in shale matrix. Clinton Shale, Ontario, Canada. ✕2. Courtesy: Academy of Natural Science, Philadelphia, Pennsylvania.

13-54. Silurian. *Dalmanites* cf. *dentatus.* A partial thorax and complete pygidium, with a lengthy spiked tail spine (telson). A cast of a trilobite shell section in a limestone matrix. Niagaran Series, Trilobite Ridge, Port Jervis, Orange County, New York. ×3.

13-55. Silurian. *Dalmanites limulurus* var. *lunatus* Lambert. A complete trilobite specimen in black shale matrix. Lockport Shale Formation, Niagaran Series, Lockport, Niagara County, New York. ×2.

13-56. Silurian. *Dalmanites limulurus lunatus* Lambert. A complete trilobite specimen (a cast of the original shell) in matrix. Part of the right side of the specimen is covered by matrix. Lockport Shales, Niagaran Series, Wrights Corners, Niagara County, New York. ×2.5.

mount. The specimen on a bit of matrix is much more interesting and looks better in a collection. Removing the specimen from matrix can be dangerous, as the specimen is likely to fall apart and no longer be of any value to a collection.

In the collecting of pyritized (or partially pyritized) trilobite specimens, such as at Bundenbach in Western Germany, many collectors have ruined fine specimens by using the crudest forms of preparation. These includes the use of "steel wool" and "wire brushes." The collectors who have rubbed steel wool or wire brushes over the "lumps" which are found in the Hunsrück shales at Bundenbach, have destroyed much fine detail of otherwise precious specimens. Such details as antennae, legs, and even some preserved soft tissues have been rubbed off by collectors who just want to see the "golden" appearance of a trilobite or an echinoid or whatever is found in the slatey shales at Bundenbach.

The proper way of preparing such slatey material is by the new air-abrasive method (a dental tool prepared by a large dental equipment company for the cleaning of teeth). These "air-dent" machines as they are commonly called, are used by museums and even a few private collectors. The method is fairly simple. A fine jet stream of air is fired at the objective (in this case, a fossil imbedded in matrix).

13-57. Middle Silurian. *Dalmanites* sp. A complete trilobite specimen, with some damage to the shell material. Specimen is in a shale matrix. Rochester Shale, Rochester, Monroe County, New York. ×3. Courtesy: Geology Department, Princeton University, Princeton, New Jersey.

13-58. Middle Silurian. *Dalmanites limulurus lunatus* Lambert. A complete trilobite specimen (see Fig. 13-57) in a shaley matrix. Rochester Shale, Rochester, Monroe County, New York. ×2.5.

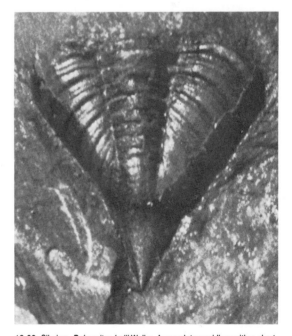

13-59. Middle Silurian. *Dalmanites limulurus lunatus* Lambert. An almost complete pygidium, missing only the "spiked" tail spine (telson). Note the flared apron on the pleural lobes. Rochester Shales, Brockport, Monroe County, New York. ×2.

13-60. Silurian. *Dalmanites halli* Weller. A complete pygidium with a short spiked tail spine. Note partial chitinous coating of the trilobite (on the pleural lobes and tail spine). Lockport Shale, Lockport, Niagara County, New York. 15 mm.

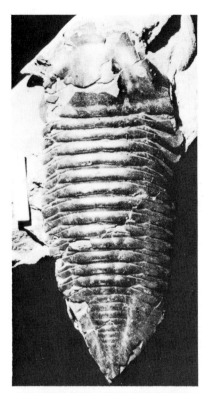

13-61. Middle Silurian. *Trimerus delphinocephalus* Green. A complete large trilobite specimen in partial matrix. Some of the chitinous coating was broken from the specimen when it was removed from the matrix. Clinton Shale, New York State. 15 cm in length.

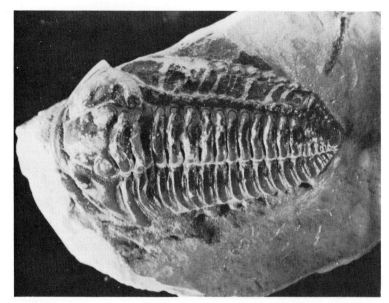

13-62. Middle Devonian. *Calymene platys* Green. An almost complete trilobite specimen in matrix. A small part of the posterior edge of the right cephalon is missing. Onondaga Shale, New York State. 6 cm in length.

13-63. Middle Devonian. *Trimerus (Dipleura) dekayi* (Green). A complete specimen of this large trilobite. Note the "shovel-nose." *Trimerus* was a bottom feeder and rooted its food from the ocean floor. Hamilton Shale, Hamilton, Madison County, New York. ×1. Courtesy: Geological Enterprises, Ardmore, Oklahoma.

13-64. Middle Devonian. *Trimerus (Dipleura) dekayi* (Green). An essentially complete (missing the tip of the snout, and part of the posterior portion of the pygidium) trilobite specimen. Flattened (probably a shell "molt") and freed from matrix. Hamilton Shale, Hubbardsville, Madison County, New York. Approx. 14 cm.

13-65. Middle Devonian. *Trimerus (Dipleura) dekayi* (Green). An interesting study of an isolated pygidium, showing the "shovel" edge (for flicking or scooping up bottom sand to find food). The shovel edge is quite evident in the lower views. Hamilton Shales, Hubbardsville, Madison County. New York. ×1.5.

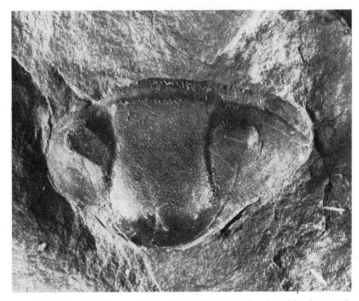

13-66. Middle Devonian. *Trimerus (Dipleura) dekayi* (Green). A three-quarter view of the cephalon of this interesting trilobite, showing the large raised eyes and the "shovel" edge of the snout. Hamilton Shale, Hubbardsville, Madison County, New York. ×1.5.

13-67. Middle Devonian. *Trimerus (Dipleura) dekayi* (Green). Another view of the above specimen (see Fig. 13-66). Looking directly down onto the cephalon (head), we can clearly see the eye mounts and the "shovel" edge snout. Hamilton Shale, Hubbardsville, Madison County, New York, ×1.5.

13-68. Late Devonian. *Trimerus (Dipleura) dekayi* (Green). A smaller specimen than usually found. Essentially complete (with the exception of the pygidium, which may be buried in an enrolled position within the matrix itself) trilobite specimen. Upper Tully Limestone Formation, Moravia Beds, Genesee Group, Moravia, Cayuga County, New York. ×1.5.

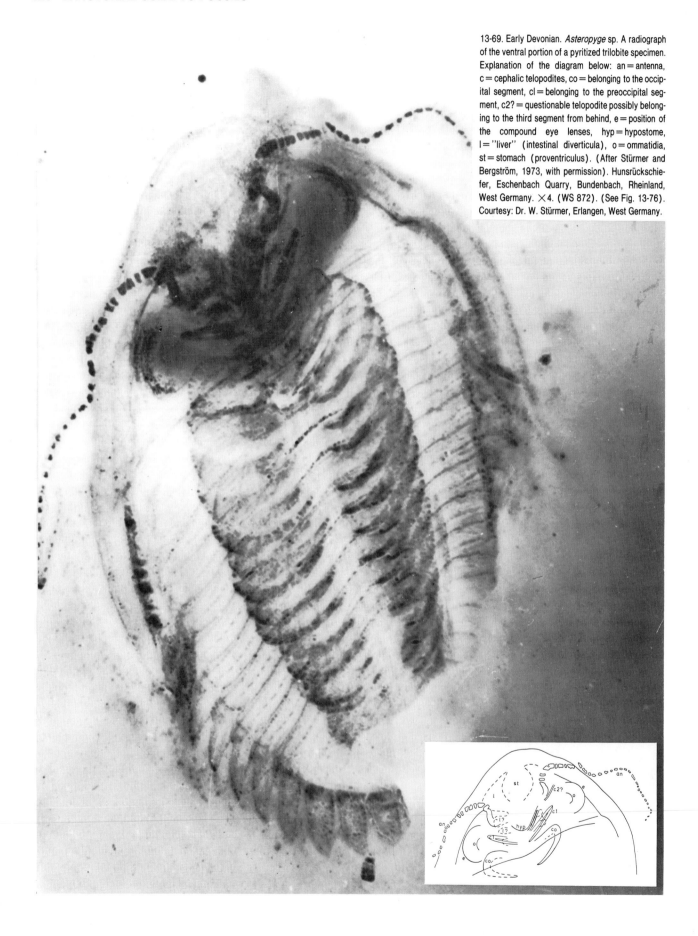

13-69. Early Devonian. *Asteropyge* sp. A radiograph of the ventral portion of a pyritized trilobite specimen. Explanation of the diagram below: an = antenna, c = cephalic telopodites, co = belonging to the occipital segment, cl = belonging to the preoccipital segment, c2? = questionable telopodite possibly belonging to the third segment from behind, e = position of the compound eye lenses, hyp = hypostome, l = "liver" (intestinal diverticula), o = ommatidia, st = stomach (proventriculus). (After Stürmer and Bergström, 1973, with permission). Hunsrückschiefer, Eschenbach Quarry, Bundenbach, Rheinland, West Germany. ✕4. (WS 872). (See Fig. 13-76). Courtesy: Dr. W. Stürmer, Erlangen, West Germany.

13-70. Middle Devonian. *Greenops boothi* (Green). An essentially complete trilobite specimen in matrix. The fringe apron on the pygidium is partially covered by shale. (See Fig. 13-43 for a comparison of the apron.) Moscow Formation, Hamilton Group, Alden, Erie County, New York. ✕2.

13-71. Middle Devonian. *Greenops boothi* (Green). An almost complete isolated cephalon showing eye mounts and the glabella. A cast in hard shale matrix. *Trimerus (Dipleura) dekayi* zone, Hamilton Shale Formation, Hubbardsville, Madison County, New York. ✕3.

13-72. Middle Devonian. *Odontocephalus aegeria* Hall. A complete trilobite specimen. Note the coalesced spines extending outward from the apron of the cephalon. This as well as a terminating crescent comprised of two spines or telsons at the end of the pygidium (not shown in this particular specimen), is diagnostic for the species. Onondaga Limestone, New York State. Approx. 10 cm. Courtesy: Geological Enterprises, Ardmore, Oklahoma.

13-73. Middle Devonian. *Odontocephalus aegeria* Hall. Another specimen of the same type of trilobite shown in Fig. 13-72. The bifurcated spine has been mechanically removed from the end of the pygidium, but the coalesced spines can still be seen at the very top of the cephalon. Onondaga Limestone, New York State. 10 cm. Courtesy: Geological Enterprises, Ardmore, Oklahoma.

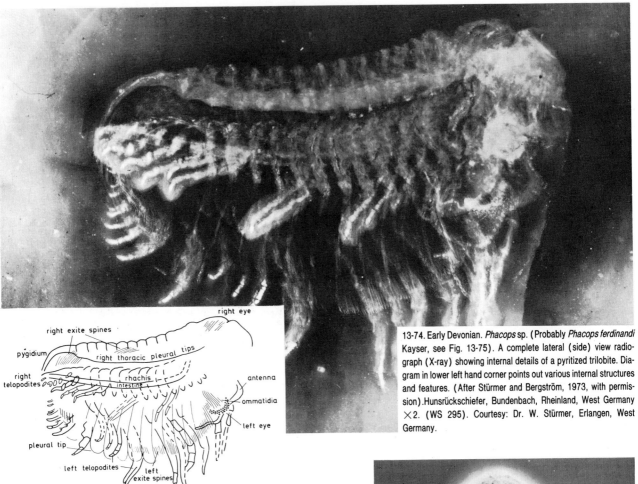

13-74. Early Devonian. *Phacops* sp. (Probably *Phacops ferdinandi* Kayser, see Fig. 13-75). A complete lateral (side) view radiograph (X-ray) showing internal details of a pyritized trilobite. Diagram in lower left hand corner points out various internal structures and features. (After Stürmer and Bergström, 1973, with permission).Hunsrückschiefer, Bundenbach, Rheinland, West Germany ×2. (WS 295). Courtesy: Dr. W. Stürmer, Erlangen, West Germany.

13-75. Early Devonian. *Phacops ferdinandi* Kayser. A complete cephalon and partial thorax of a pyritized trilobite imbedded in shale (slate) matrix. Hunsrück Shales, Gemünden, Rheinland, West Germany. ×2.

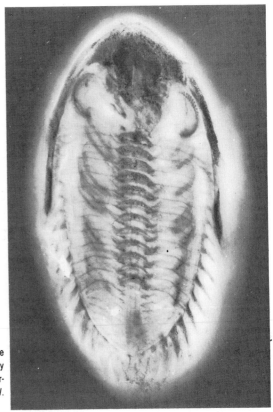

13-76. Early Devonian. *Asteropyge* sp. A radiograph of a trilobite in pyritized slate. Much internal detail is evident from this X-ray photograph. Hunsrück Shales, Bundenbach, Rheinland, West Germany. ×2. (WS 623.) (See Fig. 13-69.) Courtesy: Dr. W. Stürmer, Erlangen, West Germany.

13-77. Middle Devonian. *Phacops rana* (Green). A group of three trilobites, their shells overlapping one another. More than likely these are molted shells which were washed up against each other by the currents. Hamilton Shale, 18 Mile Creek, New York. Top specimen approx. 26 mm in length.

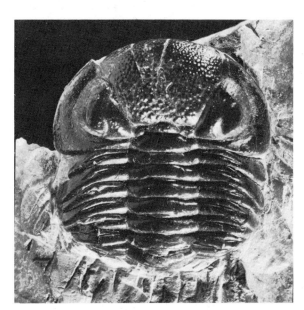

13-78. Middle Devonian. *Phacops rana* milleri (Stewart). An enrolled specimen. The posterior portion of the thorax and the pygidium are buried in the lime shale matrix. Silica Shale Formation, Medusa Limestone Quarry, Sylvania, Lucas County, Ohio. 38 mm.

13-79. Middle Devonian. *Phacops rana milleri* (Stewart). A complete trilobite specimen on a fragment of matrix. Typical of the size and preservation state of this subspecies. Silica Shale Formation, Medusa Limestone Quarry, Sylvania, Lucas County, Ohio. Approx. 4 cm in length.

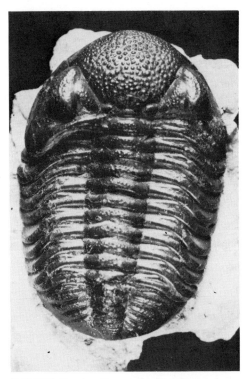

13-80. Middle Devonian, *Phacops rana milleri* (Stewart). A complete "flattened" trilobite specimen in matrix. Silica Shale Formation, Medusa Limestone Quarry, Sylvania, Lucas County, Ohio. 5 cm.

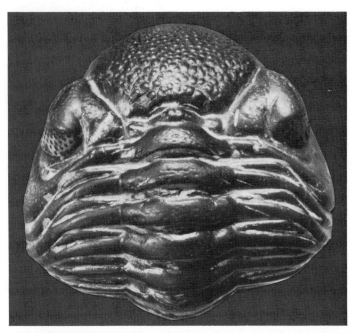

13-81. Middle Devonian. *Phacops rana milleri* (Stewart). A very fine enrolled specimen. Typical of many specimens found in the quarries. More enrolled than flattened specimens seem to be collected. Silica Shale Formation, Medusa Limestone Quarries, Sylvania, Lucas County, Ohio. ✕2.5.

A fine silicate powder is added to the air jet stream which acts like a miniature "sandblaster" by wearing down the matrix surrounding or lightly covering the specimen. It requires much patience. Hours, perhaps even days or weeks, will be needed to properly prepare the specimen. Ca e must be taken not to overwork an area with the fin. jet stream of sand particles, as the very fragile and delicate portions of the specimen can and may be worn away with the matrix.

The complete faunal listing for trilobites in the Hunsrück shales of the Lower Devonian, at Bundenbach, Rhenisch section of West Germany: *Asteropyge* sp., *Burmeisterella aculeata*, *Comura* sp., *Cornuproetus hunsrueckianus, Dipleura?* aff. *laevicauda, Homalonotus hunsrueckianus, Homalonotus* sp., *Odontochile rhenanus, Parahomalonotus planus, Phacops ferdinandi, Rhenops limbatus, Scutellum wysogorski,* and *Treveropyge drevermanni.*

The hard limestone at the Medusa Limestone Quarries near Sylvania, Lucas County, Ohio (just north of Toledo), houses some of the most beautiful and sought after trilobite specimens, *Phacops rana milleri* (Stewart) (see Figs. 13-78 through 13-83). The trilobites are hidden in massive limestone matrix, and it is necessary to use either a ten or twenty pound sledge (hammer) to break up the larger blocks of rock and to (if you are lucky!) uncover a trace of

the elusive and much prized *Phacops rana.* This species is partially pyritized and in some cases has a dark brown or blackish colored chitinous shield covering the pyritic layering. The flattened-out specimens are much more prized than the enrolled ones.

Probably the most famous of trilobite localities (now completely inaccessible to collectors) is the "Walcott" locality of the Middle Cambrian Burgess Shale at Burgess Pass, near Field, British Columbia, western Canada. Although the specimens at this locality are highly distorted due to the compaction of sediments, they are still the most desirable of specimens for the collector and student of earlier forms of trilobites. The fossils are "preserved" by a thin carbonaceous film upon the shale surface, and exhibit details of appendages as well as internal structures.

In the state of Missouri, the most common Ordovician trilobite species are *Isotelus* and *Bumastus. Phillipsia* is the most abundant species in Missouri's Mississippian rocks, while *Ameura* is the representative species for the Pennsylvanian (Late Carboniferous) of Missouri.

Trilobites have been reported from the fossil record of Wyoming.

The eastern part of Pennsylvania has *Cryptolithus* represented in the Ordovician at Swatara Gap, while *Phacops rana* and *Greenops boothi* are common to the Devonian rocks near Stroudsburg and Deer Lake. The trilobite species for the state of Virginia include: *Ampyxina scarabens, Sphaerexochus* sp., *Olenellus buttsi, Ampyx americanus, Tricrepicephalus cedarensis, Ptychoparella michaeli, Po-*

13-82. Middle Devonian. *Phacops rana milleri* (Stewart). A closeup of one of the compound lenses (eyes) of the specimen in Fig. 13-80. Note the many facettes of the lens of the eye pedestal. Silica Shale Formation, Medusa Limestone Quarry, Sylvania, Lucas County, Ohio. ×7.

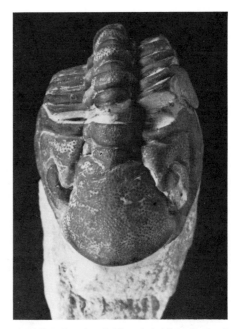

13-84. Early Devonian. *Reedops deckeri* Delo. Front end view of the cephalon and part of the thorax of this rather distinctive trilobite. See Fig. 13-85 for the complete specimen in its lateral view. Haragan Formation (Helderbergian), Hunton Group, Clarita, Coal County, Oklahoma. Approx. 2 cm.

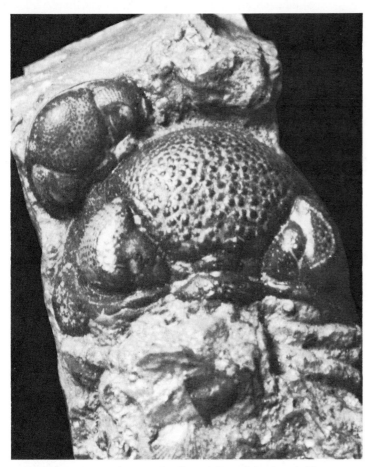

13-83. Middle Devonian. *Phacops rana milleri* (Stewart). Two cephalons in matrix. The larger of the two specimens may in fact be more complete. It appears that the thorax is covered by shale. Silica Shale Formation, Medusa Limestone Quarries, Sylvania, Lucas County, Ohio. ×4.

13-85. Early Devonian. *Reedops deckeri* Delo. Lateral (side) view of the above specimen (Fig. 13-84), showing the distinctive arched thoracic segments in the enrolled position. Note the difference in the eye pedestal, in comparision to that of *Phacops*. Haragan Formation (Helderbergian), Hunton Group, Clarita, Coal County, Oklahoma. 4 cm in length.

liella virginica, Norwoodella saffordi, and *Homagnostus bulbus (Proagnostus).*

New Jersey's typical trilobite is *Dalmanites* which extends from the Late Silurian to the Early Devonian rocks of Trilobite Ridge in Sussex County.

Indiana has trilobite representatives of the following species: *Calymene niagarensis, Dalmanites halli, Isotelus gigas,* and *Phacops rana.*

A faunal list of the trilobites of Great Britain: CAMBRIAN: *Centropleura henrichi, Eodiscus punctatus* (an agnostid), *Lotagnostus trisectus* (an agnostid), *Olenelloides armatus, Olenellus gibbosus, Olenellus lapworthi, Parabolina spinulosa, Paradoxides davidis, Peltura scarabaeoides,* and *Protolenus latouchei.* ORDOVICIAN: *Ampyx linleyensis, Angelina sedgwicki, Asaphellus homfrayi, Basilicus tyrannus, Broeggerolithus broegeri, Brogniartella bisulcata, Chasmops extensa, Cnemidopyge nudus, Corrugatagnostus sol* (an agnostid); *Dalmanitina robertsi, Diacalymene drummuckensis, Eccoptochile (Placoparina) sedgwicki, Encrinuroides sexcostatus, Euloma monile, Flexicalymene cambrensis, Flexicalymene caractaci, .Flexicalymene quadrata, Geragnostus callavei* (an agnostid), *Marrolithus favus, Neseuretus murchisoni, Ogyiocarella debuchi, Ogyiocaris selwyni, Onnia gracilis, Parabolinella triarthra, Paracybeloides girvanensis, Phaeopidina (Kloucekia) apiculata, Phillipsinella parabola, Placoparia zippei, Platycalymene duplicata, Proetus girvanensis, Pseudosphaerexochus octolobatus, Remopleurides girvanensis, Salterolithus caractaci, Selenopeltis inermis, Shumardia pusilla, Sphaerocoryphe thompsoni, Stapeleyella inconstans, Tretaspis ceroides,* and *Trinucleus fimbriatus.* SILURIAN: *Acaste downingiae* (see Fig. 13-52), *Acidaspis deflexa, Bumastus barriensis, Ca-*

lymene blumenbachi, Calymene replicata, Cheirurus bimucronatus, Dalmanites myops, Dalmanites (Delops) obtusicaudatus, Deiphon forbsi, Encrinurus onniensis, Encrinurus punctatus, Encrinurus variolaris, Phacops stokesi, Sphaerexochus mirus, and *Trimerus delphinocephalus.* DEVONIAN: *Dechenella setosa, Phacops accipitrinus, Trimerocephalus mastropthalmus,* and *Scutellum granulatum.* CARBONIFEROUS: *Brachymetopus ouralicus, Cummingella jonesi, Griffithides seminiferus, Phillipsia gemmulifera, Phillipsia laticaudata,* and *Spatulina spatulata.*

Of all the collecting localities on the North American continent (particularly, the United States of America), the state of New York seems to have the largest amount of trilobite specimens available in formations of at least three major Paleozoic time periods—the Ordovician, the Silurian, and the Devonian.

In the Chazyan of the Ordovician of New York State, the following species can be found: *Bumastus globosus,* and *Pliomerops candensis.* In the Lorraine-Eden: *Cryptolithus bellus.* In the Black River: *Bathyurus extans.* In the Middle Ordovician: *Isotelus gigas.* In the Late Ordovician: *Flexicalymene meeki.* In the Trenton Group, the following species: *Basilicus (Basiliella) barrandei, Calyptaulax (Calliops) callicephalus, Ceraurus pleuraxanthemus,* and *Cryptolithus tesselatus.*

In the Silurian Rochester Shales of New York State, we find: *Arctinurus (Lichas) boltoni, Bumastus barriensis, Dalmanites limulurus, Proetus corycaeus,* and *Trimerus (Homalonotus) delphinocephalus.*

In the Early Devonian (Oriskany and Helderbergian) of New York State, the species are: *Acidaspis tuberculata,*

13-86. Middle Devonian. *Proetus rowi* (Green). A complete trilobite specimen in matrix. Mahantango Formation, Centerfield Coral Biostrome, North of Stroudsburg, Monroe County, Pennsylvania. 4 cm.

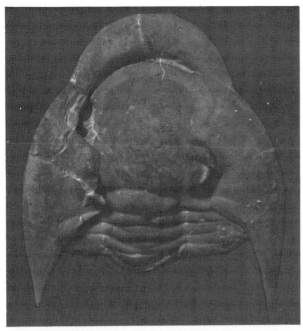

13-87. Late Pennsylvanian. *Ameura sangamonensis* (Meek and Worthen). An enrolled specimen showing the cephalon and a portion of the thorax overlapping the pygidium. Note the "shovel" edge of the pygidium indicating a "bottom feeder". Doniphan Shale, Virgil Series, Ace Hill Quarry, Plattsmouth, Cass County, Nebraska. ×6.

13-89. Late Pennsylvanian. *Ameura sangamonensis* (Meek and Worthen). Another specimen showing a complete cephalon, a partial thorax and a complete pygidium, no doubt from the same individual, in matrix. Doniphan "Limerock" Shales, Virgil Series, Ace Hill Quarry, Plattsmouth, Cass County, Nebraska. ×4.

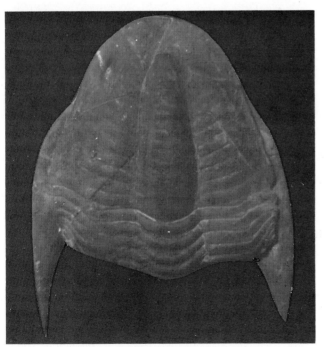

13-88. Late Pennsylvanian. *Ameura sangamonensis* (Meek and Worthen). The reverse of the above specimen (Fig. 13-87), showing the pygidium and partial thorax. The genal spines of the cephalon are extending from the other side of the specimen. Doniphan "Limestone" Shale, Virgil Series, Ace Hill Quarry, Plattsmouth, Cass County, Nebraska. ×6.

Autodetus beecheri, Cordania becraftensis, Cordania hudsonica, Dalmanites bisignatus, Dalmanites phacoptyx,

Otarion (Cyphaspis) minuscula, Phacops correlator, Phacops logani, Proetus conradi, Synphoria (Dalmanites) stemmatus, and *Synphoria (Dalmanites) stemmatus convergens.*

In the Middle Devonian (Hamilton Group) of New York State, we have: *Greenops boothi, Phacops rana, Proetus haldemani,* and *Trimerus (Dipleura) dekayi.* In the Onondaga, the following species are represented: *Odontocephalus aegeria* and *Terataspis grandis.*

Finally, in the Late Devonian, particularly the Tully Limestone, *Trimerus (Dipleura) dekayi, Greenops boothi,* and *Phacops rana* have made the transition from the Middle Devonian Hamilton Shales.

Carboniferous trilobite representations are in the rocks of the midwestern part of the United States, particularly in Iowa, Kansas and Nebraska. *Ameura* is the dominant species of the Late Carboniferous and is particularly abundant in the Virgil Series (Late or Upper Pennsylvanian) rocks in eastern Nebraska (see Figs. 13-87 through 13-89).

The trilobites died out at the end of the Carboniferous. Traces of *Ameura* and *Ditomopyge* are found in some Permian marine localities in the western states, particularly in Oklahoma. No more trilobite fossil remains have been recovered in rocks younger than the Permian, indicating strongly that the evolutionary trend for the trilobite had reached its peak. The few marine and mostly terrestrial isopods have essentially taken the place of the three-lobed arthropods—those most fascinating trilobites!

14 ARTHROPODA (Branchiopoda, Ostracoda, and Cirripedia)

In the supergroup crustacea, there are six classes of arthropods, five of which will be discussed in this chapter. The sixth class, Malacostraca (shrimps, crabs, and lobsters), will be discussed in the following chapter.

For now, we are interested in the following crustacean groups (classes): 1. Branchiopoda, including the "water fleas." 2. Ostracoda, small marine bivalved crustaceans. 3. Copepoda, small parasitic (on fishes) crustaceans ("ticks of the sea"). 4. Branchiura, similar to the copepods, also parasitic in nature. And finally, 5. Cirripedia, the barnacles.

Of the above five classes, we will concern ourselves only with the branchiopods, the ostracodes, and the cirripeds (barnacles). The copepods and branchiurids are fairly recent groups, and their occurrence to date in the fossil record is not too well known.

To begin with, we have the branchiopod ("gill-foot") crustaceans. These are tiny primitive crustaceans whose habitat is in both marine and fresh water deposits. A small percentage of modern branchiopods make their home in the oceans of the world. Most recent species are found in fresh water—in ponds, rivers, and lakes. A good example of this group in our modern freshwater areas are the "fairy shrimp," a rather ubiquitous and transparent species found in many ponds and lakes throughout the world. The "brine shrimp" live in very salty water deposits.

The branchiopods have a variable number of segments comprising their bodies, with their thoracic appendages having flattened "leaf-like" gill filaments, and the body being covered by a bivalved chitinous carapace (shell).

A typical and most common example of a fresh water branchiopod from the fossil record is the Upper Triassic species *Cyzicus (Estheria)* (see Figs. 14-1 and 14-2). Unfortunately, only the calcium carbonate impregnated shells are preserved in prehistoric fresh water lake mud shales to let us know their size and abundance. It is unlikely that their body structure and associated appendages differed radically from modern forms.

With their ornamented chitinous exoskeletons, the ostracodes are a most curious and quite ubiquitous group (class) of the crustacea. They are similar to, if not smaller than, the branchiopods, although some forms have reached sizes up to 2 centimeters. Their shells are unequal and have bulbous ornamentations of varying design on each of the valves, with protuberances at the hinge line.

There are literally hundreds of distinct species of ostracodes known, one of which, *Glyptopleurina* (see Fig. 14-6), is shown here as an example. A representative faunal listing for ostracodes is as follows: ORDOVICIAN: *Bassleratia, Dicranella, Drepanella, Jonesella, Leperditella, Milleratia, Primitia, Tetradella, Thomasatia,* and *Winchellatia*. SILURIAN: *Aparchites, Bonnemaia, Leperdita, Mastigobolbina, Welleria,* and *Zygobolba (Zygolbolbina)*. DEVONIAN: *Aechmina, Beecherella, Cytherellina, Entomozoe (Entomis), Eustephanus, Hollina, Octonaria, Parabolbina, Paraparchites, Stibus, Strepulites (Octonariella),* and *Tubulibairdia*. MISSISSIPPIAN: *Aechminella, Balantoides, Bairdiolites, Discoidella, Geisina (Perprimitia), Graphiodactylus, Healdia, Kirkbyella, Sansabella (Carboprimitia),* and *Tetratylus*. PENNSYLVANIAN: *Acratia, Bairdia, Cavellina, Coryellites, Glyptopleurina* (see Fig. 14-6), *Knightina, Moorites, Seminolites,* and *Sulcella*. PERMIAN: *Amphissites, Hollinella, Kellettina,* and *Roundyella*. JURASSIC: *Hutsonia* and *Theriosynoecum (Morrisonia)*. CRETACEOUS: *Argilloecia, Brachycythere, Cytherella, Loxoconcha, Monoceratina,* and *Paracypris*. TERTIARY: *Alatacythere, Argilloecia, Buntonia (Pyricythereis), Eocytheropteron, Hemicythere, Monoceratina, Orthonotacythere,* and Perissocytheridea.

Finally, we come to the cirripeds ("curl-foot") or barnacles, as we know them. They are attached by the base of their shells to a rock or shell surface (see Figs. 14-3 and 14-4). These are the adults of the species, sessile and encased in a rigid shell-house. Their appearance is very different from that of the other crustaceans.

Born as free swimmers, and then molting after establishing various forms in their lives, they take on an ostracode or branchiopod-like shell. Eventually they anchor themselves on a shell or rock, and, finally casting off the bivalve shell, start to secrete a series of calcareous plates to form a shortened conical shell (see Fig. 14-5). *Balanus* of the Tertiary (see Figs. 14-3 through 14-5) are well known fossil forms of cirripeds. Other known genera are *Brachylepas* and *Loriculina* of the Cretaceous.

14-1. Triassic. *Cyzicus (Estheria) ovata* (Lea). Both valves of the carapace of this most common freshwater branchiopod found in abundance in fossil lake mud shales. Lockatong Formation, Newark Series, Granton Quarry, North Bergen, Hudson County, New Jersey. × 4.

14-2. Triassic. *Cyzicus (Estheria) ovata* (Lea). Several isolated carapace sections of many individuals—possibly molted (shed) valves. Lockatong Formation, Newark Series, Granton Quarry, North Bergen, Hudson County, New Jersey. × 1.5.

14-3. Middle Miocene. *Balanus concavus* Bronn. The common "Acorn barnacle" of the Tertiary. Here it is attached to the shell of a pecten. Pungo River Marl Formation (Helvetian), Aurora, Beaufort County, North Carolina. × 1.

14-4. Middle Miocene. *Balanus concavus* Bronn. A group of acorn barnacles attached to a large pecten shell. This is representative of their colonizing a part of the ocean floor. Calvert Formation (Helvetian), Scientist's Cliffs (formerly Calvert Cliffs), Calvert County, Maryland. Reduced ½.

Inset: Another example of barnacles on a pecten. Choptank Formation, Maryland. ⅓ actual size.

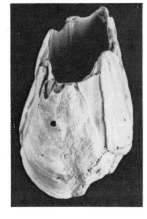

14-5. Late Miocene. *Balanus concavus* Bronn. An isolated acorn barnacle. Yorktown Formation (Upper Sahelian), Chesapeake Bay at Governors Run Road, Maryland. Half size.

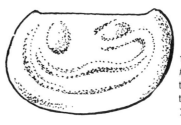

14-6. Pennsylvanian. *Glyptopleurina montifera* Coryell. One view of the right valve of the carapace of a typical Carboniferous ostracode. Desmoinesian, Northern Oklahoma. × 50.

15 ARTHROPODA (Malacostraca and Decapoda)

The class Malacostraca is a most interesting and diverse group of animals, belonging to the phylum Arthropoda, and including the crabs, crayfishes, shrimps, "sow-bugs," lobsters and others. Their geological range is Early Cambrian to Recent.

The Malacostraca are divided into five superorders. They are: the Phyllocarida (phyllocarids), a common index fossil, large with a bivalved shell and extended abdomen; the Syncarida (syncarids), animals lacking a carapace and without egg-carrying brood pouches; the Hoplocarida (hoplocarids), includes a single surviving order, the Stomatopoda or mantis shrimp; the Peracarida (peracarids), includes amphiopods and isopods (sow bugs); and the Eucarida (eucarids)—krill, shrimp, decapod shrimp, crayfish, lobsters and crabs.

The division (subclass of Moore, Lalicker, and Fischer, 1952 and currently considered as a superorder in the Treatise, 1969) Phyllocarida ("leaf-shrimps"—from the Greek Phyllo + Carid) is a primitive malacostracan whose origins can be traced back into the Cambrian Period, and which successfully survived up into the Permian Period. One species of the Phyllocarida survived into the post-Paleozoic era (specimens found in rocks of Upper (Late) Triassic age in Germany).

The Phyllocarida are quite distinctive with a large and loosely fitted body shield—a bivalved carapace (see Fig. 15-5) which is attached only to the back of the head area; a body with multiple somites (segments), with attached limbs, and an abdomen composed of at least six or seven somites; a head with eyes on stalks and antennae (see Fig. 15-1), and, most notably, a "gnathal" plate element (mandible) which the animal apparently used for feeding (see Fig. 15-3). This "gnathal" element (in reality, a part of the pincer) has a series of "toothlike" biting or crushing surfaces, almost mammallike in appearance. These so-called "gnathal" elements were originally diagnosed as the crushing teeth of primitive shark-like fishes. It was later to be proven that these structures were parts of the feeding mechanism for the phyllocarid *Echinocaris*, most commonly found in Middle Devonian rocks in the United

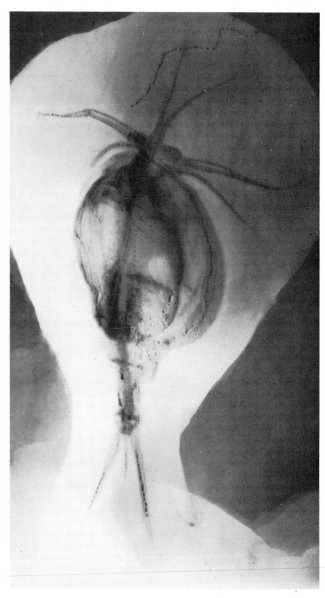

15-1. Early Devonian. *Nahecaris stuertzi* Jaekel. A radiograph of a complete phyllocarid crustacean. Note the antennae, arms, eye stalks, and three-pronged tail section extending from either side of the carapace. Hunsrück Shales, Bundenbach, Rheinland, West Germany. ×1.1. (WS4627). Courtesy: Dr. W. Stürmer, Erlangen, West Germany.

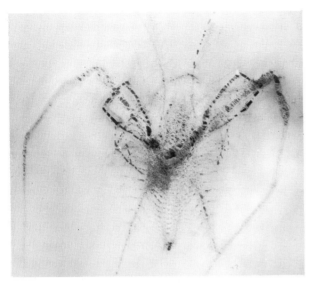

15-2. Early Devonian. *Mimetaster hexagonalis* (Gürich). A radiograph of a "spider shrimp". A bottom feeder whose habits and appearance are similar to the "spiny lobster" of our modern oceans. Hunsrück Shales, Bundenbacher Schiefer, Bundenbach, Rheinland, West Germany. ×2.5. (WS2235). Courtesy: Dr. W. Stürmer, Erlangen, West Germany.

15-5. Late Pennsylvanian. *Concavicaris sinuata* (Meek and Worthen). Both valves (one superimposed upon the other) of the carapace of this rather large phyllocarid. This crustacean was the food source of edestoid and anacanthous sharks as well as the iniopterygian and actinopterygian fishes. Queen Hill Shale, Virgil Series, Ace Hill Quarry, Plattsmouth, Cass County, Nebraska. ×1.5.

15-3. Middle Devonian. *Pseudodontichthys whitei* Skeels. A partial gnathal plate of a phyllocarid crustacean, probably belonging to the genus *Echinocaris*. The generic name *Pseudodontichthys* was assigned to these structures when they were originally thought to be the teeth of early shark-like fishes. Silica Shale Formation, Milan, Washtenaw County, Michigan. 12 mm.

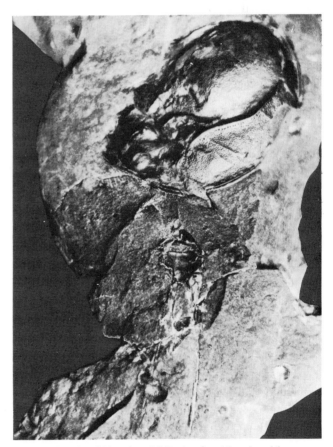

15-4. Late Pennsylvanian. *Concavicaris sinuata* (Meek and Worthen). An isolated tail section, showing three spikes (or spines) of the most common brackish-water phyllocarid from the Carboniferous. Heebner Shale, virgil Series, near Topeka, Jefferson County, Kansas. ×2.

15-6. Late Devonian. *Echinocaris socialis* Beecher. One of the valves of the carapace and the tail section of this common phyllocarid crustacean. Tully Limestone (Upper Moravian Beds), Ithaca Group, Moravia, Cayuga County, New York. ×1.5.

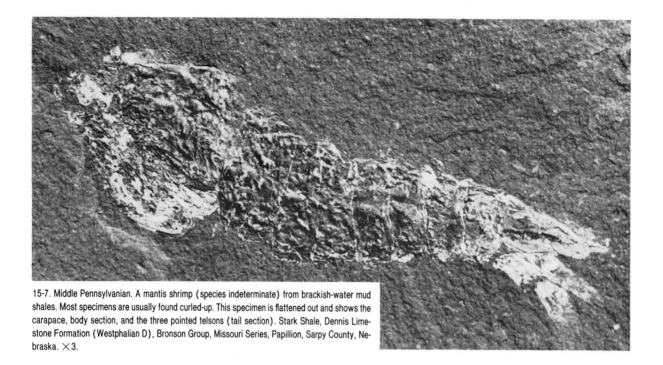

15-7. Middle Pennsylvanian. A mantis shrimp (species indeterminate) from brackish-water mud shales. Most specimens are usually found curled-up. This specimen is flattened out and shows the carapace, body section, and the three pointed telsons (tail section). Stark Shale, Dennis Limestone Formation (Westphalian D), Bronson Group, Missouri Series, Papillion, Sarpy County, Nebraska. ×3.

States in the States of Michigan and Ohio. The specimen shown in Fig. 15-3 is *Pseudodontichthys*—it should be called *Echinocaris*, but first a scientific paper will have to be published relegating *Pseudodontichthys* into synonymy with the genus *Echinocaris*.

In addition, the phyllocarids had a most distinctive trident-shaped tail appendage comprised of a series of three telson elements (see Fig. 15-4).

The division (superorder) Syncarida (syn = with, carid = shrimp) are fresh-water malacostracans without carapaces. The head and primary segment of the thoracic area are fused together. Their origins date back to the Late Mississippian of North Dakota and Montana, and only a few species have survived to the present. Their geographical distribution in fresh-water deposits today is restricted to the continent of Australia.

Palaeocaris typus Meek and Worthen, from the Pennsylvanian fresh and brackish water Braidwood fauna of the Mazon Creek area of Illinois, is a typical example of a shrimp-like syncarid. *Palaeocaris* is found in association with another syncarid *Acanthotelson stimpsoni* Meek and Worthen (see Fig. 15-22) in the Mazon Creek fossil beds. *Palaeocaris* has from twelve to thirteen body segments, its head is approximately the length of two of these body segments, with a double pair of antennae. The legs on the thoracic section are long and thin. The forward or anterior legs have no chelae. The telson or tail area is lengthy and flattened dorsoventrally.

The division (superorder) Peracarida (per = much, carid = shrimp) is a fairly large group of malacostracans

15-8. Late Jurassic. *Eryma modestiformis* (Schlotheim). A complete lobster (crayfish) in lithographic limestone matrix. Solnhofen Limestone, Eichstätt, Bavaria, West Germany. 7 cm. Courtesy: Siber and Siber Ltd., Aathal, Switzerland.

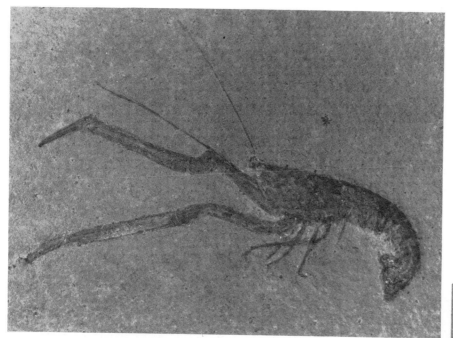

15-10. Late Jurassic. *Mecochirus longimanatus* (Schlotheim). A ventral (bottom) view of the body of a long-arm lobster showing the "fan-tail" at the back of the body. Solnhofen Limestone, Bavaria, West Germany. 5.5 cm.

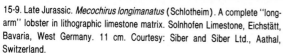

15-9. Late Jurassic. *Mecochirus longimanatus* (Schlotheim). A complete "long-arm" lobster in lithographic limestone matrix. Solnhofen Limestone, Eichstätt, Bavaria, West Germany. 11 cm. Courtesy: Siber and Siber Ltd., Aathal, Switzerland.

15-11. Late Jurassic. *Mecochirus longimanatus* (Schlotheim). A long-arm lobster with his arms folded, in a lithographic limestone matrix. Solnhofen Limestone, Solnhofen, Bavaria, West Germany. 9 cm. Courtesy: Siber and Siber Ltd., Aathal, Switzerland.

that has eight orders, three of which are listed here and are represented in the fossil record. They include the Mysidacea, the Isopoda (the "sow-bugs" or "pill-bugs"), and the Amphiopoda (marine, terrestrial and fresh-water isopod-like animals). The latter two orders are quite rare in the fossil rocord.

The peracarids are terrestrial as well as marine and fresh-water in extent. The peracarids are known for the characteristic "broodpouch," in which the females carry their live young. This brood pouch is composed of oostegites, a series of plates on the inner surface of the thoracic leg appendages.

The hoplocarids (mantis or armored shrimps, see Fig. 15-7) are, at the present time, a superorder of the Decapoda in *uncertain status* (Treatise, p. R116), and are characterized by a thin carapace connected to the first three segments of the thorax.

An interesting feature of the mantis shrimp is the stabbing motion (snap) of the mantis shrimp's dactyl—it is the fastest animal motion yet known (approximately one ten thousanth of a second). The animals have very slow reflexes; but, once triggered, the dactyl can spear prey in the

amount of time it takes a beam of light to travel across the width of Long Island, or just under 20 miles (32 kilometers).

They are not common as fossils. Nevertheless several living genera exist today in present marine environments.

The final superorder Eucarida (or "true shrimps") includes all other malacostracans not listed in the previous divisions. This final group is separated into two orders, the Euphausiacea and Decapoda. The latter includes such familiar species as lobsters, shrimps, prawns, and crabs. The former is comprised of only marine prawn-like crustaceans. We will only be concerned here in this chapter with the larger group—the Decapoda.

The earliest "lobster-like" forms include *Eryon* (see Figs. 15-12 through 15-15) from the Jurassic of Germany, while the oldest true lobster *Eryma* (see Fig. 15-8) is also represented in the Jurassic of Europe.

The lobsters, shrimps, and prawns all possess an elongated, segmented body with a light, partial armor (carapace) covering the thorax and head regions. They are also characterized by multiple pairs of limbs, and a fanlike tail. Some examples in this chapter of the various species com-

15-12. Late Jurassic. *Eryon arctiformis* (Schlotheim). Ventral (bottom) view close-up of the mouth area of this "lobster". Upper Malm, Lithographischer Schiefer, Eichstätt, Bavaria, West Germany. ×3. Courtesy: Department of Geological and Geophysical Science, Princeton University, Princeton, New Jersey.

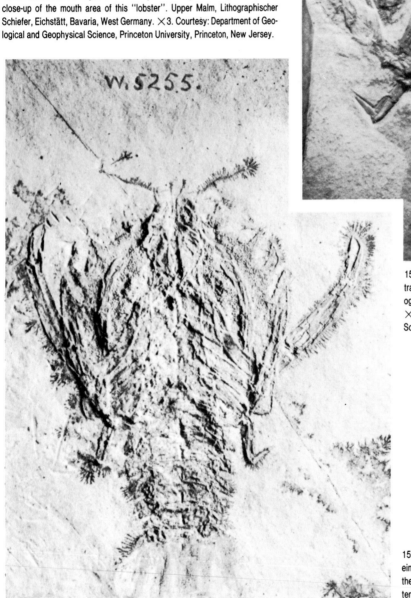

15-13. Late Jurassic. *Eryon arctiformis* (Schlotheim). A ventral (bottom) view of a complete "lobster." Upper Malm, Lithographischer Schiefer, Solnhofen, Bavaria, West Germany. ×2. Courtesy: Department of Geological and Geophysical Sciences, Princeton University, Princeton, New Jersey.

15-14. Late Jurassic. *Eryon arctiformis* (Schlotheim). A close-up of the mouth area. Lithographic Limestone, Solnhofen, Bavaria, West Germany. ×2.5. Courtesy: Department of Geology, Princeton University, Princeton, New Jersey.

15-15. Late Jurassic. *Eryon arctiformis* (Schlotheim). A complete ventral view of a "lobster." Note the dendritic patterns (branch-like streamers) extending from the outer parts of the fossil. Lithographic Limestone, Solnhofen, Bavaria, West Germany. ×1.5. Courtesy: Department of Geology, Princeton University, Princeton, New Jersey.

15-16. Late Cretaceous. Palinurid "crab" (species indeterminate). A complete ventral view of a palinurid crustacean. Cenomanian, Hadjula, Lebanon. 3.5 cm. Courtesy: Siber and Siber Ltd., Aathal, Switzerland.

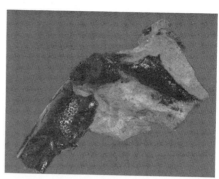

15-17. Late Cretaceous. *Hoploparia gabbi* (Pilsbry). An almost complete claw, missing the pincer. Navesink Formation (Middle Maestrichtian), Marlboro, Monmouth County, New Jersey. Approx. half size.

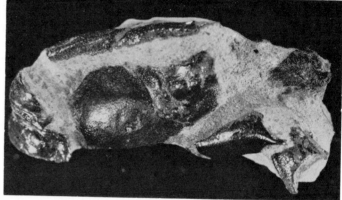

15-20. Late Cretaceous. *Pseudostacus hakelensis* (Fraas). A complete lobster with claws and antennae. Cenomanian, Hakel, Lebanon. 15 cm. Courtesy: Siber and Siber Ltd., Aathal, Switzerland.

15-18. Late Cretaceous. *Hoploparia gabbi* (Pilsbry). An almost complete crayfish-lobster. Missing parts of the claws and pincers and a part of the fan-tail. Navesink Formation (Middle Maestrichtian), Holmdel, Monmouth County, New Jersey. Approx. half size.

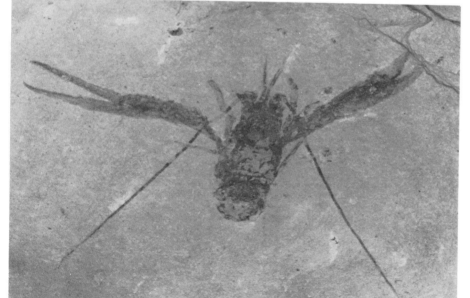

15-19. Late Cretaceous. *Hoploparia gabbi* (Pilsbry). Both pincers (one complete—in foreground, the other missing tips). Navesink Formation (Middle Maestrichtian), Holmdel, Monmouth County, New Jersey. ×1.

15-21. Late Cretaceous. *Hoploparia gabbi* (Pilsbry). An almost complete specimen. Missing only the tip of the antenna stalk (forward, to the left of the forehead), and the pincers of the claws. Navesink Formation (Middle Maestrichtian), Holmdel, Monmouth County, New Jersey, ×1.

15-22. Middle Pennsylvanian. *Acanthotelson stimpsoni* Meek and Worthen. A fairly complete shrimp in an ironstone (siderite) concretion. Braidwood Fauna (Mazon Creek), Will County, Illinois. ×1.5.

15-23. Permian. *Uronectes fimbriatus* Jordan. A complete small shrimp in red sandstone matrix. Rotliegendes (Red Beds), Lochbeunnen - Oberhof, Thüringen, East Germany. ×3.5.

prising this group are: *Mecochirus,* the "long-armed" lobster (see Figs. 15-9 through 15-11), from the Jurassic Solnhofen Limestones of Germany; several species of "lobster-like" crustaceans (see Figs. 15-16 and *Pseudostachus* 15-20) from the Late Cretaceous of Lebanon; the marine hoploparid lobsters, *Hoploparia* (see Figs. 15-17 through 15-19, and 15-21), of the Late Cretaceous of New Jersey; a small shrimp, *Uronectes* (see Fig. 15-23), from the Permian "Red-beds" of Germany; *Aeger,* a large eucarid crayfish (see Fig. 15-24), from the Jurassic of Germany; large eucarids (shrimps), *Antrimpos* (see Fig. 15-25A and B), *Dusa* (see Figs. 15-27 and 15-28), and *Drobna* (see Fig. 15-29), from the Late Jurassic of Europe; *Palinurina,* a multi-legged shrimp (eucarid) (see Fig. 15-30), from the Late Cretaceous of Lebanon; and a freshwater "long-antenna" shrimp (see Fig. 15-36) from the Middle Eocene Green River Formation of Wyoming. The true lobsters have been traced back to rocks of Jurassic age.

Pygocephalus (Anthrapalaemon) (see Fig. 15-26) is an example of a fresh-water eucarid, distinguished by a lengthy, but simple and partially serrated (at the flared-out apron), carapace covering the head and anterior thoracic segments. The head has antennae that extend from basal articulations (joints) which are wide and squarish in appearance. The abdomen is composed of six segments, and the telson (tail area) is comprised of divided flaps which fan out laterally.

Finally, there are the ubiquitous crabs. They are in a decapod suborder called Brachyura ("short-tails"). They differ from the lobsters, shrimps, and prawns in having a wider body (as opposed to the lobster's elongated one), a complete armored carapace covering the entire head and thorax, the abdomen reduced and folded underneath the cephalothorax, generally long legs, the first pair of leg appendages modified for snapping and/or crushing (see Fig. 15-42 for a closeup of the "teeth" of the chelae), short antennae, eyes on stalks, and no "fan-tail." Examples of Fossilized crabs are: *Lobocarcinus* (see Fig. 15-37) from the Late Eocene of Egypt, small shore-crabs (see Figs. 15-38 and 15-39) from the Cretaceous of Lebanon and New Jersey, and (Fig. 15-49) from the Pliocene of California; and the large "swimming" crabs: *Longysorbis* (see Fig. 15-40), *Tumidocarcinus* (see Fig. 15-43), *Harpactocarcinus* (see

15-24. Late Jurassic. *Aeger tipularis* (Schlotheim). A complete shrimp in lithographic limestone matrix. Tithonian (Malm Zeta), Eichstätt, Bavaria, West Germany. X1.5. Courtesy: Geology Department, Princeton University, Princeton, New Jersey. Inset: Another specimen of *Aeger* from the same locality.

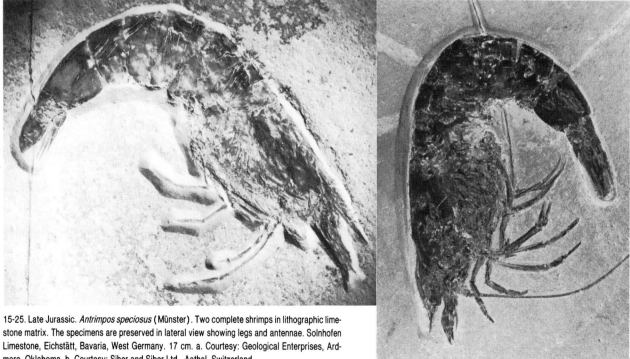

a

15-25. Late Jurassic. *Antrimpos speciosus* (Münster). Two complete shrimps in lithographic limestone matrix. The specimens are preserved in lateral view showing legs and antennae. Solnhofen Limestone, Eichstätt, Bavaria, West Germany. 17 cm. a. Courtesy: Geological Enterprises, Ardmore, Oklahoma. b. Courtesy: Siber and Siber Ltd., Aathal, Switzerland.

b

15-26. Late Jurassic. *Pygocephalus* (*Anthrapalaemon*) sp. A dorsal (top) view of a fairly complete (missing fan-tail area) shrimp with antennae, etc. Solnhofen Limestone, Schernfeld, Bavaria, West Germany, 3.5 cm.

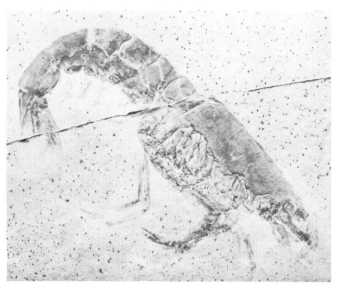

15-27. Late Jurassic. *Dusa monocera* Münster. A complete laterally preserved shrimp in lithographic limestone matrix. Missing a few legs and the antennae. Malm Zeta, Solnhofen Limestone, Solnhofen, Bavaria, West Germany. 7 cm in length.

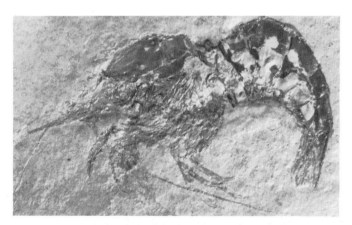

15-28. Late Jurassic. *Dusa denticulata* Balss. A complete laterally preserved shrimp in lithographic limestone matrix. Specimen has all its legs and antennae. Malm Zeta, Solnhofen Limestone, Schernfeld, Bavaria, West Germany. 6 cm.

15-29. Late Jurassic. *Drobna deformis* Münster. A complete "stretched-out" shrimp in lateral view. Malm Zeta, Solnhofen Limestone, Eichstätt, Bavaria, West Germany. 17 cm in length. Courtesy: Siber and Siber Ltd., Aathal, Switzerland.

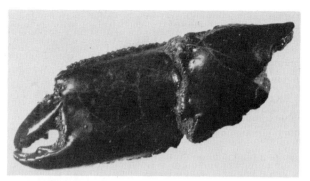

15-31. Late Cretaceous. *Protocallianassa mortoni* (Pilsbry). Complete anterior claw and pincers. Navesink Formation (Middle Maestrichtian), Holmdel, Monmouth County, New Jersey. ✕1.5.

15-30. Late Jurassic. *Palinurina longipes* Münster. A complete shrimp in dorsal (top) view with long legs and antennae in lithographic limestone matrix. Malm Zeta, Solnhofen Limestone, Eichstätt, Bavaria, West Germany. 7.5 cm. Courtesy: Siber and Siber Ltd., Aathal, Switzerland.

15-32. Late Cretaceous. *Protocallianassa faujasi*. Complete claws and pincers from the "ghost shrimp", in a chalk matrix. Maestrichtian, Maastricht, The Netherlands. 11 cm.

15-33. Late Cretaceous. *Protocallianassa mortoni* (Pilsbry). Complete set of claws (compare with Fig. 15-32), of a "ghost shrimp." The body has no chitinous material or carapace. Only the pincers and claws are found in the fossil deposits. Navesink Formation (Middle Maestrichtian), Holmdel, Monmouth County, New Jersey, ✕2.

15-34. Late Cretaceous. *Protocallianassa faujasi.* Complete claw with pincers in sandstone matrix. The claw is made up of three segments of chitinous material. The most posterior segment (right) is attached to the body. Middle Lower Campanian. Dülmen, West Germany. ×2.

15-35. Late Cretaceous. *Halymenites major* Lesquereux. The "burrow" cast of *Protocallianassa mortoni* (Pilsbry). Middle Member, Mesaverde Formation (Campanian), Washakie County, Wyoming. (Coated with magnesium oxide.) Approx. half size.

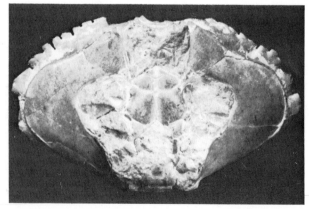

15-37. Eocene. *Lobacarcinus wurtemburgensis* Paulino. Ventral view of the carapace. Mokattum Beds, El Fayum Depression, Southwest of Cairo, Egypt. 10.5 cm.

15-36. Middle Eocene. A complete shrimp (species indeterminate) showing legs, claws and pincers, as well as antennae. Green River Formation, Kemmerer, Lincoln County, Wyoming. 16 cm. Courtesy: Siber and Siber Ltd., Aathal, Switzerland.

15-38. Late Cretaceous. Crab (species indeterminate) showing claws (legs) and carapace in chalky matrix. Cenomanian, Hadjula, Lebanon. 3.5 cm. Courtesy: Siber and Siber Ltd., Aathal, Switzerland.

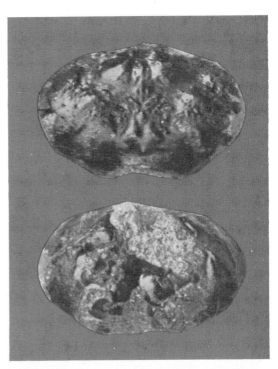

15-39. Late Cretaceous. Crab (species indeterminate). Top: dorsal view of the carapace. Bottom: ventral view. Navesink Formation (Middle Maestrichtian), Holmdel, Monmouth County, New Jersey. × 1.5.

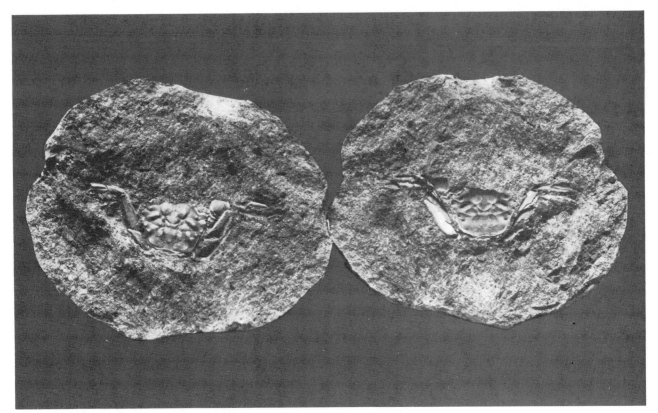

15-40. Late Cretaceous. *Longysorbis cuniculosus.* Complete crab (plate-mold and counterplate-cast) in a concretion. Spray Formation (Maestrichtian), Shelter Point near Campbell River, British Columbia, Canada. Approx. half size.

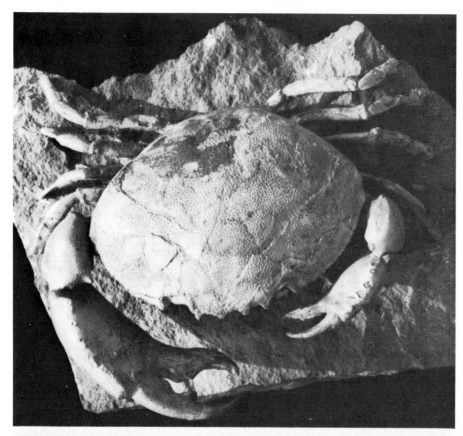

15-41. Eocene. *Harpactocarcinus punctulatus* Desmarest. Complete crab with carapace, legs, claws and pincers prepared from soft sandstone. Monte Bondone, Trento, Italy. 13 cm. Courtesy: Siber and Siber Ltd., Aathal, Switzerland.

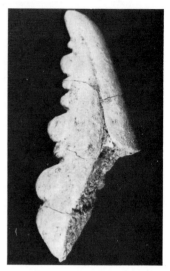

15-42. Middle Miocene. Crab pincer (species indeterminate). Helvetian. Cabrières D'Aigues, Vaucluse Department, France. ×1.5.

15-43. Miocene. *Tumidocarcinus giganteus.* Crab in the Xanthidae family in a sandstone concretion. Glen Afric Station, South Island, New Zealand. Note: Tubular structure near the "fingers" of the right cheliped (top center) is a burrow left by a worm passing through the decaying animal. 22 cm across specimen (including rock matrix).

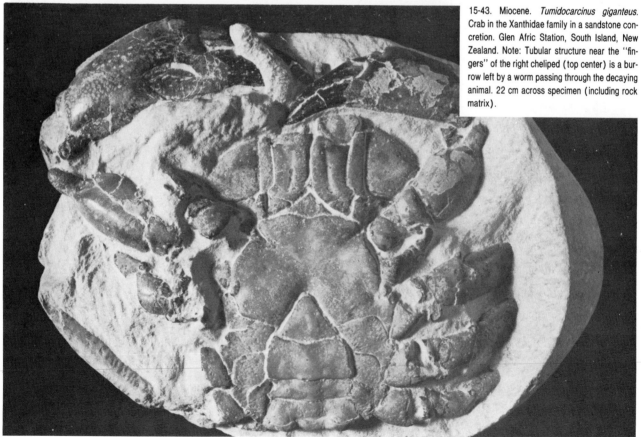

15-44. Late Eocene. *Branchioplax washingtoniana* Rathbun. Ventral (bottom) view of a crab in a limestone concretion. Twin Rivers Formation, Neah Bay, Clallam County, Washington. 7 cm.

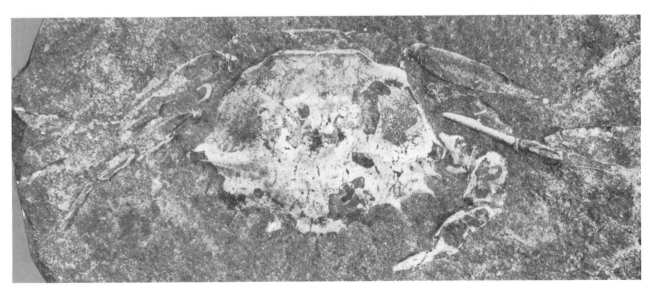

15-45. Oligocene. *Portunites alaskensis* Rathbun. A complete crab in a concretion. Washington State. 8.5 cm.

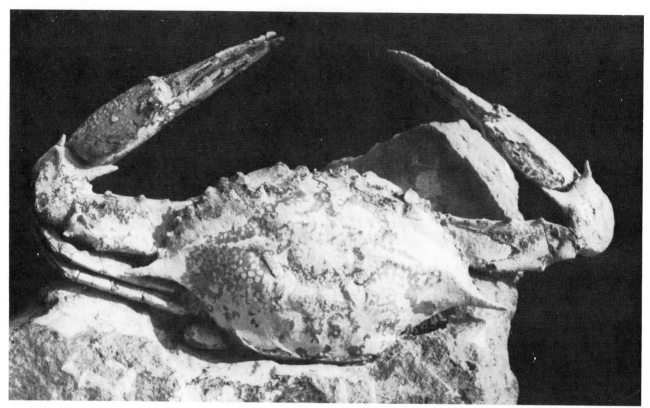

15-46. Miocene. *Portunis (Neptunis) tuberculatus*. A complete crab from diatomaceous chalk matrix. Sardinia, Italy. 16 cm. Courtesy: Siber and Siber Ltd., Aathal, Switzerland.

15-47. Late Miocene. *Coeloma* sp. Crab. Plate (cast) and counterplate (mold) of a fairly complete portunid crab. Gram, Denmark. ×1.

15-48. Late Miocene. *Coeloma* sp. A crab in a sandstone concretion. (See Fig. 15-47). Gram, Denmark. ×1.5.

15.-49. Pliocene. A shore crab (species indeterminate) in a diatomaceous matrix. Carmel Valley, California. ×2.

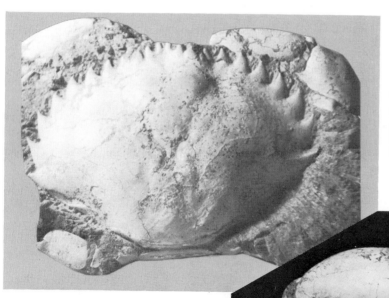

15-50. Miocene. *Necronectes proavita* (Rathbun). The dorsal shell of a swimming crab. Gatun Formation, near the Gatun Dam, Canal Zone, Panama. Approx. half size. (USNM324289). Courtesy: National Museum of Natural History (Smithsonian Institution), Washington, D.C.

15-51. Miocene. *Necronectes proavita* (Rathbun). The ventral shell of the swimming crab illustrated in Fig. 15-50. Note the strong claws and remnants of legs. Gatun Formation, Gatun Dam, Canal Zone, Panama. Approx. half size. (USNM324289). Courtesy: National Museum of Natural History, Washington, D.C.

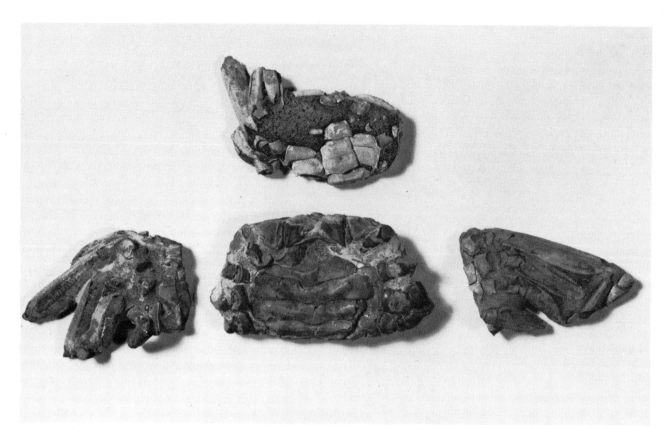

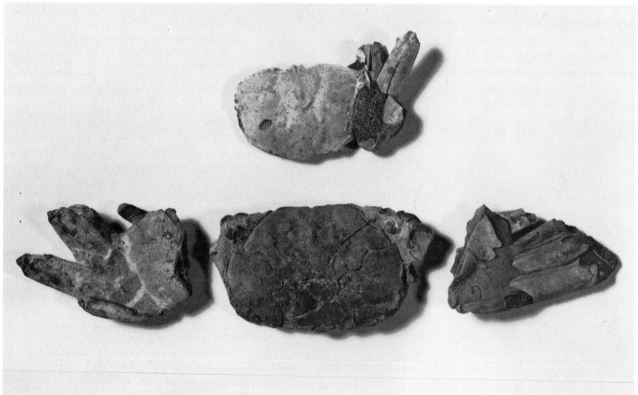

15-52. Pliocene. Crabs (species indeterminate). Top: ventral view of several individuals showing carapace (center) and legs. Bottom: dorsal view of the same specimens. Island of Guam, in the South Pacific Ocean. Approx. actual size.

15-53. Pleistocene. Portunid crabs (species indeterminate). The ventral (bottom) view of two shore crabs. Sungai Kolok, Southeast Thailand. ×2.

15-54. Pleistocene. *Macropthalmus* sp. The dorsal (top) view of a large land crab. Specimen is missing a few legs and claws. Sungai Kolok, Southeast Thailand. ×2.

Fig. 15-41), *Branchioplax* (see Fig. 15-44), *Portunites* (see Fig. 15-45), *Portunus (Neptunis)* (see Fig. 15-46), *Coeloma* (see Figs. 15-47 and 15-48), *Necronectes* (see Figs. 15-50 and 15-51), the portunid crabs (see Figs. 15-52 and 15-53), and *Macropthalmus* (see Fig. 15-54). The origins of the crabs can, at present, be traced back to the Jurassic Period. Other species of Brachyura are: *Avitelmessus* of the Cretaceous, *Callinectes* (Eocene to Recent), *Cancer* (Miocene to Recent), *Dakoticancer* (Cretaceous), *Linuparus* (Cretaceous to Recent), and *Zanthopsis* (Paleocene to Oligocene). A final note on a most common member of the infraorder Anomura: *Protocallianassa* (see Figs. 15-31 through 15-35), a primitive "soft-bodied" digging crab, left quite distinctive burrows, having dug them out of the soft bottom sands and muds of some ancient sea (see Fig. 15-35 for a cast of a typical burrow). They date from the Late Cretaceous (found in deposits of this age, throughout the world).

A note on collecting malacostracan specimens: The most interesting way to collect complete decapod crustaceans is by cracking open concretions and exposing (hopefully!) a part and counterpart (see Fig. 15-40 for example) of the actual specimen and its mold. Sandstone concretions are quite common in some localities, particularly on the beaches (at low tide) on the islands off of Washington State (Puget Sound) and British Columbia, Canada, and on the beaches of the two islands that comprise New Zealand. These sandstone concretions have eroded out of former sea-bottom sediments of Miocene through Late Cretaceous age.

16 ARTHROPODA (Insecta and Arachnida)

The insects are members of the phylum Arthropoda, and are by far the largest division of that phylum—with an extraordinary variety of species forms. The word insect comes to us from the Latin word "insectus," meaning to "cut into." This refers to the divisions of the insects' body—the head, thorax, and the abdomen (much like in the trilobites, for example, see Chapter 13). These three basic parts are themselves divided into segments, and the head in particular is tightly secured to the thorax by a hard-to-discern segment. The head is a capsule-like body region that bears the eyes, antennae, and appendages (mandibles, labia, and palps). The thoracic area is segmented in three parts (nota). The insects have three pairs of legs, one pair on each respective thoracic segment. The majority of adult forms possess wings, and these are attached to the second and third thoracic segments.

The immature, or "larval," stages of insects are quite fascinating. In fact, the life cycles of some (the dragonflies and parasitic wasps, for example) provide the substance from which horror movies are made. Some species of larvae and pupae resemble the adults of their species, while others are completely different. For example, the larval stages of the flies (Diptera), often called "maggots" (see Fig. 16-15), are wingless and do not resemble the adult fly at all.

The insects are divided into two major groups: the "winged" (Pterygota) and the "wingless" primitives (Apterygota). There are approximately thirty-six distinct orders in the former group and only about four orders in the latter.

Before discussing the insects any further, let us briefly examine the Arachnida, a subphylum of the Chelicerata (the horseshoe "crabs," see Chapter 17). The arachnids are *not* insects, as they do not possess antennae, and have only two major body segments (as opposed to the insects' three segments). The arachnids include the spiders, scorpions, ticks, and mites. The scorpions evolved from a common stock that gave rise to the extinct water dwelling eurypterids.

Early scorpions (see Fig. 16-1) emerged from the water and developed into air-breathing animals. They still have to live near water, even if only in a damp place under a rock.

Fossilized remnants of spiders can be traced back to the Carboniferous period (see Fig. 16-5) and have been found in amber, Cretaceous and Cenozoic resins, as well (see Fig. 16-48). Ticks and mites are rarely encountered as fossils, although occassionally one turns up in amber (see Fig. 16-49).

Arachnids and insects are quite rare in the fossil record. Their preservation demands unusual circumstances, primarily rapid burial without massive bacterial degradation. Occassionally, this requirement is fulfilled by some soft silt or fine-grained sediment, or even by the resinous, "gummy," secretions of plants, which eventually harden into amber.

The preservation of cockroaches (order Blattaria), "water nymphs" (see Fig. 16-6), and "dragonflies" (see Fig. 16-8) in volcanic ash shale (see Fig. 16-19 for a beetle imbedded in the shales), and in fine powdery lime sediment which eventually became lithographic limestones such as those found at Solnhofen in Bavaria (West Germany), has provided many finely preserved specimens with details of the wing veination—and even antennae and legs and certain soft parts of the animals.

In the case of amber, a resinous substance (such as the resin oozing down the tree trunk surface of a conifer or, for that matter, *Hymaenaea,* a pea tree, or the "monkey puzzle" trees of the family Araucariaceae, or even the trees of the Yew family (Taxaceae)) enveloped small insects and plant debris as they impacted on the sticky pools to entomb the animals. In some specimens the organs vanish in the course of time, but in many others they remain intact (albeit carbonized!). In several specimens, musculature, particularly of the legs, has been described, and in others (a sectioned moth from the Dominican Republic, in particular) the internal anatomy of the head capsule has

16-1. Silurian. Arachnida: Ancestral scorpion (species indeterminate). A quite rare fossil. Found in association with *Eurypterus remipes*. Fiddlers Green Dolostone, Bertie Group, Passage Gulf, Cedarville, Herkimer County, New York. ×4.5. Courtesy: S. J. Ciurca, Jr., Rochester, New York.

16-2. Middle Pennsylvanian. Palaeodictyoptera. *Eubleptus danielsi* Handlirsch. A soft-bodied, predaceous insect. Specimen is in an ironstone (siderite) concretion. Braidwood Fauna (Mazon Creek), Will County, Illinois. Approx. ×2. Courtesy: Field Museum of Natural History, Chicago.

16-3. Middle Pennsylvanian. A millipede (species indeterminate) in an ironstone (siderite) concretion. A well preserved and articulated specimen with legs showing. Braidwood Fauna (Mazon Creek), Will County, Illinois. ×1.

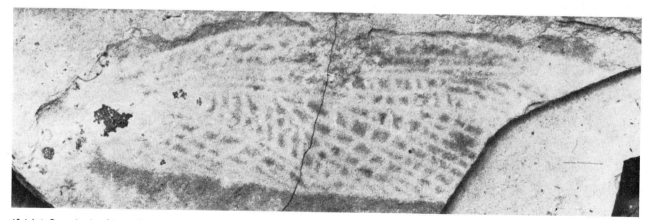

16-4. Late Pennsylvanian. Odonata: Insect wing—odonatan veination (species indeterminate) in a reddish brown shale matrix. Found in association with fishes and amphibians. Madera Shale, Manzano Mountain Range, Torrance County, New Mexico. 4 cm. Photo courtesy: Dr. David S. Berman, Carnegie Museum.

16-5. Middle Pennsylvanian. Arachnida: *Trigonomartus pustulatus* (Scudder). A ·beautifully preserved spider in a siderite concretion. Mazon Creek-Braidwood Fauna, Kankakee County, Illinois. ×3.5. Courtesy: Field Museum of Natural History, Chicago.

16-6. Pennsylvanian. Odonata: *Meganeura* sp. A swamp "water nymph." Full sculptured model of a prehistoric dragonfly. Life size dimensions with a 24″ wingspread! For the interest of the reader, the model is made of plastics: translucent polyester and fiberglass body, cellulose acetate butyrate wings, neoprene legs, polystyrene hairs and antennae. The dragonfly is found in Carboniferous coal shales in North America and elsewhere. The above model was made for the National Science Museum, Tokyo, Japan. The consultant on the project was Dr. F. M. Carpenter, Harvard University, Cambridge, Massachusetts. Courtesy: Richard Rush Studio, Inc., Chicago, Illinois.

been exposed for study. For example, Fig. 16-22 shows a water/heat swelled termite soldier. The external chiton has dissolved, leaving the entire hind gut (which *is* intact) for all to see. Figure 16-23 shows internal organs and musculature intact, as does Fig. 16-37 A and B. For references on studies of preserved musculature of amber inclusions, refer to Larson, Sven G. (1978). "Baltic amber: a paleobiological study." *Entomonograph,* Vol. 1, February 1978. Scandinavian Science Press Ltd., Klampenborg, Denmark.

In the subclass Apterygota, there are four orders: Protura (wingless and blind, delicate insects), Thysanura ("silverfish" and "bristle-tails," see Fig. 16-16), Entotrophi (wingless, active subterranean insects), and Collembola (springtails). Possibly the oldest apterygote insect group in the fossil record, Collembola-like insects were recovered from Middle Devonian rocks of Scotland.

In the subclass Pterygota, we have the following orders: Plectoptera (Ephemeroptera) (mayflies: two and four-winged delicate insects with short antennae and vestigial mouth parts); wings have a complex vein (veination) structure; *Baetis,* the small mayfly is a good example of this order, see Figs. 16-17 and 16-18; Odonata (damsel flies, water nymphs, see Fig. 16-6; and dragon flies, see Fig. 16-8); Plecoptera (salmon flies and stone flies); Orthoptera (roaches, praying mantids, sooth-sayers, and katydids, see Fig. 16-20); Dermaptera (earwigs, see Fig. 16-21), Embioptera (silk-spinning insects); Isoptera (termites, see Fig. 16-22); Psocoptera (book lice and bark lice, see Fig. 16-23); Zoraptera (zorapterans: colony insects, minute in size and terrestrial in habitat); Mallophaga (bird lice and biting lice—parasites); Thysanoptera (thrips or claw insects); Homoptera (cicadas, see Fig. 16-25, hoppers, psyllids, whiteflies, aphids, and scale insects); Hemiptera (true bugs, see Fig. 16-44, cooties and sucking lice); Neuroptera (lacewings, alderflies, hobsonflies, fishflies, snakeflies, antlions, and owlflies); Mecoptera (scorpion flies); Trichoptera (caddis flies, see Fig. 16-26); Lepidoptera (moths and butterflies, see Figs. 16-14 and 16-27); Diptera (flies, mosquitos, gnats, and midges, see Figs. 16-9 through 16-12, 16-15, 16-28 through 16-38, 16-52, and 16-54); Siphonaptera (fleas); Coleoptera (beetles and weevils—long-horned wood boring beetle, see Fig. 16-39, rove beetle, see Fig. 16-40, snout beetle or weevil, see Fig. 16-41, pin-hole borer or platypus beetle, see Fig. 16-51, and predacious water beetles, see Figs. 16-56 through 16-58); Strepsiptera (parasites of other insects: small, males with wings living "free," while larval females remain inside the bodies of their hosts); and Hymenoptera (sawflies, ichneumon flies, wasps, bees, and ants—stingless-honeybee, see Fig. 16-42, parasitic wasp see Fig. 16-43, winged ponerid ants, see Figs. 16-45 and 16-46, and wingless worker ants, see Fig. 16-47).

Most readers are familiar with the fossil insects found in Baltic Seacoast resins. This famous amber material comes from regions along the southern part of the Baltic Sea where the present day countries of East Germany, Poland, and Russia (including Lithuania, Latvia, and Estonia) have their boundaries. The amber from the Baltic territory is Late Eocene to Early Oligocene in age (approximately 40 to 45 million years old). Specimens shown here contain the following: 1. An earwig (order Dermaptera, see Fig. 16-21); 2. a cadis fly (order Trichoptera, see Fig. 16-26); 3. flies (order Diptera, see Figs. 16-29, and 16-32 through 16-36); 4. a fungus gnat (order Diptera, see Fig. 16-38); 5. a long-horned wood-boring beetle and a

snout beetle (order Coleoptera, see Figs. 16-39 and 16-41); 6. an ant (family Formicidae in the order Hymenoptera, see Fig. 16-47); 7. a spider (Arachnida, see Fig. 16-48); and finally, 8. a mite (Arachnida, see Fig. 16-49).

A brief geological history of the Oligocene Baltic amber: During the Late Eocene to Oligocene epochs, an enormous expanse of coniferous trees had grown to occupy large parts of the southernmost region of what is now the territorial land masses of the countries of Norway, Sweden and Finland, as well as west of the Ural Mountains in what is now the Soviet Union. At that time, the Baltic Sea had not yet formed its present day boundaries; the Gulf of Bothnia was a lowland, forested by large tracts of coniferous trees; and the northern boundaries of what is today Denmark were connected to Norway and Sweden, and also contained a portion of these prehistoric forests. In time, the forests vanished; but they left behind them the residues of thousands of generations of trees. Later, silts and clays carried down by rivers to the north, originating as inlets from the Barents Sea, buried these ancient woodlands in

16-7. Permian. Ichnites: Trackways of insects (species indeterminate) in a mud shale. The impressions were made in soft tidal muds and quickly covered over by drifting silt particles. Rotliegendes (Red Beds), Roter Berg at the Hornburger Sattel am Hartz, East Germany (DDR). ×1.5.

16-8. Late Jurassic. Odonata: *Cymatophlebia longialata* (Germar). A complete and beautifully preserved "dragonfly" in lithographic limestone matrix. Lithographischer Schiefer, Eichstätt, Bavaria, West Germany. 17 cm in length. Courtesy: Siber and Siber Ltd., Aathal, Switzerland.

thick sediments. While incursions of the North Sea initiated the formation of the Baltic Sea, the advancing waters covered these previously forested regions, which were now buried under hundreds of meters of clay and siltstone. The clays and silts now contained remnants of the former forest in the form of carbonized wood fragments called "lignite," and the associated "amber," which was nothing more than a plastic-like resin that oozed out from the bark of the former trees, after an injury or disease of a tree, and became hardened. The water insoluble resin became stable after several chemical alterations such as oxidation and a thorough loss of the volatile constituents of the original resin. Now, the resin (amber), having a specific gravity lighter than the lignitized wood shards, started to work its way out of the uppermost layers of its "blue-earth" environment. During storms, the pieces of amber resin would wash up along the coastal shorelines in a southeasterly direction, particularly on the shores of present day Poland (especially at Danzig), and along the western coasts of Latvia and Lithuania. There they were to be discovered by ancient man who lived along these coasts, and the amber became a "treasured" relic, which brought forth mysticisms and a form of religious fervor over articles made from this now precious fossil resin. There is no doubt that the early inhabitants of these coastal lands observed the entrapped insect remains in the amber resins, but it wasn't until the 17th and 18th centuries that the curiosity of scientifically oriented mankind saw the significance and meaning of these remarkable inclusions.

The amber resin of the Northern Highlands of the Dominican Republic: Amber ranging from Late Oligocene to Early Miocene in age has been discovered only relatively recently in the new world. A large field of amber to give challenge to the Baltic Seacoast deposits was discovered in the late part of the 19th century in the mountain range of Monte Christi in the Septrentrional Cordillera of the Northern Highlands. The Dominican Republic is on a Caribbean Sea island that it shares with its neighboring country of Haiti. The curiosity of the white settlers gave some importance and subsequent commercialization to these upland resin deposits even though, no doubt, the early paleoindians on the island must have been aware of the existence of the resin.

Many fine examples of insect inclusions have come out of these fields within the past 75 years including exceptionally well preserved representatives of the following orders: 1. A jumping bristle-tail of the order Thysanura (see Fig. 16-16); 2. a male mayfly of the genus *Baetis* (family Baetidae, order Ephemeroptera, see Figs. 16-17 and 16-18); 3. a katydid (order Opthoptera, see Fig. 16-20); 4. a termite (order Isoptera, see Fig. 16-22); 5. a bark louse (order Psocoptera, see Fig. 16-23); 6. moths (order Lepidoptera, see Fig. 16-27); 7. "flies" (order Diptera, see Figs.

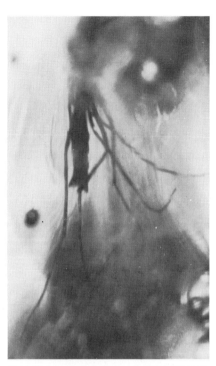

16-9. Early Cretaceous. Diptera: A "midge" in amber. This fly (see Fig. 16-10 for another view of the same specimen), as well as the other two midges on this page, was recovered in one piece of amber in the clay pits of central New Jersey. Raritan Formation, Sayreville, Middlesex County, New Jersey. ✕45. (PU88892). Courtesy: Geology Department, Princeton University, Princeton, New Jersey.

16-11. Early Cretaceous. Diptera: Another midge specimen found in the same piece of amber. Because of the angle of the insect in the amber (its refraction), it was not possible to photograph the upper portion of the insect. Raritan Formation, Sayreville, Middlesex County, New Jersey. ✕30. Courtesy: Geology Department, Princeton University, Princeton, New Jersey. (PU88892).

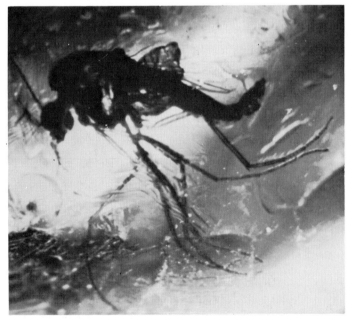

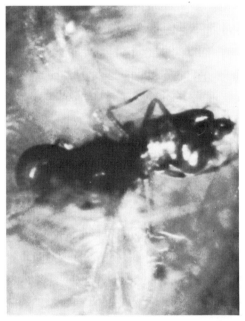

16-10. Early Cretaceous. Diptera: Another view of the midge in Fig. 16-9, showing details not as obvious as on the previous photograph. Raritan Formation, Such Clay Pits, Sayreville, Middlesex County, New Jersey. ✕30. All specimens were collected by the author in 1966. Courtesy: Geology Department, Princeton University, Princeton, New Jersey. (PU88892).

16-12. Early Cretaceous. Diptera: Another specimen of a midge in the same piece of amber collected by the author and presented to Princeton University in 1966. As of this writing, the insects remain undescribed and unstudied. Raritan Formation Sayreville, Middlesex County, New Jersey. ✕40. (PU88892). Courtesy: Geology Department, Princeton University, Princeton, New Jersey.

16-13. Paleocene. Insect wing (species indeterminate) in a shale matrix containing leaf impressions. Such finds are quite rare, and, as one can see, the details of the veination of the insect wing are quite clear. Fort Union Shale Beds, Wyoming. 13 mm in length. Courtesy: Geology Department, Princeton University, Princeton, New Jersey.

16-14. Middle Eocene. Lepidoptera: *Prodryas* sp. A beautifully preserved butterfly in a sandstone shale. The complete body and the delicate preservation of detail in the wings make this a truly remarkable specimen. Green River Formation (equivalent), Rio Blanco County, Colorado. 6.5 cm. Courtesy: Siber and Siber Ltd., Aathal, Switzerland.

16-28, 16-37, and others); 8. a rove-beetle (order Coleoptera, see Fig. 16-40); 9. a stingless honeybee (order Hymenoptera, see Fig. 16-42); 10. a parasitic wasp (order Hymenoptera, see Fig. 16-43); 11. a winged ant (family Formicidae in the order Hymenoptera, see Figs. 16-45 and 16-46); 12. a true bug that preyed upon the springtails (Collembola) (order Hemiptera, see Fig. 16-44).

Amber deposits in general are quite rare, and, within a majority of these deposits, resins with inclusions are even rarer! Typical of such deposits is the large Early Cretaceous amber beds in Middlesex County, New Jersey, on both sides of the Raritan River at such communities as Woodbridge, Parlin, and Sayreville. One small fingering of these same deposits in the lower part of the bluffs overlooking the Raritan Bay at Cliffwood Beach gave science a piece of amber with a partial colony of ants in it. This occurrence made the cover of the prestigious "Science" magazine. This author was fortunate enough to discover a single piece of amber (the size of a quarter) which contained three insect inclusions: one species of the Ceratopogonidae family and two specimens of the family Chironomidae, all in the order Diptera. These specimens are housed in the invertebrate collections of the Department of Geology, Princeton University. Hopefully, someday,

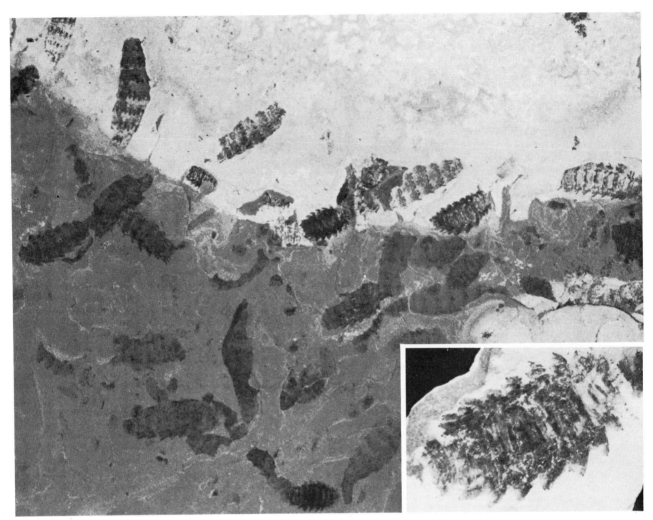

16-15. Eocene. Diptera: Family: Destridae. *Lithohypoderma ascarides*. A number of larval insects preserved as compressed chitinous films on a slab of oil shale. These large botfly larvae, or "maggots," probably infested the living and/or decaying bodies of large land mammals during the Tertiary. Inset: Isolated specimen of a dipterous larva showing "holding spines" which anchored the maggot to its host. Parachute Creek Member, Green River Formation, Uinta Basin, Uinta County, Utah. Overall shale dimensions: 21 × 30 cm. Inset: 17 mm.

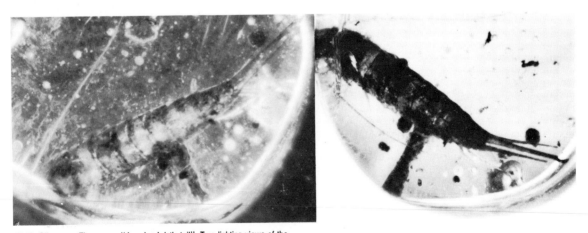

16-16. Oligocene. Thysanura: "Jumping bristle-tail". Two lighting views of the same specimen to show as much detail of the insect as possible. Specimen is in amber resin from the Northern Highlands of the Dominican Republic.

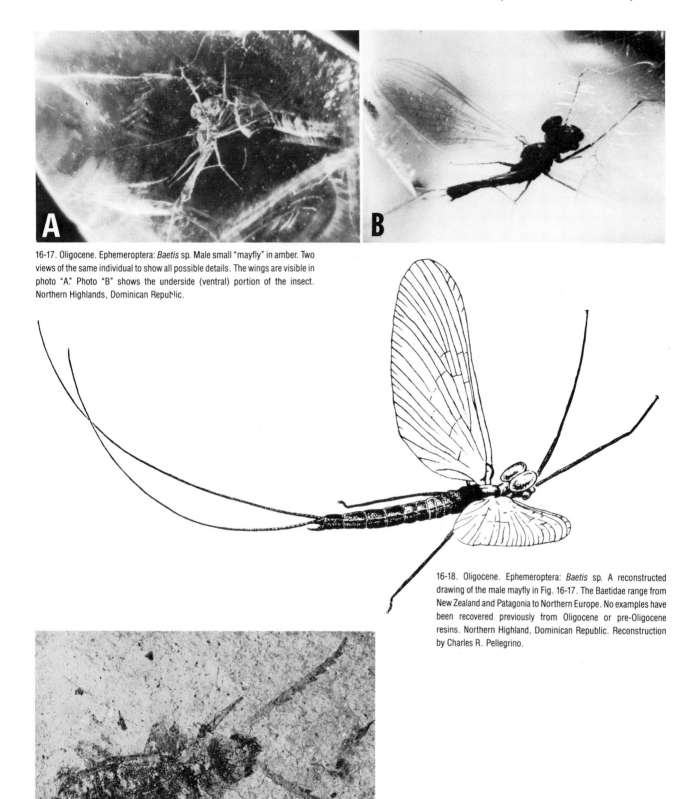

16-17. Oligocene. Ephemeroptera: *Baetis* sp. Male small "mayfly" in amber. Two views of the same individual to show all possible details. The wings are visible in photo "A." Photo "B" shows the underside (ventral) portion of the insect. Northern Highlands, Dominican Republic.

16-18. Oligocene. Ephemeroptera: *Baetis* sp. A reconstructed drawing of the male mayfly in Fig. 16-17. The Baetidae range from New Zealand and Patagonia to Northern Europe. No examples have been recovered previously from Oligocene or pre-Oligocene resins. Northern Highland, Dominican Republic. Reconstruction by Charles R. Pellegrino.

16-19. Oligocene. Coleoptera: A primitive elytra beetle with a leathery forewing and distinctive antennae in volcanic ash shale (remnants of a prehistoric lake bed) matrix. ×4. Florissant Formation, Florissant, Teller County, Colorado.

16-20. Oligocene. Orthoptera: Female "katydid" in amber. A very exceptional specimen preserved in resin (amber). Note the "air" bubbles (exhalations and regurgitations) near the legs and body cavity. The antennae are well preserved on this specimen. Northern Highlands, Domincan Republic.

16-21. Oligocene. Dermaptera: A beautifully preserved "earwig" in amber resin. A complete specimen with antennae, legs, etc. Baltic Seacoast, Poland.

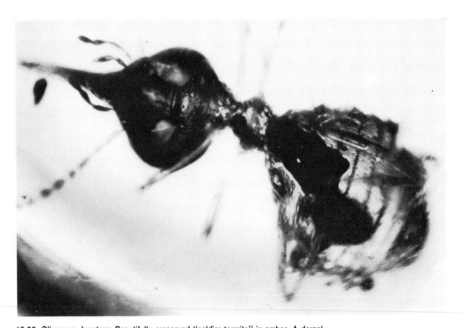

16-22. Oligocene. Isoptera: Beautifully preserved "soldier termite" in amber. A dorsal (top) view looking down on the head (left) and the body (right). The legs and antennae are deeper into the resin, and, due to the refraction of the material, these appendages appear to be out of focus. Northern Highlands, Dominican Republic.

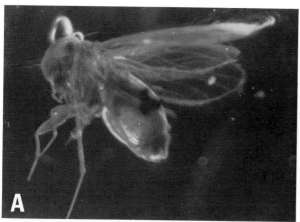

16-23. Oligocene. Psocoptera: Two lighting views of a "bark louse" in amber. A. The wings are more discernible in this illustration, as well as the eye and mouth parts. B. Body structure and appendages show up better in this photograph. Northern Highlands, Dominican Republic.

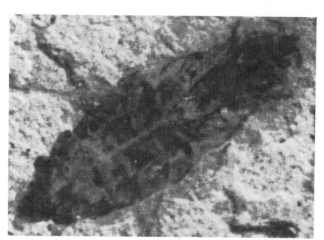

16-24. Oligocene. Hemiptera: *Naucoris* sp. "water-creeper" in volcanic ash shale. Ventral view of the body. Papier Schiefer (paper shale), Rott/Laebensbirge, Austria. 1 cm.

16-25. Oligocene. Hemiptera: A beautifully preserved cicada (locust or harvest fly) in volcanic ash shale. Preserved in lateral view with full wings. Florissant Formation, Florissant, Teller County, Colorado. ×4. Courtesy: Geological Enterprises, Ardmore, Oklahoma.

16-26. Oligocene. Trichoptera: Two lighting views of a beautifully preserved "caddisfly" in amber. Note the detail of the wings, legs and antennae in photo "B." Baltic Seacoast, Poland.

16-27. Oligocene. Lepidoptera: Lighting and positional views of two separate "moth" specimens in amber. A and B show the first moth in lateral (side) view, where one can observe the legs and antennae to their best advantage. C and D show the second moth specimen in ventral (bottom) view where the head and wings can be best observed. Both specimens are from the Northern Highlands, Dominican Republic.

someone will do a description of these unique and quite rare insects. These ants and flies are the only specimens so far collected (and reported) in the state of New Jersey. Ambers from this region are among the oldest in the world, dating back to approximately 100 million years ago, any insects contained therein perhaps once flew alongside of the dinosaurs.

More diligent collecting on the part of amateurs in these amber deposits, might eventually bring forth additional species, and a fauna might be assembled from this region.

There are many smaller deposits of amber throughout the world. In Canada there are the Late Cretaceous Cedar Lake (Manitoba) deposits of fossil resins from the monkey puzzle trees (family Araucaraceae). There are small deposits throughout North Africa, Morocco especially, and in Burma. Italy, Rumania, and many other European countries are dotted with small outcroppings of amber.

In the Pliocene amber deposits of Moa, Tanzania, in East Africa, some fine inclusions have been discovered. Among them are such species as: 1. A pin-hole borer or platypus beetle (order Coleoptera, see Fig. 16-51); and 2. a midge (order Diptera, see Fig. 16-52).

16-28. Oligocene. Diptera: A fly (species indeterminate) in amber. The insect is in a ventral (bottom) view with some of its legs folded into its body cavity. Note the "beaded" antennae. Northern Highlands, Dominican Republic.

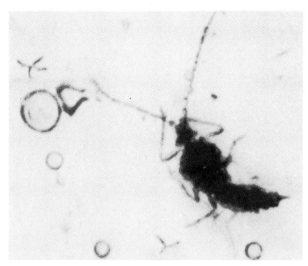

16-29. Oligocene. Diptera: A fly (species indeterminate) in amber. Note the water bubbles to the left of the specimen. The specimen is embedded deeply into the resin, and due to the refraction of the material, the specimen is in soft focus. Baltic Seacoast, East Germany (DDR).

One final note concerning amber and related material: Amber deposits were discovered in the early 1960's in the country of Lebanon in the Middle East. Unfortunately, due to the constant unrest in that region, these deposits as well as those containing Late Cretaceous fishes (see Chapters 24 and 25) have been sadly neglected. Kauri resin from North Island, New Zealand is not considered as an amber resin, and even though it does upon occassion contain insect inclusions, they are not considered as fossil (at least by the local jewelers). The Kauri resin (called "gum" locally) was at one time used in the manufacture of varnish. The resin is exuded from the Kauri pine *Agathis australis*. Kauri "gum" is presently used only for jewelry and carvings. While most of the Kauri resin is of Pliocene or Pleistocene age—even of historical times—samples have been recovered under 500 meters of sediment, and recently, New Zealand coal beds have yielded amber of Paleocene age! No insect inclusions have been reported as yet from these New Zealand Paleocene deposits.

Finally, in this chapter dealing with the varied insects and the arachnids, we cannot overlook an odd group of animals belonging to the group Myriapoda (myriapods).

16-30. Oligocene, Diptera: *Plecia pealei*, Family: Bibionidae. A march fly in volcanic ash shale matrix. As is the case with Fig. 16-31., this specimen in ventral (bottom) view is very well preserved. Note the body segments and the legs crossed over the body cavity. Florissant Formation, Florissant, Teller County, Colorado. 18 mm.

16-31. Oligocene. Diptera: *Plecia pealei*. Family: Bibionidae. Another example of a march fly in volcanic ash shale (see Fig. 16-30). The difference between this specimen and the previous one, is that the legs in this photograph are extended. Otherwise the specimens are identical. Florissant Formation, Florissant, Teller County, Colorado. 2 cm.

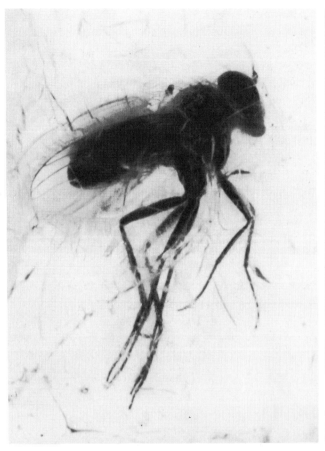

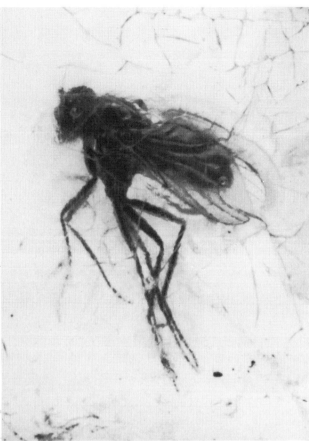

16-32. Oligocene. Diptera: Two views (front and back) of the same fly in a piece of amber. The head area is much clearer on the photograph to the right, as this view is closer to the edge of the resin. Baltic Seacoast, East Germany (DDR).

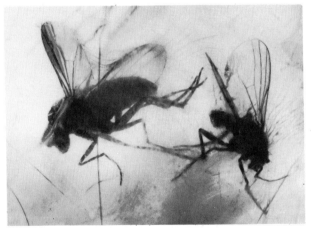

16-33. Oligocene. Diptera: Two flies in one piece of amber. Note the detail of the veination of the wings. A faceted or polished surface has to be cut near the specimens to insure a sharper image of the specimen. Baltic Seacoast, East Germany (DDR).

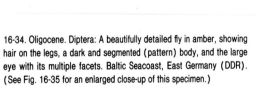

16-34. Oligocene. Diptera: A beautifully detailed fly in amber, showing hair on the legs, a dark and segmented (pattern) body, and the large eye with its multiple facets. Baltic Seacoast, East Germany (DDR). (See Fig. 16-35 for an enlarged close-up of this specimen.)

16-35. Oligocene. Diptera: A greatly enlarged photograph of the fly in Fig. 16-34, to show more detail. Note the hairs on the back and head, as well as along the body on both sides. The bristle hairs or cilia extend along the legs also. A beautifully preserved specimen in orange-colored amber resin. Baltic Seacoast, East Germany (DDR).

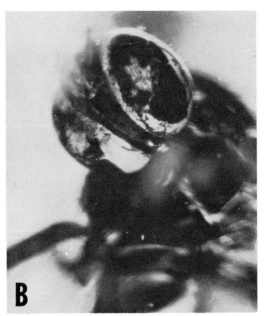

16-36. Oligocene. Diptera: Two illustrations of the same fly. A. An overall view of the insect. B. A closeup of the head area showing the lenses with their many facets. Baltic Seacoast, East Germany (DDR). A. ✕20, B. ✕30.

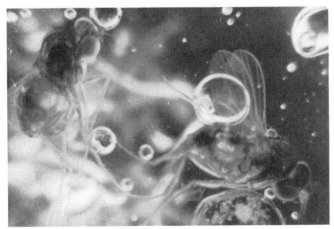

16-37. Oligocene. Diptera: Two lighting views of a male (right) and a female (left) "long-legged fly" in amber. Note the water bubbles surrounding the flies. Unfortunately these bubbles obliterate some details of the insects. These flies were apparently on their mating flight when they alighted on and became entrapped in the resin. Northern Highlands, Dominican Republic.

These are many legged invertebrates, whose modern species are all air breathers, and are among the most ancient terrestrial arthropods. They were possibly the first creatures to venture onto the land, and may well share a common ancestry with the insects.

This group includes the millipedes (also known as "thousand-leggers") in the class Diplopoda, their geological range being Silurian to Recent; and the centipedes ("the hundred-leggers") in the class Chilopoda—Carboniferous to Recent.

These "worm-like" creatures are quite rare in the fossil record. Some species of millipedes have been recovered in the Middle Pennsylvanian Period in ironstone (siderite) concretions from the Mazon Creek area of Illinois (see Fig. 16-3).

The diplopods (millipedes) are well known by their multitudinous double pairs of legs on each body segment along the lower edge of the body. The first three segments behind the head area have only single pairs of legs, while the remaining segments have legs ranging from 10 to 200 in number. The head of the millipede has antennae and a cluster of basic eyes similar to the compound type. The millipede has a chitinous exoskeleton and molts this outer shell on occassion, just like any other arthropod. The average length of millipedes is three inches (75 millemeters), but a fossil species has been found that was near eight inches (20 centimeters).

The chilopods (centipedes) are quite similar to the millipedes, but differ in that the body segments of the centipedes all have only single pairs of legs. The legs are much sturdier and are shorter than those of the diplopods. Oth-

16-38. Oligocene. Diptera: Family Fungivorinae. A "fungus gnat" in amber. Two views to show details of this specimen. Note the air bubbles at the top of photograph A. Baltic Seacoast, East Germany (DDR).

16-39. Oligocene. Coleoptera: A "long-horned wood boring beetle" in amber. A beautifully preserved specimen. Note the spurs on the legs, and the folded wings over the body. Baltic Seacoast, Poland.

16-40. Oligocene. Coleoptera: A splendid specimen of a "rove beetle" in amber. Note the hooded collar back of the head and the short antennae. Northern Highlands, Dominican Republic.

16-41. Oligocene. Coleoptera: A "snout beetle" in amber. Note the long proboscus and antennae, and the hard chitinous shell. (Photo is slightly retouched.) Baltic Seacoast, East Germany GDR. Courtesy: Academy of Natural Sciences, Philadelphia, Pennsylvania.

16-42. Oligocene. Hymenoptera: A "stingless honey bee" in amber. A beautifully preserved specimen showing stout antennae, large head (with a big eye), and its legs folded underneath its body. Northern Highlands, Dominican Republic.

16-43. Oligocene. Hymenoptera: A "parasitic wasp" in amber. The wasp rowed on its back on the slightly hardened "skin" of a resin flow until a fresh flow enveloped its body (as indicated by the flurry of wing movement which has displaced some resin around the body). Northern Highlands, Dominican Republic.

16-44. Oligocene. Hemiptera: An extremely rare "true bug" belonging to the family Enicocephalidae, which preyed upon *Collembola* (spring tails), trapped in amber. Note the large antennae and strong legs. Parts of the body and wings obliterated by air or water bubbles. Northern Highlands, Dominican Republic.

16-45. Oligocene. Hymenoptera: Family Formicidae. A "winged ponerid ant" in amber. A fine specimen with long antennae and delicate wings. The ant is slightly doubled over. Northern Highlands, Dominican Republic.

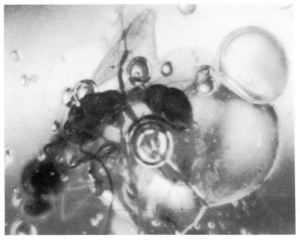

16-46. Oligocene. Hymenoptera: Family Formicidae. A "flying ponerid ant" in amber (see Fig. 16-45). The specimen is partially covered with small and large bubbles. Note the wings and the long antennae. Northern Highlands, Dominican Republic.

16-47. Oligocene. Hymenoptera: Family Formicidae. An ant in amber. In the lower left foreground is the head area with its antennae sticking out in each direction. Two legs can be seen slightly above center in the photograph. Baltic Seacoast, East Germany (DDR).

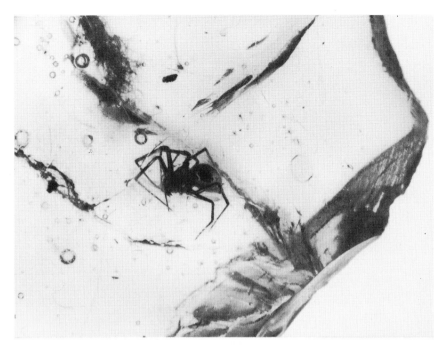

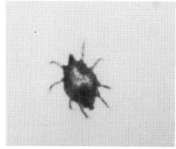

16-48. Oligocene. Arachnida: A beautifully preserved spider in amber. Note the long spindly legs and large body. The amber contains, along with the spider, pieces of pollen and other species of insects (see Figs. 16-29 and 16-49). Baltic Seacoast, East Germany (DDR).

16-49. Oligocene. Arachnida: A nicely preserved "mite" in the same piece of amber as Fig. 16-48. Its legs are splayed out and we are looking at the dorsal (top) surface. Baltic Seacoast, East Germany (DDR).

16-51. Pliocene. Coleoptera: A "pin-hole borer" or "platypus beetle" preserved in amber. Note the strange proboscus and big eye, along with the segmented (appearing to be divided in two parts) body. Moa, Tanzania (East Africa).

16-50. Late Miocene. *Libellula doris.* A group of insect larvae in volcanic ash shale. Note the splendid preservation and detail of the specimens. San Vittorio d'Alba, Italy. Overall plate dimension: 10 X 6 cm. Courtesy: Siber and Siber Ltd., Aathal, Switzerland.

16-52. Pliocene. Diptera: A very finely preserved "midge" in amber. Note the long antennae, equally long body, and beautiful wings. We are observing the specimen from the side. Moa, Tanzania.

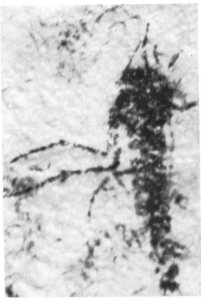

16-53. Pleistocene. Hemiptera: An insect in glacial lake bottom clay shale. Blanco Formation, Amarillo, Potter County, Texas.

16-54. Pleistocene. Diptera: A mosquito in glacial lake bottom clay shale. Blanco Formation, Amarillo, Potter County, Texas.

16-55. Pleistocene. Odonata: A dragonfly larva in glacial lake bottom clay shale. Blanco Formation, Amarillo, Potter County, Texas. 3 cm.

16-56. Pleistocene. Coleoptera: Family Dytiscidae. *Cybister explavatus.* Dorsal view of a predacious water beetle. La Brea Tar Pits, Wilshire Boulevard (Hancock Park), Los Angeles, Los Angeles County, California. ×2.

16-57. Pleistocene. Coleoptera: Family Dytiscidae. Ventral view of a predacious water beetle in peat. Note the long proboscus. It preyed upon other insects. Peat Bogs, Big Bone Lick, Northwestern Kentucky. ×2.

16-58. Pleistocene. Coleoptera: Family Dytiscidae. Dorsal view of a predacious water beetle in peat. Compare with Fig. 16-56. Peat Bogs, Big Bone Lick, Northwestern Kentucky. ×1.5.

erwise, these two animals are basically similar. The centipede also has two eye clusters (compound-like eyes) and a pair of antennae on the head. The major feature of the centipede is the modification of the leg appendages (limbs) on the first segment behind the head into fangs or poison claws!

Fossils of centipedes have been found in amber. The average length of the animal is around eight inches (20 centimeters).

Some recent species of centipedes, especially those found in and around the Sahara desert of North Africa are quite venomous.

17 ARTHROPODA (Xiphosura)

Xiphosura ("horseshoe crabs") is a group of chelicerate arthropods distinguished by a subrounded and segmented body shield design, a lengthy tail section (telson), and six pairs of appendages on the ventral side of the shield.

The anterior portion of the body shield of a xiphosuran is called the prosoma (cephalothorax), the midsection or abdomen is called the opisthosoma, and the posterior (tail region) is called the telson. The reader will note that the nomenclature for the chelicerate arthropods differs from that of the trilobites (see Chapter 13).

The shield or chitinous "shell" of the horseshoe crab is segmented transversely. The prosomal part of the shield articulates with the opisthosoma, allowing the animal to sort of "double-up," but in general it just assists the horseshoe crab in "burrowing" itself into loose sediment with a bit of flexibility.

Five pairs of ventral appendages are modified as gills and aid in the respiration of the animal. The telson may be used as a defensive weapon, but generally it is used by the animal to "right" itself when it becomes overturned on mud flats or beaches, or in very shallow water. It also serves as a rudder during swimming.

Generally, xiphosurans have compound eyes, and some species have extra vision apparatus (ocelli). The range of this group is from the Cambrian up to Recent.

The xiphosurans have three subdivisions: the Aglaspida, the Synxiphosura, and the Limulina.

The first, the Aglaspida (aglaspids) are quite primitive and not too well known from the fossil record. Only a scant half dozen or so species are known. In appearance, the aglaspids resemble the trilobites, but there is no relationship. The aglaspids have many legs (walking and/or swimming limbs), and the abdomen is separated by at least 12 segments, including the telson. The small, closely spaced compound eyes are situated high up on the dorsal portion of the prosoma (head shield). The aglaspids never attained any great size. The smallest specimens found are approximately 2.5 to 3.8 centimeters (1 to 1.5 inches), while the largest is about 23 centimeters (approximately 9 inches). The aglaspids were a short-lived line leaving behind a fossil record from the Middle to Late Cambrian. There are no pictorial representatives in this book.

The Synxiphosura (synxiphosurans) have less segments in the opisthosoma. Some species exhibit highly ornamented "fringes" at the periphery of the body segments. The synxiphosuran *Weinbergina* has been found in the "blackboard" shales ("slates") of the Rhineland in Germany. Many specimens show legs and other appendages with surprising detail (see Fig. 17-1). The synxiphosurans were also short-lived, having appeared during the Silurian and disappeared during the Devonian period.

17-1. Early Devonian. *Weinbergina opitzi* Richter. A radiograph of a complete specimen of an early xiphosuran. Note the leg appendages extending from under the body armor. Hunsrück Slate, Bundenbacher Schiefer, Bundenbach, Rheinland, West Germany ×1. (WS10433).

169

17-2. Pennsylvanian. *Euproöps danae* (Meek). A complete small xiphosuran in a clay shale. *Neuropteris* leaf impressions were in association. Clay pits, Redfield, Dallas County, Iowa. ×2.

17-3. Middle Pennsylvanian. *Palaeolimulus* sp. A complete xiphosuran in a siderite concretion. Essex Marine Fauna Mazon Creek), Braidwood, Will County, Illinois. ×2. Courtesy: Field Museum of Natural History, Chicago, Illinois.

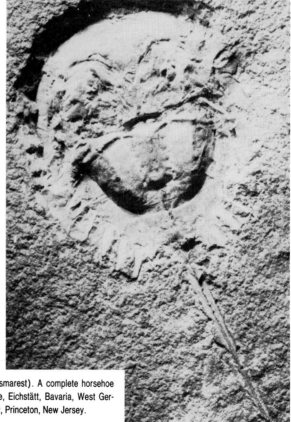

17-4. Middle Pennsylvanian. *Euproöps danae* (Meek). An essentially complete (missing tail section or telson) xiphosuran in a siderite concretion. Essex Marine Fauna (Mazon Creek), Braidwood, Will County, Illinois. ×2.

17-5. Late Jurassic. *Kouphichnium (Mesolimulus) walchi* (Desmarest). A complete horsehoe crab exoskeleton in lithographic limerock. Solnhofen Limestone, Eichstätt, Bavaria, West Germany. ×1. Courtesy: Geology Department, Princeton University, Princeton, New Jersey.

17-6. Late Jurassic. *Kouphichnium (Mesolimulus) walchi* (Desmarest). Another example of a complete horseshoe crab shield (probably a moulting) in lithographic limestone. Solenhofen Limestone, Eichstätt, Bavaria, West Germany. ×1. Courtesy: Geology Department, Princeton University, Princeton, New Jersey.

17-7. Late Jurassic. *Kouphichnium (Mesolimulus) walchi* (Desmarest). The dorsal view of an armor shield of a horseshoe crab. Specimen is missing part of its tail secton (telson). Solnhofen Limestone, Eichstätt, Bavaria, West Germany. ×1. Courtesy: Geology Department, Princeton University, Princeton, New Jersey.

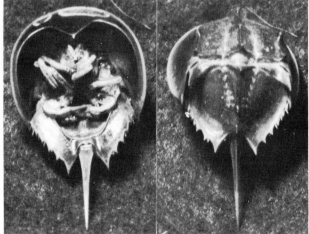

17-8. Recent. *Limulus polyphemus* (Linné). Two juvenile specimens (ventral and dorsal views) of modern horseshoe crabs for comparison. Raritan Bay, Cliffwood Beach, Middlesex County, New Jersey. ×1.

The third and final subdivision, the Limulina (limulids) constitute the majority of the xiphosurans in the fossil record. Most of the species illustrated in this chapter are of this type. The oldest limulid belongs to the Devonian Period, and it seems amazing that the present day species seems to differ little from the oldest forms.

17-9. Recent. *Limulus polyphemus* (Linné). The shed (molted) exoskeleton (or shield) of a modern adult horseshoe crab. Raritan Bay, Cliffwood Beach, Middlesex County, New Jersey. Approximately half size.

The limulids are the basic xiphosurans discussed earlier, with a three-part chitinous exoskeleton or shield made up of a prosoma, an opisthosoma, and a telson.

The fact that we find so many molted or shed exoskeletons on our present day beaches at low tide (see Figs. 17-8 and 17-9), makes us wonder if most of the fossil remains that we find in the older rock formations are molts as well. This seems to be the case with most of the trilobite specimens found.

A true mortality of one of these king or hoseshoe crabs would, if fossilized completely, show us much detail of the older species' limbs, and, in some rare cases depending upon the sort of preservation (mineralization or pyritization, etc.), even show soft tissue and other rarely preserved parts of the animal. Interestingly, a Jurassic horseshoe crab, the same species type and from the same locality as the specimens pictured in this chapter (see Figs. 17-5 through 17-7), shows the tracks made by the animal in the soft mud. The tracks terminating at the actual horseshoe crab that made the tracks! Is this an example of an animal walking along and suddenly, almost instantly, being entombed in sediment? Such graphic fossil occurrences tell us much about the animal, and also provide us with cause to reflect upon the mechanisms of fossilization itself.

Euproöps (see Figs. 17-2 and 17-4) and *Paleolimulus* (see Fig. 17-3) are good examples of limuloid xiphosurans

from the Carboniferous Period. *Kouphichnium (Mesolimulus)* (see Figs. 17-5 through 17-7) is quite common in the Jurassic of central Europe.

The xiphosurans are primarily marine to brackish in habitat, with many species favoring a littoral to shallow water environment.

Many of the older forms (primarily the synxiphosurans) have been found in "waterlime" rocks, especially during the Silurian Period, where a few species are found in association with eurypterids (see Chapter 18). The range for the limulids is from the Devonian to Recent.

NOTE: The present day *Limulus* "horseshoe crab" (see Figs. 17-8 and 17-9) is in great danger on our beaches, particularly along the New Jersey coastline. These creatures are harmless, contrary to the "old wives' tales" that they will stab you with their spikes (telsons), that they are poisonous, or that they will bite you (none of these things are true). Unfortunately, when these poor creatures venture up onto the beaches to lay their eggs, some ignorant people kill them with sticks and by tromping on their soft shells. This unpleasant news has been brought to the attention of the public by newspaper articles and the campaign started recently by the American Littoral Society, Highlands, New Jersey, to educate the public (especially the beach crowds) about the horseshoe crabs. Hopefully, these efforts will save some of these harmless and curious creatures!

18 ARTHROPODA (Eurypterida)

The order Eurypterida in the phylum Arthropoda represents the largest forms of chelicerates. All eurypterids are extinct and their fossilized remains are found in rocks of Ordovician to Permian age, with the largest variety of species found in Silurian rocks. There were both marine and freshwater (brackish) species, and most, if not all of the known species found in Silurian rocks may have been freshwater inhabitants. The dolostones of southern Ontario in Canada and New York State contain some of the most abundant eurypterid finds in the world, particularly at Passage Gulf near Herkimer, New York.

The eurypterids are crustacean-like in appearance, and are more closely related to the xiphosurans than to the malacostracans. Their chitinous exoskeletons molt as in the Xiphosura. Their overall body style is rather simplistic, and, as with the trilobites, they are constructed in three major parts: the cephalon or prosoma (head area), the abdomen or opisthosoma (body area), and the telson or terminal metasoma (tail spike). Technically speaking, the head area is actually a part (segment) of the upper body; thus these two parts are generally combined and called the cephalothorax, while the main body is called the abdomen.

Eurypterids have compound eyes (as do the trilobites) and six pairs of appendages, which are found in the head region on the ventral shield (see Fig. 18-1). The posterior pair are generally modified into (swimming) legs or paddles, e.g., *Eurypterus* (see Fig. 18-1).

In *Pterygotus* (see Fig. 18-6), the anterior pair of walking legs (chelicerae) are modified into large pincers (claw apparatus). The sexual apparatus of the eurypterid is located on the ventral shield, centered on the upper abdomen, just below the paired, paddle-bearing swimming legs (see Figs. 18-1, 18-3, and 18-4B). *Pterygotus* also has a most distinctive and large "flattened-out" telson (see Figs. 18-5 and 18-6), which may have been an aid in the animal's locomotion through the waters of its habitat.

The overall length of most eurypterids generally averaged between 10 and 25 centimeters, although *Pterygotus* attained greater lengths (see Fig. 18-5). One huge species of *Pterygotus* found in Silurian rocks near Buffalo, New York had attained a length of over 7 feet.

The eurypterids are commonly called "sea-scorpions." This is interesting as they are indeed related to the scorpions. The scorpions (see Chapter 16) evolved along with the eurypterids and there is a distinct possibility that they may have shared a common ancestry. The scorpion, which eventually evolved into an air-breathing animal (arach-

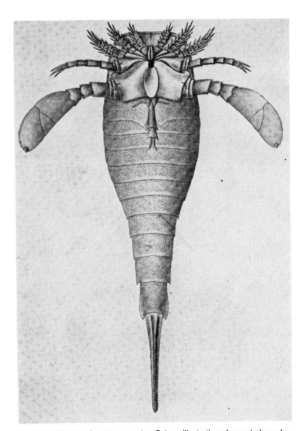

18-1. Late Silurian. *Eurypterus remipes* Dekay. Illustration of a ventral exoskeleton of a well documented species. The genital appendage and the walking legs are quite obvious on this male eurypterid. After Clarke and Ruedemann (New York State Museum 14, 1912), and from Raymond, P. E., 1939 (Fig. 28). Compare with Figs. 18-3 and 18-4B. Courtesy: New York State Museum, Albany.

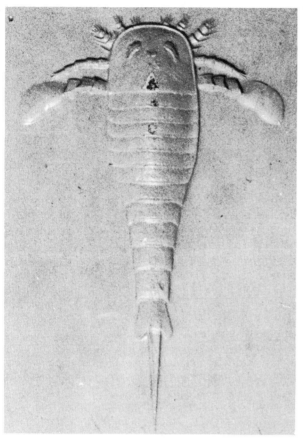

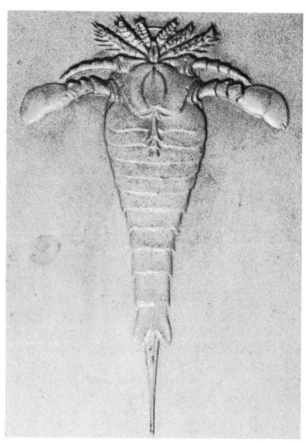

18-2. Middle Silurian. *Eurypterus fischeri* Eichwald. Plaster cast replica of the dorsal shield (exoskeleton) of an excellently preserved eurypterid. See Fig. 18-4 for comparisons. Island of Oesel, Baltic Seacoast, Northern Europe. Approx. half size. Courtesy: Geology Department, Princeton University, Princeton, New Jersey.

18-3. Middle Silurian. *Eurypterus fischeri* Eichwald. Plaster cast replica of the ventral shield (exoskeleton) of the same species illustrated in Fig. 18-2. Note the genital appendage and the walking legs at the head area. Compare with Fig. 18-4B. Island of Oesel, Baltic Seacoast, Northern Europe. Approx. half size. Courtesy: Geology Department, Princeton University, Princeton, New Jersey.

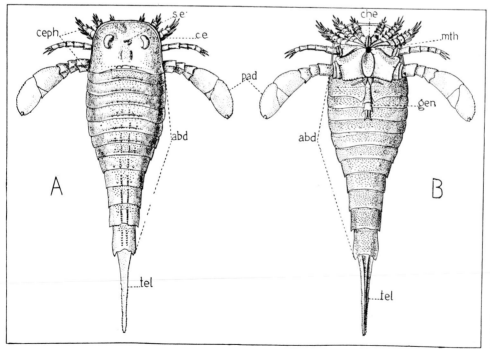

18-4. Late Silurian. *Eurypterus remipes* Dekay. A typical eurypterid. A. Dorsal view of shield (exoskeleton). B. ventral view. Abbreviations: abd = abdomen, ceph = cephalothorax, c.e. = compound eye, che = chelicerae, gen = genital appendage, mth = mouth, pad = paddle, s.e. = simple eye, and tel = telson. (After Clarke and Ruedemann, 1912, from Goldring, 1929). Courtesy: New York State Museum, Albany, New York.

18-5. Late Silurian. *Pterygotus (Acutiramus) macrophthalmus macrophthalmus.* Giant eurypterid in a fine grained dolostone block. Phelps Waterlime Member, Fiddlers Green Formation, Passage Gulf, Cedarville, Herkimer County, New York. Specimen is approximately 1.2 meters from top of carapace to the telson. Courtesy: S. J. Ciurca, Jr., Rochester, N.Y.

18-6. Late Silurian. *Pterygotus macrophthalmus cummingsi.* A reconstruction of a pterygotid (see Fig. 18-5). Note the "arachnidlike" walking legs and the lengthy claw extensions from the cephalon. Compare these extended claws with the "proboscis" of the "Tully monster" in the following chapter. After Raymond, 1939 (Fig. 29). Courtesy: New York State Museum, Albany, New York.

nida), occurs in the same rocks with eurypterids. Rare as they are in the fossil record (see Fig. 16-1), scorpions are found in the very same Silurian rocks with such forms as *Eurypterus* and *Pterygotus.*

18-7. Late Silurian. *Eurypterus remipes* Dekay. A very fine isolated cephalon carapace (head area). Note the wrinkled effect near the compound eyes, possibly indicating that this was a molting. Fiddlers Green Formation, Bertie Group, Passage Gulf, Cedarville, Herkimer County, New York. ×2.

18-8. Late Silurian. *Eurypterus lacustris* Harlan. A very fine complete eurypterid in dolostone. Bertie Group, Ft. Erie, Ontario, Canada. 6 cm.

18-9. Late Silurian. *Eurypterus remipes* Dekay. Plate and counterplate of the head and paddles (swimming legs) of a juvenile individual. Fiddlers Green Formation, Bertie Group, Passage Gulf, Cedarville, Herkimer County, New York. ×1.

18-10. Late Silurian. *Eurypterus remipes* Dekay. Another complete individual. Note the graceful curved deposition of the exoskeleton. Fiddlers Green Formation, Bertie Group, Passage Gulf, Cedarville, Herkimer County, New York. ×1. Courtesy: Geology Department, Princeton University, Princeton, New Jersey.

18-11. Late Silurian. *Eurypterus remipes* Dekay. An incomplete "molting" of a small adult eurypterid. Fiddlers Green Formation, Bertie Group, Passage Gulf, Cedarville, Herkimer County, New York. ×1. Courtesy: Geology department, Princeton University, Princeton, New Jersey.

18-12. Late Silurian. *Eurypterus remipes* Dekay. An isolated tail section with an incomplete telson (tail spike). Note the segmentation of the lower body. Fiddlers Green Formation, Bertie Group, Passage Gulf, Cedarville, Herkimer County, New York. ×2.

18-13. Late Silurian. *Eurypterus remipes* Dekay. An excellent specimen of an adult eurypterid in fine grained dolostone. Note the conchoidal breakage in the matrix surrounding the specimen. The eurypterids are usually found in these conchoidal fractures. Fiddlers Green Formation, Bertie Group, Passage Gulf, near Cedarville, Herkimer County, New York. ✕ 1.5. Courtesy: S.J. Ciurca, Jr., Rochester, N.Y.

18-14. Late Silurian. *Eurypterus remipes* Dekay. An incomplete specimen missing one swimming leg and the entire tail section. Note remnants of the walking legs near the head area. Fiddlers Green Formation, Bertie Group. Passage Gulf, Cedarville, Herkimer County, New York. ✕1.5.

18-15. Late Silurian. *Eurypterus remipes* Dekay. A very fine, complete juvenile specimen. Fiddlers Green Formation, Bertie Group, Passage Gulf, Cedarville, Herkimer County, New York. ✕1.

18-16. Late Silurian. *Eurypterus remipes* Dekay. An almost complete specimen (missing one swimming leg and a paddle). Fiddlers Green Formation, Bertie Group, Passage Gulf, Cedarville, Herkimer County, New York. ✕1. Courtesy: Geological Enterprises, Ardmore, Oklahoma.

18-17. Late Silurian. *Eurypterus remipes* Dekay. An essentially complete individual (minus the tip of the telson). Fiddlers Green Dolostone, Bertie Group, Passage Gulf, Cedarville, Herkimer County, New York. ✕1.

18-18. Late Silurian. *Eurypterus remipes* Dekay. An incomplete specimen showing only the body and head. The swimming legs and tail section are missing. Fiddlers Green Formation, Bertie Group, Passage Gulf, Cedarville, Herkimer County, New York. × 1. Courtesy: Geology Department, Princeton University, Princeton, New Jersey.

18-19. Late Silurian. *Eurypterus remipes* Dekay. An almost complete specimen. Fiddlers Green Formation, Bertie Group, Passage Gulf, Cedarville, Herkimer County, New York. × 1.

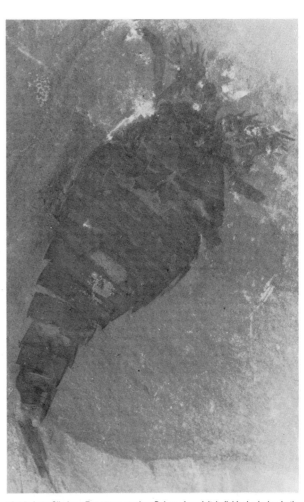

18-20. Late Silurian. *Eurypterus remipes* Dekay. An adult individual missing both swimming legs and part of the telson. Note the fine walking legs near the head area. Fiddlers Green Formation, Bertie Group, Passage Gulf, Cedarville, Herkimer County, New York. × 1.5.

18-21. Late Silurian. *Eurypterus Dekay*. A very fine example of a common eurypterid type. Specimen is missing one of its swimming legs, otherwise, it is an essentially complete individual. Fiddlers Green Formation, Bertie Group, Passage Gulf, Cedarville, Herkimer County, New York. × 1.

18-22. Late Silurian. *Eurypterus remipes* Dekay. An almost complete eurypterid specimen with a distinct walking leg showing on the right side of the carapace. Fiddlers Green Dolostone, Bertie Group, Passage Gulf, Cedarville, Herkimer County, New York. Approx. ⅔ actual size. Courtesy: Geological Enterprises, Ardmore, Oklahoma.

18-23. Late Silurian. *Eurypterus lacustris* Harlan. Two very fine exoskeletons of this second most common eurypterid species. Williamsville Formation, Bertie Group, Buffalo, Erie County, New York. ×1.5. Courtesy: S. J. Ciurca, Jr., Rochester, N.Y.

18-24. Early Devonian. *Erieopterus* sp. Cephalon (head) and partial body of a rather characteristic Early Devonian eurypterid. "H" Zone of the Upper Chrysler Formation, Southeast Syracuse, Onondaga County, New York. ×2. Courtesy: S. J. Ciurca, Jr., Rochester, N.Y.

Faunal listing of eurypterids (genera): ORDOVICIAN: *Brachyopterus, Carcinosoma, Echinognathus, Hughmilleria, Megalograptus, Mixopterus, Pterygotus,* and *Stylonurus.* SILURIAN: *Carcinosoma, Clarkeipterus, Ctenopterus, Dolichopterus, Drepanopterus, Eurypterus* (see Figs. 18-1 through 18-4, and 18-7 through 18-23), *Hughmilleria, Melbournopterus, Mixopterus, Onychopterella, Paracarcinosoma, Parahughmilleria, Pterygotus* (see Figs. 18-5 and 18-6), *Slimonia, Strobilopterus, Stylonurus, Tylopterella,* and *Waeringopterus.* DEVONIAN: *Ctenopterus, Drepanopterus, Erieopterus* (see Fig. 18-24), *Grossopterus, Hughmilleria, Lepidoderma, Mixopterus, Pterygotus, Rhenopterus, Salteropterus, Stylonurus,* and *Tarsopterella (Tarsopterus).* MISSISSIPPIAN: *Mycterops* and *Stylonurus.* PENNSYLVANIAN: *Anthraconectes, Campylocephalus, Glyptoscorpius, Mycterops,* and *Stylonurus.* PERMIAN: *Campylocephalus* and *Lepidoderma.*

Note: *Eurypterus lacustris* (see Figs. 18-8 and 18-23) does not occur in central-eastern New York State. It is known only from the Williamsville Formation, Bertie Group, of the Buffalo, New York area and adjacent Ontario, Canada (type "Bertie Waterlime"). *Eurypterus remipes* (see Figs. 18-1, 18-4, 18-7, and 18-9 through 18-22) occurs primarily in eastern new York, at a different stratigraphic horizon, in the Fiddlers Green Formation (Phelps Member), e.g., Passage Gulf, Herkimer County, New York, but does occur geographically over a larger area, perhaps even to western New York—but never associated with *E. lacustris.*

A brief history of and proceedures for the collecting of eurypterids: The earliest collecting of the fossilized remains of eurypterids was done in the famous "Old Red Sandstones" of Scotland. Early quarry workers were most curious about the strange tail sections, heads, and paddles of *Pterygotus* that they would spot while breaking up rocks in the quarry. Scientists soon became aware of the significance of these strange fossils, and careful collecting proceedures ultimately preserved many of these "stone entombed creatures" and allowed us to have knowledge of these most interesting merostomes.

The collecting of eurypterid specimens in the extremely hard dolomitic rocks of North America is quite difficult and hazardous. At Passage Gulf, a large road cut near Cedarville, New York, one would need a rock hammer to start to break up the large blocks of the hard dolostones. The rocks break conchoidally and special care should be taken by the collector to have safety goggles as well as a hard hat in the vicinity of these extremely dangerous rocks. The rock is so durable that sparks fly out from the cold chisel when struck with the hammer. It is most difficult to predict the location within the rock of the fossil specimen, because the rock breaks unevenly. It is certainly "hit or miss" when it comes to locating the thinly preserved "films" upon the surface of the freshly broken rocks. One could miss a specimen very easily by just a few precious millemeters. Pouring water on the matrix will show where a specimen is successfully exposed or even parts of specimens, but under no circumstances should you ever put a coating of shellac over any such specimen—or, for that matter, *any* fossil specimen at all! It is a bad policy for the study of the material, and is an out-of-date practice. The resulting coating will not only make it extremely difficult to study the specimen, it will also cause problems when and if a photograph is desired of the specimen. To enhance the chitinous filmy image of the eurypterid prior to photography, a light coating of glycerine may be used. A good photographer, with proper lighting, can capture an image that is most successful, and there may not be any need for any coating at all.

19 MISCELLANEOUS (Tully Monster, Etc.)

Aside from their mere presence in the fossil record, the primary significance of the animals in this chapter is that they are soft-bodied. Fossils of soft-bodied animals are geologically improbable. Soft tissues are likely to decay long before they can be sealed into rocks, and, in any event, most rocks do not form fast enough to retain an impression. But at Mazon Creek in Illinois there are literally thousands of fossils of soft-bodied creatures. It is the

19-2. Middle Pennsylvanian. *Escumasia roryi* Nitecki and Solem. Commonly called a "wye," (because of the shape of the fossilized remains which look like the letter "Y" in the alphabet). Two examples of this possible distant relative of the sea-anemone in Mazon Creek siderite concretions. Francis Creek Shale Member, Carbondale Formation (Desmoinesian/Westphalian D), Essex, Kankakee County, Illinois. Photos Courtesy: Field Museum of Natural History, Chicago.

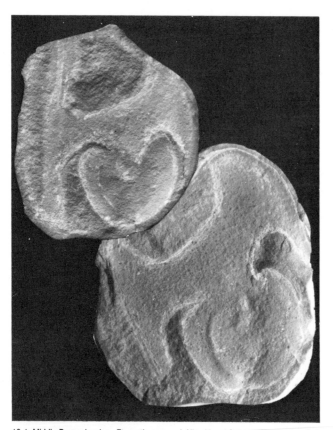

19-1. Middle Pennsylvanina. *Etacystis communis* Nitecki and Schram. Commonly called an "aitch," (because of the fossilized remains which look like the letter "H" in the Roman alphabet). Possibly a hydrozoan in the phylum Coelenterata (Foster, 1979, p. 221). Specimens are in Mazon Creek siderite concretions. Francis Creek Shale Member, Carbondale Formation (Liverpool Cyclothem, Desmoinesian, Westphalian D), Essex, Kankakee County, Illinois. Photos Courtesy: Field Museum of Natural History, Chicago.

world's most complete record of an entire fossil population. Along with jellyfish, insects, eggs, and worms that at least look somewhat familiar, there are the utterly strange Tully monster, Aitch, and Wye shown here.

There is no way of knowing for sure, but these odd orphans may have died out after Mazon Creek time. They must have had ancestors, yet, in all the known fossil record and in all the modern seas, there is no hint of anything like them. There are simply too few geological deposits capable of preserving such animals. True, there are occasional scattered records of soft worms such as *Lecthaylus* in the Silurian of Illinois (see Fig. 6-2) or the jellyfish *Brooksella*, found here and there from the Precambrian to the Ordovician. But, these are exceptions and had no soft-bodied associates.

In their own fossil collecting, many readers of this book may find concretions like those of the Mazon Creek area.

19-4. Middle Pennsylvanian. *Tullimonstrum gregarium* Richardson. A complete, large worm-like fossil (Tully monster) found in siderite concretions in the overburden of strip mines (coal region) of the Mazon Creek area of north-central Illinois. Essex Marine Facies, Will County, Illinois. Approx. half size. Courtesy: Field Museum of Natural History, Chicago.

19-3. Middle Pennsylvanina. *Palaeoxyris* sp. Possible egg case (capsule) of a chondrichthyan fish in a siderite concretion. Also considered by some to be a plant (seed). This author is of the opinion that it may in fact belong to the paleo-chimaeroid fish *Similihariotta dabasinskasi* Zangerl, which is also present in the same Mazon Creek fauna. Francis Creek Shale Member, Carbondale Formation (Westphalian D), Pit 11, Will County, Illinois. Courtesy: Field Museum of Natural History, Chicago.

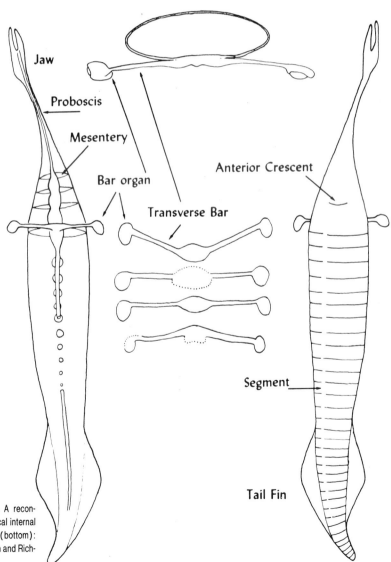

19-5. Middle Pennsylvanian. *Tullimonstrum gregarium* Richardson. A reconstruction of the Tully monster (giant wormlike form). Left: hypothetical internal anatomy. Center (top): cross-section of bar (eyes) region. Center (bottom): variations of bar organ flexure. Right: external features. From Johnson and Richardson, 1969. Courtesy: Field Museum of Natural History, Chicago.

19-6. Middle Pennsylvanian. *Tullimonstrum gregarium* Richardson. A life-size reconstruction of the Tully monster. Made under the direction of Dr. Eugene S. Richardson, Jr., Curator Geology Department, Field Museum of Natural History, Chicago, Illinois. Model prepared for the Museum of Science, Tokyo, Japan. Model constructed by Richard Rush Studio, Chicago, Illinois. Courtesy: Richard Rush Studio, Chicago.

19-7. Middle Pennsylvanian. A reconstruction of the Essex Marine fauna in Will County, Illinois, during the Middle Carboniferous (coal-forming age). Several Tully monsters swimming amongst sea floor plants (algae). One of the Tully monsters is catching a shrimp, *Kallidecthes richardsoni* Schram, while in the background other shrimps, several coelacanths, and two medusae (jelly fish) mind their own business. Courtesy: Field Museum of Natural History, Chicago.

Over the years, hundreds of collectors, like Mr. Francis Tully who found the first Tully monster, have helped to advance science by bringing their finds to the attention of scientists who can add the information to the known record.

Readers who wish identification of their concretions (nodules) should contact the Geology Department of the Field Museum of Natural History, in Chicago, Illinois.

20 PLANTS (including Algae)

Fossil aquatic plants extend far back in geologic time. Filamentous algae are known from Precambrian cherts more than 3¼ billion years old. Stromatolites (see Figs. 20-69 through 20-73), layered, calcareous masses which appear in Precambrian and later sediments, give evidence of prehistoric algal colonies. The process of deposition that formed these cauliflower-shaped masses continues today on parts of the Australian coast. Ancient seaweed (see Fig. 20-1) coexisted with eurypterids in Silurian times, as did ancient kelp (see Fig. 20-2) during the Devonian Period.

With fossil evidence of land plants beginning as far back as the Late Silurian, a well developed terrestrial flora is known to have existed since the Early Devonian. Evidence of Middle Devonian forests comes from sources such as New York State where two foot diameter tree trunks have been found in Middle Devonian sandstones. Far better known are the abundant plant fossils of the Carboniferous (Mississippian/Pennsylvanian) Period. The coal-forming forests of the Carboniferous provide a multitude of floral species and an abundance of well-preserved fossils for paleobotanists and amateurs to study. The evidence gathered from shales, often found interbedded with the coals, provides an excellent and fairly accurate picture of what the flora of these ancient forested swamps looked like. The dioramas (see Figs. 20-31, 20-32, and 20-34) and artist's renderings (see Figs. 20-30 and 20-33) give early interpretations of the habitat of the Carboniferous coal forests. More recent concepts involve a uniformly warm and wet climate as opposed to the steaming jungles depicted.

The hundreds upon hundreds of thousands of "leaves" (actually fronds, pinnae, and pinnules) of the ferns and fernlike plants, the pithy or pulpy stems and foliage of the rushes, and the fallen trunks of the lycopods or scale-trees, left such an abundance of accumulated plant debris in these ancient swampy forests that eventually peat was formed. In the process of deep burial, oxygen and eventually water were excluded, practically eliminating the decaying processes. The additional deposition that continued, resulted in heat and pressure on the buried peat, forming the coal we use today as fuel. In many parts of the world, peat itself, as well as the various coalified forms (lignite, bituminous, and anthracite), is used as fuel. Due to the heat and pressure which formed it, fossils are not easily recognizable in the coal itself. But because many leaves, stems, and trunks were preserved within the underlying and overlying shales, there is an abundant fossil flora for study.

Six basic groups of terrestrial plants are found within Carboniferous rocks. These are: Lycopodophyta (the lycopods or scale-trees); Arthrophyta (Sphenophyta-Sphenopyllales, the articulates, joint-grasses or rushes); Pterophyta (Filicophyta-filicules, the true ferns); Pteridospermophyta (the seed ferns—in which seeds evolved); Coniferophyta (Cordaitales—an extinct group of gymnospermous trees); and Coniferales (the conifers). Cycadophyta (the cycads) started to evolve during the Mesozoic Era.

The seed ferns include some of the most well-known fossil foliage such as *Neuropteris* (see Figs. 20-3 and 20-4, 20-16 through 20-20, 20-25, and 20-26); and *Alethopteris* (see Figs. 20-6, 20-7, 20-9, 20-10, and 20-25). These seed ferns are not true ferns, which reproduce from spores, but they constitute a large portion of the fossil foliage found in coal shales and the ironstone (siderite) concretions from Mazon Creek, Illinois.

The lycopods, dominated by *Lepidodendron* and *Sigillaria,* were a large floral constituent of the Carboniferous and the largest plants of the Pennsylvanian swamps. These forms were treelike in appearance and were indeed trees although unrelated to the conifers and angiosperms we know as the trees of today. Trunks of one meter in diameter are not uncommon and larger ones have been found.

The stems of the lycopods had scalelike scars (also known as leaf cushions) which were characteristic of the various form genera. Evidence obtained from large intact sections now indicates that some of the various patterns may represent different layers of the epidermis of a given tree, or even different heights on a tree. In the case of *Lepidodendron* (see Figs. 20-5 and 20-27) the leaf scars are

20-1. Silurian. Unidentified plant remains. Probably an early form of seaweed or algae. Fiddlers Green Dolomite Member, Bertie Limestone (*Eurypterus remipes* Beds), Passage Gulf Quarry, Cedarville, Herkimer County, New York. ✕ 1.5.

20-2. Late Devonian. *Thamnocladus clarkei* White. A marine alga ("feather kelp"). This algal mass grew to six feet, and along with its relative *Protaxites*, formed an ancient kelp bed beneath the waters of the Late Devonian Period. Chemung Facies, East Windsor, Broome County, New York. 65 cm in length. Courtesy: D. W. Fisher, New York State Museum, Albany.

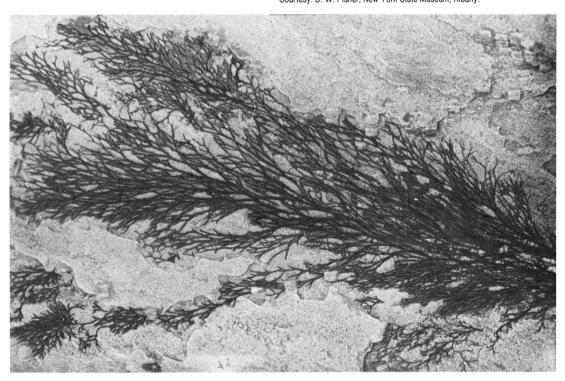

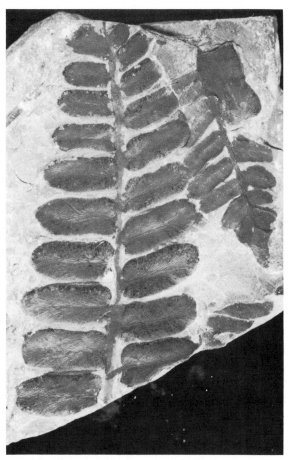

20-3. Pennsylvanian. *Neuropteris gigantea* Sternberg. Fragments of the pinna with pinnule attachments of a fernlike foliage. Carboniferous Coal Shales (Westphalian), Newcastle-on-Tyne, England. Longest Pinna 82 mm.

20-5. Mississippian. *Triphyllopteris rarinervis* Read. A sandstone block containing the leaves of a fernlike foilage. Some of the pinna have pinnule attachments. Lower Mississippian, Price Formation, Pulaski County, Virginia. Courtesy: Illinois State Museum, Springfield. Photo by R. L. Leary. Approx. half size.

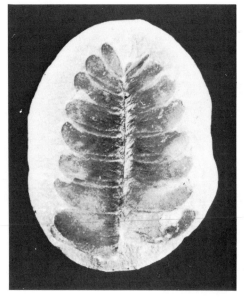

20-4. Pennsylvanian. *Neuropteris gigantea* Sternberg. The top part of the pinna of the fernlike plant in a siderite concretion. Note the upturning of the pinnules. Carboniferous (Westphalian), Cosely near Bilston, Staffordshire, England. 7 cm.

20-6. Pennsylvanian. *Alethopteris* sp. A very common Carboniferous fernlike plant. Found in association with other seed ferns such as *Neuropteris*. Note the long and sinuous pinnules (primarily attached) that distinguish this genus from the others. Coal Shales, Pottsville Formation, St. Clair, Schuylkill County, Pennsylvania. Approx. half size.

20-7. Pennsylvanian. *Alethopteris serlii* (Brongniart). A lengthy pinna of this most abundant pseudofern, surrounded by fragments of other pinnae. Coal Shales, Pottsville Formation, St. Clair, Schuylkill County, Pennsylvania. Approx. half size.

20-8. Pennsylvanian. *Mariopteris nervosa*. Several pinnae of one of the least common of the pseudoferns. The pinnules in design characteristics are reminiscent of the Mesozoic fern *Matonidium*, but ancestry is doubtful. Coal Shales, Pottsville Formation, St. Clair, Schuylkill County, Pennsylvania. Approx. half size.

20-9. Pennsylvanian. *Alethopteris* sp. Several pinnae of the common seedfern of the Eastern Coal Measures of the United States. The pinnae are silver colored on a black slaty shale. Pottsville Formation, St. Clair, Schuylkill County, Pennsylvania. Approx. half size.

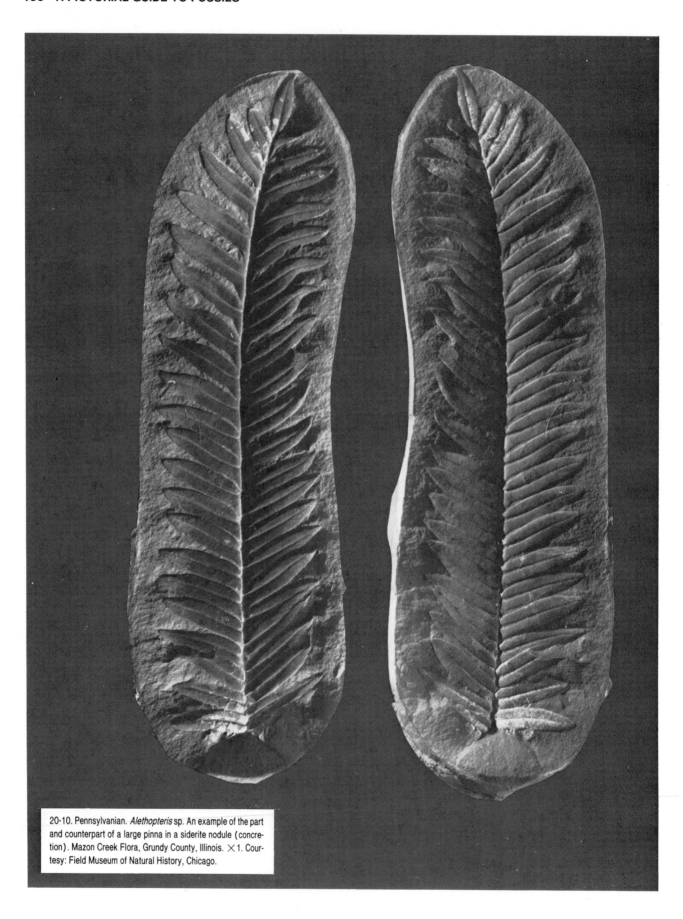

20-10. Pennsylvanian. *Alethopteris* sp. An example of the part and counterpart of a large pinna in a siderite nodule (concretion). Mazon Creek Flora, Grundy County, Illinois. ×1. Courtesy: Field Museum of Natural History, Chicago.

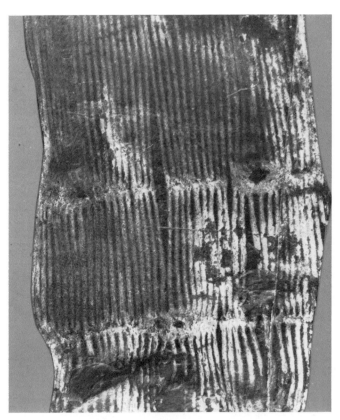

20-12. Pennsylvanian. *Calamites* sp. A fragment of the pith cast of a typical sphenophyte (rush), common in the Carboniferous (Coal Age). The branches with (their whorls of) *Annularia* (see Figs. 20-11 and 20-13) grow out and upward from the main trunk of *Calamites*. Pottsville Formation, St. Clair, Schuylkill County, Pennsylvania. ×1.5.

20-11. Pennsylvanian. *Annularia* sp. A lengthy stem of *Calamites* showing flower-shaped pinnule attachments. A fossil rush (or joint grass) which favored low swampy areas. This example is one half of a siderite concretion. Mazon Creek flora, Grundy County, Illinois. ×1. Courtesy: Field Museum of Natural History, Chicago.

20-13. Pennsylvanian. *Annularia* sp. A small branch with attached and detached whorls of the "leaves" of the scouring rush *Calamites*. Coal Shales, Florence, Cape Breton, Nova Scotia, Canada. ×1.

20-14. Pennsylvanian. *Lepidophoios* sp. A section of distinctive leaf scar impressions of a treelike lycopod not similar to *Lepidodendron* (see Figs. 20-5 and 20-25 for comparison). Westphalian, Muscatine, Muscatine County, Iowa. 114 mm.

20-15. Pennsylvanian. *Lepidostrobus* sp. A cone from the fossil lycopod *Lepidodendron*. Uncommon, but not especially rare. Pottsville Formation, Wilkes-Barre, Luzerne County, Pennsylvania. × 1.

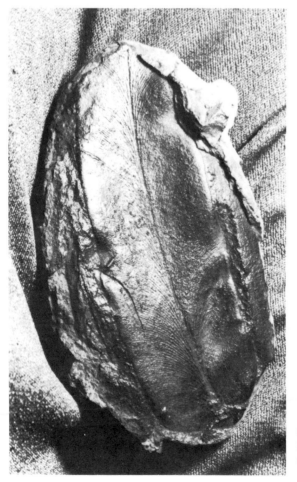

20-16. Pennsylvanian. *Neuropteris* sp. A large single pinnule in an ironstone concretion half. Coal Mines, Terre Haute, Vigo County, Indiana. × 1.

20-17. Pennsylvanian. *Neuropteris* sp. A large isolated pinnule (see Fig. 20-16) of this very common seed fern, in a siderite (ironstone) concretion half. Mazon Creek flora, Braidwood, Grundy County, Illinois. × 1.5.

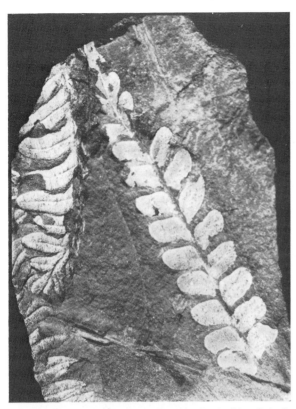

20-18. Pennsylvanian. *Neuropteris* sp. A typical seed fern fossil from Carboniferous coal shales. Pottsville Formation. St. Clair, Schuylkill County, Pennsylvania. ×1.

20-19. Pennsylvanian. *Neuropteris* sp. A very fine example of the pinna with its well developed pinnules of a Carboniferous fernlike plant. Coal Shales, Florence, Cape Breton, Nova Scotia, Canada. ×1.5.

20-20. Pennsylvanian. *Neuropteris* sp. An excellent pinna in an ironstone (siderite) concretion half. Mazon Creek flora, Grundy County, Illinois. ×1. Courtesy: Field Museum of Natural History, Chicago.

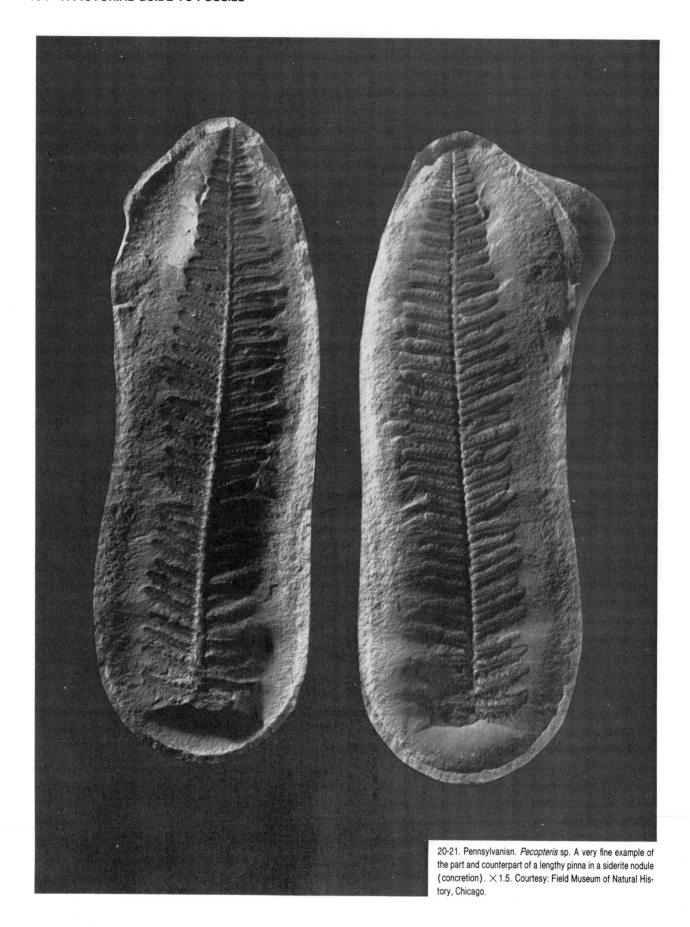

20-21. Pennsylvanian. *Pecopteris* sp. A very fine example of the part and counterpart of a lengthy pinna in a siderite nodule (concretion). ×1.5. Courtesy: Field Museum of Natural History, Chicago.

20-22. Pennsylvanian. *Senftenbergia pennaeformis.* A pinna in a siderite nodule. Note the decorative detail of the pinnules—much more ornate than in *Pecopteris.* Mazon Creek flora, Braidwood, Grundy County, Illinois. 5.5 cm.

20-23. Pennsylvanian. *Pecopteris* sp. A slender pinna in a siderite nodule. Terre Haute, Vigo County, Indiana. ✕ 1.

20-24. Pennsylvanian. *Pecopteris* sp. Part and counterpart of a pinna in an ironstone (siderite) concretion. Coal Mine, Mazon Creek flora, Braidwood, Grundy County, Illinois. ✕ 1.5.

20-25. Pennsylvanian. *Neuropteris rodgersi* (left), and *Alethopteris grandis* (right). Two pinna from different species together on the same shale fragment. Stranger Formation (Virgil Series), Franklin County, Kansas. ✕ 1.

20-26. Pennsylvanian. *Neuropteris decipiens* Lesquereux. A complete isolated pinnule from a seed fern. Note the detail of the structure of the leaf—its divergent patterns extending outward from the median line. Coal Shales, Pottsville Formation, Schuylkill County, Pennsylvania. ×1.5.

20-27. Pennsylvanian. *Lepidodendron* sp. A beautiful example of the leaf scars (or leaf cushions) of a treelike lycopod. A most decorative pattern caused by nature. Mazon Creek flora, Will County, Illinois. ×1. Courtesy: Field Museum of Natural History, Chicago.

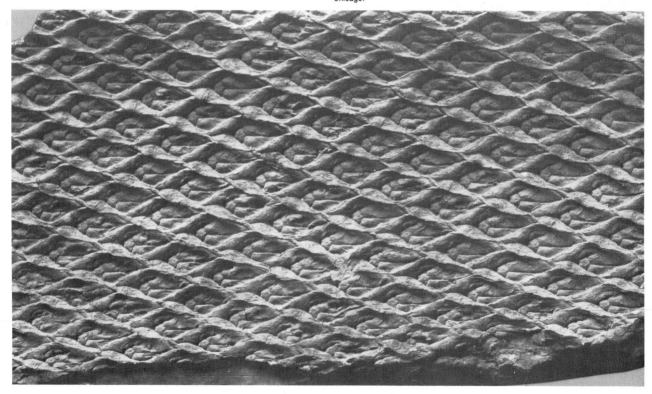

20-28. Pennsylvanian. *Sigillaria elegans* Sternberg. Coalified or "carbonized" plant remains of the outer bark impression of a lycopod tree. Coal region, Fairbury, Illinois. Courtesy: Illinois State Museum, Springfield. Photo by R. L. Leary. ✕1.

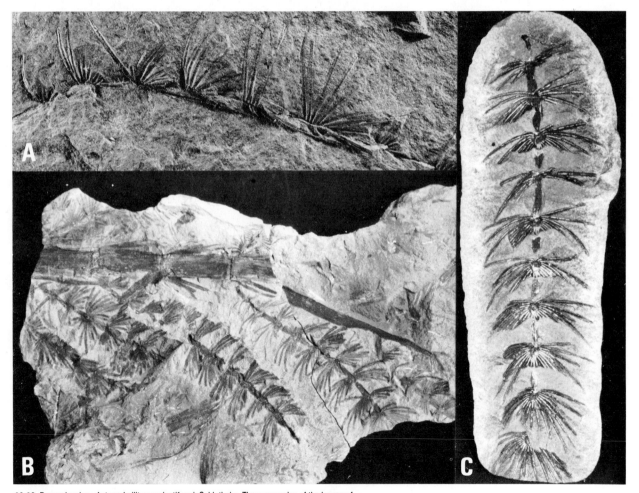

20-29. Pennsylvanian. *Asterophyllites equisetiformis* Schlotheim. Three examples of the leaves of a scouring rush similar to *Annularia* (see Figs. 20-11 and 20-13 for comparison). A. Parts of the leaf whorls are covered by shale. Coal Shales, Pottsville Formation, Schuylkill County, Pennsylvania. ✕1.5. B. A partial stem or stalk of *Calamites* (top center) with a series of leaves in association. C. A siderite concretion half showing a series of the leaves of *Asterophyllites*. B and C slightly more than half size. Francis Creek Shale, Will County, Illinois. B and C courtesy: Illinois State Museum, Springfield. Photos (B and C) by R. L. Leary.

20-30. Permian. A reconstruction of a Permian forest and swamp. Tetrapod amphibians and stem reptiles lounge in the water and on the shoreline while a giant dragonfly is hovering overhead. Along the shoreline and in the water, the rush *Calamites*, with its leaves of *Annularia* reach for the sky. While on the land, lycopods and fernlike plants grow in abundance to form a forest. Courtesy: Field Museum of Natural History, Chicago.

20-31. Pennsylvanian. A scene in a Carboniferous forest. Lycopods and fernlike trees are everywhere—later to form coal, where their fossilized remains will be packed together tightly between sandstones and shales. Reconstruction for the Hall of Geology, Wm. Penn Memorial Museum, Harrisburg, Pennsylvania. Courtesy: Richard Rush Studio, Chicago.

20-32. Pennsylvanian. An outdated diorama showing what a Carboniferous forest was thought to look like many years ago. In the foreground we see a large cockroach walking across a fallen *Lepidodendron* tree, and in the background, a tetrapod amphibian is climbing out of the water of the swamp. All around the scene are ferns and seed ferns, as well as lycopods. Courtesy: Field Museum of Natural History, Chicago.

20-33. Pennsylvanian. Another reconstruction of a Carboniferous swamp and forest. Note the treelike lycopods, the fernlike plants and the rushes *(Calamites/Annularia)* surrounding the water hole. Courtesy: Field Museum of Natural History, Chicago.

20-34. Pennsylvanian. Carboniferous coal forest. Another diorama showing a large dragonfly in the foreground (see also Fig. 16-6) surrounded by lycopods and fernlike plants. Courtesy: Field Museum of Natural History, Chicago.

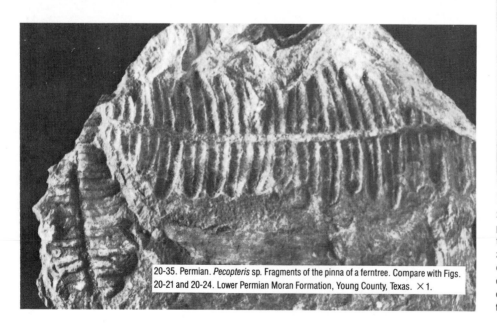

20-35. Permian. *Pecopteris* sp. Fragments of the pinna of a ferntree. Compare with Figs. 20-21 and 20-24. Lower Permian Moran Formation, Young County, Texas. ×1.

20-36. Permian. *Glossopteris* sp. A single pinnule of a rather ubiquitous broad-leaved temperate plant common to the Upper Paleozoic of the southern hemisphere. Also a characteristic species for the Gondwana sequences. New Zealand. ×2. Courtesy: Geology Museum, Victoria University, Wellington, New Zealand.

20-37. Triassic. Cross-section of a petrified tree stump (trunk) — species indeterminate. Note the annular (growth) rings. Chinle Formation, Petrified Forest National Park, Arizona. Half Size.

20-38. Triassic. A beautifully preserved agatised pine cone. Patagonia, Argentina. 5.5 cm. Courtesy: Siber and Siber Ltd., Aathal, Switzerland.

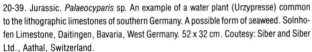

20-39. Jurassic. *Palaeocyparis* sp. An example of a water plant (Urzypresse) common to the lithographic limestones of southern Germany. A possible form of seaweed. Solnhofen Limestone, Daitingen, Bavaria, West Germany. 52 x 32 cm. Coutesy: Siber and Siber Ltd., Aathal, Switzerland.

20-40. Jurassic. *Phyllothallus* sp. Fossil seaweed from an ancient delta. Upper Malm, Lithographischer Schiefer, Schernfeld, Bavaria, West Germany. 11.5 x 13 cm.

diamond-shaped and longer than they are wide, while in *Sigillaria*, the leaf cushions are arranged in vertical rows and separated by grooves. In most species of lycopods, the leaf scars have an attachment point or nub where the leaf was affixed to the pithy part of the stalk or tree stem.

The cones of the lycopods are most interesting, and some have been found to contain spores. Although spores are abundant in many Carboniferous deposits, they are too small to be of much interest to the amateur collector. The cones of the lycopods are usually given the generic name *Lepidostrobus* (see Fig. 20-15).

Originally the parts of the lycopods, like those of many other fossil plants, were found disassociated and named separately, since at the time they were first described, their interrelationships were not known. It is not surprising that parts of the same plant were given different generic names, often referred to as form genera. For instance, *Lepidodendron* was the name given to the trunk of the lycopods, the roots were named *Stigmaria*, the leaves were named *Lepidophyllum*, the cones were named *Lepidostrobus*, and the single cone scales were named *Lepidostrobophyllum*.

The next group from the Carboniferous, the articulates or joint-grasses are represented today by *Equisetum*, the diminutive, extant horsetail. The group includes such well-known Carboniferous forms as *Calamites* (see Figs. 20-12 and 20-33) and *Sphenophyllum*. Again, as with the lycopods, different parts received different generic names. *Calamites* refers to the trunk, generally represented by the casts of the pith cavity (see Fig. 20-12), while *Annularia* refers to the flower-shaped foliage (see Figs. 20-11, 20-13, and 20-33), and *Calamostachys* refers to the cone.

The articulates used underground stems (rhizomes) to attach themselves in the soil and to propagate. The underground networks of stems ("roots") became enormously involved, taking up space and creating a veritable underground jungle.

The next group, the Filicophyta or true ferns is represented by such Carboniferous genera as *Psaronius*, a tree fern some fifty feet tall; and *Pecopteris* (see Figs. 20-21, 20-23, 20-24, and 20-35), the foliage of the tree ferns. There are at least 10,000 species of ferns in modern floras.

Less commonly found in the Carboniferous Period are

20-41. Jurassic. *Dicroidium* sp. An early Mesozic seed fern from New Zealand.
×2. Courtesy: Geology Museum, Victoria University, Wellington, New Zealand.

the Cordaitales and the Coniferales. Poorly preserved coniferous "leaf" specimens, given the artificial generic name *Walchia*, are found in Late Pennsylvanian and Early Permian strata of various places in the world. After the Carboniferous Period the newer forms, the Cycadophyta (cycads, see Fig. 20-43) and conifers became more abundant.

The angiosperms or flowering plants comprise the majority of the plants we are most familiar with today. They first appeared in the Early Cretaceous, and several excellent angiosperm floras have been found in Early Creta-

20-42. Late Cretaceous. *Ficus ceratops*. Fossilized fruit of the fig tree. A very unusual fossil. Fox Hills Sandstone, Western South Dakota. 3 cm.

20-44. Eocene. *Platanophyllum wyomingensis*. A complete leaf of a plane tree (sycamore). Chalk Bluff flora of Colorado. 10 cm. Courtesy: Geological Enterprises, Ardmore, Oklahoma.

20-43. Jurassic. *Zamites fenconis*. A complete cycad frond with pinnae attached to the rachis at right angles. Kimmeridgian, Seysel, Ain Province, France. 7 cm. Courtesy: Siber and Siber Ltd., Aathal, Switzerland.

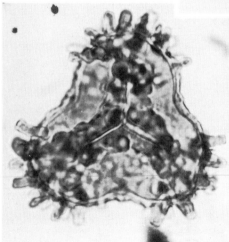

20-45. Early Cretaceous. *Trilobosporites marylandensis* Brenner. Fossil spore. Arundel Formation (Albian), Potomac Group, Clay pit near Baltimore, Maryland. × 1000.

20-46. Early Cretaceous. *Appendicisporites tricornatus* Weyland and Greifeld. Fossil spore. Arundel Formation (Albian), Potomac Group, Clay pit near Baltimore, Maryland. × 1000.

20-47. Early Cretaceous. *Trilobosporites apibaculatus* Brenner. Fossil spore. Arundel Formation (Albian), Potomac Group, Clay pit near Baltimore, Maryland. × 1000.

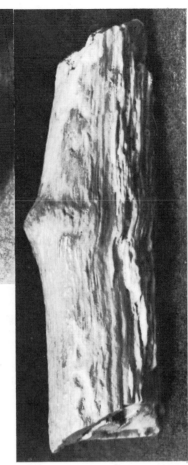

20-48. Eocene. Eden Valley petrified wood limb fragments. Note small knotholes and bark impressions. Red Canyon, South Pass, Wyoming. 8 cm.

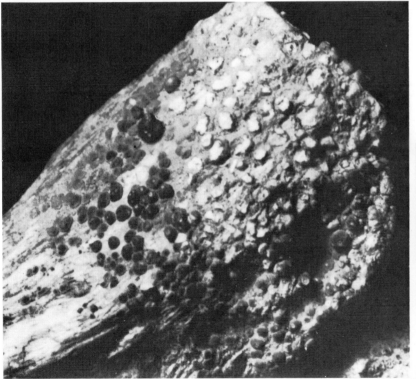

20-49. Eocene. Eden Valley petrified wood showing fossilized fungus. Very rare as a fossil. Red Canyon, South Pass, Wyoming. 2 cm.

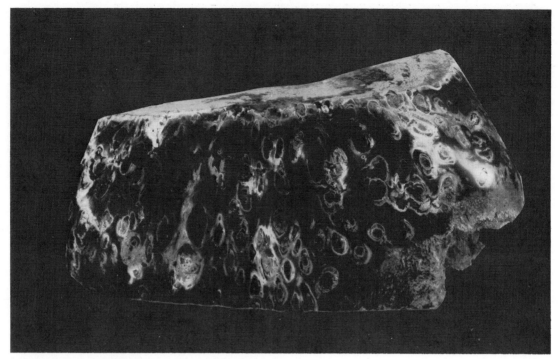

20-50. Eocene. Palm wood (species indeterminate). A polished section of a palm tree. Green River Formation, Farson, Sublette County, Wyoming. ×1.

20-51. Oligocene. *Acer* sp. A fine specimen of a fossil maple leaf. Gottschalkenberg, Switzerland. 9 cm. Courtesy: Siber and Siber Ltd., Aathal, Switzerland.

20-52. Oligocene. *Alnus carpinoides* Lesquereux. A large fossil alder leaf in volcanic ash shale. Florissant Formation, Florissant, Teller County, Colorado. ×2.

20-53. Oligocene. *Salix* sp. A beautiful specimen of a fossil willow leaf in volcanic ash shale. Florissant Formation, Florissant, Teller County, Colorado. 4 cm.

20-54. Oligocene. *Salix* sp. Another example of a fossil willow leaf in volcanic ash shale. Florissant Formation, Florissant, Teller County, Colorado. ×1. Courtesy: Brooklyn Childrens Museum, Brooklyn, New York.

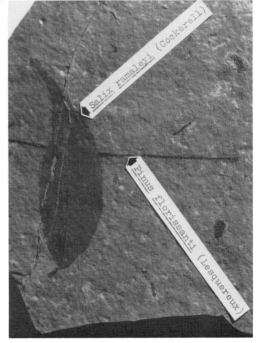

20-55. Oligocene. *Salix rameleyi* (Cockerell)—a willow leaf, and a twig from the pine tree *Pinus florissanti* (Lesquereux) in volcanic ash shale. Florissant Formation, Florissant, Teller County, Colorado. ×1.

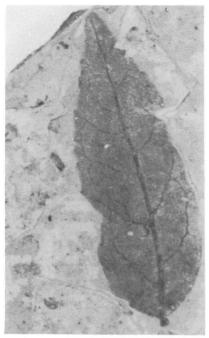

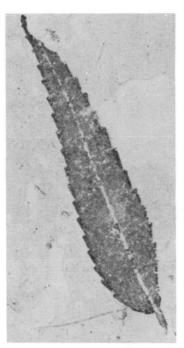

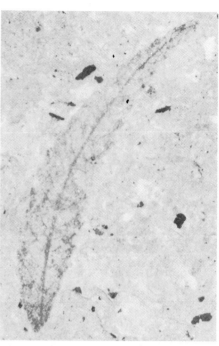

20-56. Oligocene. *Populus americanus* (Lesquereux). A fossil leaf of the poplar tree preserved in volcanic ash shale. Florissant Formation, Florissant, Teller County, Colorado. ✕ 1.

20-57. Oligocene. *Zelkova drymeja* (Lesquereux). Fossil keeki leaf in the angiosperm group preserved in volcanic ash shale. Florissant Formation, Florissant, Teller County, Colorado. ✕ 1.

20-58. Oligocene. *Zelkova drymeja* (Lesquereux). Fossil keeki leaf in volcanic ash shale. Florissant Formation, Florissant, Teller County, Colorado. ✕ 1.

20-59. Oligocene. *Zelkova drymeja* (Lesquereux). A large fossil keeki leaf in volcanic ash shale. Florissant Formation, Florissant, Teller County, Colorado. ✕ 1.5.

20-60. Oligocene. *Zelkova drymeja* (Lesquereux). A twig with three large and one small keeki leaves attached. The usual find is a solitary leaf (see Figs. 20-57 through 20-59). Leaf impressions in volcanic ash shale. Florissant Formation, Florissant, Teller County, Colorado. ✕ 1.

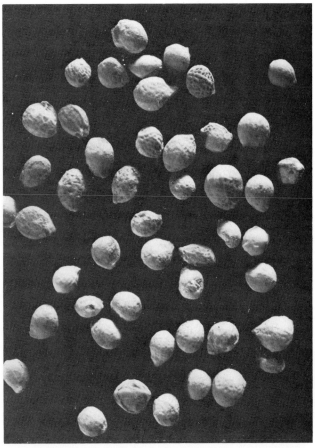

20-62. Oligocene. *Celtis occidentalis*. Numerous examples of fossilized Hackberry seeds. Orella Formation, White River Badlands, Scenic, Pennington County, South Dakota. Average seed: 4mm diam.

20-61. Oligocene. *Glyptostrobus europaeus* Heer. Fossilized branch of a water pine showing several cones at the end of the branch (top center). Chattian Stage, Preschen, Bohemia, Czechoslovakia. 11.5 cm.

20-63. Oligocene. *Nothofagus brassi*. Fossil beach leaf impression in sandstone. Note the whole leaf to the right—this is a closely related descendant (Recent) *Fagus* leaf for comparison. Fossil leaf is from the Mangles Formation, Murchison, New Zealand. The modern leaf is from New Caledonia in the South Pacific. ✕ 1. Courtesy: Geology Museum, Victoria University, Wellington, New Zealand.

20-64. Miocene. Silicified walnut. Fossil tree section showing the wood grain and annular rings (growth rings). Saddle Mountains, Washington State. ✕1.

20-66. Miocene. *Acer trilobatum tricuspidatus* (Brongniart). A fossil maple leaf. Oehningen bei Bodensee, West Germany. ×1.

20-65. Miocene. *Cocus zeylandica.* Fossilized coconut. A rare fossil preservation. Middle Miocene (Helvetian equivalent), North Island, New Zealand. 7.5 cm. Courtesy: Geology Museum, Victoria University, Wellington, New Zealand.

20-67. Miocene. *Cystoseira communis* Ung. Seaweedlike plant. Radoboj, Croatia, Jugoslavia. Less than half size.

20-68. Pleistocene. Cedar branchlet on lake bottom mud shale (see Fig. 20-74). Species indeterminate. Shiobara, Tochigi Prefecture, Japan. ×1.

20-69. Precambrian. Cryptozoon (stromatolite). A large colony of fossil algae. Note film canister for scale. Saratoga, New York. Courtesy: Monroe Community College, Rochester, New York.

20-70. Precambrian Cryptozoon (stromatolite). A large rock showing a stromatopod colony in cross-section. Saratoga, New York. Courtesy: Monroe Community College, Rochester, New York.

20-71. Precambrian. Cryptozoon (stromatolite). An isolated colony of stromatopods (algae) in a larger colony. Note pen for scale. Saratoga, New York. Courtesy: Monroe Community College, Rochester, New York.

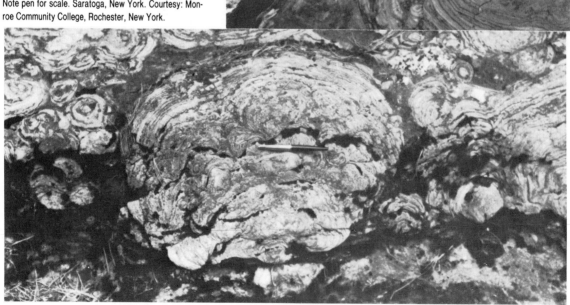

20-72. Precambrian. *Cryptozoan proliferum.* A large algal mass in a tumbled rock. Saratoga, New York. 8 cm.

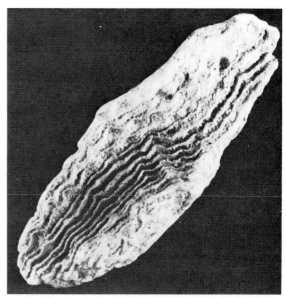

20-73. Middle Devonian. Stromatoporoid (fossil algae—species indeterminate). An isolated specimen (glacial derivation). Originally it came from the Onondaga Limestone of New York State. It was carried down by a glacier and redeposited as glacial outwash material, south of the terminal moraine in central New Jersey. × 1.

20-74. View of the Pleistocene lake bed at Shiobara, Tochigi Prefecture, Japan. See Fig. 20-68 for a specimen from these beds. The robot is now inoperative, but at one time it told the story of the fossil lake beds (the layers or stratas of which can be seen directly behind the robot). Now that the robot is no longer working as a guide, a sign (in the Japanese language) to the left tells the story of the plants and leaves found in the fossil beds. Courtesy: D. Kenney, Andover, Massachusetts.

ceous deposits of the Atlantic Coastal Plain of the United States. Examples of angiosperm fossils shown are the plaintree or sycamore (see Fig. 20-44), the maple (see Figs. 20-51 and 20-66), the alder (see Fig. 20-52), the willow (see Figs. 20-53 through 20-55), the poplar (see Fig. 20-56), the keeki (see Figs. 20-57 through 20-60), the walnut (see Fig. 20-64), and the beech (see Fig. 20-63).

Fossilized remains of plant life in general involve a number of different types of fossilization. A number of forms such as the many beautiful petrified woods have not been discussed in this chapter. The fact that leaves, seeds, branches, fern pinnules, and fruit (see Fig. 20-42, fossil fig) of plants can be found in a fossil state at all is truly a marvel.

21 PSEUDOFOSSILS

Pseudofossils, or "false-fossils", as the name of this chapter indicates, are objects or traces found in fossiliferous areas, which at first glance appear to be fossils, but are of inorganic origin.

Ripple marks (see Fig. 21-2) and mud cracks are not true fossils, however they are important as sedimentary/environmental indications and are commonly found in fossil deposits. There are many objects that fall into this category of pseudofossils. Dendrites, for example, are precipitations of manganese outlining nearby organic material, such as the bones of the fish illustrated in Fig. 25-63 (in the chapter on bony fishes). Another example is the rather deceptive "cone-in-cone" type of mineral impression found in some sedimentary bedding planes. For an authentic cone-in-cone (conularid) see Figs. 3-28 and 29 (in the chapter on corals, etc.). The cone-in-cone effect is due to an aragonitic "silex" type of mineral which usually forms as a layering above the fossiliferous zone. The effect is markedly similar to the linear ornamentation on a conularid.

Predominantly though, the false fossils are usually some form of "concretion" (see Fig. 21-1). Many concretions with remarkable organic mimicry (structure) have been recovered in or near dinosaur bone beds, particularly in the Jurassic Morrison Formation of Eastern Wyoming. These concretions have been variously identified as "gastric" or "gizzard" stones of dinosaurs and/or excretion (fecal pellets) or coprolites of sauropods found in association. Most of these concretions were formed by the growth of gypsum crystals and have no relationship to fossils at all.

21-1. Pennsylvanian. Septarian nodule. Concretion formed in Carboniferous coal beds. Power Mine, Cherokee Beds, Tebo-Mulkey Coals, Montrose, Henry County, Missouri. ×1.

21-2. Triassic. "Ripple marks" caused by gentle wave action upon the sands of the shallow part of a lake bed. Lockatong Formation, Newark Series, Granton Quarry, North Bergen, Hudson County, New Jersey. ×1.

22 PISCES (Ostracodermi / Archaic fishes)

The Ostracodermi ("shell-skinned") archaic fishes of the Ordovician/Silurian/Devonian, are the earliest and most primitive of fishes, with no known ancestors. Two subclasses make up this group of archaic fishes, the Monorhina and the Diplorhina. The subclass Monorhina has in turn, three orders: Osteostraci, Anaspida, and Galeaspida.

The order Anaspida has five families: **Jaymoytiidae, Birkeniidae (Pterygolepidae), Lasaniidae, Euphaneropsidae,** and **Endeiolepidae.** In the family **Jaymoytiidae,** *Jaymoytius* of the Middle Silurian of Europe is the only generic example. The family **Birkeniidae (Pterygolepidae)** has as its representatives: *Birkenia* and *Pterygolepis* of the Silurian of Europe. The family **Lasaniidae** is represented by *Lasanius* of the Late Silurian/Early Devonian of Europe. *Euphanerops* of the Late Devonian of North America is a good example of the family **Euphaneropsidae.** Finally, in the family **Endeiolepidae,** we have the genus *Endeiolepis* of the Late Devonian of North America.

The order Osteostraci is the second order in the subclass Monorhina and has six families: **Tremataspidae, Dartmuthiidae, Hemicyclaspidae (Ateleaspidae), Sclerodontidae, Cephalaspidae,** and **Kiaeraspidae.** Representative generic types for the Osteostraci are: *Tremataspis* of the Silurian, *Dartmuthia* (Silurian), *Hemicyclaspis* (silurian), *Sclerodus* (Silurian), *Cephalaspis* (Devonian—see Fig. 22-1), and *Kiaeraspis* (Devonian).

The third and final order of the subclass Monorhina is the Galeaspida. This order has two families: **Galeaspidae** and **Polybranchiaspidae,** with representative Devonian genera: *Galeaspis* and *Polybranchiaspis.*

The subclass Diplorhina has two orders: Heterostraci (Pteraspidomorpha) and Coelolepida (Thelodonti). In the order Heterostraci, there are ten families: **Astraspidae, Cyathaspidae, Eriptychiidae, Amphiaspidae, Corvaspidae, Pteraspidae** (see Fig. 22-1), **Traquairaspidae, Cardipeltidae, Drepanaspidae (Psammosteidae)** (see Fig. 22-1), and **Obliaspidae.** *Pteraspis* and *Drepanaspis* are good examples of the Heterostraci.

Finally, in the order Coelolepida (Thelodonti) there is

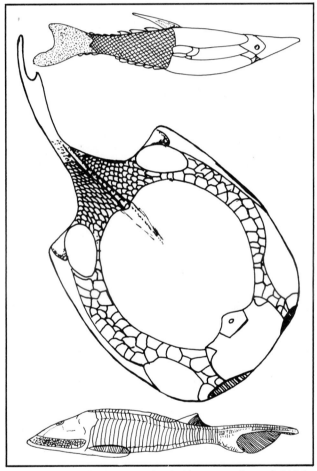

22-1. Devonian. Top: *Pteraspis rostrata* Agassiz. A heterostracan (armored) fish in lateral (side) view. Old Red Sandstone, Herefordshire, England. Slightly larger than actual size. Center: *Drepanaspis gemuendensis* Schlüter. Another heterostracan (armored) fish, in dorsal (top) view. Hunsrück Slates, Gemünden, Rheinland, West Germany. Approximately half size. Bottom: *Cephalaspis lyelli* Agassiz. A lateral (side) view of an osteostracan fish. Old Red Sandstone, Arbroath, Scotland. Slightly less than actual size. All figures are drawn from plaster casts. The author assumes responsibility for any inaccuracies in the renderings.

only one family: **Thelodontidae.** Thelodonti is represented by: SILURIAN: *Bystrowia, Coelolepis, Lanarkia, Phlebolepis,* and *Thelodus (Thelolepis).* DEVONIAN: *Coelolepis, Thelodus,* and *Turinia (Cephalopterus).*

23 PISCES (Placodermi / Arthrodires, Antiarchs, Etc.)

This chapter concludes the early or primitive fishes with a presentation of the placoderms (arthrodires and antiarchs).

The arthrodires are small to large partly armored (anterior-skull and post-cranial) primitive fishes. The larger varieties of Late Devonian arthrodires, i.e., *Dinichthys*, *Dunkleosteus* (see Fig. 23-11), and *Protitanichthys* (see Fig. 23-10) evolved from the smaller *Coccosteus* of the Middle Devonian.

These large arthrodires as represented by *Dunkleosteus* had a partial trunk-shield and head section composed of hard bony dermal armoring, with strange shearing jaws much like scissor blades. These pecular dentitions, and the fact that these arthrodires could move their heads up and down, must have made these fishes quite ferocious in the eyes of their prey.

One smaller group of the arthrodires, the ptyctodonts, may possibly have given rise to the holocephalians, particularly the chimaeriformes (see the following chapter), but this remains highly speculative. At least, they have a number of taxonomical features in common, and, since everything, or most everything, has to have an ancestor, they might eventually be proven to be archaic forms of the holocephalians.

The antiarchs or "winged-fishes" are placoderms which have strong armor (exoskeletons) on the anterior or forward part of their bodies, with soft-tissued posterior trunks or bodies (unfortunately, not well preserved in the fossil record). They may of had cartilaginous vertebral columns as well. The anterior portion of the bodies of antiarchs is made up of a series of connecting plates of dermal (bone) material, forming, in total, an outside skeleton as well as armoring for the fish. The paired fins (pectorals) are armored as well. It is these fins when extended from the thoracic armor that give the name "winged fish" to these placoderms. The most commonly known and described antiarch is *Bothriolepis* from the Late Devonian of Scaumenac Bay, Quebec, Canada. *Bothriolepis* (sometimes spelled *Bothryolepis*) (see Figs. 23-3 through 23-5) is fairly abundant in the rocks of this locality in northeastern Canada.

The following is an abbreviated listing of the genera of the various groups discussed in this chapter: Order Acanthothoraci/family **Palaeacanthaspidae:** DEVONIAN: *Dowbrowlania, Kosoraspis,* and *Palaeacanthaspis.* Family **Gemuendinidae:** DEVONIAN: *Asterosteus, Gemuendina* (see Fig. 23-9), *Hoplopetalichthys, Jagorina, Kolymaspis,* and *Radotina.* The material listed in the order Acanthothoraci was formerly listed under the order Rhenanida (Romer, 1966, p. 348R).

Order Petalichthyida/family **Macropetalichyidae:** DEVONIAN: *Macropetalichthys (Acanthaspis)* (see Fig. 23-1).

The order Arthrodira/suborder Arctolepida (Dolichothoraci)/families: **Phlyctaeniidae (Arctolepidae):** DEVONIAN: *Actinolepis, Aethaspis, Aggeraspis, Anarthraspis, Arctaspis, Arctolepis, Bryantolepis, Diadsomaspis, Elegantaspis, Heterogaspis, Huginaspis, Kujdanowiaspis, Mediaspis, Overtonaspis, Phlyctaenaspis, Polyaspis, Prescottaspis, Prosphymaspis, Rotundaspis, Simblaspis, Stuertzaspis, Svalbardaspis, Tiaraspis,* and *Wheathillaspis.* **Williamsapidae:** DEVONIAN: *Williamsaspis.* **Holonematidae:** DEVONIAN: *Anomalichthys, Aspidichthys, Deirosteus, Deveonema, Glyptaspis, Gyroplacosteus, Holonema, Megaloplax,* and *Rhenonema.* **Groenlandaspidae:** DEVONIAN: *Grazosteus* and *Groenlandaspis.*

Suborder Brachythoraci/families: **Gemuendenaspidae:** DEVONIAN: *Gemuendenaspis.* **Coccosteidae:** *Buchanosteus, Clarkosteus, Coccosteus, Dickosteus, Livosteus, Millerosteus* (see Fig. 23-6), *Oestoporus (Ostophorus?), Plourdosteus, Protitanichthys* (see Fig. 23-10), *Rhachiosteus, Taemasosteus,* and *Watsonosteus.* **Pholidosteidae:** DEVONIAN: *Malerosteus, Pholidosteus,* and *Tapinosteus.* **Heterosteidae:** DEVONIAN: *Heterosteus.* **Homostiidae:** DEVONIAN: *Angarichthys, Homostius (Homosteus), Tityosteus,* and *Tollichthys.* **Brachydeiridae:**

23-1. Devonian. *Macropetalichthys* (*Acanthaspis*) sp. A section of a pectoral fin spine of a placoderm (early arthrodire). Cedar Valley Limestone, Pints Quarry, Waterloo, Black Hawk County, Iowa. ×2.5.

23-2. Devonian. *Pterichthyodes (Pterichthys) milleri* (Agassiz). An essentially complete antiarch fish exoskeleton in a sandstone concretion. The tail section, which is missing, does not preserve due to its soft composition. Here, the entire body shield and pectoral fins (with spines) are preserved intact. Old Red Sandstone, Caithness, Scotland. ×2.

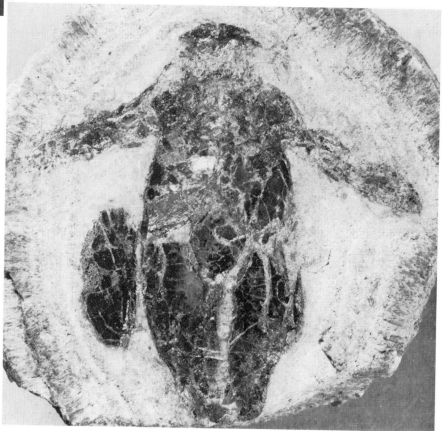

DEVONIAN: *Brachydeirus, Oxyosteus,* and *Synauchenia.* **Trematosteidae:** DEVONIAN: *Brachyosteus, Braunosteus, Cyrtosteus, Helmerosteus,* and *Trematosteus.* **Pachyosteidae:** DEVONIAN: *Enseosteus, Erromenosteus, Leiosteus, Menosteus, Microsteus, Ottonosteus, Pachyosteus, Paraleiosteus, Parawalterosteus,* and *Rhin-* *osteus.* **Titanichthyidae:** DEVONIAN: *Titanichthys.* **Selenosteidae:** DEVONIAN: *Callognathus, Gymnotrachelus, Paramylostoma, Selenosteus,* and *Stenosteus.* **Hadrosteidae:** DEVONIAN: *Hadrosteus.* **Leptosteidae:** DEVONIAN: *Leptosteus.* **Mylostomatidae:** DEVONIAN: *Dinognathus, Dinomylostoma,* and *Mylostoma.* **Dinich-**

23-3. Devonian. *Bothriolepis canadensis* Whiteaves. Dorsal and ventral views of the body armor of another antiarch in the Bothriolepidae family. Again, as was the case with *Pterichthyodes,* the soft tissue of the trunk of the body and the tail section did not fossilize. See Fig. 23-4 for a reconstruction of the tail area. Scaumenac Bay, Province of Quebec, Canada. Approx. half size (each view). Courtesy: Princeton University Museum of Natural History, Princeton, New Jersey.

23-4. Devonian. *Bothriolepis canadensis* Whiteaves. A restoration model in dorsal view, of the antiarch or "wingedfish" in Figs. 23-3 and 23-5. Courtesy: New York State Museum, Albany, New York.

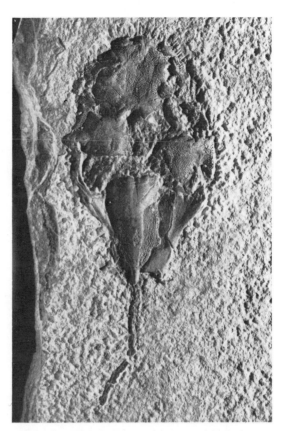

23-5. Devonian. *Bothriolepis canadensis* Whiteaves. Another ventral view of an antiarch fish's body armor or shield. See Fig. 23-3 for a similar view of the same species. Scaumenac Bay, Province of Quebec, Canada. Less than half size.

thyidae: DEVONIAN: *Dinichthys, Dunkleosteus* (see Fig. 23-11), *Eastmanosteus, Gorgonichthys,* and *Heintzichthys.*

Order Phyllolepida/family **Phyllolepidae:** DEVONIAN: *Phyllolepis.*

Order Ptyctodontida/family **Ptyctodontidae:** DEVONIAN: *Chelyophorus, Ctenurella, Eczematolepis, Goniosteus, Paraptyctodus, Ptyctodus* (see Fig. 23-8), *Rhamphodopsis,* and *Rhynchodus* (see Fig. 23-7).

Order Antiarchi/families **Asterolepidae:** DEVONIAN: *Asterolepis, Belemnacanthus, Bothriolepis* (see Figs. 23-3 through 23-5), *Byssacanthus, Cypholepis, Dianolepis, Gerdalepis, Grossilepis, Leptadolepis, Microbrachius (Microbrachium), Pterichthyodes (Pterichthys)* (see Fig. 23-2), *Sinolepis,* and *Taeniolepis.* **Remigolepidae:** DEVONIAN: *Remigolepis.* **Wudinolepidae:** DEVONIAN: *Wudinolepis.*

23-6. Devonian. *Millerosteus minor.* A crushed coccosteid arthrodire body armor (shield) in matrix. Scotland. Courtesy: Geological Enterprises, Ardmore, Oklahoma.

23-7. Devonian. *Rhynchodus major* Eastman. The upper jaw element (gnathal plate) of a ptyctodont arthrodire. 11.5 mm across specimen. Cedar Valley Limestone Formation, Pints Quarry, Waterloo, Black Hawk County, Iowa.

23-8. Devonian. *Ptyctodus compressus* Eastman. An exceptionally fine lower dental plate (gnathal element) of this primitive, small arthrodire. Cedar Valley Limestone, Pints Quarry, Waterloo, Black Hawk County, Iowa. 46 mm across the specimen.

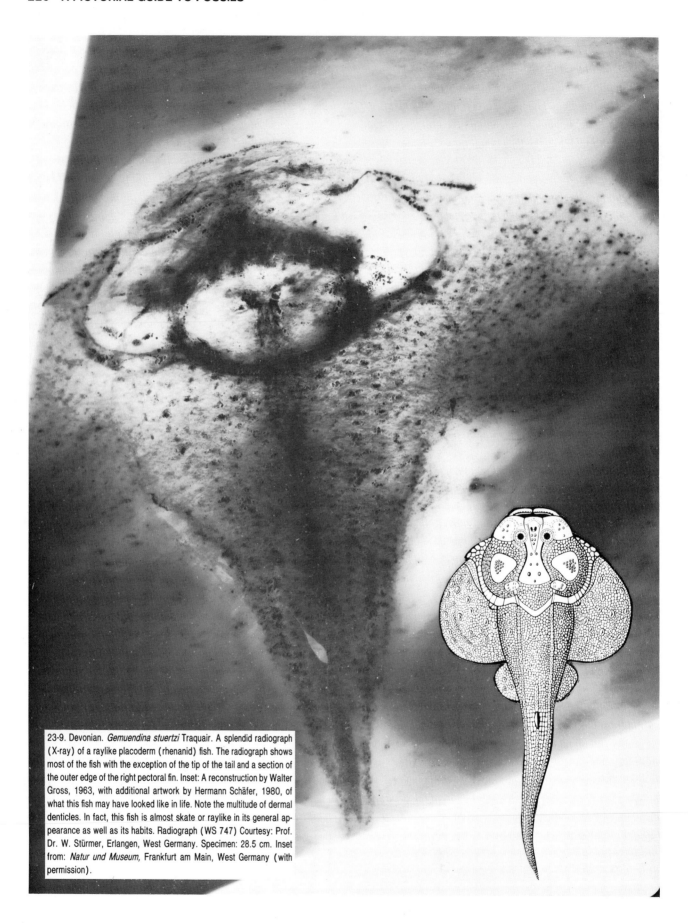

23-9. Devonian. *Gemuendina stuertzi* Traquair. A splendid radiograph (X-ray) of a raylike placoderm (rhenanid) fish. The radiograph shows most of the fish with the exception of the tip of the tail and a section of the outer edge of the right pectoral fin. Inset: A reconstruction by Walter Gross, 1963, with additional artwork by Hermann Schäfer, 1980, of what this fish may have looked like in life. Note the multitude of dermal denticles. In fact, this fish is almost skate or raylike in its general appearance as well as its habits. Radiograph (WS 747) Courtesy: Prof. Dr. W. Stürmer, Erlangen, West Germany. Specimen: 28.5 cm. Inset from: *Natur und Museum,* Frankfurt am Main, West Germany (with permission).

23-10. Devonian. *Protitanichthys* sp. (possibly *Protitanichthys* cf. *rockportensis*). Fragments of armor bone material from a large arthrodire (see 23-11). Silica Shale Formation, Milan, Washtenaw County, Michigan. × 1.5.

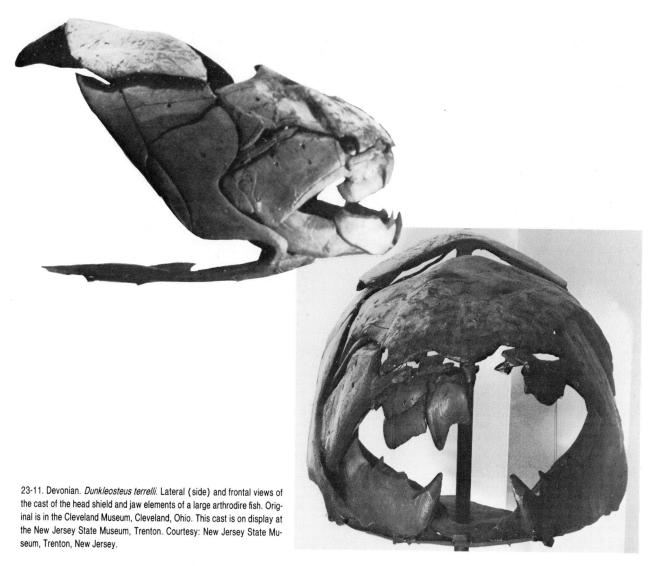

23-11. Devonian. *Dunkleosteus terrelli*. Lateral (side) and frontal views of the cast of the head shield and jaw elements of a large arthrodire fish. Original is in the Cleveland Museum, Cleveland, Ohio. This cast is on display at the New Jersey State Museum, Trenton. Courtesy: New Jersey State Museum, Trenton, New Jersey.

24 PISCES (Chondrichthyans/Sharks, Skates, Rays, Sawfishes, Chimaeras, & Iniopterygians)

In the superclass Gnathostomata (Moy-Thomas/Miles, 1971, p. 247) is Elasmobranchiomorphi (which includes, according to Moy-Thomas/Miles, 1971, p. 8, the subclass Placodermi, already discussed in the preceding chapter).

The present chapter concerns itself with the class Chondrichthyes (cartilaginous fishes), with particular attention to the infraclasses Elasmobranchii = Selachii sensu lato (sharks, skates, rays, and sawfishes); and Holocephali = Bradyodonti sensu lato (chimaeroids, etc.). The subclass Elasmobranchii includes ancient or primitive elasmobranchs and the neoselachians or modern sharks and their relatives, the skates, rays, and sawfishes. This subclass is almost certainly monophyletic.

According to Zangerl (personal communication), who agrees with a few other scientists, the geologically younger groups of elasmobranchs evolved out of the Paleozoic from the ctenacantoids, and the line leading directly to the neoselachians probably did not split off from the hybodonts (as many had previously believed), but came directly from the ctenacanths (as did the hybodonts as well).

For purposes of simplicity, the author will list the primitive sharks and their related species separately in the beginning of this chapter, and then discuss the neoselachians or modern (true) sharks. Primitive sharks and their related cartilaginous fish forms first appeared in the Late (Upper) Devonian with the earliest cladodontid form, *Cladoselache* (see Figs. 24-1 through 24-3), as well as its cousin *Ctenacanthus;* the pleuracanthid *Dittodus (Xenacanthus);* and an early hybodontid form (see Fig. 24-4, upper left corner specimen). The origins or ancestry of the primitive sharks is still being debated. Perhaps, as many believe, they evolved out of the Placodermi.

By the Mississippian, all or most of the primitive sharks and related cartilaginous fishes had evolved, including the Bradyodonti, i.e., *Deltodus, Sandalodus,* and *Cochliodus,* among others.

The classification of primitive sharks and related groups (some listings may be inaccurate due to the shifting or renaming of genera, species and even orders): Class Chondrichthyes: Subclass Elasmobranchii: Order Clado-

selachida (Cladoselachii-Pleuropterygii of Romer, 1966); Family **Cladoselachidae**: genus *Cladoselache* of the Late Devonian (see Figs. 24-1 through 24-3). Order ?Cladodontida: Family **?Cladodontidae**: genera *Cladodus* and *?Monocladodus* of the Late Devonian (*Cladodus* is now considered a nomen dubium, and as a genus, it is not recognized in the Carboniferous). The following order Symmoriida covers the Mississippian and Pennsylvanian forms previously called *Cladodus.* Hereafter they will be called cladodont in describing the style of tooth design (see Figs. 24-14 through 24-16 for example).

Order Symmoriida: Family **Symmoriidae**: PENNSYLVANIAN: *Cobelodus* (see Figs. 24-17 through 24-26), *Denaea, Stethacanthus,* and *Symmorium* (see Fig. 24-29). Note: *Stethacanthus* was formerly placed within Chondrichthyes Incertae Sedis (Romer, 1966, p. 351R).

Order Selachii (of Romer, 1966) = Cohort Euselachi (of Compagno, 1973).

Suborder Ctenacanthoidei: Family **Ctenacanthidae**: DEVONIAN: *Ctenacanthus;* MISSISSIPPIAN: *Ctenacanthus* (see Figs. 24-7 and 24-13), *Goodrichthys (Goodrichia),* and *Tristychius;* PENNSYLVANIAN: *Bandringa* (see Figs. 24-27 and 24-28), *Ctenacanthus* (see Fig. 24-54), *Dabasacanthus,* and *Phoebodus.*

Order Xenacanthida (Pleuracanthodii): Family **Xenacanthidae (Pleuracanthidae)**: DEVONIAN: *Diplodus (Dittodus),* and *Xenacanthus;* MISSISSIPPIAN: *Thrinacodus* and *Xenacanthus;* PENNSYLVANIAN: *Anodontacanthus, Compsacanthus, Triodus,* and *Xenacanthus (Pleuracanthus)* (see Figs. 24-51 through 24-53); PERMIAN: *Didymodus, Diacranodus, Hypospondylus, Orthacanthus, Triodus,* and *Xenacanthus (Pleuracanthus)* (see Figs. 24-61 through 24-70).

Suborder Hybodontoidea: Family **Coronodontidae**: DEVONIAN: *Coronodus* and *Diademodus (Tiarodontus);* PENNSYLVANIAN: *Coronodus* and *Diademodus.* Family **Hybodontidae**: DEVONIAN: *Eoörodus* and *Protacrodus;* MISSISSIPPIAN: *Coelosteus, Dicrenodus (Carcharopsis), Hybocladodus, Lambdodus, Mesodmodus,* and *Sphenacanthus;* PENNSYLVANIAN: *Dicre-*

24-1. Devonian. *Cladoselache fyleri* Newberry. An essentially complete body outline of a primitive shark. The body is in a dorsoventral position. The caudal fin (tail) section is not preserved on this specimen. Cleveland Shale, Brooklyn Heights, Cuyahoga County, Ohio. ¼ actual size.

24-2. Devonian. *Cladoselache fyleri* Newberry. Another example of a primitive shark (see Fig. 24-1) preserved in a ventral position. The impressions of the kidneys, liver, heart, stomach and intestines are preserved in this specimen. Cleveland Shale, Brooklyn Heights, Cuyahoga County, Ohio. Approx. ¼ actual size.

nodus, Echinodus, Hybodus, ?Petrodus, (see Fig. 24-59), *Sphenacanthus,* and *Styracodus (Centrodus);* PERMIAN: *Arctacanthus (Hamatus), Hybodus, Sphenacanthus, Wodnika,* and *Xystrodus;* TRIASSIC: *Acrodonchus, Acrodus, Asteracanthus (Strophodus), Carinacanthus (Hybodus), Doratodus, Hybodoconchus, Lissodus, Nemacanthus (Nematacanthus), Palaeobates, Polyacrodus,* and *Scoliorhiza;* JURASSIC: *Asteracanthus, Bdellodus, Hybodus* (see Figs. 24-74 through 24-77), *Nemacanthus, Priorybodus,* and *Pristacanthus;* CRETACEOUS: *Acrodus, Asteracanthus* (see Fig. 24-79), *Hybodus* (see Figs. 24-80 through 24-82), *Lonchidion, Orthacodus, Priorybodus,* and *Prohybodus.*

Order Eugeneodontida (Note: this is a new order erected by Zangerl (1981) for the edestoids): The super-

families of this order are Caseodontoidea and Edestoidea.

In the superfamily Caseodontoidea there are two families: **Caseodontidae** and **Eugeneodontidae.** Typical Carboniferous and Permian genera for the family: **Caseodontidae:** PENNSYLVANIAN: *Caseodus* (named after the author, see Figs. 24-33 and 24-34), *Fadenia* (see Fig. 24-35), *Romerodus* (see Figs. 24-40 through 24-43), and *Ornithoprion;* PERMIAN: *Erikodus.* In the family **Eugeneodontidae,** the following genera are listed for the Carboniferous Period: *Eugeneodus, Gilliodus,* and *Bobbodus.*

In the superfamily Edestoidea, there are two families: **Agassizodontidae** and **Edestidae.** Generic representatives of the family **Agassizodontidae** for the Carboniferous to Permian Periods: PENNSYLVANIAN: *Agassizodus*

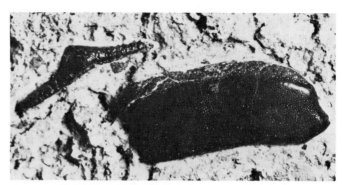

24-4. Mississippian. A primitive hybodontid (top left), and *Psammodus* sp. (lower right). Two teeth from separate orders. The hybodontid is a primitive shark, while *Psammodus* represents a member of the Bradyodonti. Burlington Limestone, Augusta, Des Moines County, Iowa. ×2.

24-3. Devonian. *Cladoselache fyleri* Newberry. An isolated lateral tooth from this early cladoselachid shark. Cleveland Shale, Brooklyn Heights, Cuyahoga County, Ohio. ×3.

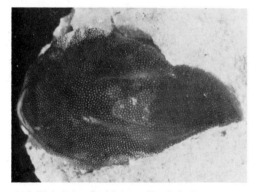

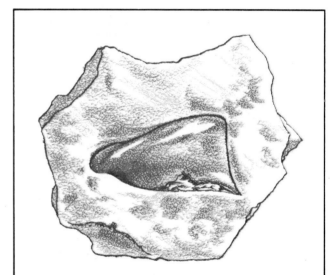

24-5. Mississippian. *Sandalodus occidentalis* Newberry and Worthen. An isolated tooth plate of a bradyodontid. Keokuk Limestone, South English, Keokuk County, Iowa. ×2.

24-6. Mississippian. *Deltodus* sp. Another example of a typical isolated dental pavement tooth (crusher plate) of a bradyodontid fish. Burlington Limestone, Oakland Mills, Burlington County, Iowa. ×1.

24-7. Mississippian. *Ctenacanthus gracillimus* Newberry and Worthen. An almost complete dorsal fin spine of an early ctenacanthid shark. Haney Member, Golconda Formation, Chester Series, Anna, Union County, Illinois. Approx. ⅔ actual size.

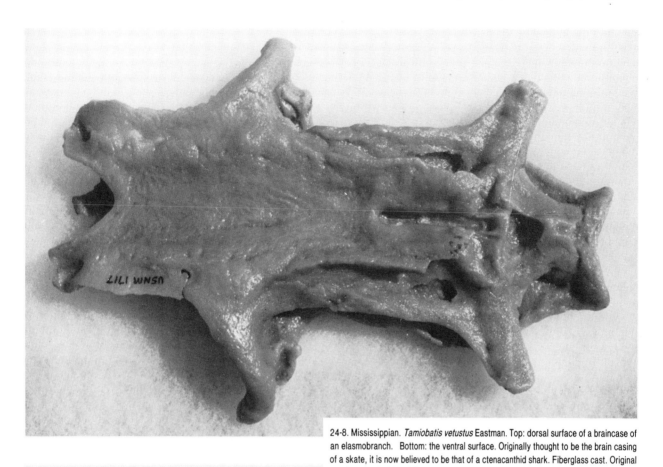

24-8. Mississippian. *Tamiobatis vetustus* Eastman. Top: dorsal surface of a braincase of an elasmobranch. Bottom: the ventral surface. Originally thought to be the brain casing of a skate, it is now believed to be that of a ctenacanthid shark. Fiberglass cast. Original specimen in the United States National Museum, Washington, DC. Powell County, Kentucky. ×1.

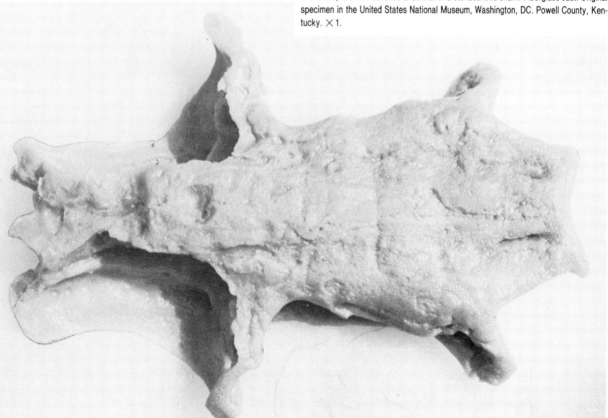

24-9. Mississippian. *Ctenoptychius* sp. A complete isolated tooth plate from a petalodont shark. Keokuk Limestone, Lee County, Iowa. ×3.5.

24-10. Mississippian. *Physonemus* sp. A fragment of a rather huge dorsal (?) fin spine of a cochliodontid cartilaginous fish. Such spines sometimes attain two feet in length. Keokuk Limestone, Kellogg, Jasper County, Iowa. ×1.

(see Figs. 24-33, 24-34, and 24-36), *Campyloprion* (see Figs. 24-37 and 24-38), and *Toxoprion;* PERMIAN: *Helicoprion* (see Figs. 24-71 and 24-72), *Parahelicoprion,* and *Sarcoprion.* The following genera comprise the family **Edestidae** Carboniferous (Mississippian/Pennsylvanian) to Permian and Triassic Periods: MISSISSIPPIAN: *Lestrodus;* PENNSYLVANIAN: *Edestus* (see Figs. 24-31 and 24-32), PERMIAN: *Helicampodus, Parahelicampodus,* and *?Syntomodus;* TRIASSIC: *Helicampodus.*

The following genera have also been attributed to the family **Edestidae** (although some of these may belong Incertae Sedis, e.g., *Physonemus* and *Listracanthus):* DEVONIAN: *Protospiraxis,* MISSISSIPPIAN: *Campodus (Lophodus),* and *Physonemus* (see Figs. 24-10 through 24-12); PENNSYLVANIAN: *Campodus, Listracanthus* (see Figs. 24-55 through 24-58), *Metaxyacanthus,* and *Protopirata (Edestodus);* TRIASSIC: *?Listracanthus* (see Fig. 24-73).

Note: Before continuing on to the classification of the "neoselachians," we must first look at the subclass Subterbranchialia (recently erected by Zangerl, 1981), which includes all chondrichthyans that are *not* sharks (elasmobranchs). This new subclass includes the Holocephali, Bradyodonti, and Chimaerina. This subclass is probably *not* monophyletic.

To begin with, in the subclass Subterbranchialia, we first have the order Iniopterygia (erected by Zangerl and Case, 1973), which includes the families **Iniopterygidae** and **Sibyrhynchidae.** These Carboniferous chondrichthyans are known only from the Late Mississippian to the Late Pennsylvanian of North America. Having (so far) been found only in the rocks of the following states: Indiana, Illinois, Iowa, Nebraska, and Montana.

The family **Iniopterygidae** includes the following genera: *Iniopteryx* (see Figs. 24-142 through 24-149) and *Promexyele* (see Figs. 24-150 through 24-152). The following genera make up the family **Sibyrhynchidae:** *Sibyrhynchus* (see Figs. 24-153 through 24-155), *Iniopera* (see

24-11. Mississippian. *Physonemus gigas* Newberry and Worthen. An essentially complete dorsal (?) fin spine (split longitudinally) in limestone matrix. Keokuk Limestone, Keswick, Keokuk County, Iowa. Approx. half size.

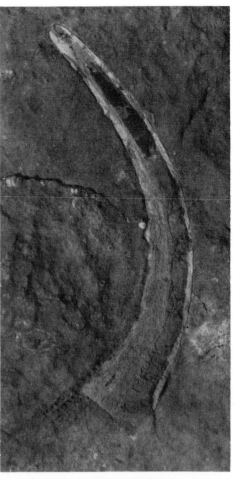

24-12. Mississippian. *Physonemus gemmatus* (Newberry and Worthen). An almost complete (missing basal portion) dorsal (?) fin spine (or cephalic hook?) of a cochliodontid cartilaginous fish. Keokuk Limestone, Keswick, Keokuk County, Iowa. ×1.

24-13. Mississippian. *Ctenacanthus* sp. The basal portion of a dorsal fin spine in association with the isolated teeth of many species of cartilaginous fishes, including: *Cladodus*, *Helodus*, *Deltodus*, and *Chomatodus*, among others. Keokuk Limestone, Keswick, Keokuk County, Iowa. Approx. half size.

24-15. Pennsylvanian. A large isolated cladodont tooth in a coprolitic mass (fecal pellet). Queen Hill Shale, Lecompton Formation, Shawnee Group, Virgil Series (Stephanian A), Ace Hill Quarry, Plattsmouth, Cass County, Nebraska. ×1.5.

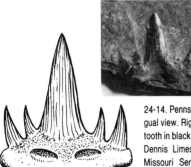

24-14. Pennsylvanian. Cladodont tooth. Left: lingual view. Right: labial view of another cladodont tooth in black sheety shale. Stark Shale Member, Dennis Limestone Formation, Bronson Group, Missouri Series, Winterset, Madison County, Iowa. ×1. (Both).

24-16. Pennsylvanian. An isolated cladodont tooth in black carbonaceous shales. Queen Hill Shale, Lecompton Formation, Shawnee Group, Virgil Series (Stephanian A), Ace Hill Quarry, Plattsmouth, Cass County, Nebraska. ×1.5.

24-17. Pennsylvanian. *Cobelodus aculeatus* (Cope). The posterior portion of the meckel's cartilage (lower jaw) of a common Carboniferous symmoriid shark. Note the articulation facet (notch) for connection with the palatoquadrate (upper jaw). No teeth are present on the upper rim of this cartilaginous jaw section. Wea Shale, Westerville Formation, Kansas City Group, Missouri Series (Westphalian D), Richfield, Sarpy County, Nebraska. ×1.5.

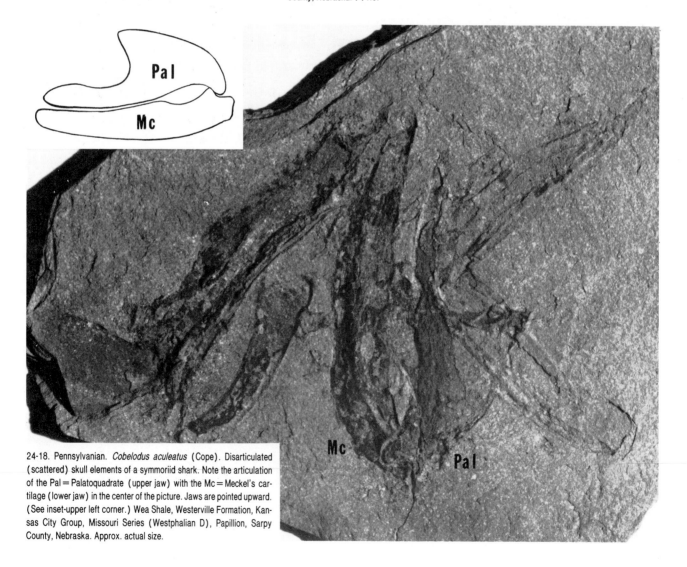

24-18. Pennsylvanian. *Cobelodus aculeatus* (Cope). Disarticulated (scattered) skull elements of a symmoriid shark. Note the articulation of the Pal = Palatoquadrate (upper jaw) with the Mc = Meckel's cartilage (lower jaw) in the center of the picture. Jaws are pointed upward. (See inset-upper left corner.) Wea Shale, Westerville Formation, Kansas City Group, Missouri Series (Westphalian D), Papillion, Sarpy County, Nebraska. Approx. actual size.

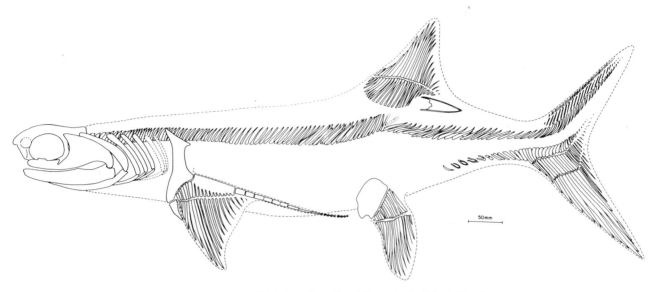

24-19. Pennsylvanian. *Cobelodus aculeatus* (Cope). A reconstruction of the skeleton of this typical anacanthous symmoriid shark. Stark Shale Member, Dennis Limestone Formation, Bronson Group, Missouri Series (Westphalian D), Papillion, Sarpy County, Nebraska. From Zangerl and Case, 1976.

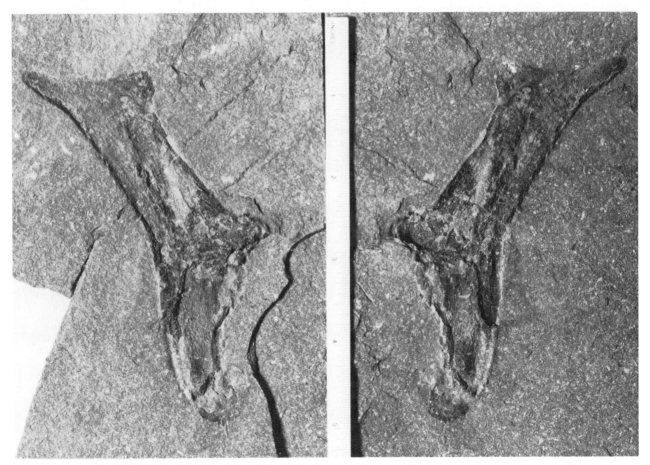

24-20. Pennsylvanian. *Cobelodus aculeatus* (Cope). The plate and counterplate of a complete scapulo-coracoid (shoulder blade) of this rather common symmoriid anacanthous shark. Black sheety shales. Wea Shale, Westerville Formation, Kansas City Group, Missouri Series (Westphalian D), Papillion, Sarpy County, Nebraska. ×1.

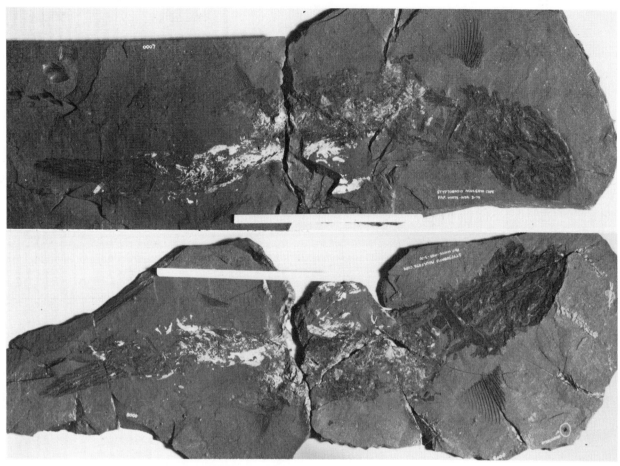

24-21. Pennsylvanian. *Cobelodus aculeatus* (Cope). The plate and counterplate of a complete shark specimen in black shale. Stark Shale Member, Dennis Limestone Formation, Bronson Group, Missouri Series (Westphalian D), Papillion, Sarpy County, Nebraska. Approx. half size.

24-22. Pennsylvanian. *Cobelodus aculeatus* (Cope). An essentially complete shark in ventral position in black sheety shale. Wea Shale, Westerville Formation, Kansas City Group, Missouri Series (Westphalian D), Papillion, Sarpy County, Nebraska. Half size.

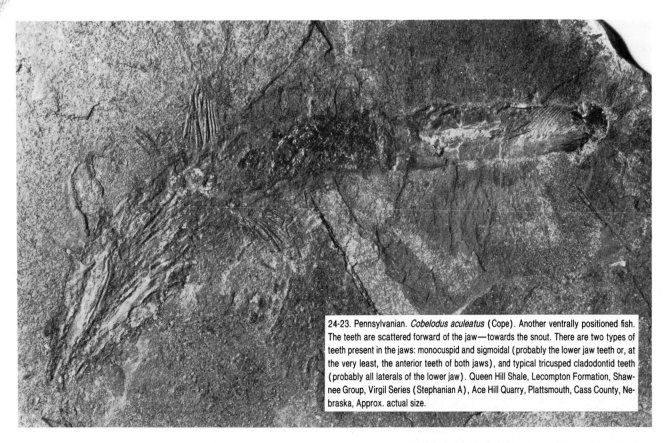

24-23. Pennsylvanian. *Cobelodus aculeatus* (Cope). Another ventrally positioned fish. The teeth are scattered forward of the jaw—towards the snout. There are two types of teeth present in the jaws: monocuspid and sigmoidal (probably the lower jaw teeth or, at the very least, the anterior teeth of both jaws), and typical tricusped cladodontid teeth (probably all laterals of the lower jaw). Queen Hill Shale, Lecompton Formation, Shawnee Group, Virgil Series (Stephanian A), Ace Hill Quarry, Plattsmouth, Cass County, Nebraska, Approx. actual size.

Figs. 24-156 through 24-158), and *Inioxyele* (see Figs. 24-159 and 24-160).

Other families comprising the subclass Subterbranchialia are as follows: **Chondrenchelyidae** with its type genus *Chondrenchelys* from the Lower Carboniferous (Mississippian) of Europe; **Polysentoridae** with its type genus *Polysentor* from the Pennsylvanian of Illinois; **Helodontidae** with its type genus *Helodus* PENNSYLVANIAN (see Fig. 24-50)—range: Devonian to Permian in North America, and the Carboniferous (Mississippian/Pennsylvanian) of Europe; **Psephodontidae** with its type genus *Psephodus*—range: Mississippian to Early Permian of North America, and the Carboniferous (Mississippian/Pennsylvanian) of Europe; **Copodontidae** with its type genus *Copodus* of the Early (Lower) Carboniferous of Europe and North America; **Deltodontidae** with its type genus *Deltodus* (see Fig. 24-6) of the Early (Lower) Carboniferous of Europe and North America; **Menaspidae** with its type genus *Menaspis* from the Late Permian of Europe; **Cochliodontidae** with its type genus *Cochliodus* from the Mississippian of North America and the Carboniferous of Europe; and finally, **Chimaeridae** ("ghostsharks" or "ratfishes"), which includes the following generic types: JURASSIC: *Brachymylus, Ganodus* and *Pachymylus;* CRETACEOUS: *Chimaera, Edaphodon,* and *Ischyodus* (see Fig. 24-113); EOCENE: *Psaliodus.* There

24-24. Pennsylvanian. *Cobelodus aculeatus* (Cope). An isolated tail fin (ventral portion) in black shale. Stark Shale Member, Dennis Limestone Formation, Bronson Group, Missouri Series (Westphalian D), La Platte, Sarpy County, Nebraska. ×1.

24-25. Pennsylvanian. *Cobelodus aculeatus* (Cope). The impression of an almost complete braincase of this symmoriid shark. The anterior (rostral) portion is missing. Stark Shale Member, Dennis Limestone Formation, Bronson Group, Missouri Series (Westphalian D), Papillion, Sarpy County, Nebraska. Slightly larger than actual size.

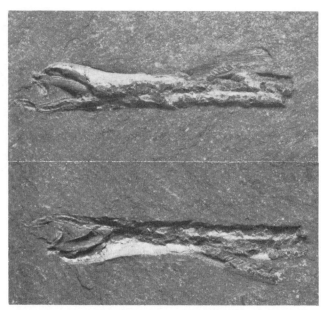

24-26. Pennsylvanian. *Cobelodus aculeatus* (Cope). The plate and counterplate of the pterygopodium or male clasper organ of this symmoriid shark. Stark Shale Member, Dennis Limestone Formation, Bronson Group, Missouri Series (Westphalian D), Papillion, Sarpy County, Nebraska. ×1.

are two other families related to the **Chimaeridae,** and these should be mentioned in passing, along with generic types: **Rhinochimaeridae:** JURASSIC: *Elasmodectes;* CRETACEOUS: *Elasmodus;* OLIGOCENE: *Amylodon.* **Callorhynchidae:** CRETACEOUS: *Callorhynchus.*

Note: The genus *Chondrenchelys* (Missippian/Pennsylvanian) was formerly placed in Elasmobranchii Incertae Sedis (Romer, 1966, p. 351L). The family and genus **Polysentoridae** *Polysentor* is newly erected by Zangerl (1979, pp. 458–459).

Incertae Sedis of Romer (1966): Order Bradyodonti. Family: **Petalodontidae:** MISSISSIPPIAN/PENNSYL-VANIAN: *Ctenoptychius* (see Fig. 24-9), *Fissodus, Janassa* (see Figs. 24-45 through 24-48), and *Petalodus* (see Fig. 24-44). Family: **Psammodontidae:** genera MISSIS-SIPPIAN/PENNSYLVANIAN: *Psammodus* (see Fig. 24-4), and *Stamiobatis* (see Fig. 24-49).

The neoselachians (modern sharks) Jurassic to recent. Order Selachii: Family **Ptychodontidae:** CRETA-CEOUS: *Hemiptychodus, Heteroptychodus, Hylaeobatis,* and *Ptychodus* (see Figs. 24-89 through 24-91). These are primitive hybodontids, which may be a bridge to the ba-toids, particularly the sting rays. Family **Paleospinacei-dae:** JURASSIC: *Paleospinax.* Family **Heterodontidae (Cestraciontidae):** TRIASSIC: *Heterodontus (Cestracion)* (see Fig. 24-78); CRETACEOUS: *Heterodontus (Cestra-cion)* and *Synechodus* (see Figs. 24-85 and 24-86); EOCENE: *Heterodontus* and *Synechodus;* MIOCENE: *Heterodontus* and *Strongyliscus.*

Suborder Hexanchoidea (Notidanoidea): Family **Hex-anchidae (Notidanidae):** CRETACEOUS: *Notorhynchus (Notidanus);* PALEOCENE: *Heptranchias (Notidanion),* and *Hexanchus;* EOCENE: *Heptranchias (Notidanion), Hexanchus,* and *Xiphodolamia;* MIOCENE: *Heptran-chias* (see Fig. 24-117), *Hexanchus* (see Fig. 24-118), and *Notorhynchus (Notidanus)* (see Fig. 24-116). Family **Chlamydoselachidae:** EOCENE: *Chlamydoselachus;* MIOCENE: *Chlamydoselachus;* PLIOCENE: *Chlamy-doselachus.*

Superorder Galeomorphii: Suborder Galeoidea (of Romer, 1966): Order Lamniformes: Family **Carchariidae** (of Romer), **Odontaspidae** (of Compagno, 1973): CRE-TACEOUS: *Hypotodus* (see Fig. 24-108), and *Odontas-pis;* PALEOCENE: *Hypotodus* and *Odontaspis;* EOCENE: *Odontaspis,* MIOCENE: *Odontaspis* (see Figs. 24-126 and 24-127); Pliocene: *Odontaspis.* Family **Pseudocarchariidae** (Recent, of Compagno, 1973); No fossils. Family **Mitsikurinidae:** CRETACEOUS: *Anom-otodon* and *Scapanorhynchus* (see Fig. 24-102); MIO-CENE: *Anomotodon.* Family **Alopiidae:** CRETA-CEOUS: *Paranomotodon;* EOCENE; *Alopias* and *Paranomotodon;* MIOCENE: *Alopecias* and *Alopias* (see

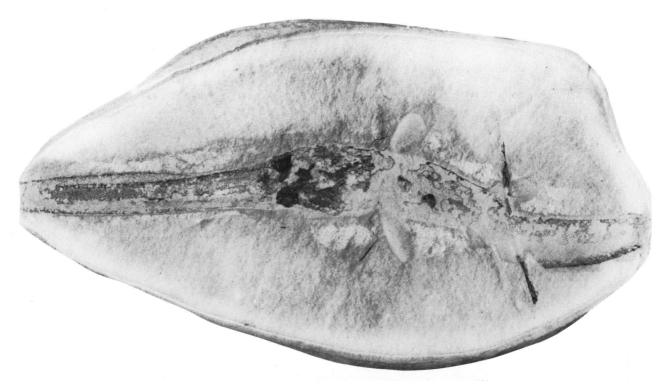

24-27. Pennsylvanian. *Bandringa rayi* Zangerl. An essentially complete ctenacanthid shark (missing a small portion of the tail and the rostrum). Fish is in a siderite concretion. Francis Creek Shale (Essex Concretion Fauna), Carbondale Formation (Westphalian D), Mazon Creek area, Grundy County, Illinois. ×2. From Zangerl, 1969. Courtesy: Field Museum of Natural History, Chicago.

24-28. Pennsylvanian. *Bandringa rayi* Zangerl. The anterior portion of the head and the entire rostrum of the fish. The specimen "runs" horizontally. The vertical mass is plant material, a woody stem of *Artisia*. Upper Kittanning Coal, Cannelton, Darlington Township, Beaver County, Pennsylvania. Approx. 16 cm. in length. From Baird, 1978. Courtesy: Princeton University Museum of Natural History, Princeton, New Jersey. (PU 19814).

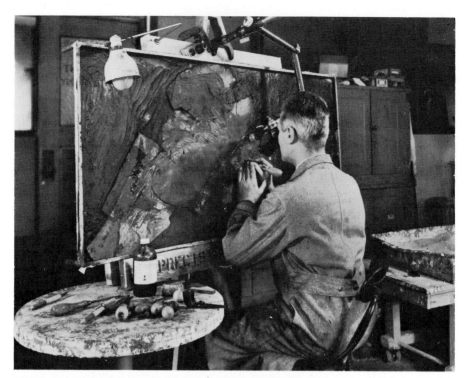

24-29. Dr. Rainer Zangerl, Curator emeritus, Department of Geology at the Field Museum of Natural History in Chicago, preparing a large *Symmorium reniforme* shark in black sheety shale. The shark is from the Mecca Quarry Shales of Western Indiana, Courtesy: Field Museum of Natural History, Chicago.

24-30. Pennsylvanian. *Orodus greggi* Zangerl. The anterior half of a primitive shark in black sheety shale. Note the paired pectoral fins, the scapulo-coracoids (shoulder blades) and the neural arches running through the middle of the shark's body. Logan Quarry Shale Member, Liverpool Cyclothem, Staunton Formation, Des Moines Series (Westphalian Upper C), Reserve Township, Park County, Indiana. From Zangerl and Richardson, 1963. Courtesy: Field Museum of Natural History, Chicago.

24-31. Pennsylvanian. *Edestus* cf. *heinrichi*. A single tooth from the symphysial whorl of a large edestid shark. Photograph is of a red rubber latex peel made from the impression of the tooth and jaw section in black tar-paper shale. Labette Black Shale, Labette Formation, Marmaton Group, Des Moines Series (Westphalian D), Madrid, Boone County, Iowa. ✕1.

24-32. Pennsylvanian. *Edestus* sp. A broken tooth from a symphysial whorl. Note the enlarged serrations on the tooth crown. Hertha Limestone, Des Moines Series, Stuart, Guthrie County, Iowa. ✕3.

24-33. Pennsylvanian. *Agassizodus variabilis* (Newberry and Worthen) (or possibly *Caseodus* sp.). An incomplete "battery" tooth (missing root) of an edestid shark. Severy Shale, Waubonsie Formation (just below the Nodaway Coal), Central Coal Mine, Clarinda, Page County, Iowa. ✕2.

24-34. Pennsylvanian. *Agassizodus variabilis* (Newberry and Worthen) (or possibly *Caseodus* cf. *eatoni* Zangerl). Two isolated battery teeth. The laterals of these two genera are very difficult to distinguish. Hertha Limestone Formation, Des Moines Series, Stuart, Guthrie County, Iowa. ✕2.5.

24-35. Pennsylvanian. *Fadenia gigas* Eaton. One complete, and one broken tooth crown on a whorl (symphysial section) of a rare species of edestid shark. Type specimen. Cherokee Shale (Lexington Coal), Lucas, Henry County, Missouri. Courtesy: University of Kansas Museum of Natural History, Lawrence, Kansas. ✕1.

24-36. Pennsylvanian. *Agassizodus variabilis* (Newberry and Worthen). A partially artic-
ulated battery of laterals (crusher teeth) in black shale matrix. Wea Shale, Westerville
Formation, Kansas City Group, Missouri Series (Westphalian D), Papillion, Sarpy
County, Nebraska.

24-37. Pennsylvanian. ? *Caseodus* sp. A
section of the jaw containing the symphysial tooth whorls and part of the battery (laterals) of crusher type teeth,
family Caseodontidae. (See Fig. 24-38
for additional views of the symphysial
whorl type teeth.) Wea Shale, Westerville Formation, Kansas City Group,
Missouri Series (Westphalian D), Papillion, Sarpy County, Nebraska. Slightly
smaller that actual size.

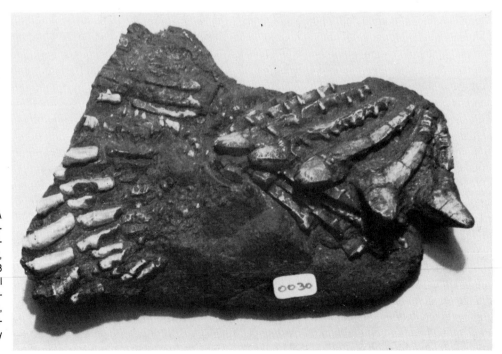

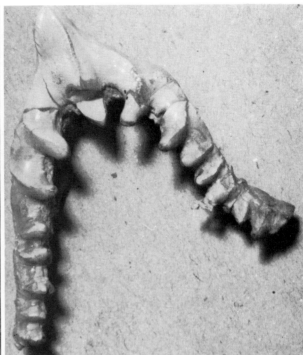

24-38. Pennsylvanian. *Campyloprion* (or *?Parahelicoprion*). A. Lateral view of a symphysial tooth whorl composed of seven teeth (in view of an agassizodontid shark. B. Isolated symphysial tooth from the whorl of a caseodontid shark. (See Fig. 24-37 for comparison.) Stark Shale Member, Dennis Limestone Formation, Bronson Group, Missouri Series (Westphalian D), Crescent, Pottawattamie County, Iowa. ×1.5.

24-39. Pennsylvanian. A complete pectoral fin of an eugeneodontid shark. Species indeterminate. Stark Shale Member, Dennis Limestone Formation, Bronson Group, Missouri Series (Westphalian D), Papillion, Sarpy County, Nebraska. ×1.5.

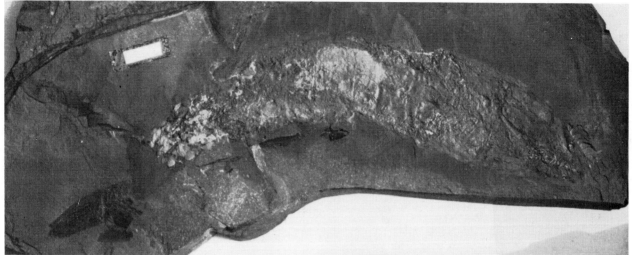

24-40. Pennsylvanian. *Romerodus* cf. *orodontus* Zangerl. Dorsoventral view of a complete shark in black sheety shale. Stark Shale Member, Dennis Limestone Formation, Bronson Group, Missouri Series (Westphalian D), Papillion, Sarpy County, Nebraska. ⅔ actual size.

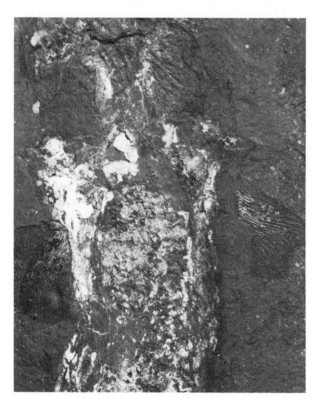

24-41. Pennsylvanian. *Romerodus* cf. *orodontus* Zangerl. A partial specimen. The tip of the head and most of the body are missing. The specimen is positioned ventrally. Wea Shale Member, Westerville Formation, Kansas City Group, Missouri Series (Westphalian D), Papillion, Sarpy County, Nebraska. ×1.

24-42. Pennsylvanian. *Romerodus* cf. *orodontus* Zangerl. Ventral view of the anterior (front) portion of a shark. Note the splendidly preserved meckel's cartilages (lower jaws) in their proper position in the head area. Stark Shale Member, Dennis Limestone Formation, Bronson Group, Missouri Series (Westphalian D), Papillion, Sarpy County, Nebraska. ×1.5.

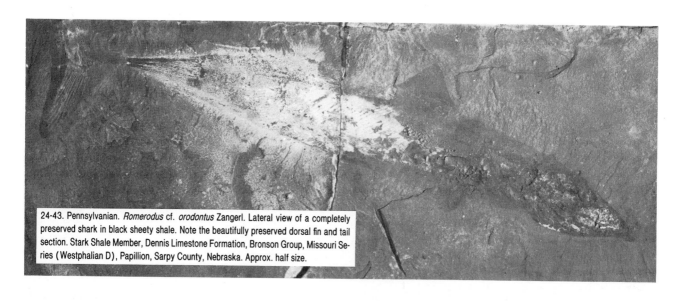

24-43. Pennsylvanian. *Romerodus* cf. *orodontus* Zangerl. Lateral view of a completely preserved shark in black sheety shale. Note the beautifully preserved dorsal fin and tail section. Stark Shale Member, Dennis Limestone Formation, Bronson Group, Missouri Series (Westphalian D), Papillion, Sarpy County, Nebraska. Approx. half size.

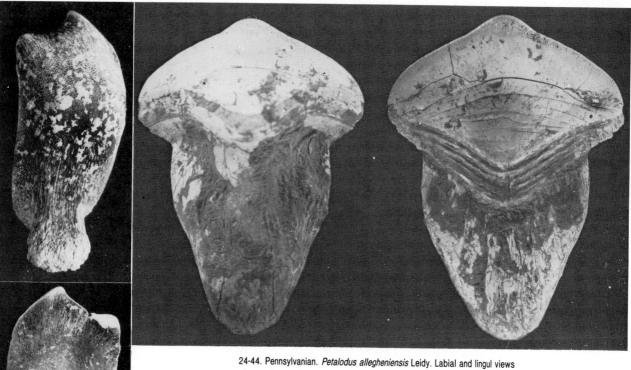

24-44. Pennsylvanian. *Petalodus allegheniensis* Leidy. Labial and lingul views of a crusher tooth plate of a possible skatelike petalodontid. South Bend Formation, Graham, Young County, Texas. ✕2.

24-45. Pennsylvanian. *Janassa bituminosa* Schlotheim. Labial and lingual views of a crusher tooth plate of a skatelike petalodontid fish. Queen Hill Quarry, Limestone Beds, Virgil Series, Rock Bluff, Cass County, Nebraska. ✕2.

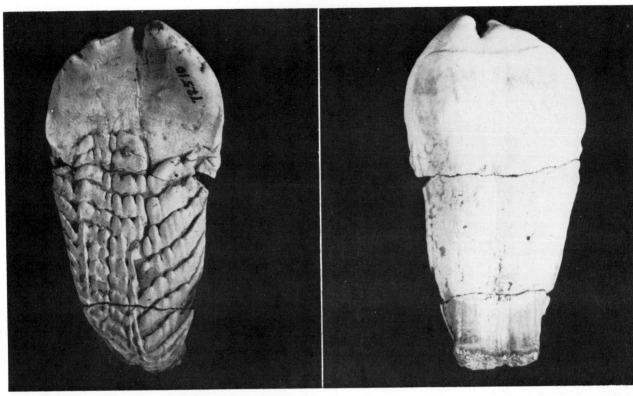

24-46. Pennsylvanian. *Janassa bituminosa* Schlotheim. Lingual and labial views of a crusher tooth plate of a skatelike fish. (See Fig. 24-47 for the tooth plate arrangement in the jaws of the fish.) Limestone Beds, Keiwitz Shale Member, Stanton Formation, Missouri Series (Westphalian D), Ash Grove Quarry, Louisville, Cass County, Nebraska. ×2.5.

24-47. Pennsylvanian. *Janassa bituminosa* Schlotheim. Reconstruction of pavement crusher teeth in jaws. (After Jaekel, 1899, from Berman, 1967, and Case, 1973.)

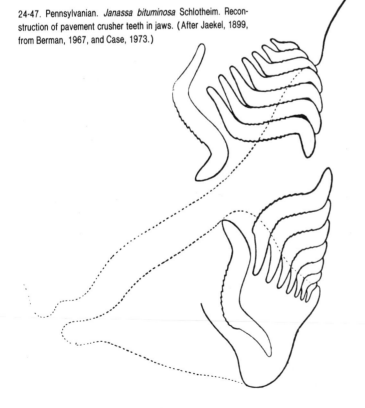

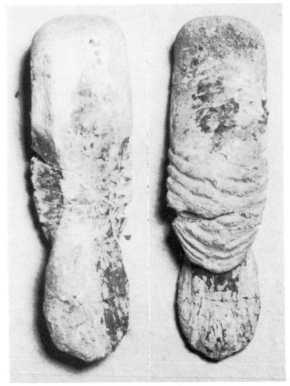

24-48. Pennsylvanian. *Janassa bituminosa* Schlotheim. Labial and lingual aspects of a pavement crusher tooth of a skatelike fish. Keiwitz Shale Member, Stanton Formation, Missouri Series (Westphalian D), Louisville, Cass County, Nebraska. ×2.5.

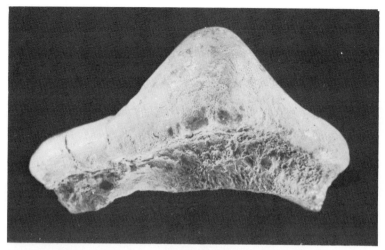

24-50. Pennsylvanian. *Helodus* sp. A pavement crusher tooth ascribed to a bradyodontid cartilaginous fish. Doniphan Shale Member, Lecompton Formation, Virgil Series (Stephanian A), Snyderville (Nehawka), Cass County, Nebraska. ×1.5.

24-49. Pennsylvanian. *Stamiobatis* sp. A large orodontid crusher tooth plate. The teeth of this genus are clearly more similar to the helodontids, hybodontids, and orodontids than to the chimaeriformes. South Bend Formation, Graham, Young County, Texas. ×1.5.

24-51. Pennsylvanian. *Xenacanthus compressus* (Newberry). Red latex peel cast of the impression of a freshwater shark tooth in tar-paper shale. Cannel below Upper Freeport Coal, Allegheny Series, Linton, Jefferson County, Ohio. ×7.

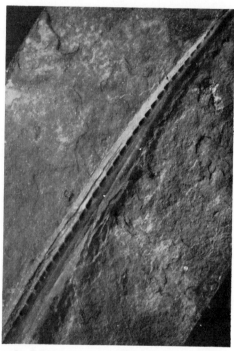

24-52. Pennsylvanian. *Xenacanthus compressus* (Newberry). Impression in tar-paper shale of a cephalic (head) spine from a freshwater shark. Cannel below Upper Freeport Coal, Allegheny Series, Linton, Ohio. ×2.

24-53. Pennsylvanian. *Xenacanthus compressus* (Newberry). A complete laterally compressed skull with postcranial elements, including a large cephalic (head) spine directed posteriorly from the back of the top of the skull roof. Cannel below Upper Freeport Coal, Allegheny Series, Linton, Jefferson County, Ohio. ×2.5.

24-54. Pennsylvanian. *Ctenacanthus* sp. A large dorsal fin spine from a ctena-canthid shark. The larger specimen is imbedded laterally in a block of massive clay-shale, and is missing the very tip and a portion of the root base of the spine. The inset, upper left corner shows another example of *Ctenacanthus* in a lateral view to show the ribs or linear pattern or grooves of the spines' outer ornamentation. Keiwitz Shale Member, Stanton Formation, Missouri Series (Westphalian D), Louisville, Cass County, Nebraska. ✕2.

24-55. Pennsylvanian. *Listracanthus hystrix* Newberry and Worthen. An isolated dermal denticle of an unknown sharklike or skate-like fish (probably the latter). The "feathery" denticle is in black fissile shale. Labette Formation, Marmaton Group, Des Moines Series (Westphalian D), Madrid, Boone County, Iowa. ×1.5.

24-56. Pennsylvanian. *Listracanthus hystrix* Newberry and Worthen. The plate and counterplate of a small group of six dermal denticles of an otherwise unknown fish. These types of denticles are at times found in association with dermal denticles similar to those of *Petrodus patelliformis* McCoy (see Fig. 24-59), and there may be some relationship. Stark Shale Member, Dennis Limestone Formation, Missouri Series (Westphalian D), Winterset, Madison County, Iowa. ×2.

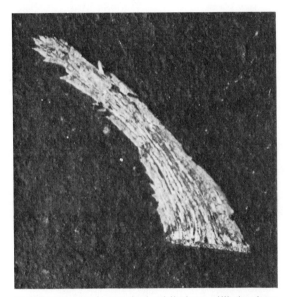

24-57. Pennsylvanian. *Listracanthus hystrix* Newberry and Worthen. A very fine single specimen of a dermal denticle of an otherwise unknown fish. Excello Shale, Carbondale Formation, Des Moines Series (Westphalian lower D), Pit 14, Peabody Coal Company Mines, Kankakee County, Illinois ×2.5.

24-58. Pennsylvanian. *Listracanthus hystrix* Newberry and Worthen. An essentially three-dimensional dermal denticle imbedded in a limestone matrix. *Listracanthus* denticles with third dimension are quite rare! Note the productid brachiopod at top center of the picture, just above the spine. Keiwitz Shale Member, Stanton Limestone Formation, Missouri Series (Westphalian D), Louisville, Cass County, Nebraska. ×5.

Fig. 24-122). Family **Cetorhinidae:** OLIGOCENE: *Cetorhinus;* EOCENE: *Cetorhinus;* MIOCENE: *Cetorhinus;* PLEISTOCENE: *Cetorhinus.* Family **Anacoracidae:** CRETACEOUS: *Anacorax, Microcorax, Pseudocorax,* and *Squalicorax* (see Figs. 24-103 through 24-105). Family **Orthacodontidae:** JURASSIC: *Orthacodus (Sphenodus);* CRETACEOUS: *Orthacodus,* EOCENE: *Orthacodus.*

Suborder Isuroidei (Family **Lamnidae** of Compagno, 1973): Family **Isuridae:** CRETACEOUS: *Cretolamna*

(see Fig. 24-109), *Cretoxyrhina,* and *Plicatolamna* (see Fig. 24-106); PALEOCENE: *Cretolamna* and *Paleocarcharodon;* EOCENE: *Paraisurus (Isurus/Oxyrhina), Lamna (Otodus),* and *Procarcharodon (Carcharodon);* MIOCENE: *Isurus (Oxyrhina)* (see Figs. 24-119 through 24-121), *Lamna,* and *Procarcharodon (Carcharodon)* (see Figs. 24-133 and 24-134); PLIOCENE: *Isurus, Lamna,* and *Procarcharodon (Carcharodon).*

Order Orectolobiformes: Family **Parascylliidae:** (Recent, of Compagno, 1973): No fossils. Family **Brachaelur-**

24-59. Pennsylvanian. *Petrodus patelliformis* McCoy. Dermal denticles of an otherwise unknown fish specimen. Found at times in association with the feathery denticles of *Listracanthus* (see Figs. 24-55 through 24-58). Hertha Limestone (*Mesolobus mesolobus* zone/clay Member), Kansas City Group, St. Charles, Madison County, Iowa. Denticles approx. × 1.

24-60. Pennsylvanian. Dermal denticle? Species indeterminate. Found in association with the teeth of *Petalodus allegheniensis* and *Janassa bituminosa*. Possibly from the skin of one of these species. Brush Creek Member, Conemaugh Formation, Albany, Athens County, Ohio. × 20.

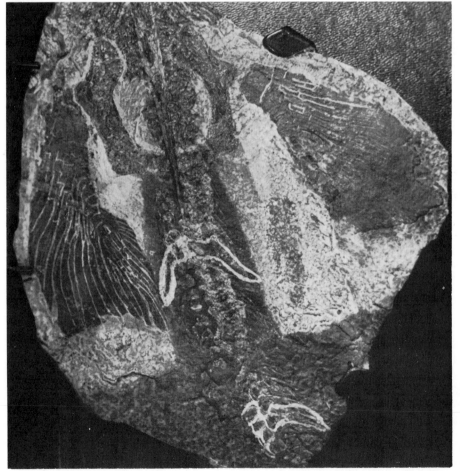

24-61. Permian. *Xenacanthus decheni* (Goldfuss). An isolated cephalic (head) spine from a large freshwater shark. Ruppersdorf bei Braunau, Bohemia, Czechoslovakia. 9 cm in length.

24-62. Permian. *Xenacanthus decheni* (Goldfuss). Dorsal view of the postcranial elements of a large freshwater shark. The elements which are preserved on this specimen are: the paired pectoral fins with ceratotrichia, the cephalic spine (top center), the scapulo-coracoids, the pectoral girdles, and spinal column. Ruprechtice near Broumov, Bohemia, Czechoslovakia. × 1.

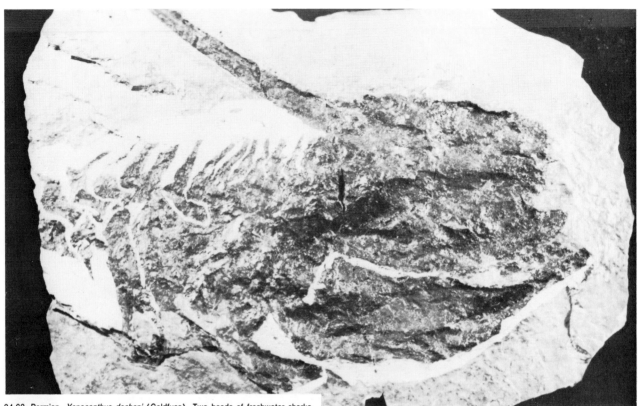

24-63. Permian. *Xenacanthus decheni* (Goldfuss). Two heads of freshwater sharks. Note the cephalic (head) spines in place back of the cranium. There are also present, the first vertebra, and a portion of the shoulder girdle and the teeth in situ (on the bottom specimen). Olivetin near Broumov, Bohemia, Czechoslovakia. ×2.5.

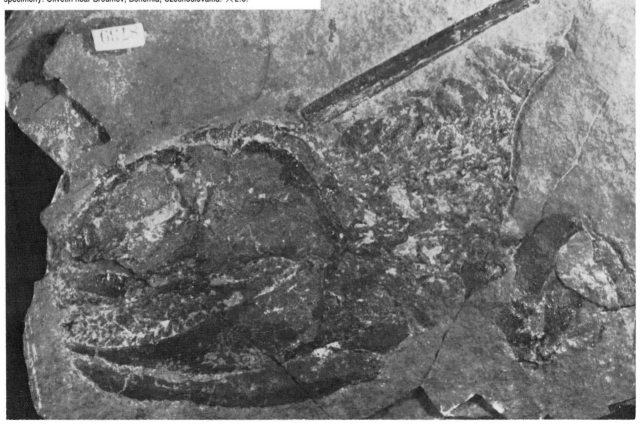

24-64. Permian. *Xenacanthus decheni* (Goldfuss). A. Lateral view of the skull, the cephalic spine, vertebral column, dorsal fin, and the pectoral girdle. B. Close-up of the Meckel's cartilage to illustrate the calcifications (prisms of cartilage), with some scattered teeth in the oral cavity (mouth area). Inner-Sudeten Depression, Lower Permian "Redbeds," Ruprechtice near Broumova, Bohemia, Czechoslovakia.

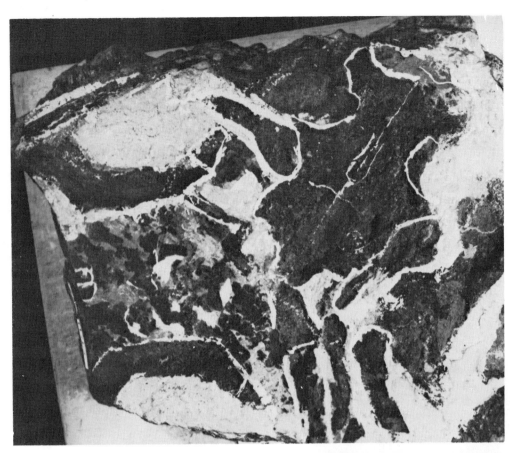

24-65. Permian. *Xenacanthus decheni* (Goldfuss). Vertical view of the braincase, jaws, visceral arches, and parts of the pectoral girdles. Inner-Sudeten Depression, Lower Permian Redbeds, Olivetin near Broumov (Oelberg bei Braunau), Bohemia, Czechoslovakia. ×1.5.

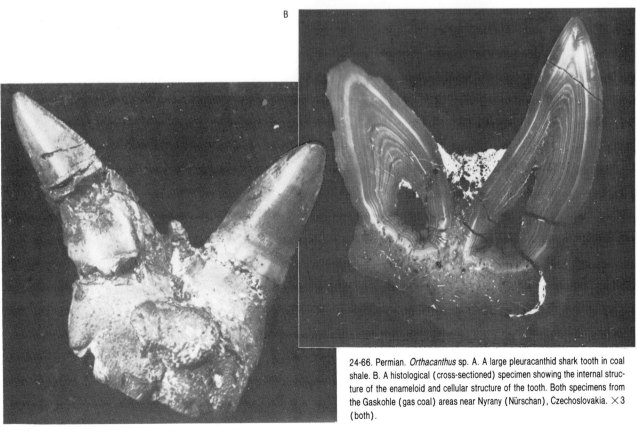

24-66. Permian. *Orthacanthus* sp. A. A large pleuracanthid shark tooth in coal shale. B. A histological (cross-sectioned) specimen showing the internal structure of the enameloid and cellular structure of the tooth. Both specimens from the Gaskohle (gas coal) areas near Nyrany (Nürschan), Czechoslovakia. ×3 (both).

24-67. Permian. *Xenacanthus* sp. A complete scapulo-coracoid element (pectoral girdle), with a portion (left) of the metapterygium (part of the pectoral fin—a support bar). Gaskohle (gas coal) shales, Nyrany (Nürschan), Czechoslovakia. ×2.

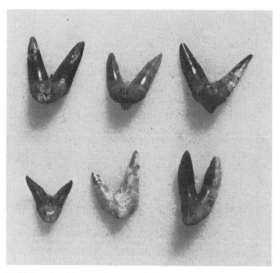

24-68. Permian. *Pleuracanthus texensis* (Cope). Isolated teeth of a pleuracanthid (freshwater) shark. All specimens are shown in their lingual views. Lower Admiral Formation, Wichita Group, Waurika, Jefferson County, Oklahoma. Approx. ×1.

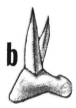

24-69. Permian. *Pleuracanthus texensis* (Cope). Three views of a typical pleuracanthid tooth: a. labial view, b. profile or lateral view, and c. lingual view. Lower Admiral Formation, Wichita Group, Rattlesnake Canyon, West of Lake Kickapoo, Archer County, Texas. ×1.5.

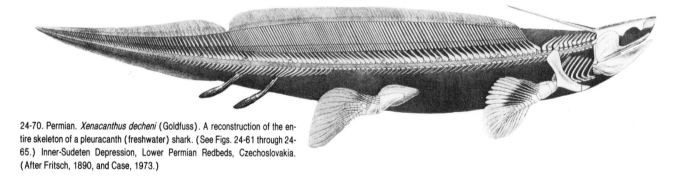

24-70. Permian. *Xenacanthus decheni* (Goldfuss). A reconstruction of the entire skeleton of a pleuracanth (freshwater) shark. (See Figs. 24-61 through 24-65.) Inner-Sudeten Depression, Lower Permian Redbeds, Czechoslovakia. (After Fritsch, 1890, and Case, 1973.)

idae: CRETACEOUS: *Brachaelurus.* Family **Orectolobidae:** CRETACEOUS: *Cretorectolobus* (see Figs. 24-84 and 24-87), *Eucrossorhinus, Orectoloboides (Archaeotriakis)* (see Fig. 24-83), and *Squatirhina.* Family **Hemiscyllidae:** CRETACEOUS: *Cantioscyllium* and *Chiloscyllium* RECENT: *Hemiscyllium.* Family **Ginglymostomatidae:** CRETACEOUS: *Ginglymostoma;* PALEOCENE: *Ginglymostoma;* EOCENE: *Ginglymostoma;* MIOCENE: *Ginglymostoma;* RECENT: *Ginglymostoma*

and *Nebrius.* Family **Rhinodontidae:** MIOCENE: *Rhiniodon.*

Order Carcharhiniformes (Family **Carcharhinidae** of Romer, 1966): Family **Scyliorhinidae:** CRETACEOUS: *Scyliorhinus;* EOCENE: *Scyliorhinus;* MIOCENE: *Scyliorhinus;* RECENT: *Scyliorhinus.* Family **Rhincodontidae (Rincodontidae** of Romer, 1966): CRETACEOUS/ RECENT: *Rhincodon (Rhineodon).* Family **Triakidae:** Tribe: Galeorhini (of Compagno, 1973): CRETACEOUS:

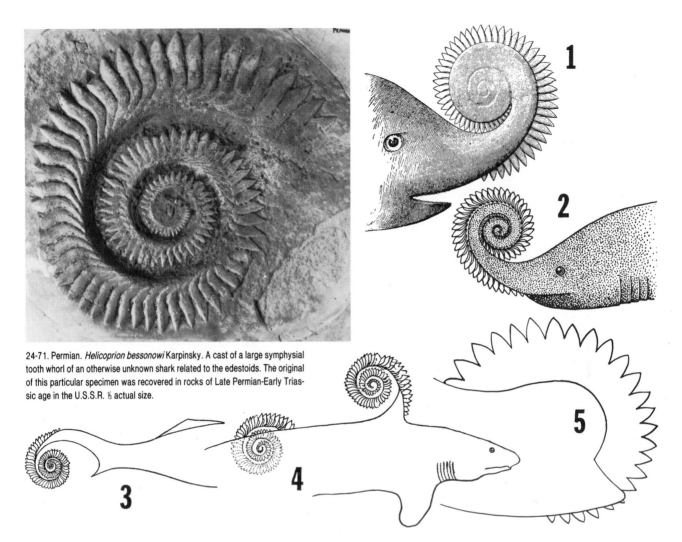

24-71. Permian. *Helicoprion bessonowi* Karpinsky. A cast of a large symphysial tooth whorl of an otherwise unknown shark related to the edestoids. The original of this particular specimen was recovered in rocks of Late Permian-Early Triassic age in the U.S.S.R. ½ actual size.

24-72. Permian. *Helicoprion bessonowi* Karpinsky. Some of the theoretical interpretations of how the giant tooth whorls fit into or "onto" the edestoid shark. 1. From inside the mouth and going around the outside of a bent-back rostrum (an interpretation in reverse, where dermal denticles become teeth when they continue into the oral cavity). 2. Another version of #1. 3. This one is a bit far-fetched: the whorl extends around towards the distal end of the caudal (tail) fin. 4. The "tooth" whorls are proposed as dorsal fin spines that curve backwards. 5. A more scientific interpretation showing the Meckel's cartilage (lower jaw) containing the entire tooth whorl. This last interpretation makes more sense than the previous four, which were at best struggling for credibility. Even though the illustrations are incongrous, they do point out a basic fact, that no one, as yet, knows the structure of the jaws that contain these peculiar tooth whorls in the edestoid, or why the teeth (if that is what they are, and every indication seems to prove that they are indeed teeth) do not shed, and just keep on growing (back onto themselves, as it were). Courtesy: Soviet Academy of Sciences, U.S.S.R. After Obruchev, 1953, Williams-Elbaum, 1973, and (illustration #5) modified from Bendix-Almgreen, 1966.

Galeorhinus; EOCENE: *Galeorhinus* and *Paragaleus* (Eugaleus);* MIOCENE: *Galeorhinus.* *Now in the family **Hemigaleidae.** Subfamily Hemipristinae: MIOCENE: *Hemipristis (Dirrhizodon)* (see Figs. 24-123 through 24-125); RECENT: *Hemipristis (Dirrhizodon)* and *Prionace (Glyphis).* Subfamily Galeocerdinae: Family **Carcharhinidae:** EOCENE: *Galeocerdo;* MIOCENE: *Galeocerdo* (see Fig. 24-128).

Subfamily Scoliodontinae: EOCENE: *Scoliodon (Loxodon**);* MIOCENE: *Scoliodon (Rhizoprionodon**).* Subfamily Carcharhininae: EOCENE: *Hypoprion;* MIOCENE: *Carcharhinus* (see Fig. 24-131), *Hypoprion, Prionace* (now considered part of the subfamily Hemipristinae), and *Negaprion.* **Loxodon and *Rhizoprionodon*

are now part of this subfamily (according to Compagno, 1973, p. 29). Family **Sphyrnidae:** MIOCENE/RECENT: *Sphyrna* (see Fig. 24-130).

Order Squaliformes: Family **Echinorhinidae:** MIOCENE: *Echinorhinus;* RECENT *Echinorhinus.* Family **Squalidae:** Subfamily Squalinae: CRETACEOUS: *Centrophoroides (Centrophorus), Centrosqualus, Cretascymnus,* and *Squalus;* PALEOCENE to RECENT: *Squalus.* Subfamily Etmopterinae: RECENT: *Aculeola, Centroscyllium,* and *Etmopterus.* Subfamily Dalatiinae: TRIASSIC/JURASSIC: *Dalatias;* MIOCENE: *Isistius.*

Order Pristiophoriformes: Family **Pristiophoridae:** OLIGOCENE: *Ikamauius;* MIOCENE: *Pliotrema* (see Fig. 24-133), and *Pristiophorus* (see Fig. 24-132).

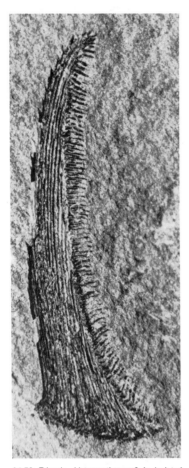

24-73. Triassic. *Listracanthus* sp.? An isolated dermal denticle of an (as yet) unknown species of elasmobranch (shark or skate). (See Figs. 24-55 through 24-58.) Vega Phroso Siltstone Member, Sulphur Mountain Formation, Lake Wapiti, British Columbia, Canada. 3 cm in length.

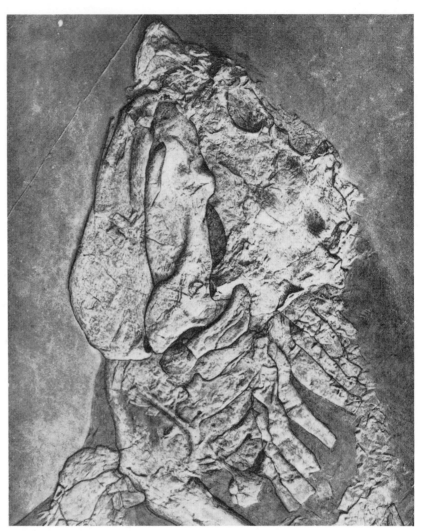

24-74. Jurassic. *Hybodus hauffianus* Koken. The skull and postcranial elements (including visceral arches, pectoral girdle, part of one pectoral fin, and some vertebrae) of a large hybodont shark. Head is directed upward towards top of page. See Fig. 24-75 for the entire skeleton of this particular specimen of shark. Liassic of Holzmaden, Germany. (After Koken, 1907, Gustav Fischer Verlag, Jena, GDR.)

24-75. Jurassic. *Hybodus hauffianus* Koken. The complete skeleton of a hybodontid shark. For detail of the head and postcranial elements elements of this specimen see Fig. 24-74. Lias, Holzmaden, Germany. (After Koken, 1907, Gustav Fischer Verlag, Jena, GDR.)

24-77. Jurassic. *Hybodus reticulatus* Agassiz. A complete dorsal fin spine. Liassic, Lyme Regis, Dorsetshire, England. Courtesy: Princeton University Museum of Natural History, Princeton, New Jersey. (PU 3202). Approx. 35 cm in length.

24-76. Jurassic. *Hybodus hauffianus* Koken. A close-up of the second (posterior) dorsal fin and the pelvic fin area of the specimen figured in 24-75. Note how deep the basal portion of the dorsal fin spine is seated into the body of the shark. It and its basal plate (to the right) come to rest upon a series of neural arch supports of the vertebral column. Lias, Holzmaden, Germany. (After Koken, 1907, Gustav Fischer Verlag, Jena, GDR.)

24-78. Jurassic. *Heterodontus (Cestracion) falcifer* (Wagner). Dorsoventrally preserved skeleton of a heterodontid shark, showing a lengthy vertebral column comprised of calcified vertebral discs (porbably the earliest form of selachian with such a preservation of its vertebral discs), the anterior (forward) and posterior (rearward) dorsal fine spines, the pectoral girdles, and the skull elements intact. Liassic, Solnhofen, Bavaria, West Germany. (After Koken, 1907, Gustav Fischer Verlag, Jena, GDR.)

24-79. Early Cretaceous. *Asteracanthus ornatissimus* Agassiz. A fragment of the dorsal fin spine of a hybodontid shark. Note the overall ornamentation of tuberculations on the spines' surface. Range: Late Triassic to Early Cretaceous. Neocomian (Hauterivian), Potton, Bedfordshire, England. ×1.5.

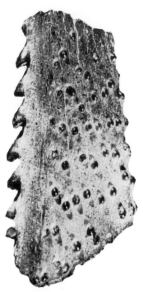

24-80. Late Cretaceous. *Hybodus* sp. #1. A fragment of the medial portion of a dorsal fin spine of a typical Late Cretaceous hybodont shark. Note the tuberculation (a possible descendancy from an *Asteracanthus* origin?). Navesink Formation (Middle Maestrichtian), Big Brook, Marlboro Township, Monmouth County, New Jersey. ×3.

24-81. Upper Cretaceous. *Hybodus montanensis* Case. A lower lateral tooth of a hybodont shark. It is rare to find the root still attached to the enameloid cusp on this species. Type specimen. Judith River Formation (Campanian), Blaine County, Montana. ×2.5.

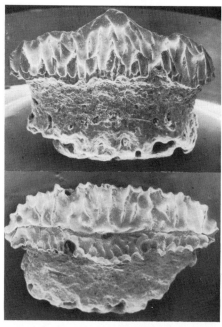

24-82. Upper Cretaceous. *Hybodus storeri* Case. Two views of a distinctive hybodont shark's tooth. Top: Lingual view, upper lateral tooth. Bottom: Occlusal (top) view of same tooth. Judith River Formation (Campanian), Blaine County, Montana. ×25. (SEM).

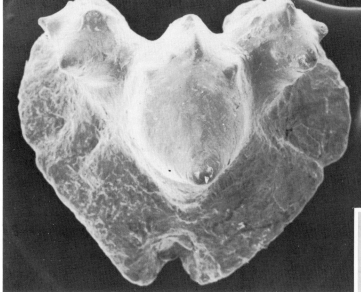

24-83. Upper Cretaceous. *Orectoloboides (Archaeotriakis) rochelleae* (Case). The occlusal (top or overhead) view of a most distinctive newly discovered shark tooth in the orectolobid family (nurse sharks). Judith River Formation (Campanian), Blaine County, Montana. ×85. (SEM).

Suborder Squatinoidei: Family **Squatinidae**: CRETACEOUS: *Squatina;* MIOCENE: *Squatina* (see Fig. 24-129). Family **Sclerorhynchidae** (Ganopristinae: **Ganopristidae** Arambourg): CRETACEOUS: *Ctenopristis, Is-*

24-84. Upper Cretaceous. *Cretorectolobus olsoni* Case. Labial view of a lower jaw lateral tooth in the orectolobid family (nurse and Wobbegong sharks). Judith River Formation (Campanian), Blaine County, Montana. ×14.

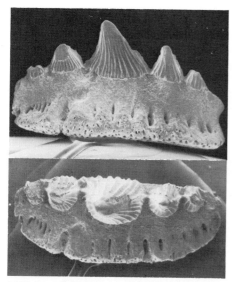

24-85. Upper Cretaceous. *Synechodus striatus* Case. Two views of a cestraciontid shark tooth. Top: Lingual view of a lower jaw lateral tooth. Bottom: An occlusal view of the same tooth. Judith River Formation (Campanian), Blaine County, Montana. ×9. (SEM).

24-86. Upper Cretaceous. *Synechodus andersoni* Case. A close-up of the rugose striations (ribbing) decoration on the labial side of a cestraciont shark tooth. Judith River Formation (Campanian), Blaine County, Montana. ×5.

24-89. Upper Cretaceous. *Ptychodus polygurus* Agassiz. Crushing tooth of a primitive ray. Note the distinctive "Fingerprint" pattern on the center of the tooth occlusal surface. Lower Chalk (Cenomanian), Sussex, England. 5 cm. Courtesy: Geological Enterprises, Ardmore, Oklahoma.

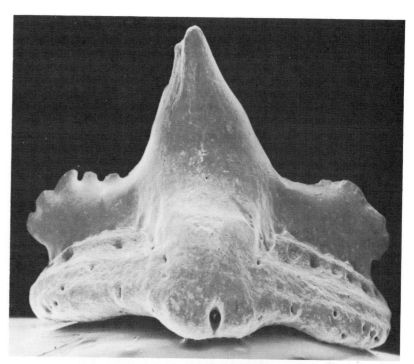

24-87. Upper Cretaceous. *Cretorectolobus olsoni* Case. Lingual view of a fossil wobbegong tooth, in the orectolobid (nurse shark) family. (See Fig. 24-84 for another example of this same species.) Judith River Formation (Campanian), Blaine County, Montana. ×21. (SEM).

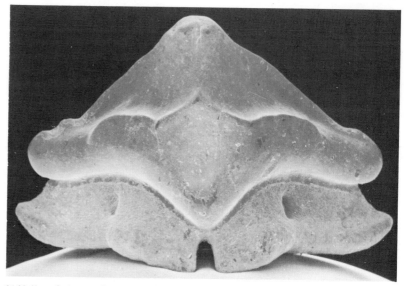

24-88. Upper Cretaceous. *Ptychotrygon blainensis* Case. Lingual view of the tooth of a ganopristine (sawfish). This is one tooth out of a mosaic pattern ("battery") of teeth that form a crushing plate. Judith River Formation (Campanian), Blaine County, Montana. ×34. (SEM).

chyrhiza (see Fig. 24-107), *Marckgrafia, Onchopristis* (see Fig. 24-101), *Onchosaurus (Dalpiazia), Peyeria, Pucapristis, Schizorhiza,* and *Sclerorhynchus.* Incertae Sedis *Ptychotrygon* (see Fig. 24-88).

Order Pristiformes: Family **Pristidae:** EOCENE: *Anoxypristis, Pristis,* and *Propristis;* MIOCENE: *Pristis.*

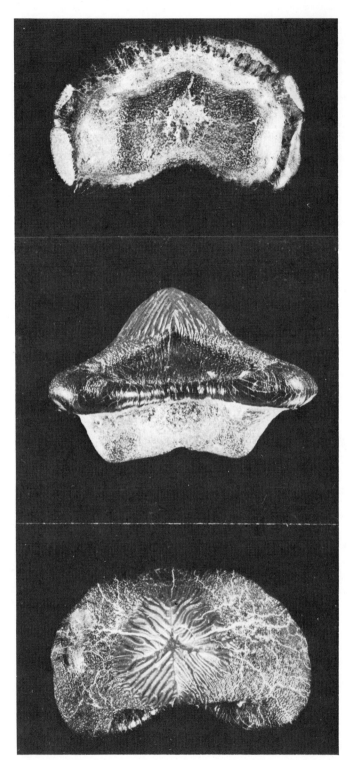

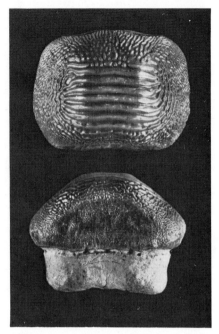

24-91. Upper Cretaceous. *Ptychodus decurrens* Agassiz. Another crusher pavement tooth from an ancient stingray. Note the similarity to the specimen in Fig. 24-89. Farmers Branch, Dallas County, Texas. × 1.

24-90. Upper Cretaceous. *Ptychodus mortoni* Mantell. Three views of a most distinctive crusher plate tooth of an ancestral stingray. Note the rugose pattern in the center of the bottom image. Top: Basal or bottom view. Center: lateral or side view. Bottom: Top or occlusal view. Eutaw Formation (Upper Santonian), Columbus, Lowndes County, Mississippi. × 1.5.

24-92. Upper Cretaceous. *Rhinobatos whitfieldi* (Hay). Ventral view of a guitarfish skate. Note the small herring (*Diplomystus*) to the left of the skate. Cenomanian, Hadjula, Lebanon. Courtesy, Siber and Siber Ltd., Aathal, Switzerland. 24 cm.

Superorder Batoidea: Order Rajiformes: Suborder Rhinobatoidei: Family **Rhinidae:** RECENT: *Rhina.* **Rhyn-**

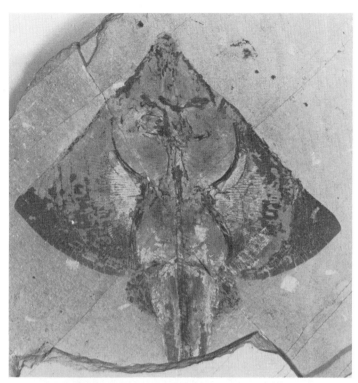

24-93. Upper Cretaceous. *Rhinobatos primarmatus* (Woodward). Another ventral (lower) view of a partial guitarfish skate (specimen is missing the lower part of its trunk and tail section). Upper Santonian, Sahel Alma, Lebanon. Courtesy: Siber and Siber Ltd., Aathal, Switzerland. 28 cm.

24-94. Upper Cretaceous. *Rhinobatos maronita* (Pictet and Humbert). A beautiful complete guitarfish skate in dorsoventral position. Note the distinctive rostrum which comes to a point at the tip of the head area. Cenomanian, Hakel, Lebanon, Courtesy: Siber and Siber Ltd., Aathal, Switzerland. 42 cm total length.

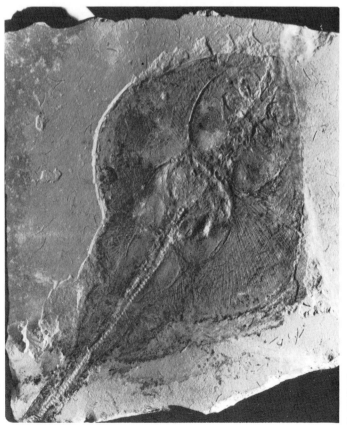

24-96. Upper Cretaceous. *Brachyrhizodus (Protorhinoptera) wichitaensis* Romer. An isolated pavement crusher chevron of a primitive cow-nose ray. The view is of the palatal position which is normally attached to the roof of the mouth. The opposing side is the crushing surface. Navesink Formation (Middle Maestrichtian), Hop Brook, Monmouth County, New Jersey. ×1.5.

24-95. Upper Cretaceous. *Rhinobatos whitfieldi* (Hay). Dorsal view of an almost complete guitarfish skate (part of lower trunk and tail section are missing). (See Fig. 24-92 for another specimen of this species.) Cenomanian, Hadjula, Lebanon. Courtesy: Geological Enterprises, Ardmore, Oklahoma. 25 cm.

24-97. Upper Cretaceous. *Cyclobatis major* Davis. A beautifully preserved ray skeleton. It is only missing a small piece of the tail. The tail section has ''thorns'' (dermal denticles) like the modern thorny skate. Cenomanian, Hakel. Lebanon. Courtesy: Siber and Siber Ltd., Aathal, Switzerland. 12 cm.

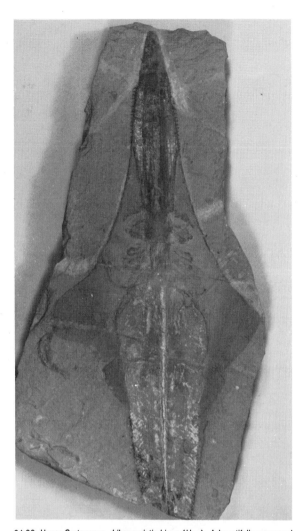

24-98. Upper Cretaceous. *Libanopristis hiram* (Hay). A beautifully preserved sawfish in dorsoventral aspect. Specimen is missing part of the lower trunk and the tail section. Note the rostral spines jutting out from either side of the rostrum. See Figs. 24-101 and 24-107 for examples of rostral spines. Cenomanian, Hadjula, Lebanon. Courtesy: Siber and Siber Ltd., Aathal, Switzerland. 38 cm in length.

24-99. Upper Cretaceous. *Centrosqualus primaevus* (Pictet). An almost complete (missing the head area) dogfish shark. Note the ''needlefish'' (*Belonorhynchus?*) just below the tail of the shark. Upper Santonian, Sahel Alma, Lebanon. 28 cm.

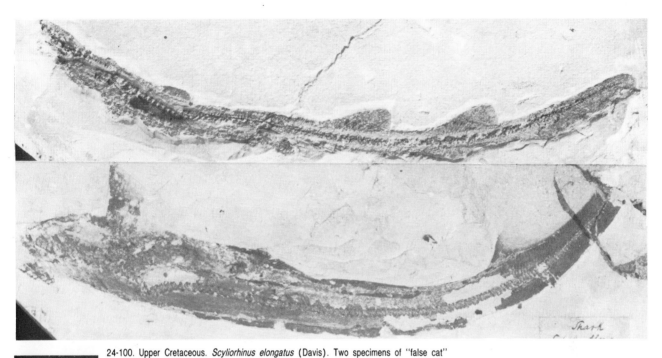

24-100. Upper Cretaceous. *Scyliorhinus elongatus* (Davis). Two specimens of "false cat" sharks. Top: A lateral view of one specimen. Bottom: Ventral view (missing the tail) of another. Upper Santonian, Sahel Alma, Lebanon. Top specimen 24 cm, Courtesy: Siber and Siber Ltd., Aathal, Switzerland. Bottom specimen 28 cm, Courtesy: Museum of Comparative Zoology, Harvard University, Cambridge, Massachusetts.

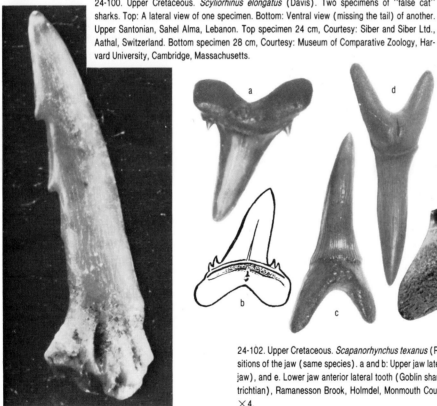

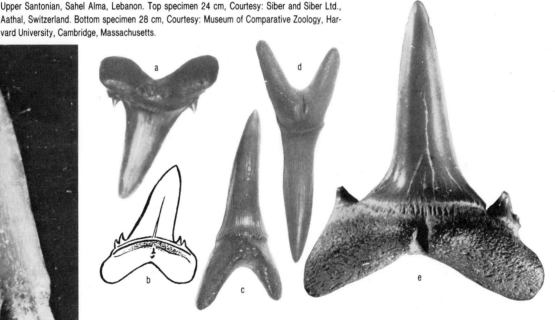

24-101. Upper Cretaceous. *Onchopristis dunklei* McNulty and Slaughter. Rostral spine of a ganopristine sawfish. Note the recurved barbs on the edge of the spine. These spines are sometimes called "rostral teeth." Their only relationship to teeth is that they are modified dermal denticles. Lower Arlington Sandstone Member, Woodbine Formation (Cenomanian), Amon Carter Airfield, Tarrant County, Texas. × 10.

24-102. Upper Cretaceous. *Scapanorhynchus texanus* (Roemer). Several teeth from different positions of the jaw (same species). a and b: Upper jaw lateral teeth. c and d: Anterior teeth (either jaw), and e. Lower jaw anterior lateral tooth (Goblin shark.) Navesink Formation (Middle Maestrichtian), Ramanesson Brook, Holmdel, Monmouth County, New Jersey. a through d × 1.5, e × 4.

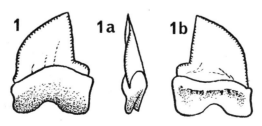

24-103. Upper Cretaceous. *Squalicorax kaupi* (Agassiz). Three views of an extinct crow shark tooth. 1: Lingual view, 1a: Lateral or side view, and 1b: Labial view. Navesink Formation (Middle Maestrichtain), Ramanesson Brook, Holmdel, Monmouth County, New Jersey. × 2.

24-104. Upper Cretaceous. *Squalicorax kaupi* (Agassiz). An isolated crow shark tooth (see Fig. 24-103) in lingual aspect. Navesink Formation (Middle Maestrichtian), Big Brook, Colts Neck Township, Monmouth County, New Jersey. ✕4.5.

24-105. Upper Cretaceous. *Squalicorax pristodontus* (Agassiz). Tooth of a larger species of crow shark. Lingual view. Navesink Formation (Middle Maestrichtian), Big Brook, Marlboro Township, Monmouth County, New Jersey. ✕1.5.

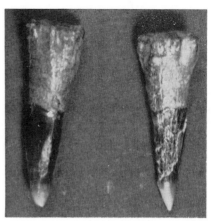

24-107. Upper Cretaceous. *Ischyrhiza mira* Leidy. Two examples of sawfish rostral spines. Navesink Formation (Middle Maestrichtian), Big Brook, Marlboro Township, Monmouth County, New Jersey. ✕1.5.

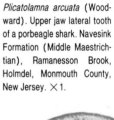

24-106. Upper Cretaceous. *Plicatolamna arcuata* (Woodward). Upper jaw lateral tooth of a porbeagle shark. Navesink Formation (Middle Maestrichtian), Ramanesson Brook, Holmdel, Monmouth County, New Jersey. ✕1.

24-108. Upper Cretaceous. *Hypotodus aculeatus* Cappetta and Case. A small, but beautifully preserved sandshark tooth. See little drawing in upper right hand corner for approximate size of the actual specimen. Mt. Laurel Sands (Early Maestrichtian), Colts Neck, Monmouth County, New Jersey.

24-109. Upper Cretaceous. *Cretolamna appendiculata* var. *lata* (Agassiz). Lingual view of a typical shark tooth found in the greensands of New Jersey. A predecessor of the mackerel shark. Navesink Formation (Middle Maestrichtian), Ramanesson Brook, Holmdel, Monmouth County, New Jersey. ✕4.

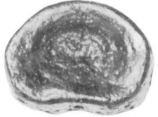

24-110. Upper Cretaceous. *Brachyrhizodus wichitaensis* Romer. The vertebral centrum of a primitive cow-nose ray. Navesink Formation (Middle Maestrichtian), Ramanesson Brook, Holmdel, Monmouth County, New Jersey. ✕1.5.

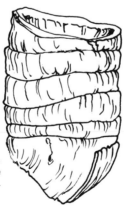

24-111. Upper Cretaceous. Coprolites or fecal pellets of sharks. Individual species indeterminate. Navesink Formation (Middle Maestrichtian), Big Brook, Marlboro, Monmouth County, New Jersey. ✕2.

24-112. Upper Cretaceous. Lamnoid shark vertebral centrum. Species indeterminate, but could belong to either *Plicatolamna* or *Cretolamna* (see Figs. 24-106 and 24-109). Merchantville Formation (Campanian), St. Georges, New Castle County, Delaware. (C. & D. Canal). ×3.

24-113. Upper Cretaceous. *Ischyodus bifurcatus* Case. Two jaw dentitions. Upper right: Upper left palatine tooth plate. Lower left: Lower left mandibular tooth plate of a chimaeroid. Note: The porous areas are the tritors or crushing contact points of the dentition. Navesink Formation (Middle Maestrichtian), Holmdel, Monmouth County, New Jersey. Palatine ×4, mandibular ×2.5.

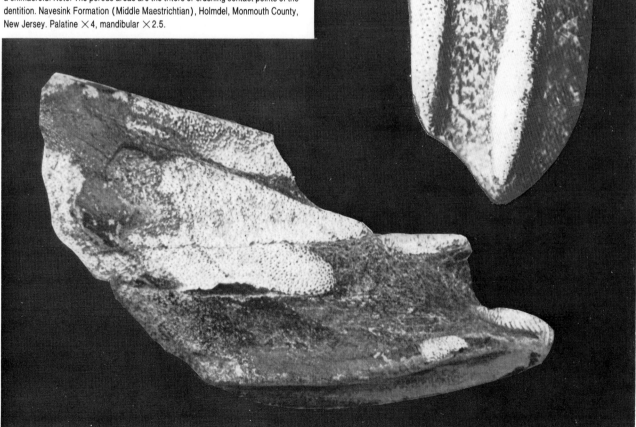

24-114. Eocene. *Myliobatis* sp. Palatal view of a large section of the lower jaw tooth plate of a duckbill ray. This is the side of the jaw that is attached to the floor of the mouth. The other side has smooth chevrons used as crushing plates. The round objects on the chevrons are bits of matrix that could not be removed from the specimen. Twiggs Clay Member, Barnwell Formation (Jacksonian), Huber, Twiggs County, Georgia. ✕ 1.5.

24-115. Eocene. *Dasyatis (Heliobatis) radians* (Marsh). Also known as *Xiphotrygon*. Two examples of freshwater skates. Note the "stinger" barbs on the tails of these skates. One of the rarer types of fossils found in the Middle Eocene Green River Formation of Wyoming. A. Specimen on a slab 18″ x 24″, Courtesy: Geological Enterprises, Ardmore, Oklahoma. B. 30 cm. Courtesy: Princeton Museum of Natural History, Princeton, New Jersey. (PU 14676).

A

B

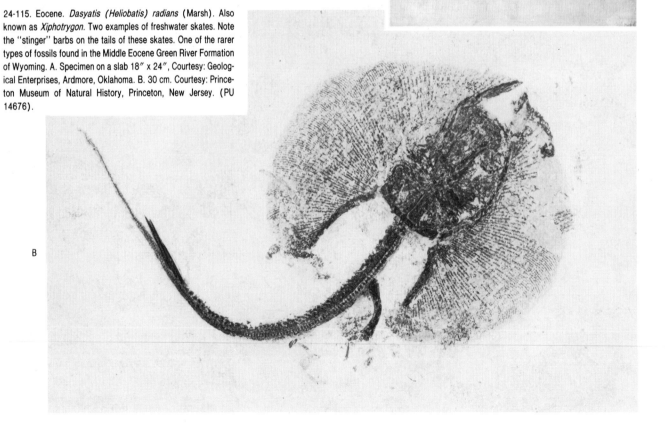

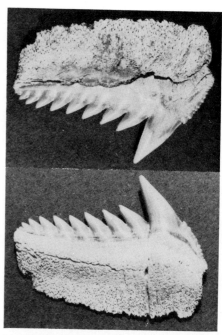

24-116. Miocene. *Notorhynchus* sp. Upper jaw lateral tooth of a fossil cow shark. Also known as the six gilled shark. Temblor Formation (Helvetian), Shark Tooth Hill, Oildale near Bakersfield, Kern County, California. ×2.

24-117. Miocene. *Heptranchias* sp. Lower jaw anterior/lateral tooth of a seven gilled cow shark. Pungo River Marl Formation (Helvetian), Aurora, Beaufort County, North Carolina. ×1.5.

24-118. Miocene. *Hexanchus* cf. *griseus* (Bonnaterre). Lower jaw lateral tooth of a cowshark. This fossil species is exactly the same as the modern species. Pungo River Marl Formation (Helvetian), Aurora, Beaufort County, North Carolina. ×3.5.

24-119. Miocene. *Oxyrhina crassa* Agassiz. (Also known as *O. benedeni*.) A rare and extinct type of mako shark tooth. It is robust and differs radically from all other known species of the isurids. Pungo River Marl Formation (Helvetian), Aurora, Beaufort County, North Carolina. ×2.

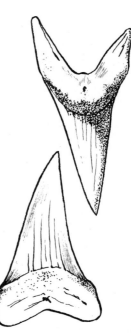

24-120. Miocene. *Isurus* cf. *oxyrhinchus* Rafinesque. Two distinctive mako shark teeth. Top: Lower jaw anterior tooth (also known as *Oxyrhina desori*). Bottom: Upper jaw anterolateral tooth (also known as *Isurus hastalis*). Pungo River Marl Formation (Helvetian), Aurora, Beaufort County, North Carolina. ×1.

24-121. Miocene. *Oxyrhina mantelli*. A distinctive isurid (mako shark) tooth with vestigial side cusplets almost becoming a lamnid. Pungo River Marl Formation (Helvetian), Aurora, Beaufort County, North Carolins. ×1.5.

24-122. Miocene. *Alopecias grandis* Leriche. Upper jaw lateral tooth of a very large thresher shark. Hawthorne Formation (Tortonian), Ashley River Phosphate Beds, North Charleston, Charleston County, South Carolina. ×1.5.

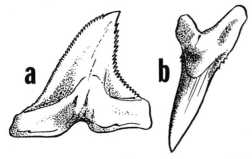

24-123. Miocene. *Hemipristis serra* Agassiz. a. Upper jaw lateral tooth. b. Lower jaw anterior tooth. Large fossil species of the present day Red sea shark. Pungo River Marl Formation (Helvetian), Aurora, Beaufort County, North Carolina. ×1.

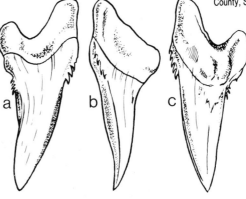

24-124. Miocene. *Hemipristis serra* Agassiz. Three views of a lower jaw anterolateral tooth. a. Labial, b. Profile, c. Lingual. Calvert Formation (Helvetian), Scientist's Cliffs (formerly Calvert Cliffs), Calvert County, Maryland. ×1.5.

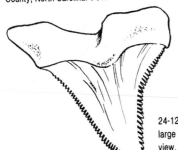

24-125. Miocene. *Hemipristis serra* Agassiz. The labial view of a rather large Red sea shark tooth. See Fig. 24-123a for the reverse (lingual) view. Pungo River Marl Formation (Helvetian), Aurora, Beaufort County, North Carolina. Slightly larger than ×1.

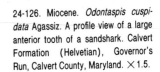

24-126. Miocene. *Odontaspis cuspidata* Agassiz. A profile view of a large anterior tooth of a sandshark. Calvert Formation (Helvetian), Governor's Run, Calvert County, Maryland. ×1.5.

24-127. Miocene. *Odontaspis* cf. *taurus* Rafinesque. A sand shark tooth with double side cusps (or denticles). Kirkwood Formation (Burdigalian), Shark River Park, Neptune, Monmouth County, New Jersey. ×1.5.

24-128a. Miocene. *Galeocerdo* cf. *cuvieri* (Leseur). Tiger shark tooth. Pungo River Marl Formation (Helvetian), Aurora, Beaufort County, North Carolina. ×1.5.

24-128b. Miocene. *Galeocerdo contortus* Gibbes. Tiger shark tooth. Calvert Formation (Helvetian), Cheasapeake Beach, Calvert County, Maryland. ×1.5.

24-128c. Miocene. *Galeocerdo aduncas* Agassiz. Tiger shark tooth. Calvert Formation (Helvetian), North Beach, Calvert County, Maryland. ×1.5.

24-129. Miocene. *Squatina occidentalis* Eastman. Two examples of the teeth of an angel shark. a. Labial view. b. Labial and profile views of another specimen. Calvert Formation (Helvetian), North Beach, Calvert County, Maryland. a. ×3, b. ×1.5.

24-130. Miocene. *Sphyrna* sp. The tooth of a hammerhead shark. Calvert Formation (Helvetian), North Beach, Calvert County, Maryland. ×1.

24-131a. Miocene. *Carcharhinus* sp. Upper jaw tooth of a dusky shark. Calvert Formation (Helvetian) North Beach, Calvert County, Maryland. ×1.5.

24-131b. Miocene. *Carcharhinus* sp. Lower jaw tooth of a dusky shark. Calvert Formation (Helvetian), North Beach, Calvert County, Maryland. ×1.5.

24-131c. Miocene. *Pterolamiops* cf. *longimanus* (Springer). Black-tip shark tooth. Pungo River Marl Formation (Helvetian), Aurora, Beaufort County, North Carolina. ×1.5.

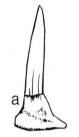

24-132. Miocene. a. *Pristiophorus* sp. Rostral spine of a "sawshark." Pungo River Marl Formation (Helvetian), Aurora, Beaufort County, North Carolina. ×1.5. b. *Pliotrema* cf. *warreni*. Cape Town Formation (Tortonian), Seapoint, Cape Province, Republic of South Africa. ×1.5. Both are rostral spines of sawsharks. Note the "serrations" on specimen b.

chobatidae: MIOCENE: *Rhynchobatus*. **Rhinobatidae:** CRETACEOUS: *Rhinobatos* (see Figs. 24-92 through 24-95); EOCENE: *Rhinobatos*, MIOCENE: *Rhinobatos*. **Platyrhinidae:** CRETACEOUS: *Protoplatyrhina;* RECENT: *Platyrhina*.

Suborder Rajoidei: Family **Rajidae:** MIOCENE: *Raja*.

Order Myliobatiformes: Superfamily Dasyatoidea: Family **Dasyatidae:** CRETACEOUS: *Dasyatis;* EOCENE: *Dasyatis* and *Trygon* (see Fig. 24-141). Family **Ptychodontidae:** CRETACEOUS: *Ptychodus* (see Figs. 24-89 through 24-91). Superfamily Myliobatoidea: Family **Myliobatidae:** CRETACEOUS: *Myledaphus* and *Rhombodus;* OLIGOCENE/EOCENE: *Aetobatus* and *Myliobatis* (see Fig. 24-114); MIOCENE: *Aetobatus* and *Myliobatis* (see Fig. 24-141). Family **Rhinopteridae:** CRETACEOUS: *Apocopodon* and *Brachyrhizodus (Protorhinoptera);* EOCENE: *Rhinoptera;* MIOCENE: *Rhinoptera*. Superfamily Mobuloidea: Family **Mobulidae:** MIOCENE: *Manta*, *Mobula*, and *Plinthicus*.

Order Torpediniformes: Superfamily Torpedinoidea: Family **Torpedinidae:** EOCENE: *Eotorpedo*. Superfamily Narcinoidea: Family **Narcinidae:** EOCENE: *Narcine* and *Pteroplatea*. **Narkidae:** RECENT: *Narke*.

Chondrichthyes (or Placodermi?)—Ichthyodorulites-Incertae Sedis of Romer, 1966: DEVONIAN: *Alienacanthus, Bulbocanthus, Cyrthacanthus,* and *Sentacanthus;*

MISSISSIPPIAN: *Acondylacanthus, Aganacanthus, Asteroptychius, Bythiacanthus, Chalazacanthus, Euctenius* (may be a part of the clasper mechanism of a pleuracanth shark), *Euphyacanthus, Glymmatacanthus, Gnathacanthus, Harpacanthus, Lispacanthus, Margaritacanthus (Euacanthus),* and *Marracanthus;* PENNSYLVANIAN: *Acondylacanthus, Asteroptychius, Euctenius, Euctenoptychius, Lepracanthus,* and *Ostracanthus;* PERMIAN: *Ancistriodus, Rapidentichthys, Thaumatacanthus,* and *Tubulacanthus.*

The distribution of fossil shark material in the USA— areas where shark remains have been recovered (list may not be absolutely complete at this time):

DEVONIAN: Northwestern New York State and northeastern Ohio (Cleveland Shales and equivalent rocks in N.Y.), and in northeastern Iowa (Cedar Valley Limestones).

MISSISSIPPIAN: Iowa (Keokuk and Burlington Limestones, and Montana (Bear Gulch Limestones).

PENNSYLVANIAN: Western Pennsylvania (Ames Limestone), eastern Ohio (Allegheny Group), central and southern Ohio (Conemaugh Group), Illinois (Mazon Creek siderite nodules), western Indiana, central and western Iowa, and eastern Nebraska (Desmoinesian, Mis-

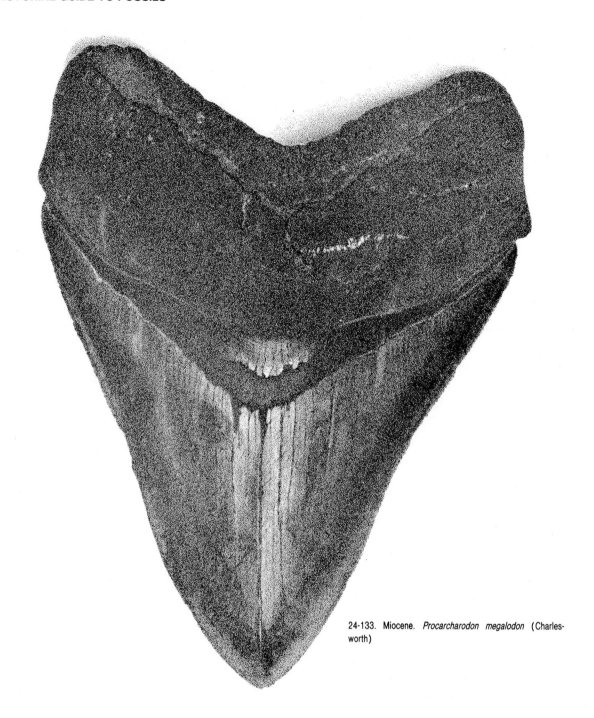

24-133. Miocene. *Procarcharodon megalodon* (Charlesworth)

sourian, and Virgilian Series of shales and limestones).

PERMIAN: Eastern Ohio (Red Bed equivalent), southeastern Nebraska (Indian Cave Sandstones—with some reworked Late Pennsylvanian material), south-central Oklahoma and north-central Texas (Wichita Group-Red Beds, also the Vale and Choaza rocks).

TRIASSIC: Eastern Pennsylvania and northern New Jersey (Lockatong Beds, only known shark remains found are of *Carinacanthus jepseni* Bryant, a hybodont shark). Triassic shark remains are quite rare in North America.

JURASSIC: No shark remains known from North America at the present time.

CRETACEOUS (Early or Lower): North-central and central Texas (Weno and Paw-paw Formations). (Late or Upper): Atlantic and Gulf Coastal Plains States: New Jersey and Delaware (Merchantville, and Mt. Laurel the Middle Maestrichtian Navesink Formations), Maryland (Late Maestrichtian/Navarroan), North and South Carolina (Peedee and Black Creek faunas—Maestrichtian), Alabama, and Mississippi (Santonian); Central USA: Arkan-

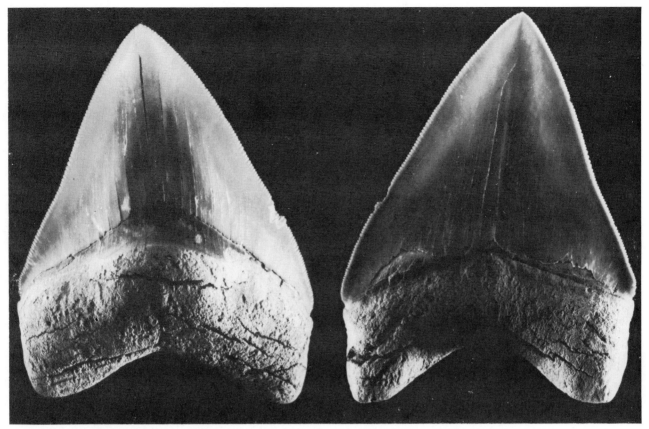

24-134. Miocene. *Procarcharodon megalodon* (Charlesworth). Left: Lingual view. Right: Labial view of a giant white shark tooth. Pungo River Marl Formation (Helvetian), Aurora, Beaufort County, North Carolina. 14 cm.

sas (Maestrichtian), Kansas (Turonian), South Dakota and Colorado (Turonian through Campanian), Texas (Cenomanian up through Late Maestrichtian), and Minnesota (Cenomanian); Western Interior (Canada): Alberta and Saskatchewan (Campanian to Lower Maestrichtian), (USA): Montana, Wyoming and northern Colorado (Judith River, Mesaverde, and Lance Formations/Campanian to Late Maestrichtian); Western USA: North-central California (Chico Local fauna/Campanian).

PALEOCENE: Central New Jersey (Danian-Hornerstown Formation), and Wyoming (Ft. Union Beds).

EOCENE: (Early or Lower): Maryland and Virginia (Brightseat and Aquia Formations/London Clay equivalents), New Jersey (Shark River/Manasquan Formations-Lutetian), and Mississippi (Bashi Marl/Paniselian). (Middle): North Carolina (Castle Hayne Limestone), east-central Texas (Mt. Selman Formation), and California/Oregon (various formations). (Late or Upper): South Carolina, Georgia, Florida, Alabama, Mississippi, Louisiana, and north-central Texas (Claibornian and Jacksonian).

OLIGOCENE: A "land" phase. No marine facies

known in this time period in the USA.

MIOCENE: (Early or Lower): North Carolina (Trent Formation/Aquitanian), and New Jersey (Kirkwood Formation-Burdigalian). (Middle): Massachusetts (Martha's Vineyard-Gay Head Cliffs/Calvert Formation equivalent/Helvetian—with some reworking of the Burdigalian), Maryland and Virginia (Calvert Formation/Helvetian), North Carolina (Pungo River Marl Formation/Helvetian, and a Calvert equivalent), California (Temblor Formation/Helvetian), and Baja California, Mexico (Santa Maria Formation/Helvetian equivalent). (Late or Upper): North Carolina (Duplin Marl-Yorktown equivalent), central Florida (Hawthorne, Citronelle, and Bone Valley sediments), and Baja California, Mexico (Tortonian).

The distribution of fossil shark material in other parts of the world (as with the previous listing, this list may not be up-to-date as of this printing):

CARBONIFEROUS: (Lower or "Mississippian" equivalent): Armagh, Ireland; and Ludlow and Bristol in England. (Upper or "Pennsylvanian" equivalent): Wirksworth and Dalkeith, England (the latter is Midlothian).

PERMIAN: Ruppersdorf bei Braunau and Nyranny (Nürschan), Czechoslovakia (Inner Sudeten Depression); and U.S.S.R.

TRIASSIC: Lake Wapiti, British Columbia, Canada

24-135. Miocene. Two views of a possible calcified duodenum (the first part of the small intestine) of a shark. Note the cartilage prisms in the interior of the left view (lower part of the specimen). Pungo River Marl Formation (Helvetian), Aurora, Beaufort County, North Carolina. ×3.5.

24-136. Miocene. Vertebral centrum possibly belonging to *Procarcharodon* or *Isurus*. Pungo River Marl Formation (Helvetian), Aurora, Beaufort County, North Carolina. ×1.

24-137. Miocene. Vertebral centrum of a carcharhinid shark. Species indeterminate. Pungo River Marl Formation (Helvetian), Aurora, Beaufort County, North Carolina. ×1.5.

(Sulphur Mountain Formation).

JURASSIC: Holzmaden, Germany; as well as Eichstätt, Daitingen, Kelheim, and Solnhofen, in Bavaria, West Germany (Liassic/Malm Zeta); Sicily, Italy (Tithonian); and Lyme Regis, Dorset, England.

CRETACEOUS: Potton, Bedfordshire, England (Hauterivian/Neocomian); Hakel and Hadjula, Lebanon (Cenomanian); Sahel Alma, Lebanon (Upper Santonian); Morocco, North Africa (Campanian and Maestrichtian); and Japan (Maestrichtian), Israel, and parts of South America.

EOCENE: Monte Bolca, Italy; Egem, Belgium (Paniselian); England and Belgium (London Clay equivalent, also in the Paris Basin of France), New Zealand, Japan, and Morocco, North Africa (Ypresian-London Clay equivalent), Egypt and Pakistan.

24-138. Miocene. A large ornate dermal denticle (actually with two denticles on the base) of a batoid (stingray). Pungo River Marl Formation (Helvetian), Aurora, Beaufort County, North Carolina. ×2.5.

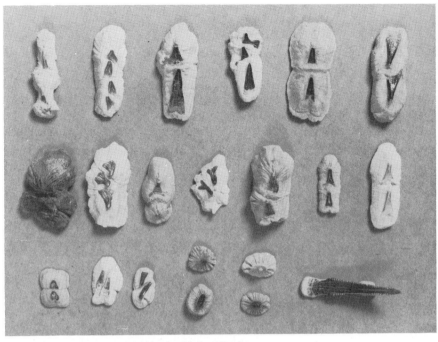

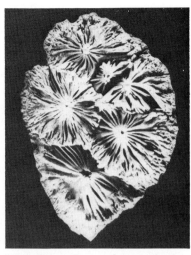

24-139. Miocene. Various dermal denticles (ossicles) attributed to thorny skates and stingrays. Species indeterminate. Bone Valley Sediments (Helvetian-Tortonian), Bartow, Polk County, Florida. ×1.

24-140. Miocene. *Trygon* sp. A compound scute or dermal ossicle of a thorny skate. The scute is made up of 6 dermal denticles fused together on one single bony base or platform. Pungo River Marl Formation (Helvetian), Aurora, Beaufort County, North Carolina. Slightly larger than actual size.

24-141. Miocene. *Trygon* and *Myliobatis*. The scutes and stingray barbs of various thorny skates and stingarees (stingrays). Bone Valley Sediments (Helvetian-Tortonian). Brewster, Polk County, Florida × 1 (Approx.).

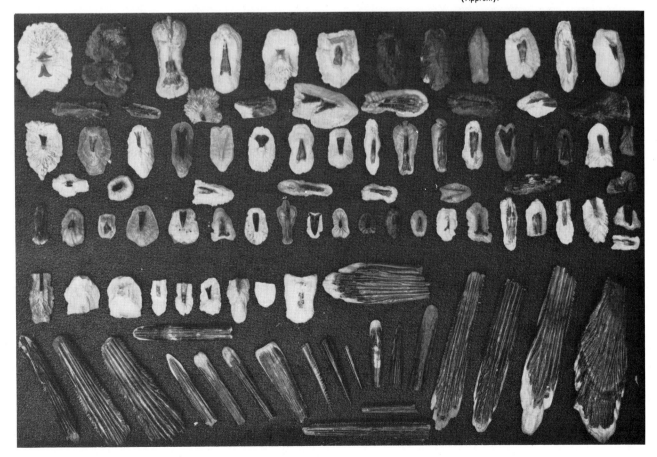

MIOCENE: Japan, New Zealand, Australia, Mexico, Belgium, Netherlands, Italy, Malta, Soviet Union, Poland, Cuba, Portugal, Spain, and France.

Particular kinds of Chondrichthyans found within various ages (list is by no means complete):

DEVONIAN: *Acmoniodus, Alienacanthus, Apateacanthus, Bulbocanthus, Cladodus, Cladoselache, Coronodus, Ctenacanthus, Cyrthacanthus, Deirolepis, Diademodus (Tiarodontus), Eoörodus, Eunemacanthus, Helodus, Hoploconchus, Ohiolepis, Phoebodus (Bathycheilodus), Proorodus, Prospiraxis, Protacrodus, Psammodus, Sentacanthus, Stethacanthus, Thoralodus, Xenacanthus (Dittodus),* and *Xenodus (Goniodus).*

MISSISSIPPIAN: *Acondylacanthus, Aganacanthus, Ageleodus, Asteroptychius, Bythiacanthus, Campodus, Chalazacanthus, Chomatodus, Chondrenchelys, Cladodus, Cochliodus, Coelosteus, Copodus, Cranodus, Ctenoptychius (Harpacodus), Cynopodius, Deltodus, Deltoptychius, Dichelodus, Dicrenodus (Carcharopsis), Diplacodus, Edestus, Erismacanthus, Eucentrurus, Euctenius, Euglossodus (Glossodus), Eunemacanthus, Euphyacanthus, Fissodus (Cholodus), Glymmatacanthus, Goodrichthys (Goodrichia), Gnathacanthus, Harpacanthus, Helodus, Hoplodus, Hybocladodus, Icanodus, Janassa, Lagarodus, Lambdodus, Lestrodus, Lispacanthus, Marracanthus, Mazodus, Menaspacanthus, Mesolophodus, Orodus, Petalodus, Petalorhynchus, Petrodus, Phoebodus, Physonemus (Xystracanthus), Platyodus, Platyxstrodus, Poecilodus, Polyrhizodus, Psammodus, Psephodus, Sandalodus, Sphenacanthus, Tamiobatis, Thrinacodus, Venustodus (Lophodus),* and *Xenacanthus (Diplodus).*

PENNSYLVANIAN: *Acondylacanthus, Agassizodus, Ageleodus, Ancistrodus, Anodontacanthus, Asteroptychius, Bandringa, Bobbodus, Campodus, Campyloprion, Caseodus, Chomatodus, Cladodus, Cobelodus, Cochliodus, Compsacanthus, Ctenacanthus, Ctenoptychius (Harpacodus), Cymatodus, Dabasacanthus, Deltodus, Deltoptychius, Dicentrodus, Dicrenodus (Carcharopsis), Echinodus, Edestodus (Protopirata), Edestus, Euctenius, Eugeneodus, Eunemacanthus, Fadenia, Gilliodus, Helodus, Janassa, Jimpholia, Lagarodus, Lepracanthus, Listracanthus, Metaxyacanthus, Monocladodus, Ornithoprion, Orodus, Ostracanthus, Petalodus (Antliodus), Petrodus, Physonemus (Xystracanthus), Platyxystrodus (Xystrodus), Polysentor, Psephodus (Aspidodus), Romerodus, Sandalodus, Similihariotta, Solenodus, Sphenacanthus, Stamiobatis, Styptobasis, Styracodus (Centrodus), Symmorium, Toxoprion,* and *Xenacanthus (Diplodus).*

PERMIAN: *Ancistrodus, Arctacanthus (Hamatus), Campodus, Cladodus, Cobelodus, Crassidonta, Ctenacanthus, Ctenoptychius, Dicentrodus, Edestus, Erikodus, Fissodus (Peltodus), Helicampodus, Helicoprion, Helidopsis, Helodus, Hybodus (Leiacanthus), Hypospondylus* (a growth stage of *Xenacanthus), Icanodus (Enniskillen), Janassa, Macrodontacanthus, Megactenopetalus, Menaspis, Orodus, Parahelicoprion, Petalodus, Physonemus (Xystracanthus), Psephodus (Aspidodus), Rapidentichthys, Sarcoprion, Sphenacanthus, Syntomodus, Thaumatacanthus, Tubulacanthus, Wodnika, Xenacanthus (Pleuracanthus),* and *Xystrodus.*

TRIASSIC: *Acrodonchus, Acrodus (Thectodus), Asteracanthus (Strophodus), Carinacanthus, Doratodus, Helicampodus, Hybodonchus, Hybodus (Leiacanthus), Lissodus, Lonchidion, Nemacanthus (Nematacanthus), Palaeobates, Polyacrodus, Raineria,* and *Scoliorhiza. Note:* The Triassic period was a time of transition towards the neoselachians (modern forms of sharks), which have come through the hybodonts from a ctenacanthoid origin in the Paleozoic. This funneling of ancestral sharks and

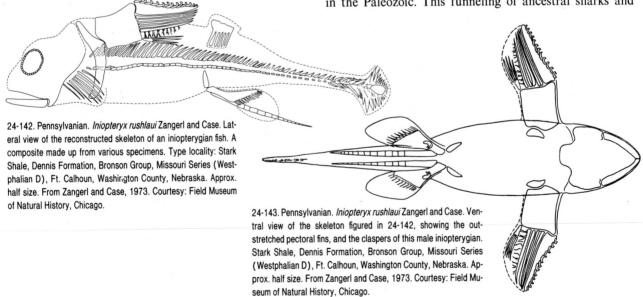

24-142. Pennsylvanian. *Iniopteryx rushlaui* Zangerl and Case. Lateral view of the reconstructed skeleton of an iniopterygian fish. A composite made up from various specimens. Type locality: Stark Shale, Dennis Formation, Bronson Group, Missouri Series (Westphalian D), Ft. Calhoun, Washington County, Nebraska. Approx. half size. From Zangerl and Case, 1973. Courtesy: Field Museum of Natural History, Chicago.

24-143. Pennsylvanian. *Iniopteryx rushlaui* Zangerl and Case. Ventral view of the skeleton figured in 24-142, showing the outstretched pectoral fins, and the claspers of this male iniopterygian. Stark Shale, Dennis Formation, Bronson Group, Missouri Series (Westphalian D), Ft. Calhoun, Washington County, Nebraska. Approx. half size. From Zangerl and Case, 1973. Courtesy: Field Museum of Natural History, Chicago.

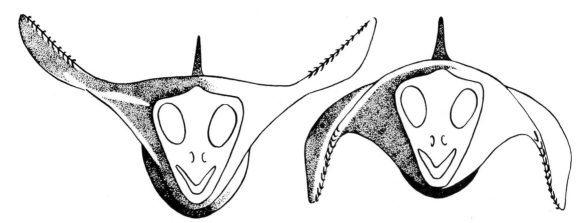

24-144. Pennsylvanian. *Iniopteryx rushlaui* Zangerl and Case. A theoretical interpretation of how this particular species of iniopterygian may have used its pectoral fins in swimming about. From Zangerl and Case, 1973. Courtesy: Field Museum of Natural History, Chicago.

24-145. Pennsylvanian. *Iniopteryx rushlaui* Zangerl and Case. A cast of a brain case of an iniopterygian. Stark Shale, Bronson Group, Dennis Limestone Formation, Missouri Series (Westphalian D), Crescent, Pottawattamie County, Iowa. ×1.5.

24-146. Pennsylvanian. *Iniopteryx rushlaui* Zangerl and Case. A lateral view of the head and postcranial elements of this rather ubiquitous iniopterygian. Note the mouth is wide open, and also note the large eye socket. Stark Shale, Dennis Formation, Bronson Group, Missouri Series (Westphalian D), Papillion, Sarpy County, Nebraska. 15 cm.

their relatives through the hybodonts proceeded into the Late Triassic Period, and by the Jurassic Period, most of the ancestral families of sharks, skates, rays, sawfishes, and chimaeroids were present in the Mesozoic and Cenozoic.

JURASSIC: *Acanthorhiza, Acrodus, Aellopos (Spathobatis), Asteracanthus, Asterodermus, Bdellodus, Belemnobatis, Brachymylus (Aletodus), Chimaeropsis, Corysodon, Crossorhinops, Cyclarthrus, Elasmodectes (Elasmognathus), ?Galeus, Ganodus (Leptacanthus), Heterodontus (Cestracion), Hexanchus (Notidanus), Hybodus, Hylaeobatis, Ischyodus (Chimaeracanthus, Lonchidion, Myriacanthus (Prognathodus), Orectolobus (Crossorhinus), Orthacodus (Sphenodus), Pachymylus, Palaeocarcharias, Palaeoscyllium, Palaeospinax, Phorcynus, Priory-*

bodus, Pristiurus, Squaloraia (Spinocorhinus), and *Squatina. Note:* The Jurassic listing is constantly being updated. Forms (genera) being found in European Jurassic marine deposits are pushing back Cretaceous generic types. The explosion of "true" or modern sharks and their related groups was indeed evident in the Jurassic Period, and many of the forms evolving out of the Triassic are with us in today's oceans.

CRETACEOUS: *Acanthoscyllium, Acrodus, Almascyllium, Ankistrorhynchus, Anomotodon, Apocopodon, Asteracanthus, Brachaelurus, Brachyrhizodus, Callorhynchus, Cantioscyllium, Centrophoroides, Centrophorus (Squalus), Centropterus, Centrosqualus, Chiloscyllium, Chimaera, Cirrhoscyllium, Cretascymnus, Cretolamna, Cretorectolobus, Cretoxyrhina (Isurus), Ctenopristis, Cy-*

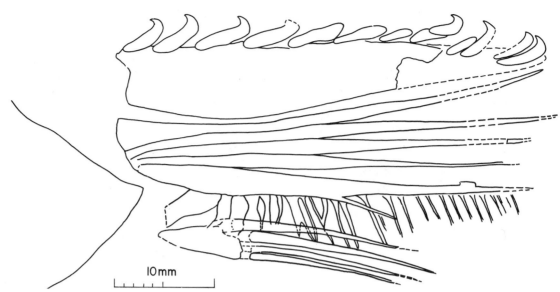

10mm

24-148. Pennsylvanian. *Iniopteryx rushlaui* Zangerl and Case. A fairly complete and articulated pelvic fin arrangement with claspers and hooks (see upper right and to the left of the cartilage plates of the fin; the hooks are broken away from the fin). Stark Shale, Dennis Formation, Bronson Group, Missouri Series (Westphalian D), Papillion, Sarpy County, Nebraska. ×3.5. ➡

24-147. Pennsylvanian. *Iniopteryx rushlaui* Zangerl and Case. Drawing of the pectoral fin of an iniopterygian fish. Note the recurved denticles on the bar, plus the transverse struts acting as an "aileron". This is a complicated pectoral fin that allows the fish to use his fins in an up and down motion (see Fig. 24-144), and to even use the "aileron" effect of its fin to slow itself down in the water. Stark Shale, Dennis Formation, Bronson Group, Missouri Series (Westphalian D), Papillion, Sarpy County, Nebraska. Zangerl and Case, 1973. Courtesy: Field Museum of Natural History, Chicago.

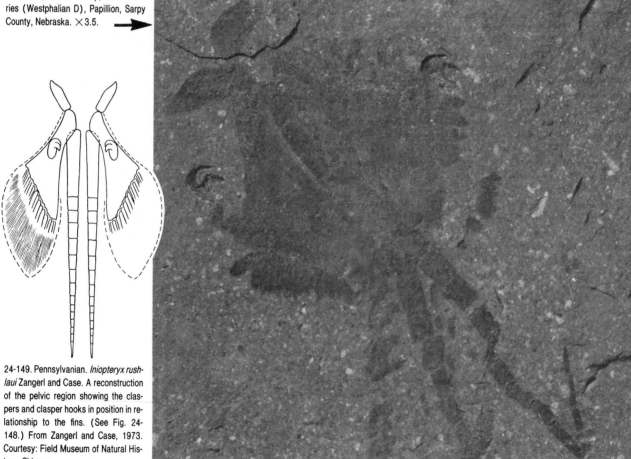

24-149. Pennsylvanian. *Iniopteryx rushlaui* Zangerl and Case. A reconstruction of the pelvic region showing the claspers and clasper hooks in position in relationship to the fins. (See Fig. 24-148.) From Zangerl and Case, 1973. Courtesy: Field Museum of Natural History, Chicago.

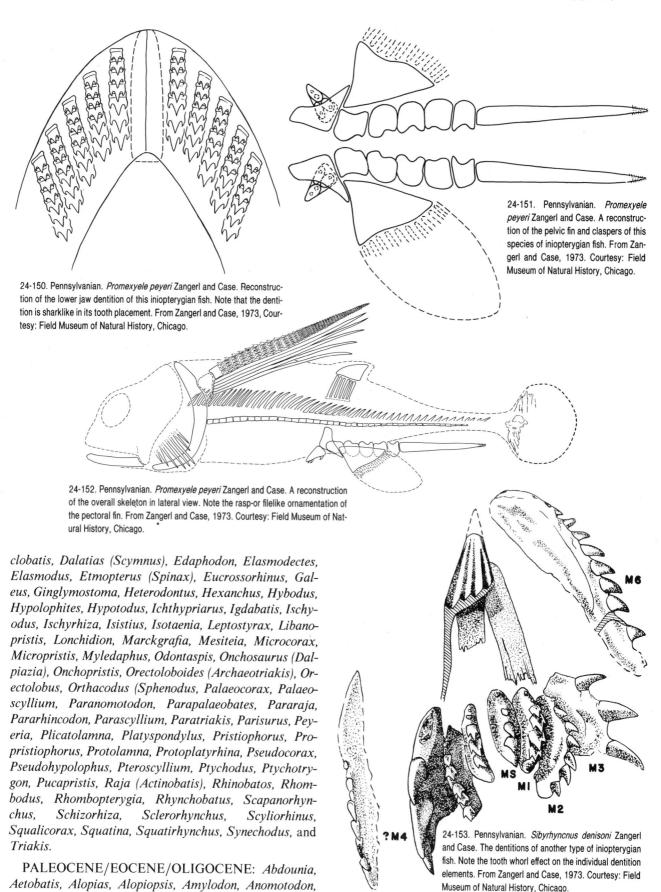

24-150. Pennsylvanian. *Promexyele peyeri* Zangerl and Case. Reconstruction of the lower jaw dentition of this iniopterygian fish. Note that the dentition is sharklike in its tooth placement. From Zangerl and Case, 1973, Courtesy: Field Museum of Natural History, Chicago.

24-151. Pennsylvanian. *Promexyele peyeri* Zangerl and Case. A reconstruction of the pelvic fin and claspers of this species of iniopterygian fish. From Zangerl and Case, 1973. Courtesy: Field Museum of Natural History, Chicago.

24-152. Pennsylvanian. *Promexyele peyeri* Zangerl and Case. A reconstruction of the overall skeleton in lateral view. Note the rasp-or filelike ornamentation of the pectoral fin. From Zangerl and Case, 1973. Courtesy: Field Museum of Natural History, Chicago.

clobatis, Dalatias (Scymnus), Edaphodon, Elasmodectes, Elasmodus, Etmopterus (Spinax), Eucrossorhinus, Galeus, Ginglymostoma, Heterodontus, Hexanchus, Hybodus, Hypolophites, Hypotodus, Ichthypriarus, Igdabatis, Ischyodus, Ischyrhiza, Isistius, Isotaenia, Leptostyrax, Libanopristis, Lonchidion, Marckgrafia, Mesiteia, Microcorax, Micropristis, Myledaphus, Odontaspis, Onchosaurus (Dalpiazia), Onchopristis, Orectoloboides (Archaeotriakis), Orectolobus, Orthacodus (Sphenodus, Palaeocorax, Palaeoscyllium, Paranomotodon, Parapalaeobates, Pararaja, Pararhincodon, Parascyllium, Paratriakis, Parisurus, Peyeria, Plicatolamna, Platyspondylus, Pristiophorus, Propristiophorus, Protolamna, Protoplatyrhina, Pseudocorax, Pseudohypolophus, Pteroscyllium, Ptychodus, Ptychotrygon, Pucapristis, Raja (Actinobatis), Rhinobatos, Rhombodus, Rhombopterygia, Rhynchobatus, Scapanorhynchus, Schizorhiza, Sclerorhynchus, Scyliorhinus, Squalicorax, Squatina, Squatirhynchus, Synechodus, and *Triakis.*

PALEOCENE/EOCENE/OLIGOCENE: *Abdounia, Aetobatis, Alopias, Alopiopsis, Amylodon, Anomotodon,*

24-153. Pennsylvanian. *Sibyrhyncnus denisoni* Zangerl and Case. The dentitions of another type of iniopterygian fish. Note the tooth whorl effect on the individual dentition elements. From Zangerl and Case, 1973. Courtesy: Field Museum of Natural History, Chicago.

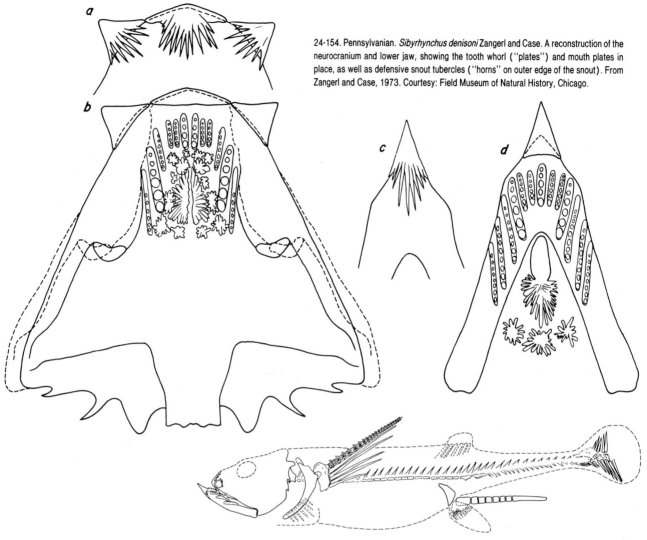

24-154. Pennsylvanian. *Sibyrhynchus denisoni* Zangerl and Case. A reconstruction of the neurocranium and lower jaw, showing the tooth whorl ("plates") and mouth plates in place, as well as defensive snout tubercles ("horns" on outer edge of the snout). From Zangerl and Case, 1973. Courtesy: Field Museum of Natural History, Chicago.

24-155. Pennsylvanian. *Sibyrhynchus denisoni* Zangerl and Case. A reconstructed lateral skeletal view, showing the snout tubercles and tooth whorls in the mouth area, as well as another type of "rasp" or "file" ornamentation on the pectoral fin. From Zangerl and Case, 1973. Courtesy: Field Museum of Natural History, Chicago.

Anoxypristis, Aprionodon, Archaeomanta, Cetorhinus, Chimaera, Dalatias, Dasyatis, Echinorhinus, Edaphodon, Elasmodus, Eotorpedo, Eugaleus, Galeocerdo, Galeorhinus, Galeus, Hemipristis, Heterodontus, Heptranchias, Hexanchus, Hypolophites, Hypolophodon, Hypoprion, Ikamauius, Isistius, Isurus, Lamna (Otodus), Mesiteia, Mustelus, Myliobatis, Narcine, Narcopterus, Nebrius, Negaprion, Notorhynchus, Odontaspis, Orthacodus, Paleocarcharodon, Paragaleus, Pararhincodon, Parisurus, Physodon, Physogaleus, Platyrhina, Prionodon, Pristis, Procarcharodon, Protogaleus, Pteroplatea, Raja, Rhinobatos, Rhinoptera, Rhynchobatus, Scoliodon, Scyliorhinus, Sphyrna, Squalus, Squatina, Squatirhina, Squatiscyllium, Synechodus, Taeniura, Torpedo, Triakis, Trigonodus, Trygonorrhina, Urolophus, and *Xiphodolamia.*

MIOCENE/PLIOCENE/RECENT: *Aetobatis, Alopias, Anomotodon, Anoxypristis, Aprionodon, Carcharias, Carcharhinus, Cetorhinus, Chiloscyllium, Clamydoselachus, Dasyatis, Echinorhinus, Etmopterus, Galeocerdo, Galeorhinus, Galeus, Ginglymostoma, Hemipristis, Heptranchias, Heterodontus, Hexanchus, Hypoprion, Isistius, Isurus (Oxyrhina), Lamiostoma, Lamna, Manta, Mobula, Mustelus, Myliobatis, Negaprion, Notorhynchus, Odontaspis, Paragaleus, Paratodus, Physodon, Physogaleus, Plinthicus, Pliotrema, Pristiophorus, Pristis, Procarcharodon (Carcharodon), Pteromylaeus, Pteroplatea, Raja, Rhincodon, Rhinobatos, Rhinoptera, Rhynchobatus, Rhizoprionodon, Scoliodon, Scyliorhinus, Sphyrna, Squalus, Squatina, Strongyliscus,* and *Trygon.*

Additional copy from this chapter can be found in the Addenda section (p. 495).

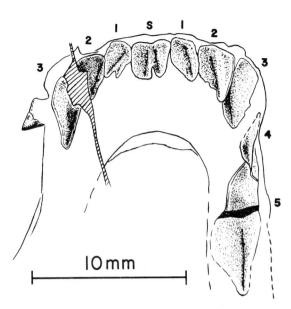

24-156. Pennsylvanian. *Iniopera richardsoni* Zangerl and Case. Lower jaw reconstruction of an iniopterygian with an unusual dentition. S = symphysial tooth; numbers 1 through 5 indicate teeth comparable to incisors, premolars and molars. From Zangerl and Case, 1973. Courtesy: Field Museum of Natural History, Chicago.

24-157. Pennsylvanian. *Iniopera richardsoni* Zangerl and Case. Starlike or "snowflake" denticles outlining two pouches which were attached to the pectoral fin (see Fig. 24-158). It is assumed that an inklike substance was squirted out from these pouches when this particular iniopterygian was in danger. Stark Shale, Dennis Formation, Bronson Group, Missouri Series (Westphalian D), Ft. Calhoun, Washington County, Nebraska. × 1.

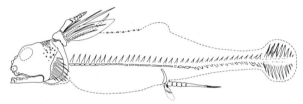

24-158. Pennsylvanian. *Iniopera richardsoni* Zangerl and Case. A reconstruction of the skeleton in lateral view, showing rasp or file type (shorter than previous species) ornamentation on the pectoral fin, plus the "ink pouches" under the rasps. From Zangerl and Case, 1973. Courtesy: Field Museum of Natural History, Chicago.

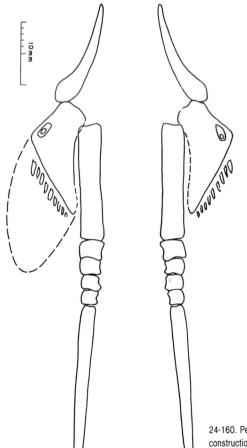

24-160. Pennsylvanian. *Inioxyele whitei* Zangerl and Case. A reconstruction of the pelvic fins and claspers of this species of iniopterygian. From Zangerl and Case, 1973. Courtesy: Field Museum of Natural History, Chicago.

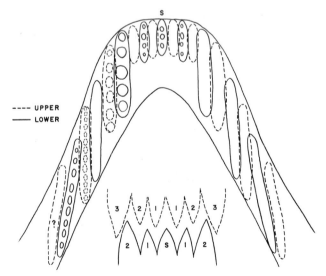

24-159. Pennsylvanian. *Inioxyele whitei* Zangerl and Case. An interpretation of the dentition of the jaws of this iniopterygian. We have mammalianlike dentition in the jaws (see Figs. 24-153, 154, and 156). From Zangerl and Case, 1973. Courtesy: Field Museum of Natural History, Chicago.

25 PISCES (Osteichthyes/Bony fishes)

This chapter deals with Osteichthyes, "bony fishes," as opposed to the cartilaginous fishes (sharks, skates, rays, etc.) of the previous chapter. Also included in this chapter are the crossopterygians and dipnoans (as prescribed by Romer, 1966).

It is too large a task to list all of the known fossil fishes in the fossil record; so the reader will have to forgive the author for leaving out a few for the sake of brevity. It will be simpler for the author to proceed with orders or groups of fish types along with their family listings. In the long run, this will be most helpful to the reader. Select families within an order will be listed, while the remaining orders, suborders and families where possible will be mentioned without species representation, again, for the sake of brevity.

Order Cyclostomata (lampreys, hagfishes, and slime eels)/suborder Petromyzontoidei/family **Petromyzontidae:** PENNSYLVANIAN: *Mayomyzon* (see Fig. 25-1). Suborder Myxinoidei/family **Myxinidae (Bdellostomatidae-Eptatretidae)** RECENT: No fossil representatives.

Order Dipnoi (lungfishes): Family **Dipnorhynchidae.** Family **Dipteridae:** DEVONIAN representatives: *Chirodipterus, Conchodus (Cheirodus), Dipteroides, Dipterus* (see Figs. 25-6 and 25-7), *Grossipterus, Paradipterus, Pentlandia,* and *Rhinodipterus.* Family **Phaneropleuridae:** DEVONIAN representatives: *Fleurantia, Jarvikia, Nielsenia, Phaneropleuron,* and *Scaumenacia (Canadipterus).* Family **Ctenodontidae:** MISSISSIPPIAN: *Ctenodus* and *Tranodus,* PENNSYLVANIAN: *Ctenodus.* Family **Sagenodontidae:** MISSISSIPPIAN: *Straitonia;* PENNSYLVANIAN: *Sagenodus* (see Fig. 25-8) and *Megapleuron;* PERMIAN: *Sagenodus* (see Fig. 25-9) and *Megapleuron.* Family **Uronemidae:** MISSISSIPPIAN: *Uronemus (Ganopristodus).* Family **Conchopomidae:** PENNSYLVANIAN: *Conchopoma (Conchiposis);* PERMIAN: *Conchopoma (Conchiopsis.* Family **Ceratodontidae:** TRIASSIC: *Gosfordia, Microceratodus, Paraceratodus,* and Ptychoceratodus; JURASSIC: *Ceratodus (Hemictenodus)* (see Fig. 25-10); CRETACEOUS: *Cera-todus (Hemictenodus)* (see Fig. 25-11), and *Neoceratodus (Epiceratodus).* Family **Lepidosirenidae:** PENNSYLVANIAN: *Gnathorhiza;* PERMIAN: *Gnathorhiza;* EOCENE: *Protopterus (Protomalus);* MIOCENE: *Lepidosiren (Ampibichthys.*

Order Crossopterygii (crossopterygians and coelacanths): Family **Osteolepidae:** DEVONIAN: *Bogdanovia, Canningius, Glyptopomus (Glyptognathus), Gyroptychius (Diplopterax)* (see Fig. 25-12), *Latvius, Megistolepis, Osteolepis (Pleiopterus), Panderichthys, Thaumatolepis,* and *Thurius;* MISSISSIPPIAN: *Megalichthys (Carlukeus)* (see Fig. 25-18); PERMIAN: *Ectosteorhachis.* Family **Rhizodontidae:** DEVONIAN: *Eusthenodon, Eusthenopteron* (see Fig. 25-13), *Litoptychius, Platycephalichthys, Rhizodopsis (Characodus), Sauripterus,* and *Tristichopterus;* MISSISSIPPIAN: *Rhizodus* (see Fig. 25-19); PENNSYLVANIAN: *Rhizodopsis (Characodus)* (see Fig. 25-14) and *Rhizodus* (see Fig. 25-20).

Subclass Sarcopterygii of the class Osteichthyes/order Crossopterygii/suborder Rhipidistia, superfamily Holoptychoidea (Porolepiformes)/families: **Holoptychidae, Porolepidae,** and **Onychodontidae.**

Family **Holoptychidae:** DEVONIAN: *Glyptolepis, Holoptychus, Lacognathus,* and *Pseudosauripterus.* Family **Porolepidae:** DEVONIAN: *Porolepis (Gyrolepis).* Family **Onychodontidae:** DEVONIAN: *Onychodus* (see Figs. 25-15 through 25-17), and Strunius.

Suborder Coelacanthini (Actinistia): (Coelacanths): Family **Euporosteidae.** Family **Diplocercidae:** DEVONIAN: *Chagrinia, Dictyonosteus, Diplocercides,* and *Nesides;* MISSISSIPPIAN: *Rhabdoderma (Conchiopsis);* PENNSYLVANIAN: *Rhabdoderma (Conchiopsis)* (see Fig. 25-21), and *Synaptotylus.* Family **Coelacanthidae:** PENNSYLVANIAN: *Coelacanthus (Hoplopygus);* PERMIAN: *Coelacanthus (Hoplopygus),* and *Spermatodus;* TRIASSIC: *Axelia, Coelacanthus (Hoplopygus), Diplurus* (see Figs. 25-22 through 25-29), *Graphiurichthys*

25-1. Pennsylvanian. *Mayomyzon pieckoensis* Bardack and Zangerl. A primitive lamprey, and the first fossil lamprey discovered and described. The picture shows only ⅓ of the actual fish fossil. The fish was found in its entirety in an ironstone (siderite) concretion. This view shows the head and postcranial areas. Type specimen (FMNH PF5687). Francis Creek Shale (Essex Marine concretion fauna), Carbondale Formation (Mazon Creek), Will County, Illinois. From Bardack and Zangerl, 1968. Courtesy: Field Museum of Natural History, Chicago.

25-2. Pennsylvanian. *Acanthodes bridgei.* A nicely preserved acanthodian fish in limestone matrix. Kansas. Courtesy: Geological Enterprises, Ardmore, Oklahoma. 9.5 cm.

25-3. Pennsylvanian. An immature acanthodian fish, probably *Acanthodes* sp., in brown shale matrix. Note the fin spines. Madera Shale, Sandia, Bernalillo County, New Mexico. 3.2 cm.

(Graphiurus), Heptanema, Moenkopia, Mylacanthus, Piveteauia, Sassenia, Scleracanthus, Sinocoelacanthus, Whitea, and *Wimania (Leioderma);* JURASSIC: *Bunoderma, Coccoderma (Kokkoderma), Cualabaea, Libys,* and *Rhipis;* CRETACEOUS: *Macropoma (Eurycormus), Macropomoides,* and *Mawsonia.* Family **Laughiidae:** TRIASSIC: *Holophagus (Undina),* and *Laughia.* Family **Latimeriidae:** RECENT: *Latimeria.*

Subclass Actinopterygii (Actinopterygians), Order Palaeonisciformes (paleoniscoids): Family **Cheirolepidae.** Family **Stegotrachelidae:** DEVONIAN: *Kentuckia* (see Figs. 25-30 through 25-33), *Moythomasia, Orvikuina,* and *Stegotrachelos.* Family **Tegeolepidae.** Family **Rhabdole-** pidae. Family **Rhadinichthyidae.** Family **Carbovelidae.** Family **Canobiidae.** Family **Cornuboniscidae:** MISSISSIPPIAN: *Cornuboniscus* (see Fig. 25-41). Families: **Cryphiolepidae. Holuriidae. Cosmoptychiidae. Styracopteridae. Ptgopteridae. Elonichthyidae. Acrolepidae. Coccocephalichthyidae. Haplolepidae:** PENNSYLVANIAN: *Haplolepis (Parahaplolepis)* (see Fig. 25-35), and *Pyritocephalus* (see Fig. 25-35). **Amblypteridae:** PERMIAN: *Amblypterus* (see Fig. 25-46). **Commentryidae. Palaeoniscidae:** TRIASSIC: *Turseodus* (see Figs. 25-49 and 25-50). **Aeduellidae:** PERMIAN: *Aeduella* (see Fig. 25-45), and *Westollia (Lepidopterus).* **Dicellopygidae. Boreolepidae. Birgeriidae. Scanilepidae. Centrolepidae.**

25-4. Pennsylvanian. Acanthodian fin spines in matrix. Madera Shale, Sandia, Bernalillo County, New Mexico. Courtesy: Geology Department, University of Oklahoma, Norman. ×2.4. U0106258.

25-5. Mississippian. Acanthodian tail and / or fin spine drag marks in the bottom muds (sandstone matrix). Hortons Bluff Formation, Blue Beach, Hants County, Nova Scotia, Canada. ×1.

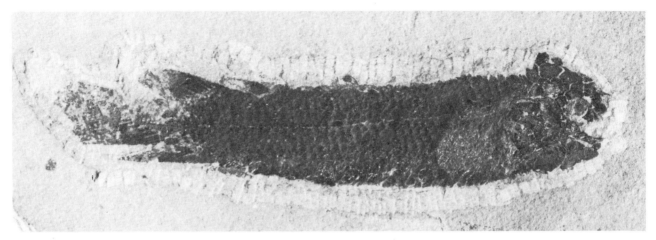

25-6. Devonian. *Dipterus valenciennesi* Sedgwick and Murchison. A nearly complete skeleton of a lungfish. Middle Old Red Sandstone, Caithness, Scotland. Approx. half size.

25-7. Devonian. *Dipterus* sp. A drawing of a lungfish from the Middle Old Red Sandstone of northern Scotland. Very few fossil specimens are found in complete condition. A re-construction is necessary to show what the fish may have looked like in life. Half size.

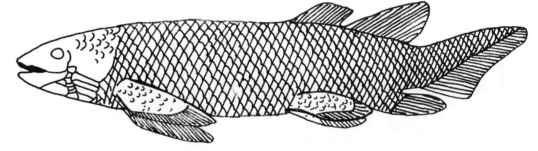

25-8. Pennsylvanian. *Sagenodus* sp. An isolated upper dental plate of a lungfish. South Bend Formation, Graham, Young County, Texas. ✕1.5.

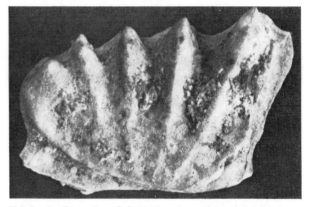

25-9. Permian. *Sagenodus* sp. An isolated lower jaw tooth plate of a lungfish. Lower Permian of Young County, Texas. ✕ 3.

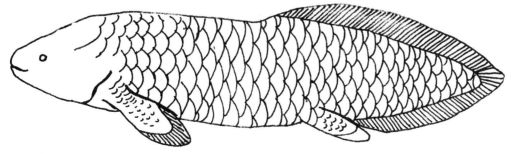

25-10. Jurassic. *Ceratodus (Hem-ictenodus) formosus.* A drawing of a reconstruction based upon fossil remains found in New South Wales, Australia. Approx. half size.

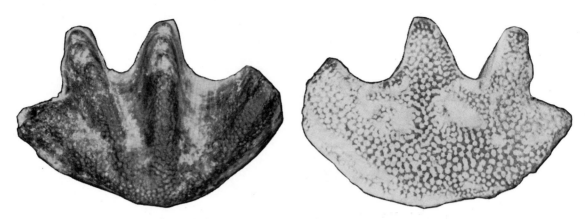

25-11. Cretaceous. *Ceratodus (Hemictenodus)* sp. Dorsal and ventral views of an isolated dental plate of a lungfish. Trinity Sands Formation (Albian), Forestburg, Montague County, Texas. ×2.

25-12. Devonian. *Gyropychius (Diplopterax) agassizi* McCoy. A complete rhizodont crossopterygian fish. Middle Old Red Sandstone, Orkney, Scotland. 35 cm. Courtesy: Siber and Siber Ltd., Aathal, Switzerland.

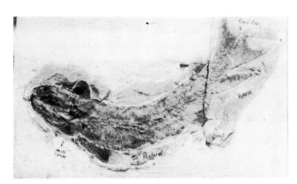

25-13. Devonian. *Eusthenopteron foordi* Whiteaves. A completely preserved fish in sandstone. Scaumenac Bay, Province of Quebec, Canada. ⅓ actual size. Courtesy: Rutgers University Museum, New Brunswick, New Jersey.

25-14. Pennsylvanian. *Rhizodopsis (Characodus)* sp. The scale from a rhizodont crossopterygian fish. Tebo-Mulkey Coal Shales, Montrose, Henry County, Missouri. ×3.

25-15. Devonian. *Onychodus sigmoides* Newberry. An arch of presymphysial teeth from the lower jaw of a large crossopterygian fish. Columbus Limestone (Middle Devonian), Delaware, Delaware County, Ohio. ×2.

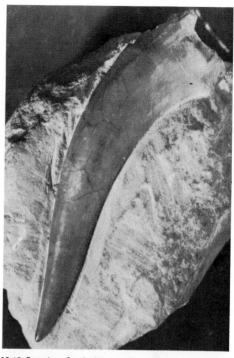

25-16. Devonian. *Onychodus sigmoides* Newberry. A single tooth from the presymphysial section of the lower jaw of a crossopterygian fish. (See Figs. 25-15 and 25-17.) Columbus Limestone Formation, Columbus, Franklin County, Ohio. 6.5 cm.

Coccolepidae.

Suborder Platysomoidei (Platysomatid fishes): Family **Platysomidae**: PENNSYLVANIAN: *Platysomus* (see Figs. 25-39 and 25-44). Family **Amphicentridae**: PENNSYLVANIAN: *Amphicentrum* (see Figs. 25-37 and 25-38). Families: **Bobastraniidae. Dorypteridae. Tarrasiidae. Ptycholepidae. Pholidopleuridae. Luganoiidae. Redfieldiidae (Catopteridae)**: TRIASSIC: *Redfieldia (Catopterus)* (see Fig. 25-47), and *Synorichthys* (see Fig. 25-48). **Perleididae (Colobondontidae). Cleithrolepidae. Peltopleuridae. Platysiagidae. Cephaloxenidae.** and **Aethodontidae.**

Order Acipensiformes (long-snouted, partly cartilaginous (chondrostean) fishes, including sturgeons): Families: **Phanerorhynchidae**: PENNSYLVANIAN: *Phanerorhynchus* (see Fig. 25-43). **Errolichthyidae. Chondrostidae. Saurichthyidae. Acipenseridae.** and **Polyodontidae.**

Infraclass Holostei (Holostean fishes). Order Semionotiformes: Family **Semionotidae (Lepidotidae)**: TRIASSIC: *Semionotus* (see Fig. 25-51); JURASSIC: *Dapedius* (see Fig. 25-68). Family **Lepisosteidae**: EOCENE: *Lepisosteus (Lepidosteus)* (see Figs. 25-74, and 25-113 through 25-115).

Order Pycnodontiformes (pycnodonts). Family **Pycnodontidae**: JURASSIC: *Gyrodus (Stromateus)* (see Figs. 25-67 and 25-69), and *Proscinetes (Microdon)* (see Fig. 25-66); CRETACEOUS: *Anomaeodus (Pycnodus)* (see Figs. 25-71 and 25-72), *Eomesodon* (see Fig. 25-70), and *Ancistrodon (Stephanodus)* (see Figs. 25-100 and 25-101).

25-17. Devonian. *Onychodus sigmoides* Newberry. Another single tooth from the presymphysial region, this time showing partial internal structure. Columbus Limestone Formation, Columbus, Franklin County, Ohio. 6 cm.

25-18. Mississippian. *Megalichthys (Carlukeus)* sp. An isolated scale from a crossopterygian fish—probably *Megalichthys (Carlukeus) hibberti*. Upper Carboniferous, Wakefield, Yorkshire, England. ✕2.

25-20. Pennsylvanian. *Rhizodus* sp. An isolated fish scale probably belonging to this crossopterygian fish. Cannel below the Upper Freeport Coal, Allegheny Series, Linton Mine Tailings, Jefferson County, Ohio. ✕1.5.

Order Amiiformes, suborder Amioidei (bowfins): Family **Caturidae:** JURASSIC: *Caturus* (see Figs. 25-61 and 25-63). **Amiidae:** JURASSIC: *Urocles (Megalurus)* (see Fig. 25-60); EOCENE: *Amia* (see Fig. 25-109 and 25-112); OLIGOCENE: *Amia* (see Fig. 25-126). **Macrosemiidae. Pachycormidae:** CRETACEOUS: *Protosphyraena* (see Fig. 25-80).

Order Aspidorhynchiformes: Family **Aspidorhynchidae:** JURASSIC: *Aspidorhynchus* (see Figs. 25-64 and 25-65), CRETACEOUS: *Belonostomus* (see Figs. 25-62, 25-75, 25-102 through 25-104).

Order Pholidophoriformes: Family **Pholidophoridae:** JURASSIC: *Pholiodophorus* (see Fig. 25-53).

Infraclass Teleostei (Teleostean fishes), Order Leptolepiformes: Family **Leptolepidae:** JURASSIC: *Leptolepis* (see Figs. 25-52, 25-54 through 25-58); CRETACEOUS: *Cearana* (see Figs. 25-105 through 25-107), and *Tharrias.*(See Fig. 25-108).

Superorder Elopomorpha, Order Elopiformes, Suborder Elopidei: Family **Elopidae:** JURASSIC: *Thrissops* (see Fig. 25-59). **Apsopelicidae. Pachyrhizodontidae:** CRETACEOUS: *Pachyrhizodus* (see Fig. 25-80).

Suborder Albuloidei: Family **Albulidae:** CRETACEOUS: *Paralbula* (see Figs. 25-76 and 25-77).

Order Anguilliformes (Apodes) (eels, etc.). No representatives of this order in this book.

25-19. Mississippian. *Rhizodus hibberti* (Agassiz). A tooth fragment, partially restored (dotted lines indicate missing parts), of a crossopterygian fish. Carboniferous, Lochgelly, Fife, Great Britain (England). ✕6.5.

25-21. Pennsylvanian. *Rhabdoderma (Conchiopsis) exiguum* Eastman. A complete skeleton of a coelacanth fish in half of a siderite concretion. Essex Marine Fauna (Mazon Creek), Braidwood, Will County, Illinois. ×2. Courtesy: Field Museum of Natural History, Chicago.

25-22. Triassic. *Diplurus newarki* (Bryant). A complete coelacanth fish in a curved position. Lockatong Formation, Newark Series, Granton Quarry, North Bergen, Hudson County, New Jersey. Half size.

25-23. Triassic. *Diplurus newarki* (Bryant). A nicely preserved coelacanth fish in hard mud-shale. Lockatong Formation, Newark Series, Granton Quarry, North Bergen, Hudson County, New Jersey. ×2.

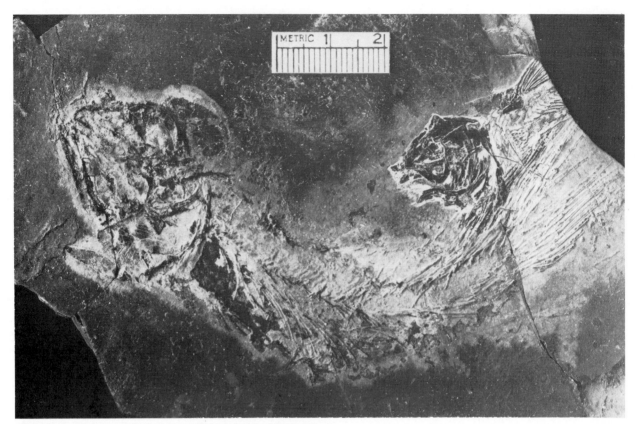

25-24. Triassic. *Diplurus newarki* (Bryant). A complete dorsoventrally preserved skeleton of a coelacanth fish. Note the opercula flattened-out on either side of the skull. A smaller coelacanth fish can be seen (skull and postcranial elements) to the upper right of the main fish's tail. Lockatong Formation, Newark Series, Granton Quarry, North Bergen, Hudson County, New Jersey.

25-25. Triassic. *Diplurus newarki* (Bryant). A splendidly preserved skeleton (lateral view) of a coelacanth fish. Lockatong Formation, Newark Series, Granton Quarry, North Bergen, Hudson County, New Jersey. × 1.5.

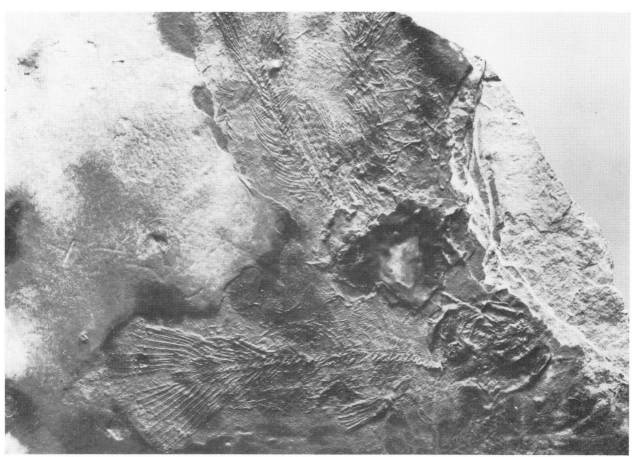

25-26. Triassic. *Diplurus newarki* (Bryant). Two skeletons of coelacanth fishes in mud-shales. Note the scales showing up just beneath the head of the top fish. Lockatong Formation, Newark Series, Granton Quarry, North Bergen, Hudson County, New Jersey. ×2.

25-27. Triassic. *Diplurus newarki* (Bryant). An isolated operculum (gill plate cover) of a coelacanth fish. Lockatong Formation, Newark Series, Granton Bergen, Hudson County, New Jersey. ×5.5.

25-28. Triassic. *Diplurus newarki* (Bryant). A skull and postcranial elements of a coelacanth fish. Lockatong Formation, Newark Series, Princeton, Mercer County, New Jersey. Slightly smaller than actual size.

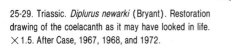

25-29. Triassic. *Diplurus newarki* (Bryant). Restoration drawing of the coelacanth as it may have looked in life. ×1.5. After Case, 1967, 1968, and 1972.

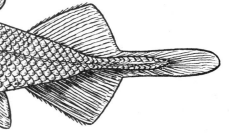

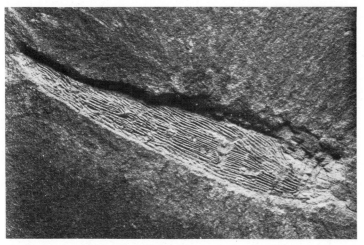

25-30. Devonian. *Kentuckia hlavini* Dunkle. The lower jaw of a paleoniscoid fish. Cleveland Shales, Cleveland, Cuyahoga County, Ohio. ✕3.

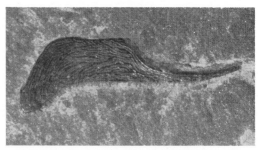

25-3l. Devonian. *Kentuckia hlavini* Dunkle, The (maxillary) upper jaw of a paleoniscoid fish. (See Fig. 25-30 for the lower jaw.) Cleveland Shales, Brooklyn Heights, Cuyahoga County, Ohio. ✕2.

25-32. Devonian. *Kentuckia hlavini* Dunkle. An isolated scale from a paleoniscoid fish. Cleveland Shales, Brooklyn Heights, Cuyahoga County, Ohio. ✕15.

25-33. Devonian. A gular plate (throat bone) of a paleoniscoid fish (species indeterminate, but possibly *Kentuckia*). Silica Shales, Milan, Washtenaw County, Michigan. ✕2.

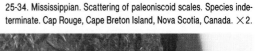

25-34. Mississippian. Scattering of paleoniscoid scales. Species indeterminate. Cap Rouge, Cape Breton Island, Nova Scotia, Canada. ✕2.

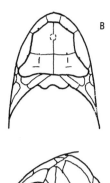

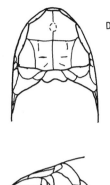
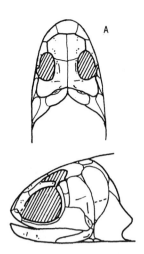

25-35. Pennsylvanian. Haplolepid (paleonisciformes) fishes: A. *Pyritocephalus lineatus*. B. *Haplolepis corrugata*. C. *Haplolepis (Parahaplolepis) tuberculata* (formerly: *Eurylepis tuberculatus* Newberry). D. *Haplolepis ovoidea*. All of the above skulls (top row: dorsal view of the skull roof, bottom row: lateral (side) view of the skulls) are haplolepid fishes from the Allegheny series, of Linton, Jefferson County, Ohio. From Baird, 1962, with permission. Drawings by D. Baird.

25-36. Pennsylvanian. Skeleton (covered by scales) of a paleoniscoid fish (species indeterminate) in coal shale. Severy Shale Member, Wabaunsee Formation (Virgilian), Evans Coal Mine, Clarinda, Page County, Iowa. × 1. Courtesy: Iowa State Historical Society, Des Moines, Iowa.

25-37. Pennsylvanian. *Amphicentrum* sp. A complete platysomoid chondrostean fish in black tar-paper shales. Stark Shale, Missouri Series, Winterset, Madison County, Iowa. Slightly less than half size. Courtesy: Iowa State Historical Society, Des Moines, Iowa.

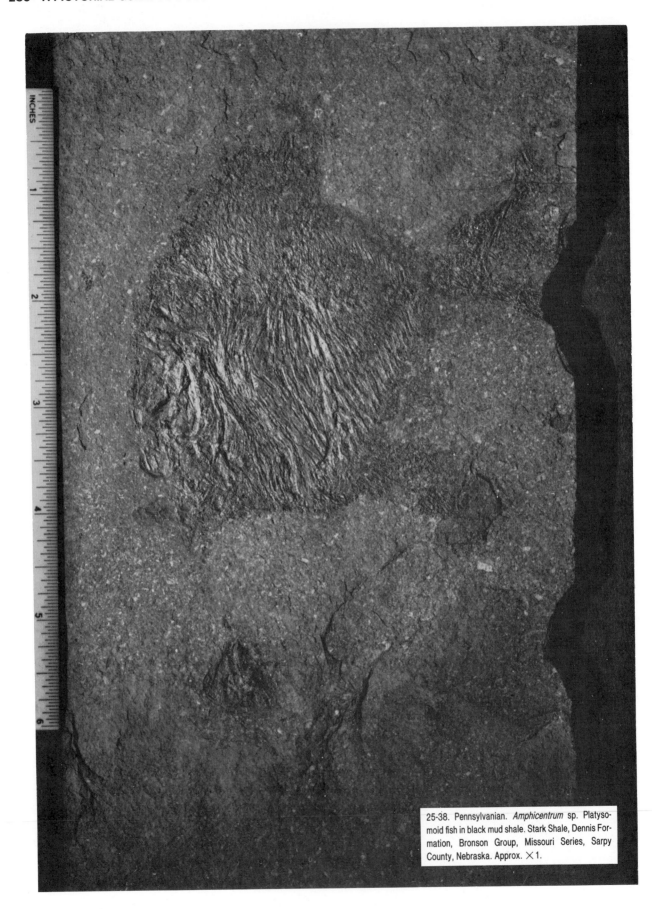

25-38. Pennsylvanian. *Amphicentrum* sp. Platyso-moid fish in black mud shale. Stark Shale, Dennis Formation, Bronson Group, Missouri Series, Sarpy County, Nebraska. Approx. ✕ 1.

25-39. Pennsylvanian. *Platysomus circularis* (Newberry and Worthen). A complete platysomoid fish in a half of a siderite concretion. Essex Marine Fauna (Mazon Creek), Will County, Illinois. ×1.5. Courtesy: Field Museum of Natural History, Chicago.

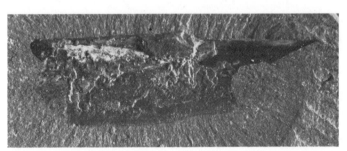

25-40. Pennsylvanian. Paleoniscoid fish scale in black shale. Stark Shale, Dennis Limestone Formation, Bronson Group (Missourian), Ft. Calhoun, Washington County, Nebraska. ×2.

25-41. Mississippian. *Cornuboniscus* sp. The complete skeleton of a paleoniscoid fish from the Carboniferous of Nova Scotia, Canada. 8 cm. Courtesy: Geological Enterprises, Ardmore, Oklahoma.

25-42. Pennsylvanian. A large paleoniscoid gular plate in black shale. The scale is outlined on the shale with white paint, as it is difficult to spot the scale on the black matrix. Stark Shale, Dennis Formation, Bronson Group (Missourian), Ft. Calhoun, Washington County, Nebraska. ×1.5.

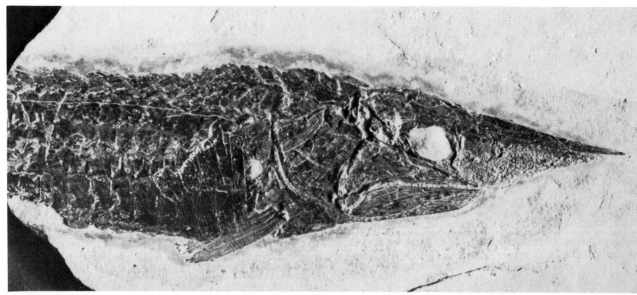

25-43. Pennsylvanian. *Phanerorhynchus* sp. The front half of the body of a long-snouted chondrostean fish of the order paleonisciformes. Fishes of this type may have given rise to the order acipensiformes. Madera Formation, Manzano Mountains, Bernalillo County, New Mexico. ×2. From Cavender and Case, 1973.

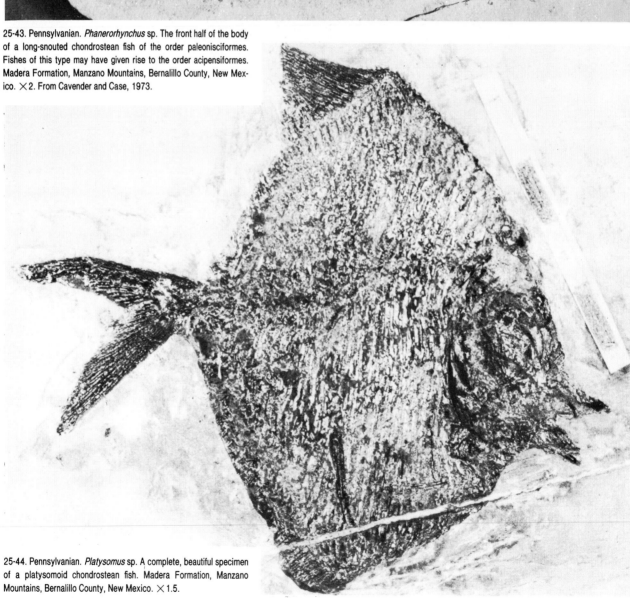

25-44. Pennsylvanian. *Platysomus* sp. A complete, beautiful specimen of a platysomoid chondrostean fish. Madera Formation, Manzano Mountains, Bernalillo County, New Mexico. ×1.5.

25-45. Permian. *Aeduella blainvillei.* Two paleoniscoid fish on a slab of coal shale. Massif Central, France. Slightly less than actual size. Courtesy: Geological Enterprises, Ardmore, Oklahoma.

25-46. Permian. *Amblypterus macropterus.* A nicely preserved paleoniscoid fish from the Lower Permian Rothliegendes (Red Beds). Kreuznach, Germany. 17 cm Courtesy: Siber and Siber Ltd., Aathal, Switzerland.

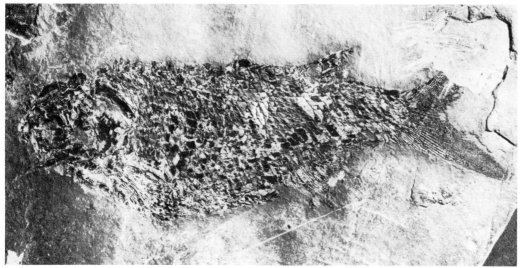

25-47. Triassic. *Redfieldia (Catopterus)* sp. *(Redfieldius redfieldi).* A skeleton of a catopterid paleonisciforme fish in lake bottom mud shales. Stockton Formation, Durham, Middlesex County, Connecticut. ⅔ actual size.

25-48. Triassic. *Synorichthys* sp. Disarticulated scales of a paleoniscoid fish in lake bottom mud shales. Lockatong Formation, Newark Group, Granton Quarry, North Bergen, Hudson County, New Jersey. ×1.

25-49. Triassic. *Turseodus* sp. A section of articulated body scales of a paleoniscoid fish. Lockatong Formation, Newark Group, Granton Quarry, North Bergen, Hudson County, New Jersey. ×2.

25-50. Triassic. *Turseodus acutus.* Complete lower jaws and scattered upper jaws and skull elements of this paleoniscoid fish. Lockatong Formation, Newark Group, Granton Quarry, North Bergen, Hudson County, New Jersey. ×1.5.

Order Notacanthiformes (Lyopomi) Same as above.

Superorder Clupeomorpha, Order Clupeiformes, Suborder Denticiptoidei (no representatives). Suborder Clupeoidei: Family **Clupeidae:** CRETACEOUS: *Diplomystus* (see Figs. 25-81 through 25-83); EOCENE: *Diplomystus* (see Figs. 25-120, 25-122 through 25-124, *Knightia* (see Figs. 25-116, 25-119 through 25-121, 25-125 and 25-138); MIOCENE: *Xyne* (see Fig. 25-151); PLIOCENE: *Ganolytes* (see Fig. 25-163). Family **Chirocentridae** (no representatives).

Superorder Osteoglossomorpha, Order Osteoglossiformes, Suborder Ichthyodectoidei: Family **Ichthyodectidae:** CRETACEOUS: *Ichthyodectes* (see Fig. 25-80) and *Xiphactinus (Portheus)* (see Fig. 25-80). **Saurocephalidae:** CRETACEOUS: *Saurocephalus* (see Figs. 25-78 and 25-79). **Thryptodontidae (Plethodontidae):** CRETACEOUS: *Bananogmius (Anogmius).*

Suborder Osteoglossoidei: Family **Osteoglossidae:** EOCENE: *Phareodus (Dapedoglossus)* (see Figs. 25-116 through 25-118). **Pantodontidae** (no representatives).

Suborders Mormyriformes, Notopteroidei, and Salmonoidei (no representatives).

The following suborders also have no representation in this work: Argentinoidei, Galaxiodei, Esocoidei (Haplomi), Stomiatoidei, Alepocephaloidei, and Bathylaconoidei.

Suborder Myctophoidei: Family **Enchodontidae:** CRETACEOUS: *Cimolichthys (Empo)* (see Fig. 25-80), *Enchodus* (see Figs. 25-84 through 25-90), and *Eurypholis* (see Fig. 25-91). Families: **Tomognatidae. Ipnopidae. Paralepidae (Sudidae). Omosusidae. Aleposauridae. Anopteridae. Evermannellidae. Scopelarchidae. Scopelosauridae. Myctophidae (Scopelidae). Neoscopelidae. Cheirothricidae:** CRETACEOUS: *Exocoetoides* (see Fig. 25-92). **Dercetidae.**

The following suborders have no representation in this book: Cetomimoidei, Ateleopodoidei, Giganturoidei, Mir-

25-51. Triassic. *Semionotus* sp. A complete holostean fish in lake bottom mud (sandstone). Brunswick Formation, Newark Group, above Hook Mt. Basalt, Boonton Reservoir, Boonton, Morris County, New Jersey. Approx. ×1. Courtesy: Princeton University Museum of Natural History, Princeton, New Jersey.

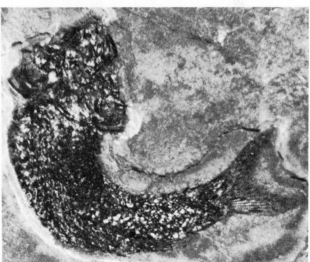

25-53. Jurassic. *Pholidophorus americana*. A holostean fish in sandstone. Dorsoventrally preserved (note the spread-out opercula reminiscent of the preservation of the fish in Fig. 25-24). Todilto Formation, Guadalupe County, New Mexico. ×1. Courtesy: Black Hills Institute of Geological Research, Hill City, South Dakota.

25-52. Jurassic. *Leptolepis sprattiformis* Agassiz. A complete teleostean fish in lithographic limestone. Solnhofen Limestone (Liassic Malm Zeta), Solnhofen, Bavaria, West Germany. ×1.

25-54. Jurassic. *Leptolepis schoewi*. A nicely preserved teleostean fish in sandstone. Todilto Formation, Guadalupe County, New Mexico. ×1. Courtesy: Black Hills Institute of Geological Research, Hill City, South Dakota.

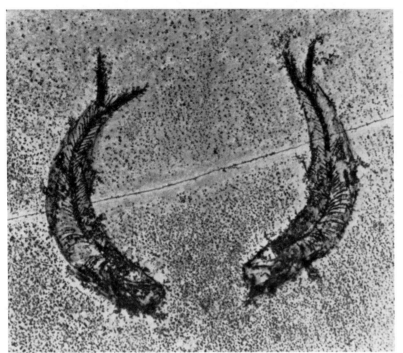

25-55. Jurassic. *Leptolepis sprattiformis* Agassiz. Two nicely preserved teleostean fishes in lithographic limestone. Malm Zeta (Liassic), Solnhofen, Bavaria, West Germany. 7 cm. Courtesy: Siber and Siber Ltd., Aathal, Switzerland.

25-56. Jurassic. *Leptolepis sprattiformis* Agassiz. A matrix covered teleost specimen. Malm Zeta (Liassic), Solnhofen Limestone, Eichstätt, Bavaria, West Germany. ✕ 1.

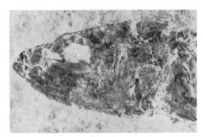

25-57. Jurassic. *?Leptolepis* sp. A complete lateral view of the head of a teleostean fish common to the lithographic limestones of Southern Germany. Solnhofen, Bavaria, West Germany. ✕ 1.5.

25-58. Jurassic. *Leptolepis knorri* Agassiz. A teleostean fish. The fish is curved around upon itself in death. Upper Malm (Liassic), Solnhofen Plattenkalk, Langenheim, Bavaria, West Germany. ✕ 2.

appinnatoidei, Gonorynchoidei, Chanoidei, Characoidei, and Gymnotoidei.

Order Ctenothrissiformes: Family **Ctenothrissidae:** CRETACEOUS: *Ctenothrissa* (see Fig. 25-98).

Suborder Cyprinoidei: Family **Cyprinidae:** MIOCENE: *Leuciscus* (see Fig. 25-162).

Order Siluriformes (Nematognathi) (catfishes): Families: **Diplomystidae. Ictaluridae (Ameiuridae):** EOCENE: *Hypsidoris* (see Figs. 25-127 and 25-128). **Bagridae (Porcidae-Mystidae). Craniglanididae. Siluridae. Schilbeidae and Pangasidae. Amblycipitidae. Amphiliidae. Akysidae. Sisoridae (Bagariidae). Clariidae. Heteropneustidae. Chacidae. Olyridae. Malapteruridae. Mochokidae (Syn-**odontidae). **Ariidae (Tachysuridae):** EOCENE: *Rhineastes (Astephas)* (see Fig. 25-129). **Doradidae. Auchenipteridae. Aspredinidae (Bunocephalidae). Plotosidae. Pimelodontidae. Ageneiosidae. Hypopthalmidae. Helogeneidae. Cetopsidae. Stegophilidae (Pygidiidae-Eretmophilidae-Trichomycteridae). Callichthyidae. and Loricariidae.** Superorder Paracanthopterygii, Suborder Aphredoderoidei: Family **Aphredoderidae:** EOCENE: *Amphiplaga* (see Figs. 25-131 and 25-132), *Asineops* (see Fig. 25-130), and *Erismotopterus* (see Figs. 25-133 and 25-134).

The following suborders are not represented in this book: Percopsoidei, Lophiodei, Antennarioidei, Cera-

25-59. Jurassic. *Thrissops salmoneus* (Blainville). A beautifully preserved elopiform teleost fish in lithographic limestone. Malm Zeta (Lias), Solnhofen Plattenkalk, Eichstätt, Bavaria, West Germany. 18 cm. Courtesy: Siber and Siber Ltd., Aathal, Switzerland.

25-60. Jurassic. *Urocles (Megalurus) polyspondylus* (Wagner). A finely preserved amiid fish in lithographic limestone. Quarry workers call it the "broom fish." Solnhofen Plattenkalk, Malm Zeta (Liassic), Eichstätt, Bavaria, West Germany. 20 cm. Courtesy: Siber and Siber Ltd., Aathal, Switzerland.

25-61. Jurassic. *Caturus furcatus* (Agassiz). A splendidly preserved amioid fish in lithographic limestone. Solnhofen Limestone, Eichstätt, Bavaria, West Germany. 14 cm. Courtesy: Siber and Siber Ltd., Aathal, Switzerland.

25-62. Jurassic. *Belonostomus* sp. A superbly preserved garfish in lithographic limestone. Note the ammonoid in the lower right hand corner. Solnhofen Formation, Daitingen, Bavaria, West Germany. 43 cm. Courtesy: Siber and Siber Ltd., Aathal, Switzerland.

25-63. Jurassic. *Caturus furcatus* (Agassiz). Another example of an amioid fish (see Fig. 25-61). Note that this specimen is outlined by manganese dendrites (the little "treelike" branches all around the specimen). Solnhofen Limestone, Solnhofen, Bavaria, West Germany. 31 cm. Courtesy: Siber and Siber Ltd., Aathal, Switzerland.

25-64. Jurassic. *Aspidorhynchus acutirostris* Blainville. A spectacular example of an elongated fish in lithographic limestone. Solnhofener Plattenkalk (Liassic), Eichstätt, Bavaria, West Germany. 61 cm. Courtesy: Siber and Siber Ltd., Aathal, Switzerland.

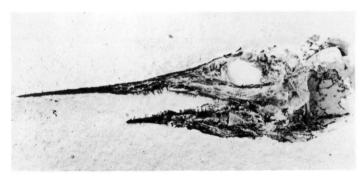

25-65. Jurassic. *Aspidorhynchus* sp. The head of a long-snouted fish in lithographic limestone. Solnhofener Plattenkalk (Liassic), Solnhofen, Bavaria, West Germany. 11 cm.

25-66. Jurassic. *Proscinetes (Microdon) radiatus* (Agassiz). An almost complete (missing tail section and posterior portion of the haemal and neural arches) pycnodontid fish in a fragment of limestone. Purbeck Beds (Portlandian), Swanage, Dorset, England. 10 cm. Courtesy: Siber and Siber Ltd., Aathal, Switzerland.

25-67. Jurassic. *Gyrodus (Stromateus) macrophthalmus* Agassiz. Upper jaw (right side) of a pycnodont fish. Liassic, Kelheim, Bavaria, West Germany. ×5.5.

25-68. Jurassic. *Dapedius politus* (Leach). A remarkably preserved semionotid fish. Lyme Regis, Dorset, England. Slightly less than actual size. (PU 3219.) Courtesy: Princeton University Museum of Natural History, Princeton, New Jersey.

25-69. Jurassic. *Gyrodus (Stromateus) macropthalmus* Agassiz. A complete pycnodont fish. (See Fig. 25-67 for part of the dentition of this species.) Tithonian, Sicily, Italy. 4.5 cm.

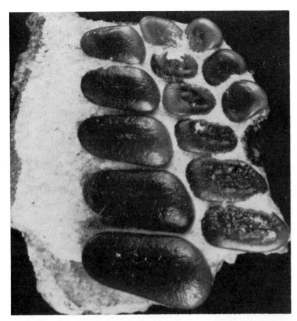

25-70. Early Cretaceous. *Eomesodon* sp. A fragment of one of the jaw plates of a pycnodontid fish. Trinity Sands Formation (Albian), Forestburg, Montague County, Texas. ×30.

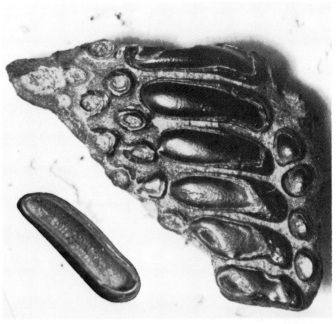

25-71. Cretaceous. *Anomaeodus (Pycnodus) phaseolus* (Hay). A section of dentition from the lower jaw of a pycnodontid fish. Note the isolated tooth plate in its ventral view (lower left corner). Navesink Formation (Middle Maestrichtian), Big Brook, Marlboro Township, Monmouth County, New Jersey. ×2.

25-72. Cretaceous. *Anomaeodus (Pycnodus) phaseolus* (Hay). Section of a dentition from the upper jaw of a pycnodontid fish. (See Fig. 25-71 for the lower jaw.) Navesink Formation (Middle Maestrichtian), Holmdel, Monmouth County, New Jersey. ×1.5.

25-73. Cretaceous. A fragment of scales (squamation) of a garfish. Species indeterminate. Chalk Beds of England. ×2.

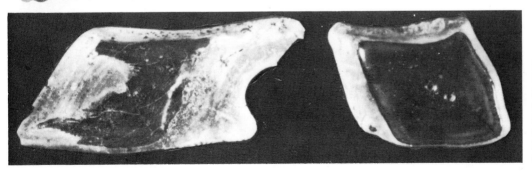

25-74. Cretaceous. Two isolated garfish scales. Species possibly Lepisosteus (Lepidosteus), the garpike. Teapot Sandstone Member, Mesaverde Formation (Late Campanian), Big Horn Basin, Washakie County, Wyoming. ×3.

25-75. Cretaceous. A complete, curled-up aspidorrhynchid fish—possibly *Belonostomus* (Vinctifer) sp. The mandible is longer than that of *Aspidorhynchus*. Cenomanian, Hadjula, Lebanon. 7 cm. Courtesy: Siber and Siber Ltd., Aathal, Switzerland.

25-76. Cretaceous. *Paralbula casei* Estes. Almost complete basibranchial dentition plate of an albulid fish. (Species named after the author.) Judith River Formation (Campanian), Blaine County, Montana. ×2.

25-77. Cretaceous. *Paralbula casei* Estes. Section or fragment of the basibranchial tooth plate of an albulid fish (type species named after the author). Navesink Formation (Middle Maestrichtian), Tributary Brook, Holmdel, Monmouth County, New Jersey. ×4.5

25-78. Cretaceous. *Saurocephalus* sp. A fragment of the jaw bone of an osteoglossomorph fish. Navesink Formation (Middle Maestrichtian), Big Brook, Marlboro, Monmouth County, New Jersey. ×1.

25-79. Cretaceous. *Saurocephalus* sp. A large section of the dentary of an osteoglossomorph fish. Maestrichtian, Robin-Their, Belgium. ×1. Courtesy: Musée du Silex, Eben-Emael, Belgium.

25-80. Cretaceous. Fragments of bony fishes from the Kansas Chalk Beds, i.e, jaw sections, vertebral centra, and fin support bones: 1. *Protosphyraena* sp. (spine fragment), 2. *Cimolichthys (Empo)* sp. (dentary fragment), 3. *Cimolichthys (Empo) semianceps* (palatine section), 4. *Ichthyodectes* sp. (vertebral column fragment), 5. *Xiphactinus audax* (hypural bone fragment), 6. *Pachyrhizodus* sp. (dentary section), 7. *Ichthyodectes* sp. (vertebral column fragment), 8. *Cimolichthys (Empo) semianceps* (mandible section), 9. *Cimolichthys (Empo) nepoelica* (ethmoid section), 10. *Ichthyodectes* sp. (vertebral column section at the hypural), 11. *Ichthyodectes* sp. (vertebra), 12. *Xiphactinus audax* (?neural arch section), 13. *Gillicus arcuatus* (matching mandible sections upside down in photo), and 14. *Ichthyodectes* sp. (vertebra). All material from the Niobrara Chalk Formation of Kansas. Approx. ⅔ actual size.

25-81. Cretaceous. *Diplomystus* sp. A. A single clupeid (herring) fish in lake bottom sandstone. Cenomanian, Hadjula, Lebanon. B. A group of *Diplomystus* (herrings) in light buff sandstone. Cenomanian, Mt. Lebanon, Hakel, Lebanon. A. ✕ 1.5. B. Half size.

A

B

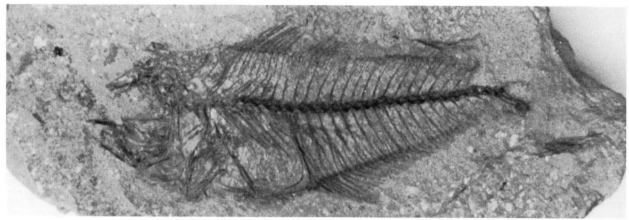

25-82. Cretaceous. *Diplomystus* sp. A clupeid (herring) fish skeleton in chalk (missing tail section). Cenomanian, Misburg bei Hannover, West Germany. ×1.5.

25-83. Cretaceous. *Diplomystus brevissimus* Blainville. A rather "puffed-up" clupeid (herring). Cenomanian, Hadjula, Lebanon. 7 cm. Courtesy: Siber and Siber Ltd., Aathal Switzerland.

tioidei, Muraenolepidoidei, Gadoidei, Ophidioidei, Zoarcoidei, Macrouroidei, Exocoetoidei, Cyprinodontoidei, Atherinoidei, Stephanoberycoidei, Polymixiodei.

Suborder Dinopterygoidei: Family **Aipichthyidae:** CRETACEOUS: *Aipichthys* (see Figs. 25-94 through 25-97).

Suborder Berycoidei: Families: **Diretmidae. Kosorogateridae. Anoplogasteridae. Berycidae:** CRETACEOUS: *Holopteryx* (see Fig. 25-93). **Monocentridae. Anomalopidae. Holocentridae.**

The following order and suborders are not represented in this work: Order: Zeiformes. Suborders: Lampridoidei, Veliferoidei, Trachipteroidei, and Styleophoroidei.

Suborder Gasterosteoidei: Family **Gasterosteidae:** PLIOCENE: *Gasterosteus* (see Figs. 25-164 and 25-165).

Family **Macrorhamphidae (Centriscidae):** Miocene: *Centriscus (Amphisyle)* (See Fig. 25-152).

The following orders and suborders are not represented in this book: Orders: Channiformes, Synbranchiformes, Scorpaeniformes, Dactylopteriformes, and Pegasiformes. Suborders: Aulostomoidei, Syngnathoidei, Scoraenoidei, Hexagrammoidei, Platycephaloidei, Hoplichthyoidei, Congiodoidei, and Cottoidei.

Order Perciformes, Suborder Percoidei: Families: **Centropomidae. Serranidae:** OLIGOCENE: *Dapalis (Smerdis)* (see Figs. 25-110 and 25-111); EOCENE: *Cyclopoma* (see Fig. 25-139). **Plesiopidae. Pseudoplesiopidae. Anisochromidae. Acanthoclinidae. Glaucosmidae. Therapomidae. Banjosidae. Kuhlidae. Gregoryinidae. Centrarchidae. Priacanthidae. Apogonidae (Cheilodipteridae).**

25-84. Cretaceous. *Enchodus ferox* Leidy. The dentary section of a rather ubiquitous myctophoid fish. Navesink Formation (Middle Maestrichtian), Hop Brook, Colts Neck, Monmouth County, New Jersey. ×2.

25-85. Cretaceous. *Enchodus* sp. Tooth in fragment of dentary bone. Campanian, Chico Local Fauna, Oroville, Butte County, California. ×1.5.

25-86. Cretaceous. *Enchodus ferox* Leidy. Two views of a palatine tooth of a myctophoid fish. Navesink Formation (Middle Maestrichtian), Ramanesson Brook, Holmdel, Monmouth County, New Jersey. ×1.5.

Percidae: EOCENE: *Mioplosus* (see Figs. 25-136 through 25-138), *Perca* (see Fig. 25-135). **Sillaginidae. Branchiostegidae (Latilidae). Labracoglossidae. Lactaridae. Menidae:** EOCENE: *Mene* (see Fig. 25-140). **Sparidae:** EOCENE: *Sparnodus* (see Fig. 25-141); MIOCENE: *Sargus* (see Fig. 25-159). **Scatophagidae:** EOCENE: *Scatophagus* (see Fig. 25-149). **Pomacentri-**dae: EOCENE: *Priscacara* (see Figs. 25-144 through 25-147). **Sphyraenidae:** MIOCENE: *Sphyraena* (see Fig. 25-154). **Labridae:** EOCENE: *Phyllodus* (see Figs. 25-142 and 25-143). **Trichiuridae:** EOCENE: *Trichiurides* (see

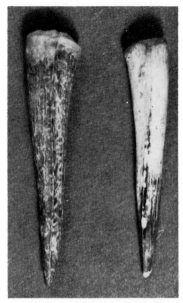

25-87. Cretaceous. *Enchodus* sp. Two isolated mandibular teeth (detached from sockets of jaw). Navesink Formation (Middle Maestrichtian), Ramanesson Brook, Holmdel, Monmouth County, New Jersey. ×1.5.

25-88. Cretaceous. *Enchodus* sp. Top: palatine bone with a fragment of the tooth. Bottom: two views of a complete tooth from a palatine. Merchantville Formation (Late Santonian), Chesapeake and Delaware Canal, St. Georges, New Castle County, Delaware. ×1.

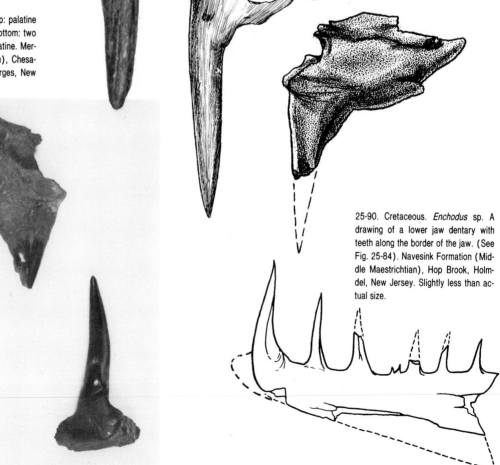

25-89. Cretaceous. *Enchodus ferox* Leidy. A large upper jaw palatine bone with complete tooth. Navesink Formation (Middle Maestrichtian), Hop Brook, Holmdel, Monmouth County, New Jersey. ×2.

25-90. Cretaceous. *Enchodus* sp. A drawing of a lower jaw dentary with teeth along the border of the jaw. (See Fig. 25-84). Navesink Formation (Middle Maestrichtian), Hop Brook, Holmdel, New Jersey. Slightly less than actual size.

25-91. Cretaceous. *Eurypholis boissieri* (Pictet). A nicely preserved myctophoid fish. Upper Santonian, Sahel Alma, Lebanon. 19 cm. Courtesy: Siber and Siber Ltd., Aathal, Switzerland.

25-92. Cretaceous. *Exocoetoides minor* (Davis). A very fine dorsoventrally preserved cheirothricid fish. Upper Santonian, Sahel Alma, Lebanon. 4.5 cm. Courtesy: Siber and Siber Ltd., Aathal, Switzerland.

25-93. Cretaceous. *Holopteryx* sp. A partial berycid fish in chalk. The head and tail are missing. The body shows squamation. Chalk Beds, Blue Bell Hill, Chatham, Kent, England. ×1.5.

25-94. Cretaceous. *Aipichthys* sp. (probably *A. formosus*). A nicely preserved carangid (horse mackerel) fish. Upper Santonian, Sahel Alma, Lebanon. 5 cm. Courtesy: Geological Enterprises, Ardmore, Oklahoma.

Fig. 25-148). *Xiphiidae:* EOCENE: *Cylindracanthus* (see Fig. 25-150); MIOCENE: *Xiphiorhynchus* (see Fig. 25-155). **Diodontidae:** MIOCENE: *Diodon* (see Fig. 25-158).

Subclass Acanthodii/order Acanthodiformes/family **Mesacanthidae:** DEVONIAN: *Mesacanthus.* Family **Acanthodidae:** DEVONIAN-PERMIAN: *Acanthodes*

25-95. Cretaceous. *Aipichthys vellifer* Steindachner. A very fine skeleton of a horse-mackerel fish. Upper Santonian, Sahel Alma, Lebanon. 10.5 cm. Courtesy: Geological Enterprises, Ardmore, Oklahoma.

25-96. Cretaceous. *Aipichthys* sp. A horse mackerel fish skeleton. Sediment is marine limestone. Upper Santonian, Sahel Alma, Lebanon. 7.5 cm. Courtesy: Siber and Siber Ltd., Aathal, Switzerland.

25-97. Cretaceous. *Aipichthys formosus.* A beautifully preserved horse mackerel fish. Upper Santonian, Sahel Alma, Lebanon. 3.5 cm. Courtesy: Siber and Siber Ltd., Aathal, Switzerland.

25-98. Cretaceous. *Ctenothrissa vexilifer* Pictet. The skeleton of a perch-like fish. Cenomanian, Hadjula, Lebanon. 5 cm. Courtesy: Geological Enterprises, Ardmore, Oklahoma.

(see Figs. 25-2 through 25-5). Order Climatiformes/family **Climatiidae**: DEVONIAN: *Climatius.*

Osteichthyes, the bony fishes: The myriad species in the world of fossil fishes could not possibly fit into this one volume—let alone this particular chapter. What the author has done here, is to present the reader with typical fish skeletons and isolated fish parts which are commonly found by collectors and scientists. The chapter starts out with Dipnoi or the lungfishes. Historically, these lung-

fishes are most important, for they and their associated species of crossopterygians and coelacanths gave rise to the amphibians (see the following chapter).

The lungfishes are able to survive outside of water, and, in some modern species, can roll themselves up in mud-ball "cocoons." The dampness of the mud-balls enables the lungfish to survive during drought seasons. These fishes are transported to laboratories, for study, in mud-balls as well. The most usual fossil of a lungfish that you might find in a fossil deposit is part of the jaw or the isolated upper or lower jaw tooth plates (see Figs. 25-8, 25-9, and 25-11). The whole skeleton of these fishes is rather rare (see Figs. 25-6, 25-7, and 25-10).

The crossopterygians or "fringe-finned ganoids" are represented in the fossil rocord by two suborders: Rhipidistia and Coelacanthini (Actinistia). In the Rhipidistia there are two basic superfamilies: the Osteolepidoidea (Osteolepiformes) and the Holoptychoidea (Porolepiformes). In appearance, many species of the Crossopterygii resemble the amphibians, and may well be the ancestral stock from whence they came.

The coelacanths are most curious. For a long time it was thought that this was an extinct lineage, until, in the year 1938, a scientist discovered that the coelacanth still existed in the oceans. A fisherman netted one off the east coast of

25-99. Cretaceous. A fancy perch (species indeterminate). Upper Santonian, Sahel
Alma, Lebanon. 15 cm. Courtesy: Siber and Siber Ltd., Aathal, Switzerland.

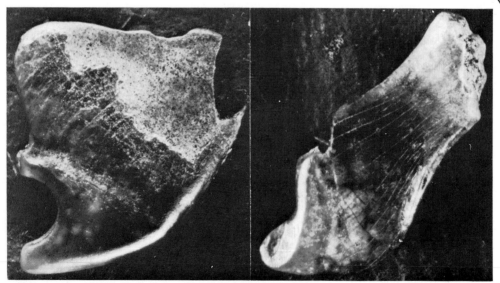

25-100. Cretaceous. *Stephanodus (Ancistrodon)* sp. Two specimens of the "nibbling"
teeth of a sclerodermid fish. Navesink Formation (Middle Maestrichtian), Ramanesson
Brook, Holmdel Monmouth County, New Jersey. ×5.

25-101. Cretaceous. *Stephanodus
(Ancistrodon)* sp. Nibbling teeth of
a sclerodermid fish. Navesink Formation (Middle Maestrichtian),
Big Brook, Marlboro, Monmouth
County, New Jersey. Photo specimen. ×4.

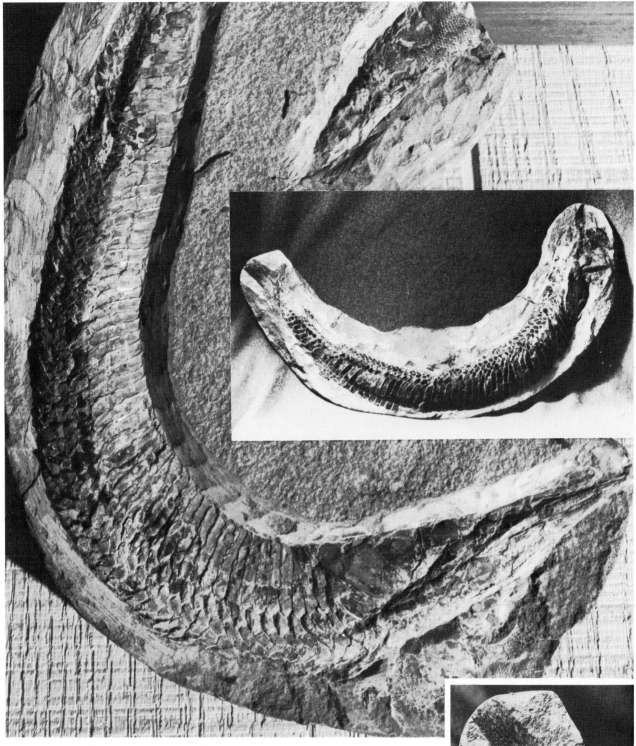

25-102. Cretaceous. *Belonostomus (Vinctifer) comptoni* Agassiz. An essentially "three-dimensional" fish specimen—a lengthy garfish in a concretion. Inset: Another specimen of the same species. Note the lengthy scales. Série Ararife, Santana Formation, Chapada do Ararife, Regiao do Céara, Brazil. 35 cm.

25-103. Cretaceous. *Belonostomus (Vinctifer) comptoni* Agassiz. Close-up of the tail region of the specimen in the inset of Fig. 25-102. Série Ararife, Santana Formation, Chapada do Ararife, Regiao do Céara, Brasil.

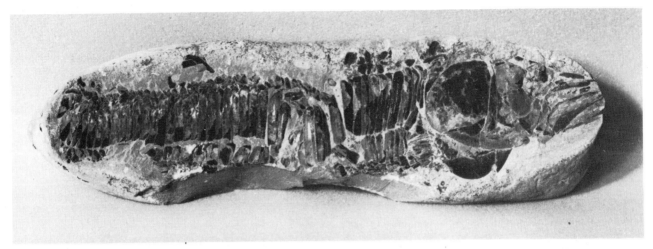

25-104. Cretaceous. *Belonostomus (Vinctifer) comptoni* Agassiz. Two-thirds of a garfish body. This specimen is missing its tail section and part of its rostrum or snout. Série Ararife, Santana Formation, Chapada do Ararife, Regiao do Céara, Brazil. Approx. half size.

25-105. Cretaceous. *Cearana* sp. A three-dimensional fish in a concretion. This specimen is missing its tail section. Série Ararife, Santana Formation, Chapada do Ararife, Regiao do Céara, Brazil. Approx. half size.

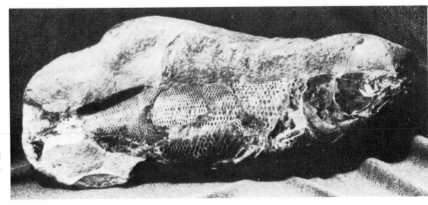

25-106. Cretaceous. *Cearana* sp. (See Figs. 25-105 and 25-107.) This specimen is complete. Série Ararife, Santana Formation, Chapada do Ararife, Regiao do Céara, Brasil. Approx. ⅓ actual size.

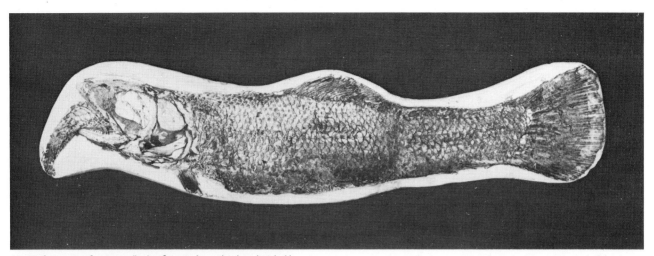

25-107. Cretaceous. *Cearana* swallowing *Cearana*. A complete large leptolepid fish in a concretion (in three dimensional state of preservation). The large fish has ingested a smaller fish of the same species, and this had probably caused its death. Série Ararife, Santana Formation, Chapada do Ararife, Regiao do Céara, Brasil. 78 cm. Courtesy: Siber and Siber Ltd., Aathal, Switzerland.

25-108. Cretaceous. *Tharrias ciraripis.* Part and counterpart of another leptolepid fish, showing, this time, the internal structures (vertebral centra, etc.) of the fish. Série Ararife, Santana Formation, Chapada do Ararife, Regiao do Céara, Brasil.

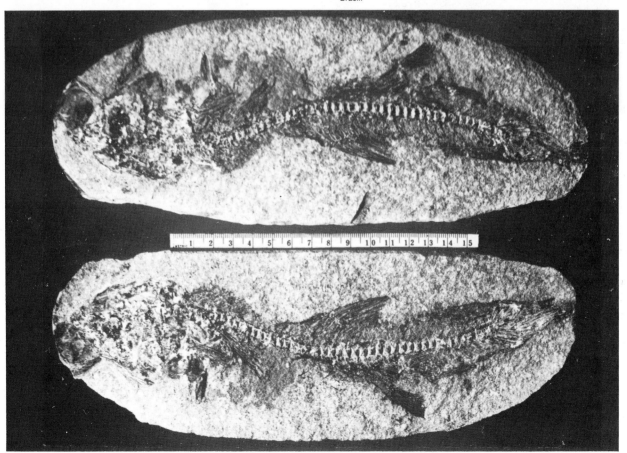

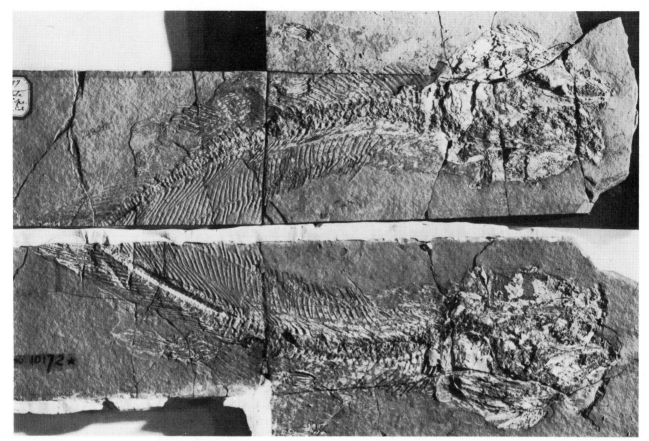

25-109. Oligocene. *Amia scutata* Cope. The plate and counterplate of a bowfin fish. Florissant, Teller County, Colorado × 1. Courtesy: Princeton University Museum of Natural History, Princeton, New Jersey.

South Africa, and now we have *Latimeria,* a "living fossil."

Fossil coelacanths were never too common, with the exception of the ubiquitous Triassic form *Diplurus* (see Figs. 25-22 through 25-29), which constitutes the largest population in some eastern (USA) fossil lake mudshales, namely the Lockatong Formation of New Jersey. The largest group of Paleozoic fishes found in the fossil record is Paleonisciformes or the paleoniscoids. This group comprises many of the types of scale fishes found in the Mississippian through the Permian, with a large population in the Pennsylvanian. Species such as: *Paleoniscus* (Fig. 25-36), *Amphicentrum* (Figs. 25-37 and 25-38), *Platysomus* (Figs. 25-39 and 25-44), *Cornuboniscus* (Fig. 25-41), and *Amblypterus* (Fig. 25-46) are quite common to certain Carboniferous formations throughout the world. Not all the skeletons may be intact (articulated), and most finds are of disarticulated or scattered scales and jaw parts, see, for example, Fig. 25-34. A complete skeleton is usually the exception to the rule.

The Mesozoic brought a very large array of species of bony fishes, many of which are the ancestral stock of today's ocean fauna.

25-110. Oligocene. *Dapalis (Smerdis) minutus* (Agassiz). A well preserved perch fish. Aix-en-Provence, France. Half size.

The Cretaceous, in particular, included a most interesting variety of species including the earliest gars, pycnodonts, and clupeids, among many others. Many of our modern forms of sharks arrived during the Cretaceous as well. A larger discussion on these latter fishes is offered the reader in the preceeding chapter.

The Eocene is by far the time for early "freshwater" fishes, as seen by the myriad species found within the fossil lake beds of the vast sediments of the Green River Formation of Wyoming, Utah, and Colorado. In these beds can be found all manner of fishes from the common herrings (*Diplomystus* and *Knightia,* see Figs. 25-121 through 25-125), to perches (*Mioplosus,* see Figs. 25-136

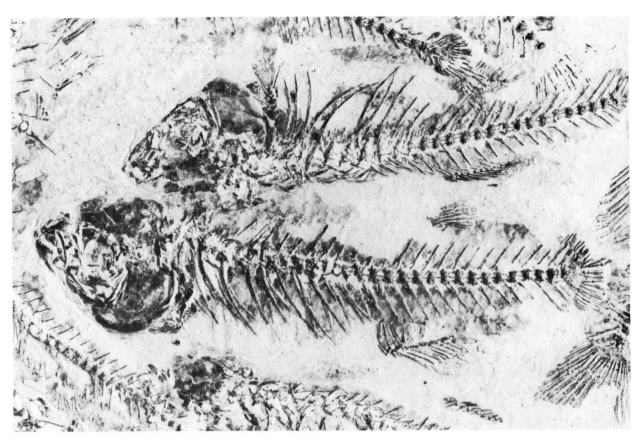

25-111. Oligocene. *Dapalis (Smerdis) minutus* (Agassiz). Two complete perch fish among several other fragments of the same species. Aix-en-Provence, France. ×1.5.

25-112. Eocene. *Amia kehreri* Andreae. A complete amiid (bowfin) fish in a coal-ball concretion. Messel bei Darmstadt, Germany. 22 cm. Courtesy: Geological Enterprises, Ardmore, Oklahoma.

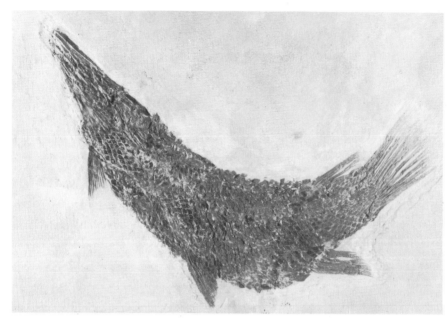

25-113. Eocene. *Lepisosteus cuneatus* (Cope) (also known as *Lepidosteus*). A complete garfish (garpike) in lake bottom mud shale. Fossil Butte Member, Green River Formation, Sweetwater County, Wyoming. 66.5 cm. Courtesy: Geological Enterprises, Ardmore, Oklahoma.

25-114. Eocene. *Lepisosteus (Lepidosteus) simplex* (Leidy). A beautifully preserved garpike fish in lake bottom mud shale. Fossil Butte Member, Green River Formation, Fossil, Lincoln County, Wyoming. 75 cm. Courtesy: Princeton University Museum of Natural History, Princeton, New Jersey.

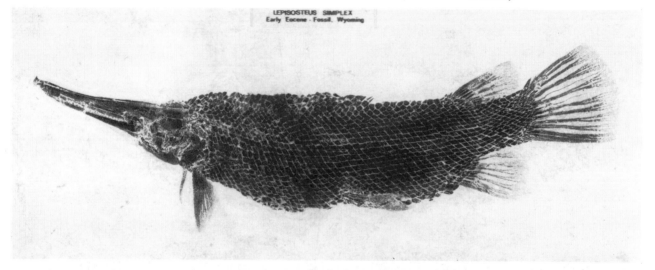

25-115. Eocene. *Lepisosteus (Lepidosteus) strausi.* A splendid garpike fish in oil-schist shale. Messel bei Darmstadt, West Germany. 24 cm. Courtesy: Siber and Siber Ltd., Aathal, Switzerland.

25-116. Eocene. *Phareodus* and *Knightia*. A beautifully preserved osteoglossid fish with a little clupeid (herring) lying just above its head. Fossil Butte Member, Green River Formation, Kemmerer, Lincoln County, Wyoming. 20 cm. Courtesy: Black Hills Institute of Geological Research, Hill City, South Dakota.

25-117. Eocene. *Phareodus testis* (Cope). A superb example of this osteoglossid fish. Fossil Butte Member, Green River Formation, Ulrich Quarry, Diamondville, Lincoln County, Wyoming. 13 cm.

25-118. Eocene. *Phareodus* sp. Dorsal view of the braincase and a partial jaw section. Fossil Butte Member, Green River Formation, Kemmerer, Lincoln County, Wyoming. ✕1.5.

25-119. Eocene. *Knightia eocaena* Jordan. A small specimen of this rather ubiquitous clupeid (herring). Fossil Butte Member, Green River Formation, Kemmerer, Lincoln County, Wyoming. ✕1.5.

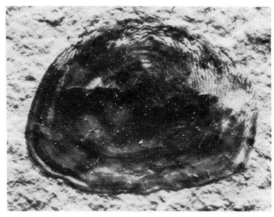

25-120. Eocene. An isolated scale of a clupeid (herring) belonging to either *Knightia* or *Diplomystus*. Fossil Butte Member, Green River Formation, Kemmerer, Lincoln County, Wyoming. 2 mm.

25-121. Eocene. *Knightia* sp. A school of herrings of one of the most common fishes in the Middle Eocene. Fossil Butte Member, Green River Formation, Fossil, Lincoln County, Wyoming. Avg. fish about 8.5 cm. Courtesy: Geological Enterprises, Ardmore, Oklahoma.

through 25-138) and catfishes (*Hypsidoris,* see Figs. 25-127 and 25-128, and *Rhineastes,* see Fig. 25-129). Freshwater stingrays (*Heliobatis,* see Fig. 24-115) have also been found in these shales and sandstones, as well as crocodiles, snakes, turtles, bats, and birds!

25-122. Eocene. *Diplomystus dentatus* Cope. A large clupeid fish in lake bottom mud shale. Fossil Butte Member, Green River Formation, Kemmerer, Lincoln County, Wyoming. 46 cm.

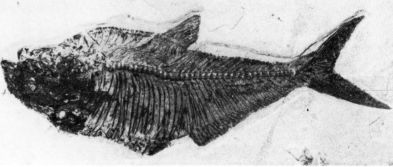

25-123. Eocene. *Diplomystus dentatus* Cope. A nicely preserved clupeid (herring) fish in lake bottom mud shale. Fossil Butfe Member, Green River Formation, Uinta County, Wyoming. ✕1.5

25-124. Eocene. *Diplomystus dentatus* Cope. Two fine examples of clupeid (herring) fishes. Fossil Butte Member, Green River Formation, Kemmerer, Lincoln County, Wyoming.

25-125. Eocene *Gosiutichthys parvus* Grande. Multiple fish burial in lake bottom mud shale. Fossil Butte Member, Green River Formation, La Barge, Lincoln County, Wyoming. 1 fish is 5 cm. Courtesy: Siber and Siber Ltd., Aathal, Switzerland.

25-126. Eocene. *Amia (Kindleia) fragosa* (Jordan). A beautifully preserved large bowfin fish in lateral view. Fossil Butte Member, Green River Formation, Kemmerer, Lincoln County, Wyoming. 35 cm in length. Courtesy: Siber and Siber Ltd., Aathal, Switzerland.

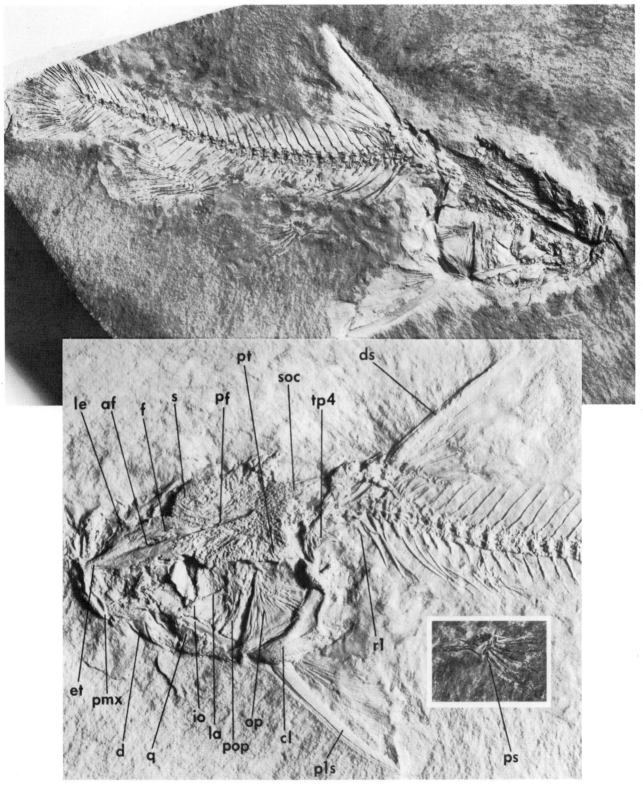

25-127. Eocene. *Hypsidoris farsonensis* Lundberg and Case. Top: a dorsal view of the skull roof and a lateral view of the body of an ictalurid catfish. Laney Shale Member, Green River Formation, Farson, Sublette County, Wyoming. ×1. Bottom: labels for the various bones in the skull: af = anterior fontanelle, cl = cleithrum, d = dentary, ds = dorsal spine, et = supraethmoid, f = frontal, io = intraorbitals, la = levator palatini crest on hyomandibular, le = lateral ethmoid, op = opercle, pf = posterior fontanelle, pmx = premaxilla, pop = preopercle, pt = pterotic, pls = pectoral spine, q = quadrate, rl = first rib, s = sphenotic, soc = supraoccipital, and tp4 = expanded transverse process of fourth vertebra. From Lundberg and Case, 1970, *Jour. Paleo.*, vol. 44. no. 3.

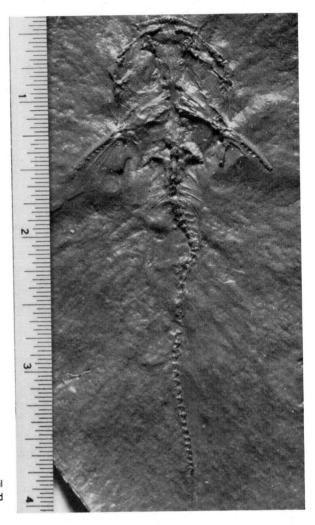

25-128. Eocene. *Hypsidoris farsonensis* Lundberg and Case. Left: original specimen in oil shale. Right: a red colored rubber latex peel made from the original. Ventral view of an ictalurid catfish. Laney Shale Member, Green River Formation, Farson, Sublette County, Wyoming.

The rarer fishes from the Green River Formation (Eocene beds) are: the garpike (*Lepisosteus,* see Figs. 25-113 and 25-114), the osteoglossid fish (*Phareodus,* see Figs. 25-116 through 25-118), *Asineops* (Fig. 25-130), *Amphiplaga* (Figs. 25-131 and 25-132), and *Erismotopterus* (Figs. 25-133 and 25-134), the latter three fishes are particularly rare.

There are two basic types of fossil preservation in the Green River Formation. The most common type is the "chalky" matrix found in most outcroppings, particularly at Fossil Lake and Kemmerer, Lincoln County, Wyoming. In this matrix, the skeleton of the fish is preserved intact with all its bones and, sometimes, scales. The "oil shales" are the other type of preservation. These are the harder rocks found up in Sublette County, and Shirley Basin, Carbon County, Wyoming. The best examples of these "hard-rock" fishes are found in the Farson area. These rocks, when split open, show a plate and a counterplate of the specimen. The bones are gone, and there is a natural mold (for example, see *Hypsidoris,* the ictalurid catfish, in Figs. 25-127 and 25-128). A rubber latex peel (bottom of

Fig. 25-127, and the right hand picture of Fig. 25-128) can be made from these oil-shale specimens. Thus, a cast can be made from the "mold." These casts give us much information—the shape and detail of missing bones and teeth, for example.

Note: The Green River fossil beds are essentially closed to collecting now. Basically, the land and its geological treasures must be preserved for future generations. Most of the fossiliferous areas of the Green River Formation in Wyoming are now part of a National Park called "Fossil Buttes."

The Eocene was also a time for many marine fishes as well as the freshwater species of the Green River Basin. In the seas from Midway time (Paleocene) up through the Wilcoxian/Claibornian/Jacksonian of the Eocene, many types of wrasses (*Phyllodus,* see Figs. 25-142 and 25-143) and breams (*Sparnodus,* see Fig. 25-141) were nibbling at coral reefs, while gars and bowfins were gliding silently and stealthfully through calm estuary waters in search of

25-129. Eocene. *Astephus antiquus* (Leidy), an ictalurid (catfish), and *Diploymstus,* a clupeid (herring) in lake bottom mud shale. Fossil Butte Member, Green River Formation, Kemmerer, Lincoln County, Wyoming. ✕1. Courtesy: Black Hills Institute of Geological Research, Hill City, South Dakota.

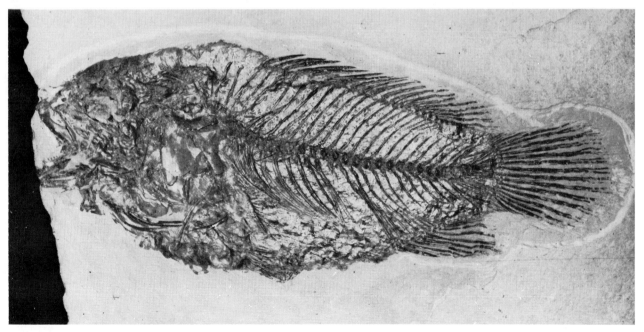

25-130. Eocene. *Asineops squamifrons* Cope. One of the larger of the freshwater fishes. Fossil Butte Member, Green River Formation, Kemmerer, Lincoln County, Wyoming. ✕1. Courtesy: Black Hills Institute of Geological Research, Hill City, South Dakota.

25-131. Eocene. *Amphiplaga brachyptera* Cope. Another species of typical "trout-perch" fishes from the Fossil Butte Member of the Green River Formation of Wyoming. × 1. Courtesy: Black Hills Institute of Geological Research, Hill City, South Dakota.

25-132. Eocene. *Amphiplaga brachyptera* Cope. Another specimen of this unusual "trout-perch" fish from the Fossil Butte Member of the Green River Formation of Wyoming. × 1. Courtesy: Geological Enterprises, Ardmore, Oklahoma.

25-133. Eocene. *Erismatopterus levatus* (Cope). Another type of freshwater fish from the Fossil Butte Member of the Green River Formation of Wyoming. × 1.5.

25-134. Eocene. *Erismatopterus levatus* (Cope). Another specimen of the third species of freshwater fishes indigenous to the Fossil Butte Member of the Green River Formation of Wyoming. × 1.5.

25-135. Eocene. *Perca* sp. A perch fish in lake bottom mud shale. Monte Bolca, Italy. × 1.

25-136. Eocene. *Mioplosus labracoides* Cope. The skeleton of a rather large perch fish in lake bottom mud shale. Fossil Butte Member, Green River Formation, Kemmerer, Lincoln County, Wyoming. ✕1.5. Courtesy: Geological Enterprises, Ardmore, Oklahoma.

25-137. Eocene. *Mioplosus labracoides* Cope. A beautifully preserved perch fish in lake bottom mud shale. Fossil Butte Member, Green River Formation, Ulrich Quarry, Diamondville, Lincoln County, Wyoming, 24 cm.

25-138. Eocene. *Mioplosus* (a perch) swallowing the herring *Knightia*. Fossil Butte Member, Green River Formation of Wyoming. × 1. Courtesy: Princeton University Museum of Natural History, Princeton, New Jersey.

25-139. Eocene. *Cyclopoma* sp. A rather large perch in lateral view. Monte Bolca, Italy. 40 cm. Courtesy: Siber and Siber Ltd., Aathal, Switzerland.

smaller prey; in the oceans, the fierce barracuda (*Sphyraena,* see Fig. 25-154) chased after the pompanos and perch-like fishes.

The Miocene found all of the same species and more in its oceans, bays and estuaries. By now, the croakers or sea basses ("drum fishes") were nibbling on the Miocene cor-

als along with the porcupine fishes (*Diodon,* see Fig. 25-158), while the barracuda and the giant "great white shark" ruled the watery expanses.

The Miocene was also the time when sailfishes and swordfishes started their evolutionary trend towards today's modern oceanic forms.

25-140. Eocene. *Mene rhombea* Volta. Large deep-bodied fish. Monte Bolca, Italy. 23 cm. Courtesy: Siber and Siber Ltd., Aathal, Switzerland.

25-141. Eocene. *Sparnodus vulgaris.* A beautifully preserved seabream in lake bottom mud shale. Monte Bolca, Italy. 20 cm. Courtesy: Geological Enterprises, Ardmore, Oklahoma.

25-142. Eocene. *Phyllodus toliapicus* Agassiz. A complete pharyngeal mouth plate of a wrasse fish. Aquia Formation (London Clay equivalent-Ypresian), Belvidere Beach, Prince Georges County, Virginia. ×1.5.

25-143. Eocene. *Phyllodus toliapicus* Agassiz. An almost complete pharyngeal mouth plate of a wrasse fish (see Fig. 25-142). Aquia Formation (Ypresian). Marlborough Point, Potomac River, Prince Georges County, Virginia. ✕ 1.5.

25-144. Eocene. *Priscacara liops* Cope. A fine example of the subtropical freshwater chromid fish. Fossil Butte Member, Green River Formation, Ulrich Quarry, Diamondville, Lincoln County, Wyoming. 15.2 cm.

25-145. Eocene. *Priscacara serrata* Cope. A subtropical chromid fish with no known living relative. Fossil Butte Member, Green River Formation, Ulrich Quarry, Diamondville, Lincoln County, Wyoming. Slightly less than half actual size.

25-146. Eocene. *Priscacara serrata* Cope. A subtropical freshwater fish of the family Chromidae. Fossil Butte Member, Green River Formation, Kemmerer, Lincoln County, Wyoming. ½ actual size.

25-147. Eocene. *Priscacara liops* Cope. A slightly disarticulated subtropical freshwater chromid fish from the lake bottom mud shales. Fossil Butte Member, Green River Formation, Kemmerer, Lincoln County, Wyoming. Half size.

25-148. Eocene. *Trichiurides* sp. An isolated tooth from a scombroid fish. Twiggs Clay Member, Barnwell Formation (Jacksonian), Huber, Twiggs County, Georgia. ×25.

25-149. Eocene. *Scatophagus frontalis* Agassiz. A beautifully preserved perciform fish. Monte Bolca, Italy. 10 cm. Courtesy: Siber and Siber Ltd., Aathal, Switzerland.

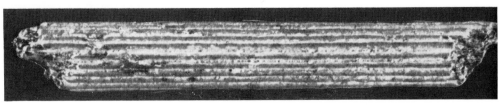

25-150. Eocene. *Cylindracanthus acus.* Fragment of a rostrum of the blociid fish (similar in appearance, but not related to the gars). Twiggs Clay Member, Barnwell Formation (Jacksonian), Huber, Twiggs County, Georgia. ×3.5.

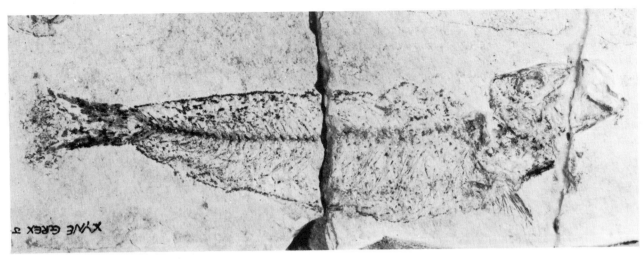

25-151. Miocene. *Xyne grex* Jordan and Gilbert. Shoal herrings recovered in diatomaceous rocks. Monterey Formation, Lompoc, Santa Barbara County, California. 17.5 cm.

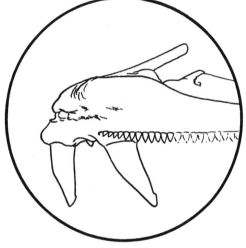

25-152. Miocene. *Centriscus (Amphisyle) heinrichsi* (Heckel). A fine example of a laterally preserved skeleton of a trumpet fish. Ancona, Italy. 11 cm long. Courtesy: Siber and Siber Ltd., Aathal, Switzerland.

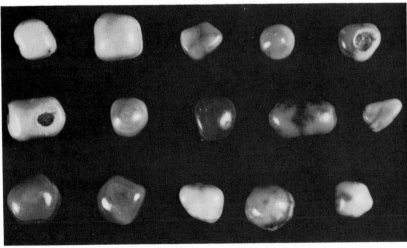

25-153. Miocene. Isolated teeth of a croaker or "drum" fish in the sciaenid group of percoids. Burdigalian, Port-du-Bouc near Fos-sur-Mer, Bouches-du-Rhone Departement, France. Avg. size 4 mm.

25-154. Miocene. *Sphyraena* sp. Bottom: isolated incisor tooth with bony root of a fossil barracuda (see drawing above for position of the tooth in the jaw). Pungo River Marl Formation (Helvetian), Aurora, Beaufort County, North Carolina. ×2.

25-155. Miocene. Two views of a swordfish rostrum tip, probably: *Xiphiorhynchus* sp. Top: ventral or lower side of rostrum. Bottom: dorsal or upper side of the rostrum. Pungo River Marl Formation (Helvetian), Aurora, Beaufort County, North Carolina. 20 cm.

25-156. Miocene. Typical garfish scales (ganoin ivory) found in the phosphate pits of Polk County, Florida. (See Fig. 25-74 for comparison.) Tortonian, Palmetto Washer, Brewster, Polk County, Florida. ×1.5.

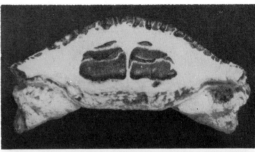

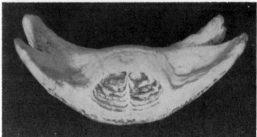

25-158. Miocene. *Diodon circumflexus* Leriche. Upper and lower jaws of a fossil porcupine fish. Pungo River Marl Formation Helvetian), Aurora, Beaufort County, North Carolina. ×1.5.

25-157. Miocene. *Polymerichthys nagurai* Uyeno. Newly discovered fossil deep-sea fish. Helvetian, Horaiji-san, Minami-sidara, Aichi-ken, Japan. ×1. Courtesy: Dr. Y. Hasegawa, Yokohama University.

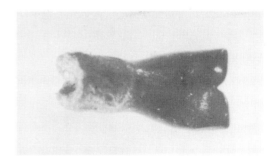

25-159. Miocene. *Sargus* sp. Incisor tooth of a sea bream. Note the typical "chisellike" biting edge of the tooth. Pungo River Marl Formation (Helvetian), Aurora, Beaufort County, North Carolina. ×22.

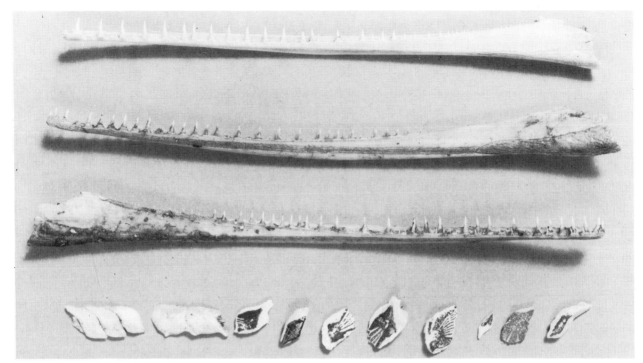

25-160. Recent. Garfish jaws and scales (with the exception of the last 8 scales at the lower right in the bottom row which are Miocene in age). Modern material from Lake Monroe, Enterprise, Volusia County, Florida. The fossil garfish scales are from the Bone Valley Sediments (Phosphate Pits), Polk County, Florida. Approx. ×1.

25-161. Miocene. Various actinopterygian (bony fishes) vertebrae. Species indeterminate. Pungo River Marl Formation (Helvetian), Aurora, Beaufort County, North Carolina. ×1.

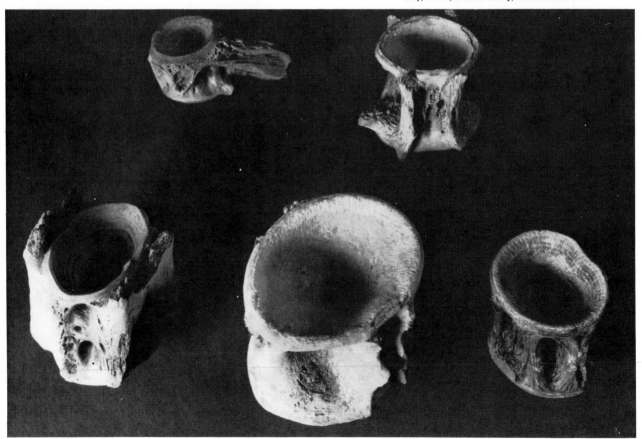

25-162. Miocene. *Leuciscus tarsiger* Troschel. A cyprinid (carp) fish. Rott bei Bonn, West Germany. ×1.5.

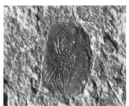

25-163. Pliocene. *Ganolytes (Dirasias)* sp. An isolated clupeid (herring) fish scale. Note the interesting ornamentation (pattern) on the scale. Gaviotta Beach, Santa Barbara County, California. ×2.

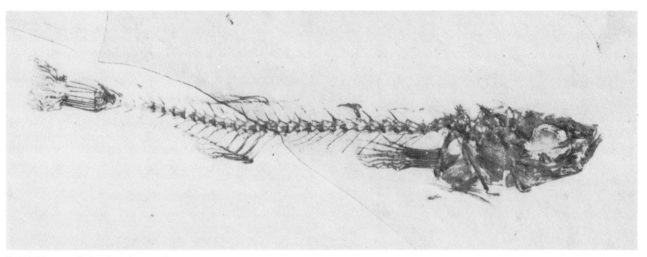

25-164. Pliocene. *Gasterosteus doryssus.* A complete fossil "stickleback" fish in diatomaceous shale. Lyon County, Nevada. ×2.

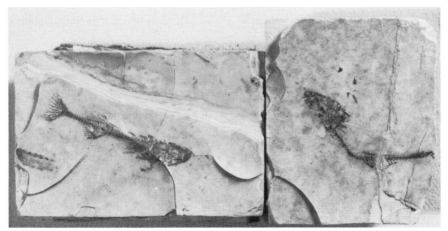

25-165. Pliocene. *Gasterosteus doryssus.* Two examples of stickleback fishes in diatomaceous shales. Lyon County, Nevada. Slightly smaller than actual size.

26 AMPHIBIA (Amphibians)

The class amphibia is made up of three major subclasses: Labyrinthodontia, Lepospondyli, and Lissamphibia.

In the subclass Labyrinthodontia, there are three orders: Ichthyostegalia, Temnospondyli, and Anthracosauria. The first order Ichthyostegalia is comprised of three families: **Elpistostegidae, Ichthyostegidae:** DEVONIAN: *Ichthyostega* (see Fig. 26-1) and *Ichthyostegopsis,* and the family **Otocratiidae.**

The order Temnospondyli has three suborders: Rhachitomi, Stereospondyli, and Plagiosauria. In the order Rhachitomi, there are these superfamilies with their individual families and some genera: Loxommatoidea, family **Loxommatidae.** Edopoidea, families **Edopidae, Dendrererpetontidae, Cochleosauridae,** and **Colosteidae.** Trimerorhachoidea, families **Trimerorhachidae:** PENNSYLVANIAN: *Lafonius* (see Figs. 26-2 and 26-3), *Saurerpeton* (see Figs. 26-4 and 26-5), **Dvinosauridae.** Superfamily Eryopoidea, families: **Eryopidae:** PERMIAN: *Eryops* (see Figs. 26-14 through 26-17), *Onchiodon (Branchiosaurus)* (see Figs. 26-24 through 26-27), **Dissorophidae:** PENNSYLVANIAN: *Eoscophus* (see Fig. 26-6); PERMIAN: *Cacops* (see Fig. 26-18), **Trematopsidae:** PERMIAN: *Trematops* (see Fig. 26-19), **Parioxyidae, Zatracheidae, Archegosauridae, Melanosauridae** and **Intasuchidae.** Superfamily Trematosauroidea, families **Trematosauridae** and **Rhytidosteidae.** In the suborder Stereospondyli: superfamily Rhinesuchoidea, families **Rhinesuchidae, Lydekkerinidae** and **Uranonocentrodontidae.** Superfamily Capitosauroidea, families **Benthosuchidae** and **Capitosauridae.** Superfamily Metoposaurioidea, family **Metosauridae.**

In the suborder Plagiosauria, are the families **Peltobatrachidae** and **Plagiosauridae.**

The order Anthracosauria has four suborders: Schizomeri, Diplomeri, Embolomeri, and Seymouriamorpha.

In the suborder Schizomeri, the family **Pholidogasteridae.** In the suborder Diplomeri, the family **Diplovertebrontidae.** In the suborder Embolomeri, families **Anthracosauridae:** PENNSYLVANIAN: *Anthracosaurus* (see Figs. 26-7 and 26-8), and **Cricotidae.** In the suborder Sey-

mouriamorpha, the families **Seymouridae:** PERMIAN: *Seymouria* (see Figs. 26-20 and 26-22), **Kotlassiidae, Discosaurisidae, Chronisuchidae, Tseajaiidae,** and **Diadectidae.**

In the subclass Lepospondyli, there are three orders: Nectridea, Aistopoda, and Microsauria.

In the order Nectridea, the following families are represented: **Urocordylidae, Lepterpetontidae,** and **Keraterpetontidae:** PERMIAN: *Diplocaulus* (see Figs. 26-21 and 26-23).

In the order Aistopoda, the families **Phlegethontidae:** PENNSYLVANIAN: *Phlegethontia* (see Fig. 26-10), **Ophiderpetontidae:** PENNSYLVANIAN: *Ophiderpeton* (see Fig. 26-9). In the order Microsauria, the families **Adelogyrinidae, Molgophidae:** PENNSYLVANIAN: *Mogolphis* (see Fig. 26-12), **Lysorophidae, Microbrachidae, Gymnarthridae,** and **Tuditanidae.**

In the subclass Lissamphibia, there are two superorders: Salientia and Candata.

In the superorder Salientia, there are two orders: Proanura and Anura. There is one family in the first order: **Triadobatrachidae.**

In the order Anura (frogs), there are in turn, two suborders: Archaeobatrachia and Neobatrachia.

In the suborder Archaeobatrachia, are the families **Notobatrachidae, Leiopelmatidae (Ascaphidae)** and **Discoglossidae (Bombinidae).**

In the order Neobatrachia, the families **Pelobatidae, Pelodytidae, Leptodactylidae, Bufonidae, Pseudidae, Hylidae:** MIOCENE: *Hyla* (see Fig. 26-26), **Centrolenidae, Atelopodidae, Ranidae:** MIOCENE: *Palaeobatrachus* (see Fig. 26-27), **Rhacophoridae (Polypedatidae, Microhylidae (Brevicipitidae),** and **Phyrnomeridae.**

In the suborder Caudata, there are two orders: Urodela and Apoda.

In the order Urodela, the families are: **Cryptobranchidae, Hynobiidae, Scapherpetontidae, Ambystomatidae, Salamandridae, Amphiumidae, Plethodontidae, Batrachosauroididae, Proteidae,** and **Sirenidae.**

In the order Apoda, there is only one family: **Caeciliidae.**

26-1. Devonian. *Ichthyostega* sp. A labyrinthodont amphibian from the Upper-most Devonian of Greenland. A reconstructed model. Courtesy: Richard Rush Studio, Chicago, Illinois.

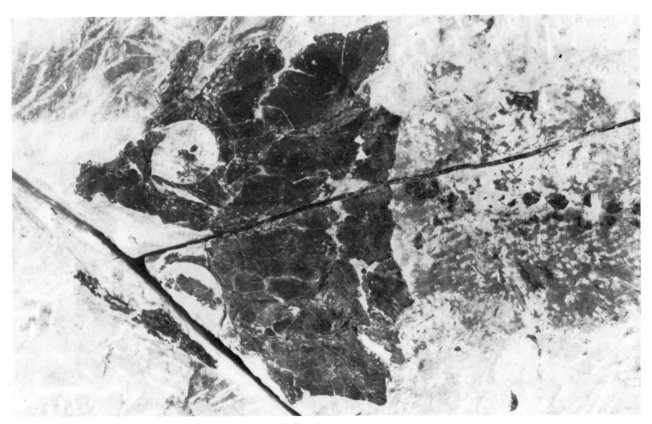

26-2. Pennsylvanian. *Lafonius lehmani* Berman. The skull and partial vertebral column of a trimerorhachid labyrinthodont amphibian. Madera Formation, Manzano Mountains, Bernalillo County, New Mexico. Approx. 2 cm. across width of skull. Photo courtesy: Dr. David S. Berman, Carnegie Museum, Pittsburgh, Pennsylvania.

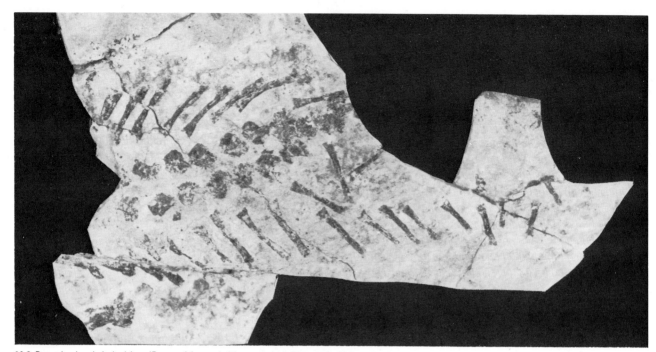

26-3. Pennsylvanian. *Lafonius lehmani* Berman. A fragment of the counterpart of the area back of the skull of a trimerorhachid labyrinthodont amphibian (see Fig. 26-2), showing a series of paired neural arch (halves) and ribs in association. Madera Formation, Manzano Mountains, Bernalillo County, New Mexico. Approx. length of shale fragment, 3 cm. Photo courtesy: Dr. David S. Berman, Carnegie Museum, Pittsburgh, Pennsylvania.

26-4. Pennsylvanian. *Saurerpeton* cf. *S. obtusum* (Cope). A trimerorhachoid amphibian in a siderite concretion (nodule). Mazon Creek fauna, Braidwood, Will County, Illinois. Courtesy: Field Museum of Natural History, Chicago.

26-5. Pennsylvanian. *Saurerpeton* cf. *S. obtusum* (Cope). A close-up of the head and postcranial elements of the trimerorhachoid amphibian in Fig. 26-4. Mazon Creek fauna, Braidwood, Will County, Illinois. Courtesy: Field Museum of Natural History, Chicago.

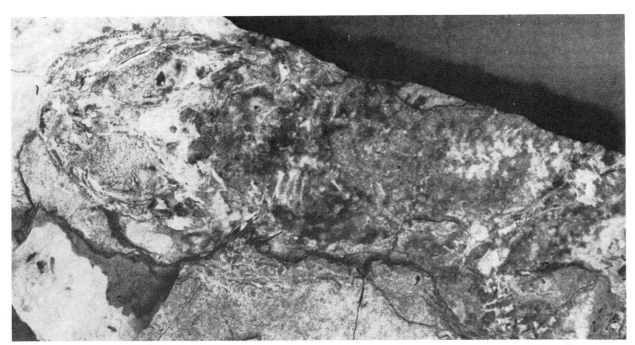

26-6. Pennsylvanian. *Eoscophus lockardi*. A dissorphid amphibian. A skull and partial body in fissile shale from the Late Carboniferous of Kansas. 11.5 cm. Courtesy: Geological Enterprises, Ardmore, Oklahoma.

26-7. Pennsylvanian. *Anthracosaurus* sp. An embolomere amphibian. A life-size model in clay. Model was prepared for and is presently in the "walk thru" restoration of a Pennsylvanian Period coal forest, Wm. Penn Memorial Museum, Harrisburg, Pennsylvania. Consultant on the model: Dr. Donald Baird, Princeton University. Courtesy: Richard Rush Studio, Chicago, Illinois.

26-8. Pennsylvanian. *Anthracosaurus* sp. An embolomere amphibian. Side view of the clay model in Fig. 26-7, showing additional details and especially the total length of the reconstructed animal. Courtesy: Richard Rush Studio, Chicago, Illinois.

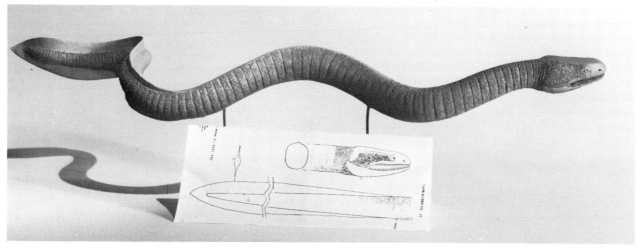

26-9. Pennsylvanian. *Ophiderpeton* sp. A lepospondyl (aïstopod) amphibian. A life size model of a snakelike amphibian from the Carboniferous Period. On display at the Geology Hall, Wm. Penn Memorial Museum, Harrisburg, Pennsylvania. Consultant: Dr. Donald Baird, Princeton University. Courtesy: Richard Rush Studio, Chicago, Illinois.

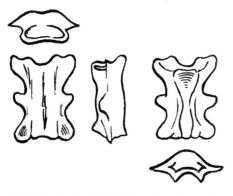

26-10. Pennsylvanian. *Phlegethontia* sp. Five views of a Late Carboniferous lepospondyl (aïstopod) amphibian vertebra. Indian Cave Sandstone, Peru, Nemaha County, Nebraska. 22 mm in length.

26-11. Pennsylvanian. *Ophiderpeton amphiuminum* (Cope). The skull and postcranial elements of an aïstopod amphibian in a siderite concretion. Francis Creek Shale, Carbondale Formation (Middle Pennsylvanian), Mazon Creek fauna, Pit 11, Grundy / Will Counties, Illinois. 3.5 cm. Specimen from the H. & T. Piecko Collection. D. Baird photo.

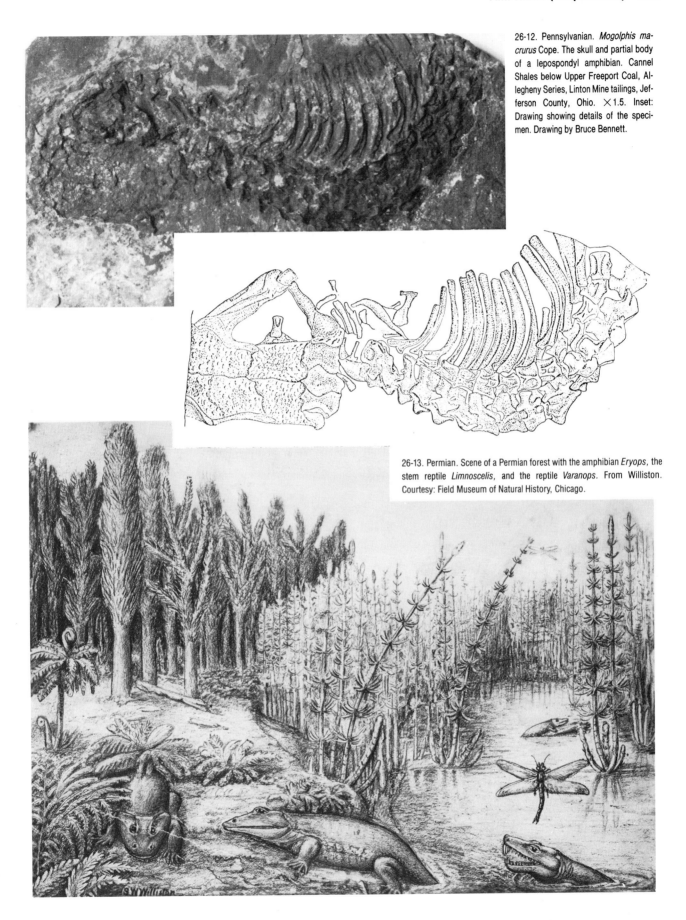

26-12. Pennsylvanian. *Mogolphis ma-crurus* Cope. The skull and partial body of a lepospondyl amphibian. Cannel Shales below Upper Freeport Coal, Allegheny Series, Linton Mine tailings, Jefferson County, Ohio. ×1.5. Inset: Drawing showing details of the specimen. Drawing by Bruce Bennett.

26-13. Permian. Scene of a Permian forest with the amphibian *Eryops*, the stem reptile *Limnoscelis*, and the reptile *Varanops*. From Williston. Courtesy: Field Museum of Natural History, Chicago.

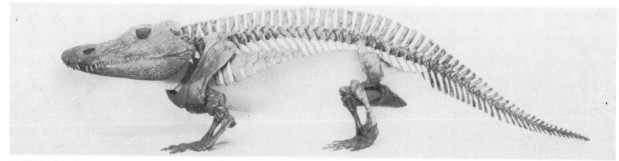

26-14. Permian. *Eryops megacephalus* Cope. A mounted skeleton of a large rhachitomous amphibian from the Early Permian red beds of north Texas. Courtesy: Museum of Comparative Zoology, Harvard University, Cambridge, Massachusetts.

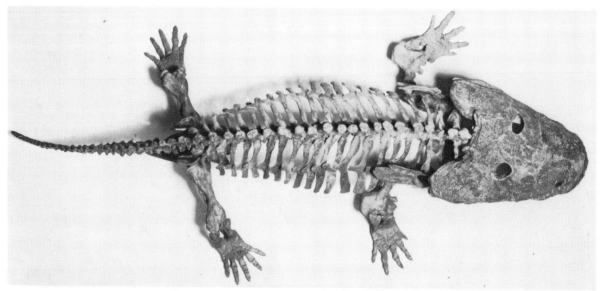

26-15. Permian. *Eryops megacephalus* Cope. A dorsal view of the skeleton of one of the larger amphibians from the Early Permian "red beds" of west Texas. Courtesy: Field Museum of Natural History, Chicago.

26-16. Permian. *Eryops megacephalus* Cope. Another view of the skeleton of the rhachitomous amphibian in Fig. 26-15. Note the massive head on the specimen. Courtesy: Field Museum of Natural History, Chicago.

From 1868 to 1885, Edward Drinker Cope described a rich fauna of amphibians (plus reptiles and fish) recovered from Pennsylvanian coal shale in a mine at Linton, Ohio. This single site yielded skeletons of the labyrinthodonts *Megalocephalus, Macrerpeton, Gaudrya, Colosteus, Erpetosaurus, Saurerpeton, Amphibamus, Stegops, Eusau-*
ropleura and *Anthracosaurus;* the lepospondyls *Ophiderpeton, Phlegethontia, Sauropleura, Ptyonius, Ctenerpeton, Diceratosaurus, Keraterpeton, Tuditanus, Odonterpeton, Mogolphis* and *Cocytinus;* and the reptiles *Anthracodromeus, Cephalerpeton* and *Archaeothyris.* The site is still producing fossils, but not as it was in the heyday of Cope.

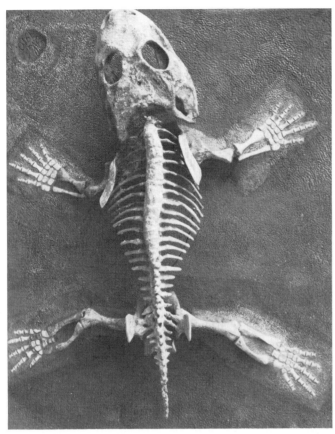

26-17. Permian. *Cacops aspidophorus* Williston. A skeleton of the dissorophid amphibian from the Early Permian red beds of Texas, a dorsal view. Courtesy: Field Museum of Natural History, Chicago.

26-18. Permian. *Trematops milleri* Williston. Dorsal view of the skeleton of a rhachitomous amphibian. Arroyo Formation, Wilbarger County, Texas. Courtesy: Field Museum of Natural History, Chicago.

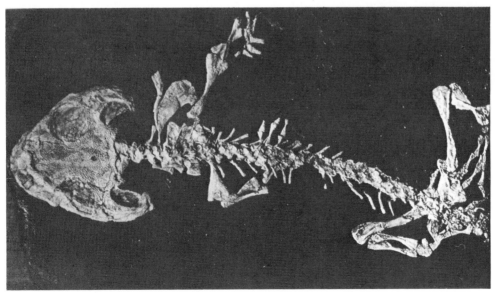

26-19. Pennsylvanian. *Amphibamus lyelli* (Wyman). An almost complete (missing tail section) skeleton of a dissorophid amphibian from the Middle Pennsylvanian (Carboniferous). Upper Freeport Coal, Linton, Jefferson County, Ohio. Length of specimen, approx. 12 cm. Photo by D. Baird.

26-20. Permian. *Seymouria baylorensis* Broili. Reconstruction of a stegocephalian amphibian. At right, a primitive reptile similar to *Labidosaurus* (see Fig. 27-5). Seymour, Baylor County, Texas. Courtesy: Field Museum of Natural History, Chicago.

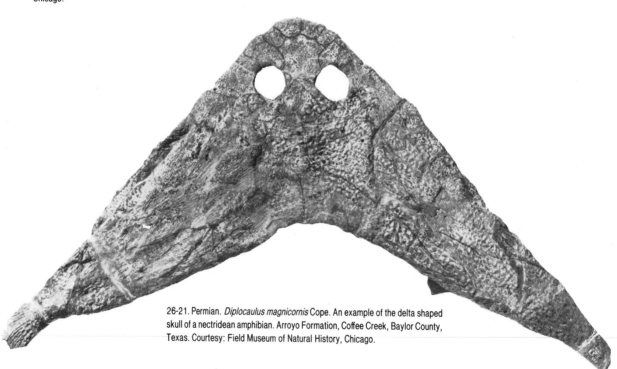

26-21. Permian. *Diplocaulus magnicornis* Cope. An example of the delta shaped skull of a nectridean amphibian. Arroyo Formation, Coffee Creek, Baylor County, Texas. Courtesy: Field Museum of Natural History, Chicago.

26-22. Permian. *Onchiodon (Branchiosaurus) amblystomus* (Credner). A rhachitomous stegocephalian amphibian in reddish colored shale, with skin impression and body outline. Rothliegendes (red beds). Bad Kreuznach, West Germany. 18 cm. Courtesy: Siber and Siber Ltd., Aathal, Switzerland.

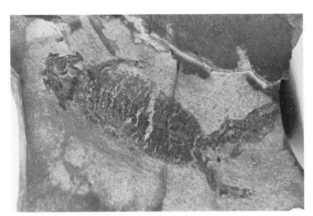

26-23. Permian. *Onchiodon (Branchiosaurus) amblystomus* (Credner). Another example of a completely preserved (skin impression and body outline) rhachitomous stegocephalian amphibian from the red beds of Odernheim Platz, Rhenish Prussia, East Germany. Courtesy: Field Museum of Natural History, Chicago.

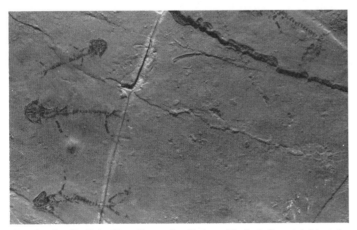

26-24. Permian. *Onchiodon (Branchiosaurus) amblystomus* (Credner). Several skeletons of a rhachitomous stegocephalian amphibian from the Lower Permian Rothliegendes (red beds) of Niederhässlich bei Dresden, East Germany (DDR). ×1. Courtesy: Geological Enterprises, Ardmore, Oklahoma.

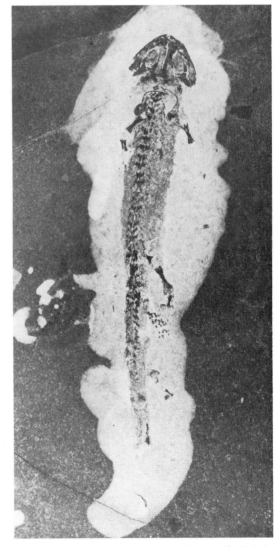

26-25. Permian. *Onchiodon (Branchiosaurus) amblystomus* (Credner). An almost complete (missing hands and feet) rhachitomous stegocephalian amphibian from the famous red beds of Bad Kreuznach, West Germany. Courtesy: Siber and Siber Ltd., Aathal, Switzerland. 7.5 cm.

26-26. Miocene. A beautiful example of a tree-frog caught in resin (amber). Dominican Republic. Courtesy: Geological Enterprises, Ardmore, Oklahoma. ×2.

26-27. Miocene. *Palaeobatrachus luedecki* Wolterstorff. A beautifully preserved, dorsoventrally flattened skeleton of a frog in lignite. Lignite Beds, Markersdorf, Czechoslovakia (CCCR). ×1.5.

26-28. Miocene. A disarticulated anuran (frog, species indeterminate) skeleton in lignite shales. Posterior portion of the body essentially intact (missing feet), while the head area is crushed. Bonn, West Germany. ×2.

27 REPTILIA (Early Stem Reptiles)

The stem reptiles arose from the labyrinthodont amphibians (see preceding chapter) in the Paleozoic, near the beginning of the Pennsylvanian period, and flourished on into the Mesozoic era. Various groups of reptiles (treated in subsequent chapters) branched off the "stem" and evolved into the higher orders. Two of these, in turn, gave rise to the birds and the mammals.

The unique feature that permitted the success of the early stem reptiles, and is retained by present-day reptiles and birds, was their ability to lay shelled eggs and live in a terrestrial habitat. Although the early reptiles probably favored a life environment near some body of water (where they found abundant food in the form of insects, amphibians and fishes), they were able to exploit the resources of the land and were no longer dependent on water for breeding and egg-laying as their amphibian ancestors had been. With this advantage, the reptiles were able to diversify and occupy environments, including the sea and the air, that were denied to the amphibians. In the Class Reptilia, we have the following subclasses: Anapsida, Lepidosauria, Archosauria, Euryapsida (Synaptosauria of Romer), Ichthyopterygia, and Synapsida.

In the Subclass Anapsida, we have the following orders: Cotylosauria, Mesosauria, and Chelonia.

In the Order Cotylosauria we have the suborders Captorhinomorpha and Procolophonia. These are the two suborders we will be discussing in this particular chapter. The orders Mesosauria and Chelonia, as well as the Ichthyosaurs, will be discussed in the following chapter. The latter are in the subclass Euryapsida, but are placed in the following chapter as a "varied" reptilian type.

The following families comprise the suborder Captorhinomorpha (the captorhinomorph reptiles): **Romeriidae, Limnoscelidae, Captorhinidae,** and the questionable **Bolosauridae.**

In the Family **Romeriidae,** we have the following Paleozoic genera: PENNSYLVANIAN: *Archerpeton, Cephalerpeton* (see Figs. 27-1 through 27-3), *Hylonomus (Fritschia), Paleothyris,* and *Anthracodromeus.* PERMIAN: *Melanothyris, Paracaptorhinus, Protorothyris,* and *Romeria.*

In the Family **Limnoscelidae,** we have the Pennsylvanian forerunner *Romeriscus* (see Fig. 27-4), and the Permian representative *Limnoscelis* (see Figs. 20-30 and 26-14).

In the Family **Captorhinidae,** we have the following genera, all represented in the Permian: *Captorhinikos, Captorhinoides, Captorhinus (Ectocynodon), ?Chamasaurus, Geocatogomphius, Hecatogomphius, Kahneria, Labidosaurikos, Labidosaurus* (see Fig. 27-4), *Pleuristion, ?Puercosaurus, Rothioniscus (Rothia),* and *?Sphenosaurus (Palaeosaurus).*

The final family of the suborder Captorhinomorpha is the questionable **Bolosauridae,** with its only Permian representative, *Bolosaurus.*

In the suborder Procolophonia (procolophonoid reptiles), there are three superfamilies: Procolophonoidea, Pareiasauroidea, and Millerosauroidea.

The following two families are represented in the superfamily Procolophonoidea: **Nyctiphruretidae** and **Procolophonidae.**

In the Family **Nyctiphruretidae,** we have the following genera, all represented in the Permian: *Barasaurus, ?Broomia, Nyctiphruretus,* and *Owenetta.*

In the Family **Procolophonidae,** we have the following genera represented in the Permian and Triassic: PERMIAN: *Anomoidon, ?Estheriophagus,* and *?Microcenemus (?Tichvinskia);* TRIASSIC: *Candelaria, Hypsognathus* (see Fig. 27-8), *Koiloskiosaurus, Leptopleuron (Telepeton), Myocephalus, Myognathus, Neoprocolophon, Paoteodon, Phaanthosaurus, Procolophon* (see Fig. 27-7), *Santaisaurus, Sclerosaurus (Aristodesmus), Sphodrosaurus,* and *Spondylolestes.*

The following two families are represented in the superfamily Pareiasauroidea: **Rhipaeosauridae** and **Pareiasauridae.**

In the Family **Rhipaeosauridae,** we have the following genera, all represented in the Permian: *Leptoropha, Parabradysaurus,* and *Rhipaeosaurus.*

In the Family **Pareisauridae,** we have the following genera, all represented in the Permian: Anthodon, *Brady-*

27-1. Pennsylvanian. *Cephalerpeton ventriarmatum* Moodie. A cotylosaurian reptile skeleton in the suborder Captorhinomorpha, family Romeriidae. A skeleton (missing the rear end and tail section) in a siderite concretion. Photo is of a red latex peel showing the bones and teeth as a casting. Francis Creek Shale above Morris (No. 2) Coal, Carbondale Formation (Westphalian D), Mazon Creek, Grundy County, Illinois. From Carroll and Baird, 1972. Photo by D. C. Stager. 23 cm.

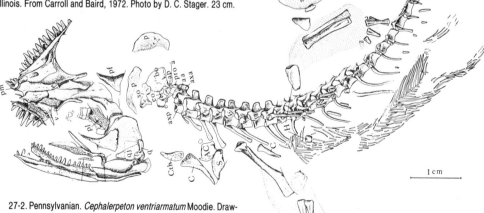

27-2. Pennsylvanian. *Cephalerpeton ventriarmatum* Moodie. Drawing of the skeleton in Fig. 27-1. Yale Peabody Museum specimen. From Carroll and Baird, 1972.

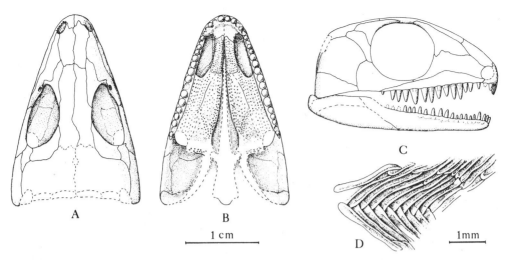

27-3. Pennsylvanian. *Cephalerpeton ventriarmatum* Mooaie. Restoration of the skull of the cotylosaurian reptile illustrated in Figs. 27-1 and 27-2. A. Dorsal view. B. Palatal view. C. Lateral view. D. Detail of the ventral scales in ventral view (partially schematic). × 10. From Carroll and Baird, 1972.

27-4. Pennsylvanian. *Romeriscus periallus* Carrol and Baird. Partial skeleton of the earliest limnoscelid amphibian-reptile. Type specimen. Port Hood, Nova Scotia, Canada. Courtesy: Princeton University Museum of Natural History, Princeton, New Jersey. Photo by D. Baird.

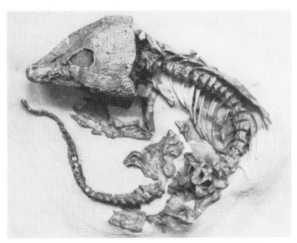

27-5. Permian. *Labidosaurus hamatus* Cope. Cast of the skeleton of a captorhinomorph reptile. Original in the Smithsonian Institution, Washington, D.C. Clear Fork Group, Texas. Courtesy: Rutgers University, New Brunswick, New Jersey.

saurus (Brachypareia) (see Figs. 27-6 through 27-8, and 27-10), *Elginia, Embrithosaurus (Dolichopareia), Parasaurus, Pareiasaurus* (see Fig. 27-9), *Scutosaurus (Amalitzkia)* (see Fig. 27-5), and *Shihtienfenia.*

There is one family represented in the superfamily of Millerosauria, **Millerettidae.**

The genera represented in the Family **Millerettidae,** all in the Permian Period, are as follows: *?Elliotsmithia, ?Heleosaurus, Milleretoides, Milleretta (Millerina), Millerettops, Millerosaurus,* and *Nanomilleretta.*

27-6. Permian. *Diadectes tenuitectus* Case. Skeletal mount exhibited at Harvard University. Arroyo Formation, Baylor County, Texas. Courtesy: Museum of Comparative Zoology, Harvard University, Cambridge, Massachusetts. Photo by D. Baird.

27-7. Triassic. *Procolophon trigoniceps* Owen. Dorsal view of the skull of an anapsid reptile. Karroo Series (Early Triassic). Republic of South Africa. Specimen in the collection of the University of California at Berkeley. Approx. length of skull, 5.5 cm. Photo by P. E. Olsen.

27-8. Triassic. *Hypsognathus fenneri* Gilmore. Dorsal view of the skull of a reptile related to *Procolophon* (see Fig. 27-7). New Haven Arkosic Sandstone (Late Triassic), Connecticut. Yale Peabody Museum specimen. Approx. width of skull, 10.5 cm. Photo by P. E. Olsen.

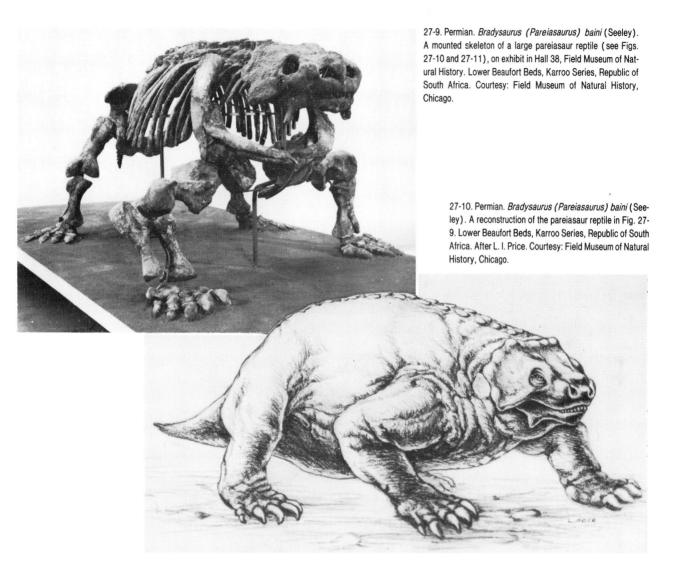

27-9. Permian. *Bradysaurus (Pareiasaurus) baini* (Seeley). A mounted skeleton of a large pareiasaur reptile (see Figs. 27-10 and 27-11), on exhibit in Hall 38, Field Museum of Natural History. Lower Beaufort Beds, Karroo Series, Republic of South Africa. Courtesy: Field Museum of Natural History, Chicago.

27-10. Permian. *Bradysaurus (Pareiasaurus) baini* (Seeley). A reconstruction of the pareiasaur reptile in Fig. 27-9. Lower Beaufort Beds, Karroo Series, Republic of South Africa. After L. I. Price. Courtesy: Field Museum of Natural History, Chicago.

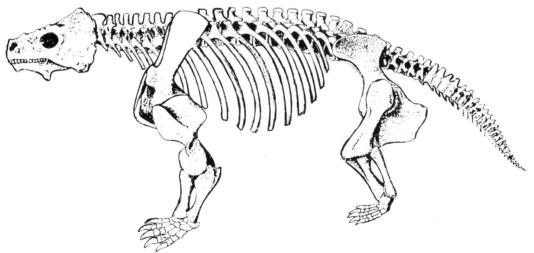

27-11. Permian. *Bradysaurus (pareiasaurus) serridens* (Owen). A drawing of the lateral (side) view of a pareiasaurian reptile. Lower Beaufort Beds, Karroo Series, Republic of South Africa. Courtesy: South African Museum, Capetown, Republic of South Africa.

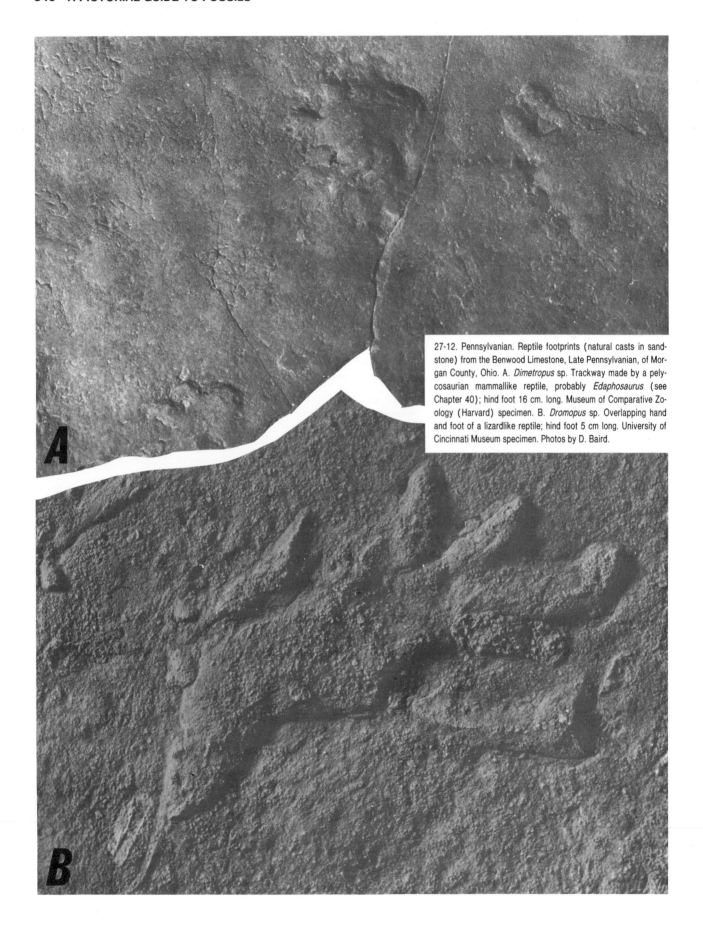

27-12. Pennsylvanian. Reptile footprints (natural casts in sandstone) from the Benwood Limestone, Late Pennsylvanian, of Morgan County, Ohio. A. *Dimetropus* sp. Trackway made by a pelycosaurian mammallike reptile, probably *Edaphosaurus* (see Chapter 40); hind foot 16 cm. long. Museum of Comparative Zoology (Harvard) specimen. B. *Dromopus* sp. Overlapping hand and foot of a lizardlike reptile; hind foot 5 cm long. University of Cincinnati Museum specimen. Photos by D. Baird.

28 REPTILIA (Varied types)

This chapter deals with turtles, mesosaurs, and ichthyosaurs. The order Ichthyosauria belongs to the subclass Ichthyopterygia. The orders Chelonia (turtles) and Mesosauria (aquatic reptiles) belong to the subclass Anapsida. The mesosaurs and turtles are carried over from the previous chapter. As the title of the chapter indicates, these are groups of reptiles having numerous forms or types. They are, therefore, presented in a separate chapter.

These groups had their ancestry in the stem reptiles (see preceding chapter). The mesosaurs were a shortlived group confined to the Permian; the turtles and ichthyosaurs first appeared in the Triassic.

The order Chelonia (turtles) is a large order with at least 4 suborders and 8 superfamilies. The suborders are as follows: Proganochelydia, Amphichelydia, Cryptodira, Pleurodira, and the questionable Eunotosauria.

In the suborder Proganochelydia, there is one superfamily, Proganochelyoidea, with one family represented, **Proganochelyidae.** The Triassic genera *Proganochelys* and *Proterochersis* are representatives of this family.

In the suborder Amphichelydia, there are two superfamilies: Pleurosternoidea and Baenoidia.

In the suborder Pleurosternoidea, the following families are represented: **Pleurosternidae, Plesiochelyidae, Thalassemyidae, Sinemydidae,** and **Apertotemporalidae.**

Representative genera of the above five families: **Pleurosternidae:** JURASSIC: *Platychelys, Pleurosternon* and *Stegochelys.* CRETACEOUS: *Desmemys, Glyptops,* and *Helochelys.* **Plesiochelyidae:** JURASSIC: *Craspedochelys* and *Plesiochelys.* **Thalassemyidae:** JURASSIC: *Changisaurus, Eurysternum (Achelonia), Pelobatochelys,* and *Tropidemys.* CRETACEOUS: *Pygmaeochelys* and *Sontiochelys.* **Sinemydidae:** JURASSIC: *Manchurochelys* and *Sinemys (Cinemys).* **Apertotemporalidae:** CRETACEOUS: *Apertotemporalis* and *Chitracephalus.*

In the superfamily Baenoidea, there are four families with the following representative genera: **Neurankylidae:** CRETACEOUS: *Boremys.* **Baenidae:** CRETACEOUS: *Baena, Chengyuchelys, Chisternon,* and *Macrobaena.* **Meiolaniidae:** CRETACEOUS: *Niolamia.* EOCENE: *Crossochelys.* **Eubaenidae:** CRETACEOUS: *Eubaena.*

In the suborder Cryptodira, there are five superfamilies: Testudinoidea, Cheloniodidea, Dermocheloidea, Carettochelyoidea, and Trionychoidea.

In the superfamily Testudinoidea, there are three major families: **Dermatemydidae, Chelydriidae,** and **Testudinidae.** Representative genera in the above families: **Dermatemydidae:** JURASSIC: *Sinochelys;* CRETACEOUS: *Adocus, Agomphus,* and *Basilemys.* **Chelydridae:** EOCENE: *Gafsochelys;* MIOCENE: *Acherontemys, Chelydrops,* and *Macroclemys (Macrochelys);* PLIOCENE: *Chelydra* and *Kinosternum (Cinosternum).* **Testudinidae (Emydidae):** EOCENE: *Achilemys;* OLIGOCENE: *Gopherus* and *Stylemys* (see Figs. 28-8 and 28-9); PLIOCENE: *Testudo* (see Fig. 28-10).

In the superfamily Chelonioidea, the families **Toxochelyidae, Protostegidae,** and **Cheloniidae** have the following representative genera: **Toxochelyidae:** CRETACEOUS: *Peritresius* (see Figs. 28-3 and 28-4), and *Prionochelys* (see Fig. 28-5). **Protostegidae:** CRETACEOUS: *Archelon* (see Fig. 30-23), and *Protostega.* **Cheloniidae:** CRETACEOUS: *Allopleuron* (see Figs. 28-6 and 28-7), *Chelonia (Chelone), Glyptochelone,* and *Rhinochelys;* EOCENE: *Eochelone.*

In the superfamily Dermocheloidea, family **Dermochelyidae:** EOCENE: *Cosmochelys;* MIOCENE: *Dermochelys.*

In the superfamily Carettochelyoidea, family **Carettochelyidae:** EOCENE: *Allaeochelys;* MIOCENE: *Carettochelys.* In the superfamily Trionychoidea, family **Trionychidae:** JURASSIC: *Sinaspiderites;* PLEISTOCENE: *Chitra.*

The suborder Pleurodira ("side-neck" turtles), family **Pelomedusidae** has the following representative genera: CRETACEOUS: *Amblypeza, Apodichelys, Bothremys,* and *Podocnemis* (see Figs. 28-1 and 28-2); OLIGOCENE: *Anthracochelys, Cyclochelys,* and *Pelomedusa.* In the family **Chelyidae:** OLIGOCENE: *Chelodina.*

Finally, the questionable suborder Eunotosauria, family **Eunotosauridae:** *Eunotosaurus* (genus dubium) (PERMIAN).

28-1. Cretaceous. *Podocnemis barberi* Zangerl. Dorsal view of a pleurodire (side-neck) turtle shell from the Mooreville Chalk of Alabama. Type specimen. Courtesy: Field Museum of Natural History, Chicago.

28-2. Cretaceous. *Podocnemis barberi* Zangerl. Ventral view of the specimen figured in 28-1. Mooreville Chalk of Alabama. Courtesy: Field Museum of Natural History, Chicago.

28-3. Cretaceous. *Peritresius ornatus* (Leidy). Lateral (side) view of the right side of the carapace of a toxochelyid (saw) turtle. New Egypt Formation (Maestrichtian), Inversand Marl Pits, Sewell, Gloucester County, New Jersey. Courtesy: New Jersey State Museum, Trenton, New Jersey.

28-4. Cretaceous. *Peritresius ornatus* (Leidy). Dorsal view of the carapace specimen figured in 28-3. A giant marine (sea) turtle. NJSM 11051. Maximum length of the carapace 58 cm, width 53 cm. Courtesy: New Jersey State Museum, Trenton, New Jersey.

28-5. Cretaceous. *Prionochelys* cf. *P. nauta* Zangerl. Three views of an almost complete 10th peripheral bone from the carapace of a "saw turtle." Navesink Formation (Middle Maestrichtian), Lumberton, Burlington County, New Jersey. From Baird and Case, 1966, *Jour. Paleo.*, v. 40. Slightly less than actual size. Courtesy: Princeton University, Princeton, New Jersey.

28-6. Cretaceous. *Allopleuron hoffmani* Baur. The lower jawbone in matrix of a large marine turtle. Maastricht, Holland. × 1.

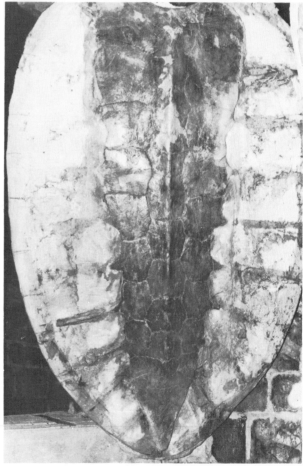

28-7. Cretaceous. *Allopleuron hoffmani* Baur. A reconstructed carapace of a large marine turtle. Maastricht, Holland. Courtesy: Musée du Silex, Eben-Emael, Belgium.

The ichthyosaurs ("fish-lizards") are in the subclass Ichthyopterygia, which has five families in the order Ichthyosauria. The families, and representative genera, are as follows: **Mixosauridae:** TRIASSIC: *Mixosaurus.* **Omphalosauridae:** TRIASSIC: *Grippia* and *Omphalosaurus.* **Shastasauridae:** TRIASSIC: *Shastasaurus.* **Ichthyosauridae:** JURASSIC: *Brachypterygius, Ichthyo-*

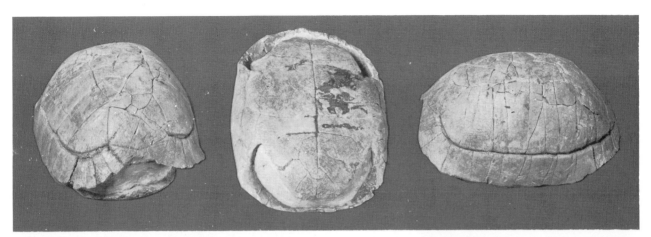

28-8. Oligocene. *Stylemys nebrascensis* Leidy. Three views of the complete shell (carapace and plastron) of a common turtle from the Badlands of South Dakota. Courtesy: Brooklyn Children's Museum, Brooklyn, New York. ⅕ actual size.

28-9. Oligocene. *Stylemys nebrascensis* Leidy. Another view (¾) of the complete shell (carapace and plastron) of the common Badlands turtle. White River Badlands, Scenic, Pennington County, South Dakota. Courtesy: Geological Enterprises, Ardmore, Oklahoma. ⅛th actual size.

28-10. Pliocene. *Testudo* cf. *T. orthopygia*. Complete skeleton of a land tortoise from the Valentine Formation, Broadwater, Nebraska. Courtesy: Field Museum of Natural History, Chicago.

28-11. Jurassic. *Ichthyosaurus communis* Conybeare. The skull of an ichthyosaur ("fish-lizard"). In the same exhibit case, there are smaller skeletons and a complete paddle (hand) of other specimens of ichthyosaurs. Lyme Regis, Dorset, England. Courtesy: Field Museum of Natural History, Chicago.

saurus (see Figs. 28-11 through 28-13), *Macropterygius, Myopterygius,* and *Opthalmosaurus* (see Figs. 28-14 and 28-17). **Stenopterygiidae:** JURASSIC: *Stenopterygius* (see Figs. 28-15 and 28-16); CRETACEOUS: *Platypterygius.*

The mesosaurs (small aquatic reptiles) are the most recent of subclass Anapsida. The sole family **Mesosauridae** has only one representative in the Permian; *Mesosaurus* (see Figs. 28-19 through 28-21). These aquatic "lizards" are found in Permian rocks in South Africa and South America (Brazil).

28-12. Jurassic. *Ichthyosaurus campylodon* Carter. An isolated tooth from an ichthyosaur. Mill Road Cementworks, Cambridge, England. × 1.

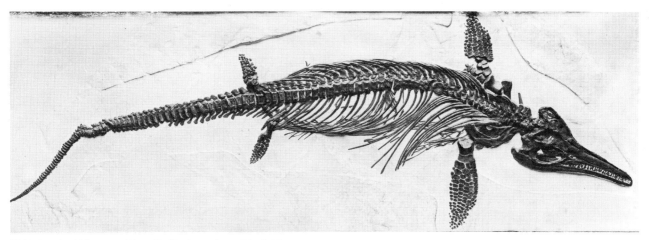

28-13. Jurassic. *Ichthyosaurus intermedius* Conybeare. A complete skeleton from the Liassic of Würtemberg, Germany. Courtesy: Field Museum of Natural History, Chicago.

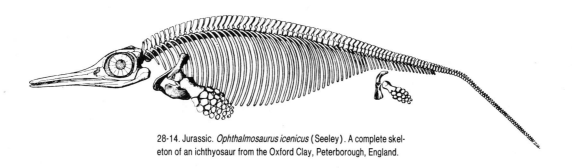

28-14. Jurassic. *Ophthalmosaurus icenicus* (Seeley). A complete skeleton of an ichthyosaur from the Oxford Clay, Peterborough, England.

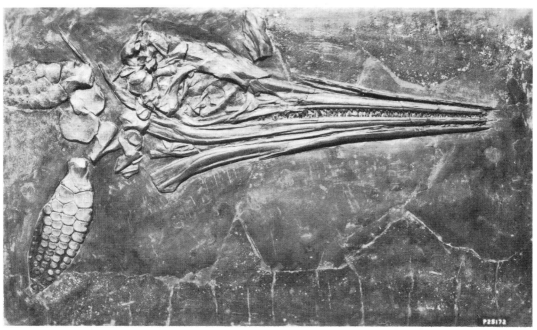

28-15. Jurassic. *Stenopterygius quadricissus* Quenstedt. The skull and paddle bones of an ichthyosaur. Ohmden, Würtemberg, Germany. Courtesy: Field Museum of Natural History, Chicago.

28-16. Jurassic. *Stenopterygius quadricissus* Quenstedt. A complete skeleton of an ichthyosaur showing skin outline preserved. Lias Epsilon, Holzmaden, Germany. Courtesy: Field Museum of Natural History, Chicago.

28-17. Jurassic. An ichthyosaur vertebral centrum (species indeterminate, but possibly belonging to *Opthalmosaurus*). Oxford Clay, Ramsay, near Peterborough, Northants, England. ×1.

28-18. Jurassic. Ichthyosaur vertebral centra (possibly belonging to *Ichthyosaurus communis* Conybeare). Lyme Regis, Dorset, England. ×1.

28-19. Permian. *Mesosaurus brasiliensis* McGregor. An essentially complete skeleton of an aquatic reptile that inhabited freshwater ponds and lakes in South America. Irati Formation, Sao Paulo, Brazil. Courtesy: Black Hills Institute of Geological Research, Hill City, South Dakota.

28-20. Permian. *Mesosaurus brasiliensis* McGregor. The skeleton of a primitive aquatic reptile. Irati Formation, Sao Paulo, Brazil. Courtesy: Black Hills Institute of Geological Research, Hill City, South Dakota.

28-21. Permian. *Mesosaurus brasiliensis* McGregor. A complete skeleton of a South American freshwater reptile. Irati Formation, Sao Paulo, Brazil. (Siegel Collection.)

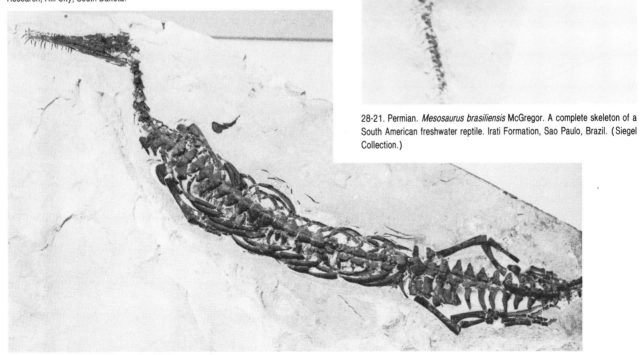

29 REPTILIA (Euryapsida)

This chapter deals with three groups of swimming reptiles—the plesiosaurs, nothosaurs and placodonts—that belong to the subclass Euryapsida.

The plesiosaurs were stout-bodied reptiles having oar-shaped limbs with which they rowed themselves through the water. They had flattened, platelike bones in the shoulder and hip girdles (see Fig. 29-5) plus a stout ventral armor of bony rods that served to protect the belly when these reptiles hauled out onto a beach to lay their eggs. Plesiosaurs were particularly abundant in the Jurassic and early Cretaceous, but they tapered off to extinction at the end of the Cretaceous time.

Within the plesiosaur suborder we see two different adaptive types: the small-headed, long-necked elasmosaurs that fed mainly on fishes (see Fig. 29-5), and the large-headed, short-necked pliosaurs that could cope with larger prey (such as sea turtles) and that included the largest marine predators of their time, reaching lengths over 40 feet in the Australian genus *Kronosaurus*. Even fragmentary remains of plesiosaurs can be recognized by the characteristic teeth and vertebral centra, for example, those of the pliosaur *Cimoliasaurus* (see Fig. 29-3) that is fairly common in the Upper Cretaceous greensand marls of the Atlantic coastal plain of the United States.

The placodonts (see Fig. 29-8) of the Triassic were rather manatee-shaped reptiles, some having a turtlelike body armor. Their blunt, pavement teeth were adapted for crushing shellfish.

The nothosaurs were small aquatic reptiles, similar in habitat, although not related, to the mesosaurs of the preceding chapter. They had webbed hind feet and teeth (although on a smaller scale) very similar to the plesiosaurs (particularly *Cimoliasaurus*, see Fig. 29-3), and vertebral centra similar to the lepidosaurian reptile *Champsosaurus* (see Fig. 30-1).

These various small aquatic (swimming or crawling reptiles) such as *Mesosaurus* (Chapter 28), *Pachypleurosaurus* (this chapter), and *Tanytrachelos* (Chapter 30) are most curious, since they are all essentially similar in their general appearance and habitat, although none are related to the others.

In the subclass Euryapsida (Synaptosauria) there are three orders: Araeoscelidia (Protosauria), Sauropterygia (the nothosaurs and plesiosaurs), and Placodontia. The order Araeoscelida consists of primitive lizardlike terrestrial reptiles, with two definite and three probable families: **Araeoscelidae:** PERMIAN: Araeoscelis. **?Protosauridae:** PERMIAN: *Protorosaurus*. TRIASSIC: *Trachelosaurus*. **Tanystropheidae:** TRIASSIC: *Tanystropheus* and *Tanytrachelos* (see Figs. 30-2 and 30-3). **?Weigeltisauridae:** PERMIAN: *Coelurosauravus*. **Trilophosauridae:** TRIASSIC: *Trilophosaurus*, and *Tricuspisaurus*.

The order Sauropterygia contains the suborder Nothosauria, with the families (all genera TRIASSIC): **Nothosauridae:** *Nothosaurus*. **Pachypleurosauridae:** *Pachypleurosaurus* (see Fig. 29-7), and **Simosauridae:** *Simosaurus*.

The suborder Plesiosauria has two families: **Pistosauridae** and **Cymatosauridae**. Typical generic representatives are: **Pistosauridae:** TRIASSIC: *Pistosaurus*, and **Cymatosauridae:** TRIASSIC: *Cymatosaurus* sp.

In the superfamily Plesiosauroidea (Dolichodeira), there are three families: **Plesiosauridae:** JURASSIC: *Archaeonectrus*, *Cryptocleidus (Apractocleidus)*, *Muraenosaurus*, *Picrocleidus*, and *Plesiosaurus* (see Fig. 29-6); CRETACEOUS: *?Plesiosaurus* (see Fig. 29-4). **Thaumatosauridae:** JURASSIC: *Eurycleidus*, *Seelyosaurus*, *Simolestes*, and *Thaumatosaurus (Enigmatosaurus)*. **Elasmosauridae:** CRETACEOUS: *Alzadasaurus* (see Fig. 29-5), *Aphrosaurus*, *Brancasaurus*, *Elasmosaurus*, *Fresnosaurus*, *Hydralmosaurus*, *Hydrotherosaurus*, *Leuspondylus*, *Mauisaurus*, *Morenosaurus*, *Scanisaurus*, *Styxosaurus*, *Thallassomedon*, and *Woolungasaurus*.

In the superfamily Pliosauroidea (Brachydeira), there are three families: **Pliosauridae:** JURASSIC: *Liopleurodon (Ischyrodon)*, *Peloneustes*, *Pliosaurus (Chelonosaurus)*, and *Stretosaurus*. CRETACEOUS: *Brachauchenius*, *Gymocetus*, and *Kronosaurus*. **Polycotylidae:** CRETACEOUS: *Cimoliasaurus (Symoliosaurus)* (see Fig. 29-3), *Discosaurus*, *Embaphias*, *Piptomerus*, *Piratosaurus*, *Polycotylus*, *Polyptychodon*, *Scarrisaurus*, *Taphrosaurus*, and *Trinacromerum (Dolichorhynchus)*.

29-1. Jurassic. Ichthyosaurs ("fish-lizards") leaping out from the water in an effort to catch a fish for their dinner, while long-necked plesiosaurs are sharing a freshly caught fish. An ocean scene during the Jurassic Period. Painted by Charles R. Knight. Courtesy: Field Museum of Natural History, Chicago.

29-2. Jurassic. An isolated plesiosaur vertebral centrum. Oxford Clay (Liassic), Peterborough, England. ✕ 1.

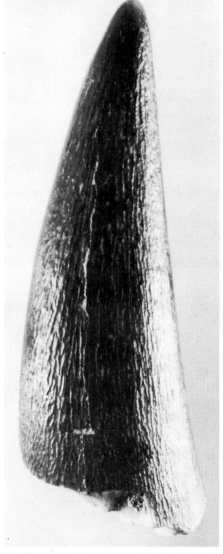

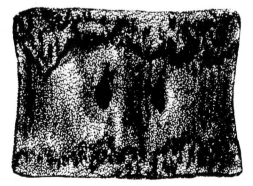

29-3. Cretaceous. *Cimoliasaurus magnus* Leidy. The ventral portion of a pliosaur vertebral centrum. Note the two distinctive large foramina (holes) found on the ventral face of these vertebrae. Navesink Formation (Middle Maestrichtian), Ramannesson Brook, Holmdel, Monmouth County, New Jersey. ✕ 1. From D. C. Parris, 1974. Princeton University Museum specimen.

29-4. Cretaceous. *Plesiosaurus mauritanicus* Arambourg. An isolated large tooth fragment (anterior end or portion) of a plesiosaur. Maestrichtian of Beni-Idir, Oulad Abdoun Basin, Phosphate region of Morocco. ✕ 2.5.

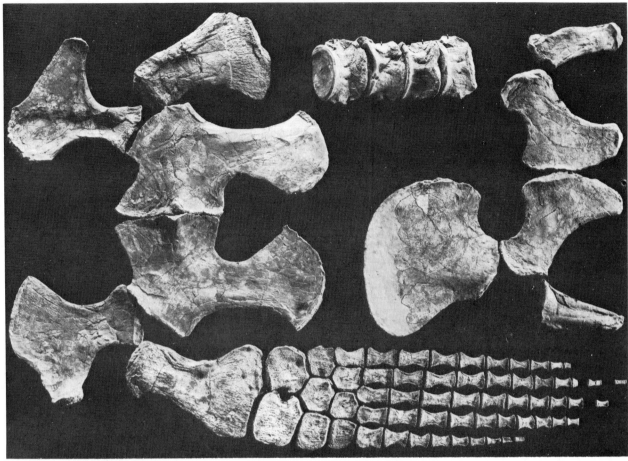

29-5. Cretaceous. *Alzadasaurus riggsi* Welles. The paddles, vertebral centra, and pelvic and pectoral girdles of an elasmosaur (plesiosaur group) from Montana. ×0.1. Courtesy: Field Museum of Natural History, Chicago.

29-6. Jurassic. *Plesiosaurus dolichodeirus* Conybeare. A complete skeleton of a large plesiosaur in shale matrix. Lyme Regis, Dorset, England. Courtesy: Academy of Natural Sciences, Philadelphia, Pennsylvania.

29-7. Triassic. *Pachypleurosaurus edwardsi* Cornalia. An almost complete skeleton (missing the tail section) of a nothosaurian swimming reptile. Keuper of Besano, Italy. 16 cm. Courtesy: Siber and Siber Ltd., Aathal, Switzerland.

29-8. Triassic. *Placodus gigas* Agassiz. Three-fourths view of an almost complete skeleton of an aberrant euryapsid reptile which specialized in eating molluscs. Note the flat crushing teeth in its jaws. Muschelkalk (Middle Triassic), Steinsfurt bei Heidelberg, Germany. Senckenberg Museum photo (from Drevermann, 1933).

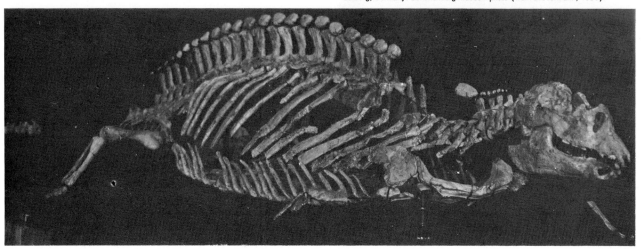

Leptocleididae: CRETACEOUS: *Dolichorhynchops, Leptocleidus,* and *Peyerus.* **Plesiosauria Incertae Sedis:** JURASSIC: *Eurysaurus, Hexatarsostinus, Megalneusaurus,* and *Pantosaurus;* CRETACEOUS: *Aptychodon, Oligosimus, Orophosaurus,* and *Uronautes.*

In the order Placodontia, there are four families: **Hel-** **veticosauridae:** TRIASSIC: *Helveticosaurus.* **Placodontidae:** TRIASSIC: *Paraplacodus* and *Placodus.* **Placochelyidae (Cyamodontidae):** TRIASSIC: *Cyamodus, Placochelys, Psephoderma, Psephosaurus,* and *Saurosphargis.* **Henodontidae:** TRIASSIC: *Henodus* sp.

30 REPTILIA (Lepidosauria)

This chapter deals with the Lepidosaurians—reptiles with two openings in the temporal region of the skull separated by a bar formed by the postorbital and squamosal bones. Temporal openings developed independently in the cheek regions of various reptile groups to accommodate the muscles used in biting, and the different configurations are useful in classification.

The primitive stem-reptiles (see Chapter 27) and their present-day descendants, the turtles, show an anapsid condition without a complete "window" in the temple (although various embayments in the skull roof and lower margin of the cheek may serve the same purpose).

In the previous chapter, the reptiles discussed had a euryapsid condition with a single opening high up on the skull roof. The diapsid or two-opening condition, seen in lepidosaurians, also occurs in the dinosaurs and their relatives (see subsequent chapters). Still another condition, called synapsid, characterizes the mammallike reptiles (see Chapter 40), which have a single temporal opening in the same position as the lower opening of the diapsids.

The lepidosaurians include: the eosuchians, squamatans, and rhynchocephalians. The eosuchians (order Eosuchia) were small, varied reptilian forms which may have given rise to the more advanced archosaurians (see following chapter). Representative of the Eosuchia in this chapter: *Champsosaurus* (see Fig. 30-1).

The order Eosuchia is comprised of four suborders: Younginiformes, Choristodera, Thalattosauria, and Procertiformes.

The suborder Younginiformes has in turn two representative families: **Younginiidae** with *Youngina* from the Late Permian of South Africa, and **Tangasauridae** with *Tangasaurus* from the Late Permian of East Africa.

There is only one family in the suborder Choristodera: **Chamosauridae** with *Champsosaurus* (see Fig. 30-1) as a representative genus from the Upper Cretaceous of North America.

As with the above order, there is only one family represented in the suborder Thalattosauria: **Thalattosauridae** with *Thalattosaurus* from the Middle Triassic of North America.

Finally, we have the suborder Prolacertiformes with two families represented: **Prolacertidae** with *Prolacerta* from the Early Triassic of South Africa and **Tanystropheidae** with *Tanystropheus* and *Tanytrachelos* (see Figs. 30-2 and 30-3) as representative genera of the Middle to Late Triassic of North America.

The squamatans (order Squamata) are varied forms of lizards and snakes, and some marine lizard forms as well, e.g., the mosasaurs.

The order Squamata has two suborders: Lacertilia (lizards) and Ophidia (serpents or snakes). The suborder Lacertilia contains six infraorders: Eolacertilia, Iguania, Nyctisauria (Gekkota), Leptoglossa (Scincomorpha), Annulata (Amphisbaenia), and Diploglossa. In turn, the infraorder Diploglossa has two superfamily representatives: Anguoidea and Varanoidea (Platynota).

The suborder Ophidia (Serpentes) has three superfamilies: Typhlopoidea (Scolecophidia), Booidea (Henophidia), and Colubroidea (Caenophidia).

The following is a condensed faunal listing for the suborder Lacertilia:

Infraorder Eolacertilia, family **Kuehneosauridae**: *Icarosaurus* (see Fig. 30-4) and *Kuehneosaurus* from the Late Triassic of North America.

Infraorder Iguana, families **Bavarisauridae**: *Bavarisaurus* from the Late Jurassic of Europe; **Euposauridae**: *Euposaurus* from the Late Jurassic of Europe; **Iguanidae** with *Iguana* from the Pleistocene of the West Indies; **Agamidae**: *Agama*—Eocene/Europe; and **Chameleontidae**: *Chamaeleo*—Pleistocene/S.W. Asia, and *Mimeosaurus*—Late Cretaceous of Eastern Asia.

Infraorder Nyctisauria (Gekkota), families **Ardeosauridae**: JURASSIC: *Ardeosaurus;* **Broiliasauridae**: JURASSIC: *Broiliasaurus;* and **Gekkonidae**: EOCENE: *Cadurcogekko.* MIOCENE: *Gerandogekko.*

Infraorder Leptoglossa, families **Xantusidae**: PLEISTOCENE: *Xantusia;* **Teiidae**: CRETACEOUS: *Chamops.* OLIGOCENE: *Teius.* MIOCENE: *Dracaena;* **Scincidae**: CRETACEOUS: *Sauriscus;* **Lacertidae**: MIOCENE: *Lacerta;* and **Cordylidae**: MIOCENE: *Gerrhosaurus.*

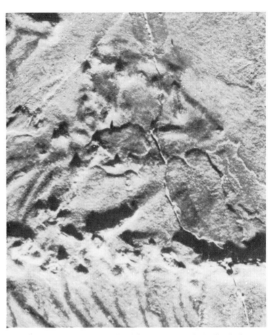

30-1. Cretaceous. *Champsosaurus* sp. Four views of a single vertebral centrum of a small reptile. Teapot Sandstone Member, Mesaverde Formation (Late Campanian), Big Horn Basin, Washakie County, Wyoming. ×1.5.

30-2. Triassic. *Tanytrachelos ahynis* Olsen. An aquatic reptile. A close-up of the head (covered by silicified mud-shales) (see Fig. 30-3). Lockatong Formation, Newark Series, Granton Quarry, North Bergen, Hudson County, New Jersey. ×2.

30-3. Triassic. *Tanytrachelos ahynis* Olsen. The skeleton (missing a small portion of the tail) of a swimming reptile (see Fig. 30-2). Lockatong Formation, Newark Series, Granton Quarry, North Bergen, Hudson County, New Jersey. ×1.5.

30-4. Triassic. *Icarosaurus siefkeri* Colbert. A reconstructed model of a gliding reptile from the Lockatong Formation, Newark Series, Granton Quarry, North Bergen, Hudson County, New Jersey. Model prepared for the Wm. Penn Memorial Museum, Harrisburg, Pennsylvania (Geology Hall). Consultant: Dr. Donald Baird, Princeton University Geology Museum. Courtesy: Richard Rush Studio, Chicago.

30-5. Permian. *Dromopus lacertoides* (Geinitz). An ichnite or "footprint" of a small reptile (see also Fig. 27-12B). Kuhles Tai (Cold Valley), Thüringen, DDR (East Germany). ✕1.5.

30-6. Triassic. Skull of an unnamed rhynchocephalian, very similar to the living *Sphenodon* of New Zealand. New Haven Arkose, Meriden, Connecticut. ✕2.5. Courtesy: Princeton University, Princeton, New Jersey.

30-7. Cretaceous. *Clidastes* sp. An isolated humerus bone of a mosasaurine reptile. Taylor Formation (Campanian), North Sulphur River, Ladonia, Fannin County, Texas. Approx. half size.

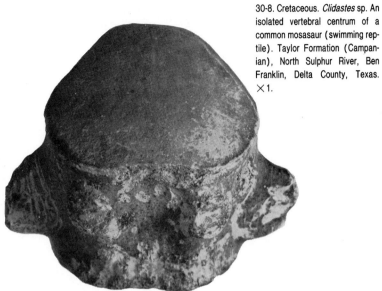

30-8. Cretaceous. *Clidastes* sp. An isolated vertebral centrum of a common mosasaur (swimming reptile). Taylor Formation (Campanian), North Sulphur River, Ben Franklin, Delta County, Texas. ×1.

30-9. Cretaceous. *Clidastes* sp. Three isolated vertebral central of a mosasaur. Taylor Formation (Campanian), North Sulphur River, Ben Franklin, Delta County, Texas. ×1.

30-10 Cretaceous. *Clidastes* sp. A fragment of the dentary (lower jaw) of a mosasaur reptile. The tooth roots are the only part of the dentition preserved. The points (crowns or caps) are missing. Taylor Formation (Campanian), North Sulphur River, Pecan Gap, Delta County, Texas. ×1.

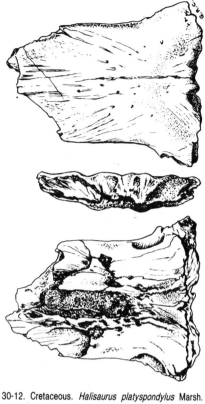

30-11. Cretaceous. *Halisaurus platyspondylus* Marsh. Three views of the frontal bone of a rather rare mosasaur. a. Ventral view. b. Dorsal view. c. Basal view. (See Fig. 30-12.) Navesink Formation (Middle Maestrichtian), Lumberton, Burlington County, New Jersey. From Baird and Case, 1966, *Jour. Paleo.,* v. 40. Courtesy: Princeton University, Princeton, New Jersey. Approx. ⅔ actual size.

30-12. Cretaceous. *Halisaurus platyspondylus* Marsh. Drawing of the frontal bone in Fig. 30-11. Top: dorsal view. Center: basal view. Bottom: ventral view. Compare with Fig. 30-11. Navesink Formation (Middle Maestrichtian), Lumberton, Burlington County, New Jersey. From Baird and Case, 1966. Courtesy: Princeton University, Princeton, New Jersey. Approx. ⅔ actual size of specimen.

30-14. Cretaceous. *Mosasaurus conodon* Marsh. An isolated mosasaur tooth cap. Navesink Formation (Middle Maestrichtian), Big Brook, Marlboro, Monmouth County, New Jersey. ×1.5.

30-13. Cretaceous. *Mosasaurus conodon* Marsh. Two views of a single tooth in a jaw fragment. Note the depression in the root area of the illustration at left, this is the area for the replacement tooth. Navesink Formation (Middle Maestrichtian), Holmdel, Monmouth County, New Jersey. ×1.

30-16. Cretaceous. *Mosassaurus hoffmani.* A close-up of the sclerotic (eye) ring from the mosassaur skull in the top view of Fig. 30-15. Courtesy: Musée du Silex, Eben-Emael, Belgium. × 1.

30-15. Cretaceous. *Mosasaurus hoffmani.* Three photos showing Monsieur Robert Garcet of Temple Eben-Ezer, Eben-Emael, Belgium, preparing skulls of mosassaurs. Top: Preparing a large mosasaur skull. Note the sclerotic (eye) ring (see Fig. 30-16 for a more detailed view of the ring). Center: Mr. Garcet uncovering a small mosasaur skeleton. Bottom: Preparation of a large head of a mosasaur. Note the size of the teeth. Courtesy: Musée du Silex, Eben-Emael, Belgium.

30-17. Cretaceous. *Mosasaurus conodon* Marsh. A lower jaw fragment containing two teeth. Navesink Formation (Middle Maestrichtian), Holmdel, Monmouth County, New Jersey. Approx. half size.

Infraorder Annulata, family **Amphisbaenidae:** PALEOCENE: *Oligodontosaurus.* EOCENE: *Jepsibaena.* OLIGOCENE: *Changlosaurus.* PLEISTOCENE: *Tro-*

gonophis. In the infraorder Diploglossa, superfamily Anguoidea, families **Anguidae:** CRETACEOUS: *Gerrhonotus.* PALEOCENE: *Haplodontosaurus.* OLIGOCENE: *Anguis.* EOCENE: *Ophisauriscus.* Superfamily Varanoidea, families **Necrosauridae:** PALEOCENE: *Necrosau-*

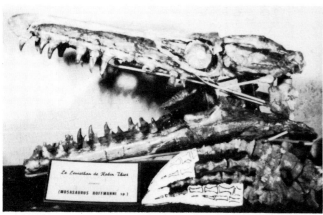

30-18. Cretaceous. *Mosasaurus hoffmani.* Two views of mosasaur heads with complete dentitions intact. Maestrichtian, Robin-Their, Belgium. Courtesy: Musée du Silex, Eben-Emael, Belgium.

30-19. Cretaceous. Mosasaurus maximus Cope. A complete skull with a partial vetebral column. Note the close similarity to *Mosasaurus hoffmani.* Inversand Marl Pits, Sewell, Glouchester County, New Jersey. Courtesy: New Jersey State Museum, Trenton, New Jersey.

30-20. Cretaceous. *Mosasaurus maximus* Cope. Another view of the skull in Fig. 30-19. This head is hanging over the 2nd floor stair landing of the New Jersey State Museum in Trenton. New Egypt Formation (Late Maestrichtian), Sewell, Gloucester County, New Jersey. Courtesy: New Jersey State Museum, Trenton.

30-21. Cretaceous. *Platycarpus coryphaeus* Cope. Skull of a mosasaur from the Niobrara Chalk Formation of Western Kansas. Courtesy: Field Museum of Natural History, Chicago.

rus; **Parasaniwidae:** CRETACEOUS: *Parasaniwa;* **Helodermatidae:** OLIGOCENE: *Heloderma;* **Varanidae:** CRETACEOUS: *Pachyvaranus* and *Palaeosaniwa.* EOCENE: *Saniwa.* MIOCENE: *Varanus;* **Aigialosauridae:** CRETACEOUS: *Aigialosaurus;* **Dolichosauridae:** CRETACEOUS: *Dolichosaurus;* **Mosasauridae:** CRETACEOUS: *Clidastes* (see Figs. 30-7 through 30-10), *Globidens* (see Fig. 30-22), *Halisaurus* (see Figs. 30-11 and 30-12), *Mosasaurus* (see Figs. 30-13 through 30-20), *Platycarpus (Platecarpus)* (see Fig. 30-21), and *Tylosaurus* (see Fig. 30-22); **Paleophidae:** EOCENE: *Palaeophis* (see Fig. 30-25). NOTE: this family should rightfully be listed in the suborder Ophidia. The genus is a snake, *not* a lizard. **Simoliophidae:** CRETACEOUS: *Simoliophis.*

In the suborder Ophidia, superfamily Typhlopoidea, families **Typhlopidae** and **Leptotyphlopidae.** Superfamily Booidea, families **Dinilysiidae; Aniliidae;** and **Boidae.** Superfamily Colubroidea, families: **Archaeophidae; Colubridae; Elapidae; Hydrophiidae;** and **Viperidae.**

Finally, we have the order Rhynchocephalia, which has five families: **Sphenodontidae** (see Fig. 30-6); **Rhynchosauridae:** *Mesosuchus, Parasuchus,* and *Rhynchosaurus,* from the Early Triassic of South Africa, South America, and Europe, respectively; **Sapheosauridae (Saurodontidae); Claraziidae;** and **Pleurosauridae.** The Rhynchocephalia are extinct except for one genus, the tuatara *(Sphenodon),* that survives in New Zealand today.

30-22. Cretaceous. *Globidens aegyptiacus* Zdansky. Two views; Top: occlusal view, Bottom: lateral view, of a tooth of a shell-crushing mosasaur. Maestrichtian, South of Khouribga, Oulad Abdoun Basin, Morocco. ✕1.5.

30-23. Cretaceous. *Tylosaurus* sp. This painting by Charles R. Knight depicts a huge aquatic lizard in the mosasaurine family chasing *Archelon,* the giant turtle, while overhead a flock of *Pteranodon* (flying reptiles) are awaiting any bits of food left over from the forthcoming battle between the two water reptiles. Scene is during Middle Cretaceous time in what is now the State of Kansas. Courtesy: Field Museum of Natural History, Chicago.

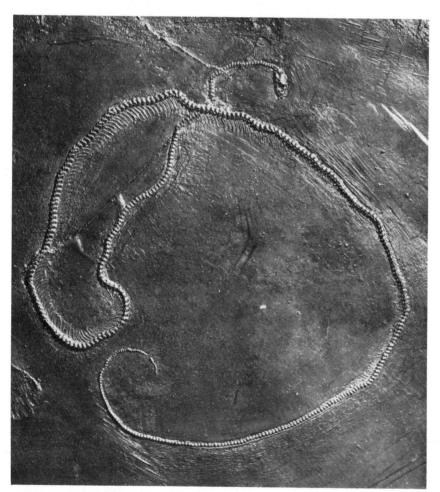

30-24. Eocene. Fossilized snake (cast)—species indeterminate. Messel, Germany. Courtesy: Geological Enterprises, Ardmore, Oklahoma.

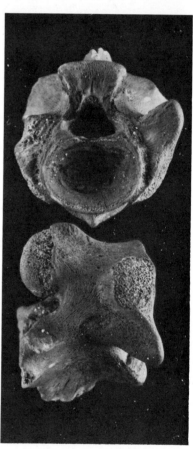

30-25. Eocene. Two views of an isolated vetebral centrum of *Palaeophis* sp., a water python. Twiggs Clay Member, Barnwell Formation (Jacksonian), Huber, Twiggs County, Georgia. ×2.

31 REPTILIA (Ruling types)

The ruling reptiles or archosaurians are a varied group that include the thecodonts (both quadrupedal and bipedal) and their descendants, the crocodilians, pterosaurians, and dinosaurs.

The archosaurians evolved from the eosuchians (see preceding chapter) and became bipedal as modifications were made in their pelvic girdle and limb elements, which enabled them to walk and run much more smoothly than their primitive predecessors.

The archosaurians were carnivorous, as evidenced by their simplified, sharp, needlelike teeth and strong jaw bones. Many of the thecodonts show a diapsid condition in their skull structure. The thecodonts were undoubtedly the precursors of the later dinosaurs of the Jurassic/Cretaceous.

The order Thecodontia in the subclass Archosauria has four distinct suborders: Proterosuchia, Pseudosuchia, Aetosauria, and Phytosauria.

A condensed faunal listing follows for the families of the above suborders of the order Thecodontia:

In the suborder Proterosuchia, families: **Chasmatosauridae:** PERMIAN: *Archosaurus.* TRIASSIC: *Chasmatosaurus;* **Erythrosuchidae:** TRIASSIC: *Cuyosuchus, Erythrosuchus,* and *Shansisuchus.*

In the suborder Pseudosuchia, families: **Euparkeriidae:** TRIASSIC: *Euparkeria;* **Erpetosuchidae:** TRIASSIC: *Dibothrosuchus, Hesperosuchus, Parringtonia, Saltoposuchus,* and *Strigosuchus;* **Teleocrateridae:** TRIASSIC: *Teleocrater;* **Elastichosuchidae:** TRIASSIC: *Elastichosuchus;* and **Prestosuchidae:** TRIASSIC: *Prestosuchus.*

In the suborder Aetosauria, family: **Aetosauridae (Stagonolepidae):** TRIASSIC: *Aetosauroides, Aetosaurus, Argentinosuchus, Desmatosuchus, Ebrachosuchus, Stagonolepis, Stegomus* (see Figs. 31-1 and 31-2), and *Typothorax.*

In the suborder Phytosauria, family **Phytosauridae:** TRIASSIC: *Mystriosuchus, Nicrosaurus (Phytosaurus),* and *Rutiodon (?Clepsysaurus)* (see Fig. 31-3).

The crocodilians (crocodiles and alligators) evolved from the thecodonts, most probably from aetosaur/phytosaur origins. The earliest crocodilians appeared before the end of the Triassic, and have remained successful, dominant land, water, and swamp dwellers even to the present. The order Crocodilia has five suborders: Protosuchia, Archaeosuchia, Mesosuchia, Sebecosuchia, and Eusuchia.

A condensed faunal listing follows for the families of the above suborders of the order Crocodilia:

In the suborder Protosuchia, families: **Sphenosuchidae:** TRIASSIC: *Sphenosuchus;* **Protosuchidae:** JURASSIC: *Protosuchus* and *Stegomosuchus.*

In the suborder Archaeosuchia, families **Notochampsidae:** TRIASSIC: *Erythrochampsa, Microchampsa,* and *Notochampsa;* **Proterochampsidae:** TRIASSIC: *Proterochampsa.*

In the suborder Mesosuchia, families: **Teleosauridae:** JURASSIC: *Aelodon, Gavialinum, Mycterosuchus, Pelagosaurus, Steneosaurus* (see Fig. 31-12), *Teleidosaurus,* and *Teleosaurus.* CRETACEOUS: *Heterosaurus;* **Pholidosauridae:** JURASSIC: *Crocodilaemus, Peipehsuchus, Petrosuchus, Pholidosaurus,* and *Sunosuchus.* CRETACEOUS: *Dyrosaurus, Suchosaurus,* and *Teleorhinus.* EOCENE: *Rhabdognathus, Rhabdosaurus,* and *Sokotosaurus;* **Atoposauridae:** JURASSIC: *Alligatorellus, Alligatorium, Atoposaurus, Hoplosuchus,* and *Shantungosuchus;* **Goniopholidae:** JURASSIC: *Goniopholis, Hsisosuchus, Machimosaurus,* and *Oweniasuchus.* CRETACEOUS: *Baharijodon, Doratodon, Itasuchus, Paralligator, Pinacosuchus, Pliogonodon, Polydectes,* and *Symptosuchus;* **Notosuchidae:** CRETACEOUS: *Notosuchus* and *Uruguaysuchus;* **Metriorhynchidae:** JURASSIC: *Dakosaurus, Geosaurus,* and *Metriorhynchus.* CRETACEOUS: *Capellineosuchus* and *Enaliosuchus.*

In the suborder Sebecosuchia, families **Baurusuchidae:** CRETACEOUS: *Baurusuchus* and *Cynodontosuchus;* **Sebecidae:** CRETACEOUS: *Peirosaurus.* PALEOCENE: *Sebecus.* In the suborder Eusuchia, families: **Hylaeochampsidae:** JURASSIC: *Hylaeochampsa;* **Stomatosuchidae:** CRETACEOUS: *Chiayusuchus* and *Stomatosuchus;* **Gavialidae:** MIOCENE: *Gavialis;* **Cro-**

31-1. Triassic. *Stegomus arcuatus* Marsh. A pseudosuchian thecodont reptile. A model prepared for the Wm. Penn Memorial Museum, Geology Hall. Consultant: Dr. Donald Baird, Princeton University. Courtesy: Richard Rush Studio, Chicago.

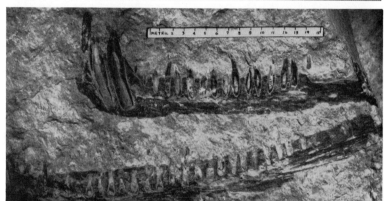

31-3. Triassic. *Rutiodon (?Clepsysaurus) carolinensis* Emmons. Two dentaries of a phytosaurian reptile. Both the left and right dentaries of the same individual are present in the photograph. Lockatong Formation, Newark Supergroup, Blue Bell, Pennsylvania. Courtesy: Academy of Natural Sciences, Philadelphia.

31-2. Triassic. *Stegomus arcuatus jerseyensis.* A fragment of the body armor tail area of a thecodont pseudosuchian reptile from the Passaic Formation of New Jersey. Approx. half size. Courtesy: New Jersey State Museum, Trenton. (Type specimen.)

31-4. Jurassic. *Rhamphorhynchus,* a flying reptile soaring above and hovering near some cycads. A scene in the Jurassic of Germany. Painting by Charles R. Knight, Courtesy: Field Museum of Natural History, Chicago.

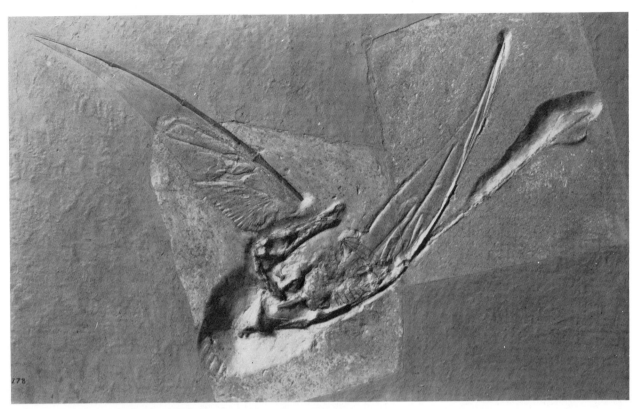

31-5. Jurassic. *Rhamphorhynchus phyllurus*. Type specimen of the pterosaurian reptile showing a partial body, a head, wings, and the distinctive vane-tip tail. In collection of the Peabody Museum, Yale University. Solnhofen, Bavaria, West Germany. (See Fig. 31-6 for detailed drawing of this specimen.) Courtesy: Field Museum of Natural History, Chicago.

31-6. Jurassic. *Rhamphorhynchus phyllurus*. A drawing to show details of the illustration in Fig. 31-5. Courtesy: Field Museum of Natural History, Chicago.

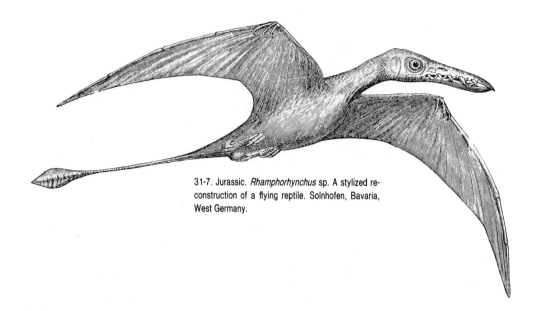

31-7. Jurassic. *Rhamphorhynchus* sp. A stylized reconstruction of a flying reptile. Solnhofen, Bavaria, West Germany.

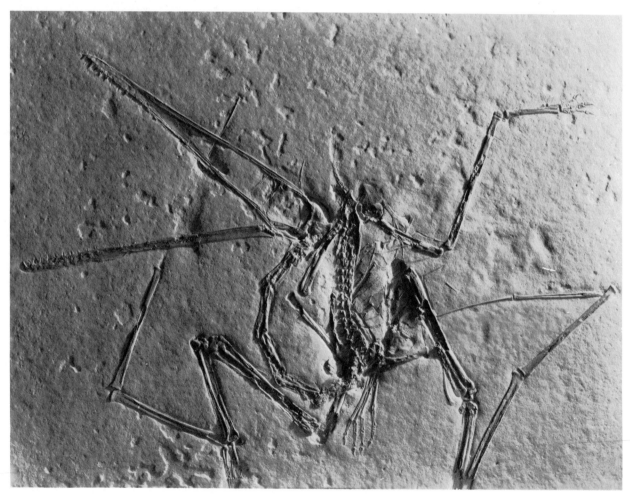

31-8. Jurassic. *Pterodactylus antiquus* Sömmering. A cast of a completely articulated skeleton of a flying reptile. Solnhofen, Bavaria, West German. Courtesy: Field Museum of Natural History, Chicago.

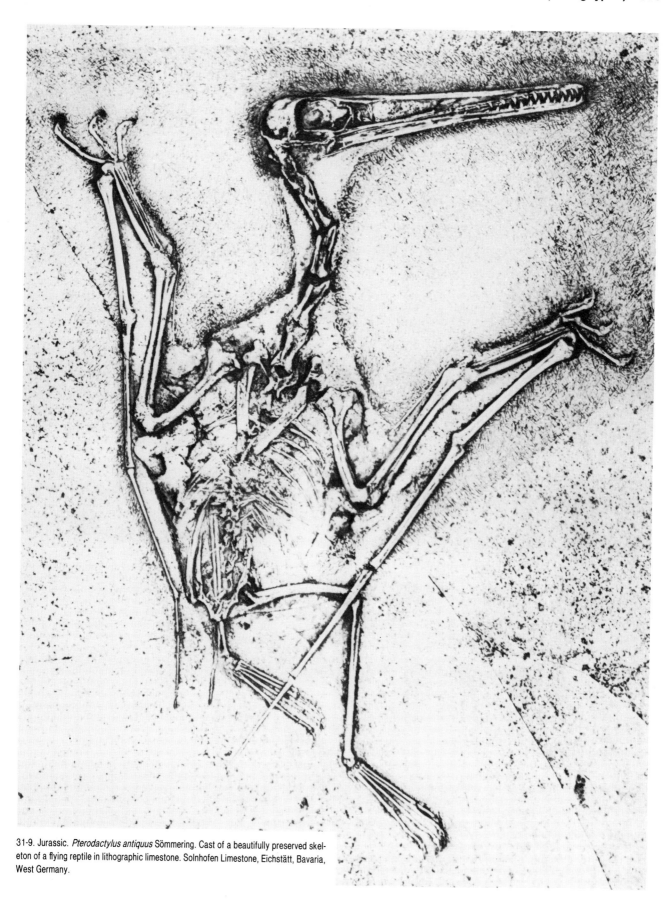

31-9. Jurassic. *Pterodactylus antiquus* Sömmering. Cast of a beautifully preserved skeleton of a flying reptile in lithographic limestone. Solnhofen Limestone, Eichstätt, Bavaria, West Germany.

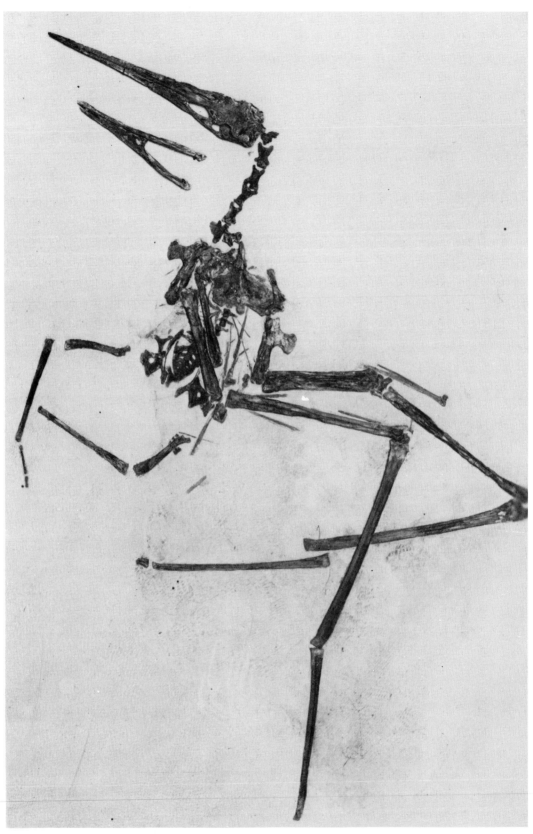

31-10. Cretaceous. *Nyctosaurus gracilis* Marsh. Disarticulated skeleton of a flying reptile similar to *Pteranodon*. Collected in the Niobrara Formation chalk beds of Western Kansas. Courtesy: Field Museum of Natural History, Chicago.

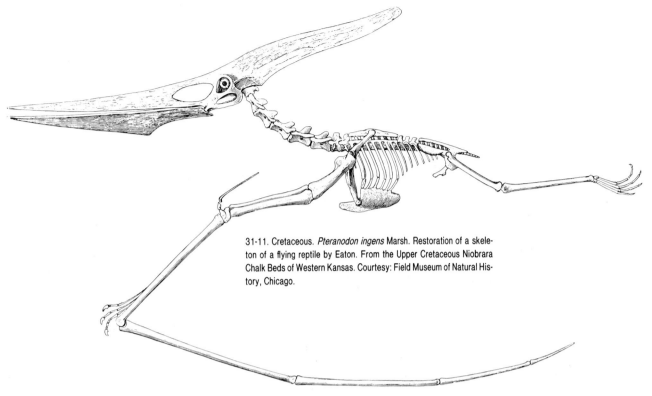

31-11. Cretaceous. *Pteranodon ingens* Marsh. Restoration of a skeleton of a flying reptile by Eaton. From the Upper Cretaceous Niobrara Chalk Beds of Western Kansas. Courtesy: Field Museum of Natural History, Chicago.

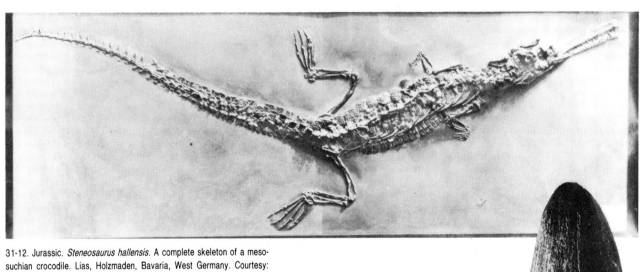

31-12. Jurassic. *Steneosaurus hallensis*. A complete skeleton of a mesosuchian crocodile. Lias, Holzmaden, Bavaria, West Germany. Courtesy: Field Museum of Natural History, Chicago.

codylidae: CRETACEOUS: *Allodaposuchus, Bottosaurus, Brachychampsa, Crocodylus, Deinosuchus (Phobosuchus)* (see Fig. 31-13), *Eotomistoma, Holops, Leidyosuchus, Prodiplocynodon,* and *Thoracosaurus (Sphenosaurus)*. PALEOCENE: *Allognathosuchus, Caiman, Ceratosuchus* (see Fig. 31-14), *Leidyosuchus* (see Fig. 31-15), and *Necrosuchus*. EOCENE: *Asiatosuchus, Brachyuranochampsa, Caimanosuchus, Dollosuchus, Eocenosuchus, Eosuchus, Kentisuchus, Lianghusuchus, Limnosaurus, Megadontosuchus, Orthogenysuchus, Or-*

31-13. Cretaceous. *Deinosuchus (Phobosuchus)*. An isolated tooth cap (crown) of a giant dinosaur-eating crocodile. Peedee Formation (Late Maestrichtian), Giddensville, Duplin County, North Carolina. ×1.5.

31-14. Paleocene. *Ceratosuchus burdoshi* Schmidt. Reconstruction of the head of the "horned crocodile" from the Paleocene of Colorado. Courtesy: Field Museum of Natural History, Chicago.

31-15. Paleocene. *Leidyosuchus riggsi* Lambe. The dorsal view of the skull of a Paleocene crocodile from Colorado. Courtesy: Field Museum of Natural History, Chicago.

thosaurus, Pristichampsus, Procaimanoidea, Tienosuchus, Tomistoma, and *Weigeltisuchus.* OLIGOCENE: *Alligator, Balanerodus,* and *Hispanochampsa.* MIOCENE: *Charactosuchus, Euthecodon, Gavialosuchus, Maroccosuchus* (see Figs. 31-16 through 31-18), and *Mourasuchus.* PLIOCENE: *Leptorrhamphus* and *Proalligator.* PLEISTOCENE: *Colossoemys.*

Finally, we come to the pterosaurs (flying reptiles) who dominated the skies of the Jurassic and Cretaceous. These peculiar flying creatures had sharp, pointed teeth, leatherlike wings, and in the case of the form *Rhamphorhynchus,* a lengthy tail ending in a "paddle" shape (see Figs.

31-4 through 31-7). Some forms like *Pteranodon* (see Fig. 31-11) had a long and slender head region.

The order Pterosauria has two suborders: Rhamphorhynchoidea and Pterodactyloidea.

A condensed faunal listing follows for the families of the above suborders of the order Pterosauria: In the suborder Rhamphorhynchoidea, families: **Dimorphodontidae:** JURASSIC: *Campylognathoides, Dimorphodon, Parapsicethalus, Rhampocephalus,* and *Scaphognathus;* **Rhamphorhynchidae:** JURASSIC: *Dermodactylus, Dorygnathus, Odontorhynchus,* and *Rhamphorhynchus* (see Figs. 31-4 through 31-7); **Anurognathidae:** JURAS-

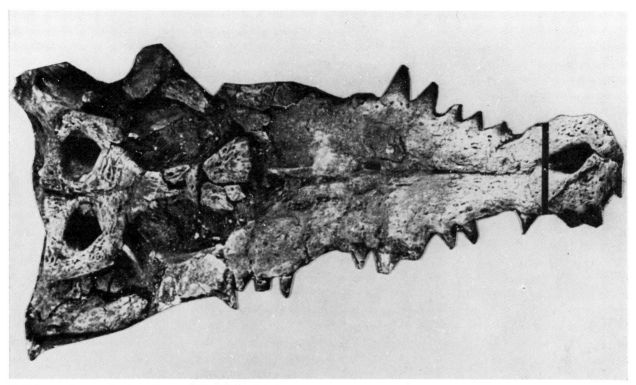

31-16. Eocene. *Maroccosuchus zennaroi* Jonet and Wouters. The palatal view of the skull of a crocodile. Ypresian, Oued-Zem, Oulad Abdoun Basin, Kingdom of Morocco, North Africa.

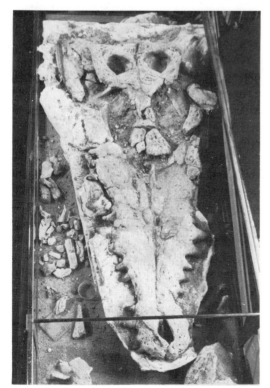

31-17. Eocene. *Maroccosuchus zennaroi* Jonet and Wouters. Another view of the crocodile skull figured in 31-16. Ypresian, Oued-Zem, Oulad Abdoun Basin, Morocco.

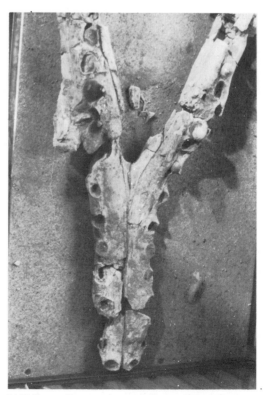

31-18. Eocene. *Maroccosuchus zennaroi* Jonet and Wouters. A dorsal view of the two lower jaws of the crocodile figured in 31-16 and 31-17. Ypresian, Oued-Zem, Oulad Abdoun Basin, Morocco.

31-19. Modern. A typical Nile River crocodile from North Africa. From an old wood engraving.

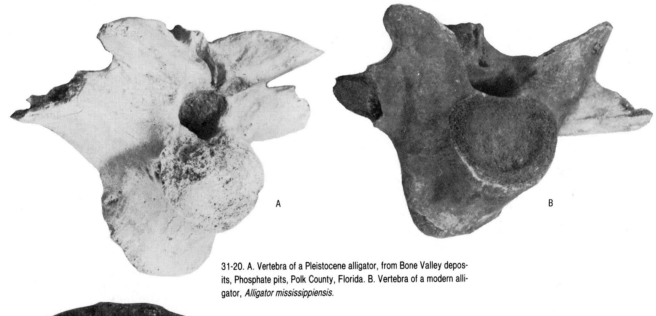

31-20. A. Vertebra of a Pleistocene alligator, from Bone Valley deposits, Phosphate pits, Polk County, Florida. B. Vertebra of a modern alligator, *Alligator mississippiensis*.

31-21. Recent. *Alligator mississippiensis*. An isolated dorsal scute (from the back of the body) of a lake alligator. Lake Monroe, Enterprise, Volusia County, Florida.

SIC: *Anurognathus* and *Batrachognathus*.

In the suborder Pterodactyloidea, families **Pterodactylidae:** JURASSIC: *Belonochasma, Ctenochasma, Gnathosaurus,* and *Pterodactylus* (see Figs. 31-8 and 31-9); **Ornithocheridae (Nyctosauridae or Pteranodontidae):** CRETACEOUS: *Criorhynchus (Coloborhynchus), Dsungaripterus, Nyctosaurus (Nyctodactylus)* (see Fig. 31-10), *Ornithocheirus, Ornithodesmus, Pteranodon* (see Fig. 31-11), and *Titanopteryx.*

32 DINOSAURIA (Saurischia-Theropoda)

The dinosaurs ("terrible-lizards") are broken up into two basic orders, the Saurischia ("lizard-hipped") and the Ornithischia ("bird-hipped").

In these first two chapters on dinosaurs, we are concerned with the suborder Theropoda ("three-toed") of the order Saurischia. In the suborder Theropoda there are the infraorders Coelurosauria and Carnosauria (see the following chapter). In turn, the infraorder Coelurosauria consists of three recognized and two questionable families. They are: **Procompsognathidae,** which includes the following generic types (all from the Triassic Period): *Avipes, Coelophysis, Dolichosuchus, Halticosaurus, ?Loukousaurus, Podokesaurus, Procompsognathus (Pterospondylus), Saltopus, Scleromochlus, ?Spinosuchus, Triassolestes,* and *Velocipes.* **?Segisauridae,** with the genus *Segisaurus* from either the Late Triassic or Early Jurassic of North America. **Coeluridae (Coelurosauridae/Compsognathidae):** JURASSIC: *Agrosaurus, Caudocoelus, Coelurus, Ornitholestes* (see Fig. 32-1), *Compsognathus, Elaphrosaurus, ?Hallops,* and *Sinocoelurus;* CRETACEOUS: *Aristosuchus, Brasileosaurus, Calamospondylus (Calamosaurus), Coeluroides, Compsosuchus, ?Dromaeosaurus, Elaphrosaurus, Jubbulpuria, Laevisuchus, Paronychodon, Saurornithoides* (see Fig. 32-2), *Thecocoelurus (Thecospondylus),* and *Velociraptor* (see Fig. 32-3). **Ornithomimidae:** CRETACEOUS: *?Betasuchus, Cheirostenotes (Chirostenotes), Macrophalangia, Ornithomimus (Coelosaurus and Struthiomimus)* (see Figs. 32-4 and 32-5), *Ornithomimoides,* and *Oviraptor.* **?Caenagnathidae,** with its Late Cretaceous genus: *Caenagnathus.*

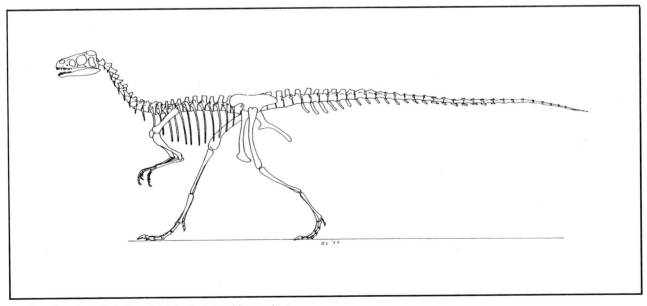

32-1. Upper Jurassic. *Ornitholestes* sp. A skeleton of a theropod dinosaur about 2 meters long, from the Morrison Formation near Medicine Bow, Wyoming. Drawing by Ken Carpenter, University of Colorado, Boulder.

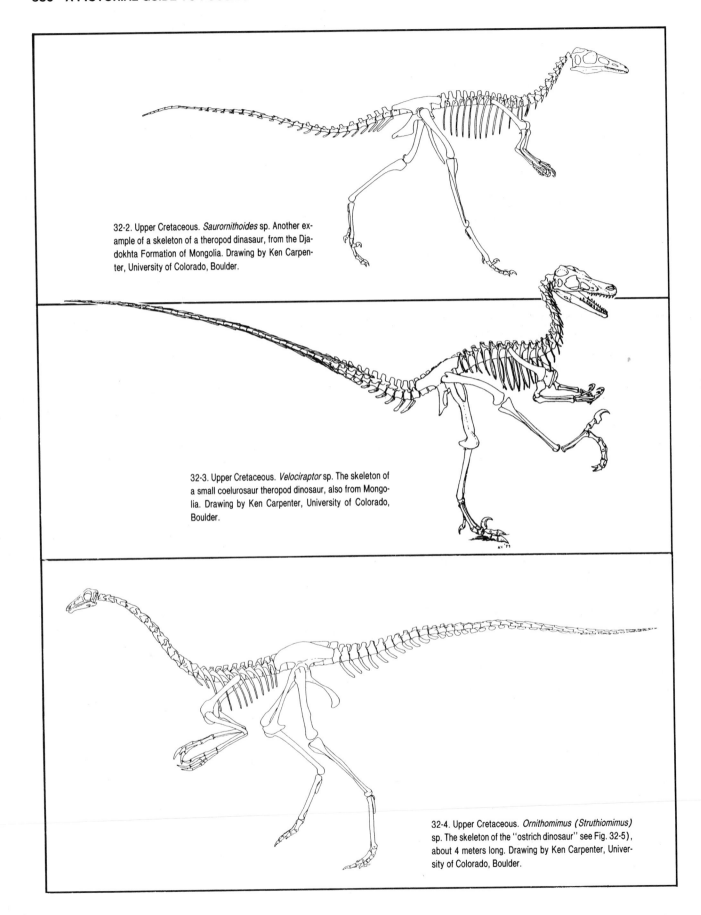

32-2. Upper Cretaceous. *Saurornithoides* sp. Another example of a skeleton of a theropod dinasaur, from the Djadokhta Formation of Mongolia. Drawing by Ken Carpenter, University of Colorado, Boulder.

32-3. Upper Cretaceous. *Velociraptor* sp. The skeleton of a small coelurosaur theropod dinosaur, also from Mongolia. Drawing by Ken Carpenter, University of Colorado, Boulder.

32-4. Upper Cretaceous. *Ornithomimus (Struthiomimus)* sp. The skeleton of the "ostrich dinosaur" see Fig. 32-5), about 4 meters long. Drawing by Ken Carpenter, University of Colorado, Boulder.

32-5. Upper Cretaceous. *Ornithomimus* (*Struthiomimus*) sp. A model by the sculptor Louis Paul Jones of the "ostrich dinosaur" of North America and east Asia (see Fig. 32-4). Inset to the right shows a close-up of the head area. The model is on exhibit at the New Jersey State Museum, Trenton, New Jersey. Courtesy: New Jersey State Museum.

33 DINOSAURIA (Saurischia-Carnosauria)

The larger "flesh-eaters" (the carnosaurs) evolved along with the much smaller coelurosaur carnivores. Many of these large carnivores had strong, massive bodies, with hollow bones, and quite a few of them attained great size. *Allosaurus (Antrodemus)* (see Figs. 33-1 through 33-3) is an example of a carnivorous dinosaur which attained a great length of up to 35 feet (from snout to the tip of the tail). Other dinosaurs such as *Gorgosaurus* (see Fig. 33-9) and *Tyrannosaurus* (see Figs. 33-11 and 33-12) later attained even greater lengths.

Not very much is known of Early Cretaceous carnivores, but there is no doubt that the origins for such large carnivores as *Allosaurus, Gorgosaurus,* and *Tyrannosaurus* are to be found in the Late (Upper) Jurassic and Early (Lower) Cretaceous.

Classification of the carnosaurs: The infraorder Carnosauria is comprised of four families. They are as follows: **Ornithosuchidae, Megalosauridae, Spinosauridae,** and **Tyrannosauridae (Deinodontidae).**

The generic listing for the above families is as follows: **Ornithosuchidae:** TRIASSIC: *Ornithosuchus (Dasygnathus).* **Megalosauridae:** JURASSIC: *Allosaurus (Antrodemus)* (see Figs. 33-1 through 33-3), *Ceratosaurus* (see Fig. 33-4), *Chienkosaurus, Eustreptospondylus (Streptospondylus), Macrodontophion, Megalosaurus (Magnosaurus), Metriacanthosaurus, Proceratosaurus,* and *Sarcosaurus;* CRETACEOUS: *Allosaurus (Antrodemus), Bahariasaurus, Carcharodonsaurus, Chilantaiosaurus, Embasaurus, Inosaurus,* and *Megalosaurus (Magnosaurus).* **Spinosauridae:** CRETACEOUS: *Acrocanthosaurus, Altispinax,* and *Spinosaurus* (see Fig. 33-5). **Tyrannosauridae (Deinodontidae):** CRETACEOUS: *Albertosaurus (Gorgosaurus* and *Deinodon)* (see Figs. 33-6 through 33-10), *Alectrosaurus, Chingkankonsaurus, Dryptosaurus, Genyodectes (Loncosaurus), Majungasaurus, Orthogoniosaurus (Indosaurus), Prodeinodon, Szechusanosaurus, Tarbosaurus* (see Fig. 33-13), and *Tyrannosaurus (Dynamosaurus)* (see Figs. 33-11 and 33-12).

Notes concerning a few of the carnivorous dinosaurs: Allosaurus, also known as *Antrodemus* (a *synonym* or re-jected taxonomic name—the name *Allosaurus* being current usage), preyed particularly upon the swamp dwellers such as *Apatosaurus, Camarasaurus,* and *Diplodocus,* among others, but was also one of the many bipedal terrestrial carnivores that preyed upon other species of carnosaurs. *Allosaurus* had claws (talons) on all four feet that, along with its long, sharp, serrated teeth, must have made it a formidable aggressor in battle (see Figs. 33-1 through 33-3). It was indeed a terror of its time.

The metabolism of these giant carnivores was rather high, so *Allosaurus* and his relatives had to eat an enormous amount of food to keep healthy and stay alive. In life, *Allosaurus* must have weighed upwards of two tons or even more, and attained a body length of from thirty to thirty-five feet.

The Morrison Formation (Jurassic) supplies us with many complete and disarticulated skeletons of *Allosaurus,* as well as other species or allied forms. Many exceptional skeletons come from the Como Bluff quarries in the southeastern part of Wyoming. The geological range for *Allosaurus* is Late (Upper) Jurassic to Early (Lower) Cretaceous.

Albertosaurus (Gorgosaurus), Tarbosaurus, and *Tyrannosaurus* were even more ferocious carnivores than *Allosaurus.* Larger and much more powerful, these Cretaceous carnosaurs took the dominant place that *Allosaurus* once held, and they jointly ruled on up into the Late Cretaceous in their particular territories. Skeletons of *Albertosaurus* are found in the Belly River Formation of western Canada. Many fine skulls (see Fig. 33-9) and articulated skeletons (see Figs. 33-8 and 33-10) have been unearthed in the province of Alberta in Canada. *Albertosaurus* has also been reported from eastern Asia and the Soviet Union.

Tarbosaurus (see Fig. 33-13) of the Soviet Union and eastern Asia is yet another of the larger carnivores which roamed the flood plains and savannahs looking for prey. *Tarbosaurus* had double claws (as did *Albertosaurus* and *Tyrannosaurus*) with recurved talons on its shortened arms.

33-1. Upper Jurassic. *Allosaurus (Antrodemus)* sp. Two views of a mounted skeleton on exhibit at the Geology Museum of Princeton University. Top: Side view of the carnivorous dinosaur showing its 9.3 meter length. Left: A frontal view showing the animal's massive jaws and sharp teeth. Morrison Formation, near Cleveland, Utah. Courtesy: Princeton University, Princeton, New Jersey.

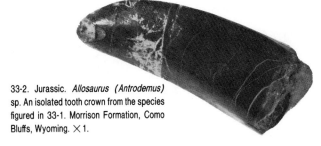

33-2. Jurassic. *Allosaurus (Antrodemus)* sp. An isolated tooth crown from the species figured in 33-1. Morrison Formation, Como Bluffs, Wyoming. ×1.

Tyrannosaurus rex (see Figs. 33-11 and 33-12) is an excellent example of the "terrible-lizards." Larger and heavier than its predecessor *Allosaurus,* and its contemporaries, *Albertosaurus* and *Tarbosaurus,* its teeth were larger, sharper and recurved, and as with *Tarbosaurus* and *Gorgosaurus,* it only had two claws armed with talons on the end of each arm.

Smaller carnosaurs related to those mentioned above are *Deinonychus* ("terrible-claws") which was discovered only fairly recently in Montana in 1964 by Dr. John Os-

33-3. Jurassic. *Allosaurus (Antrodemus)* sp. A skeleton of the carnivorous dinosaur shown in Fig. 33-1. Drawing by Ken Carpenter, University of Colorado, Boulder.

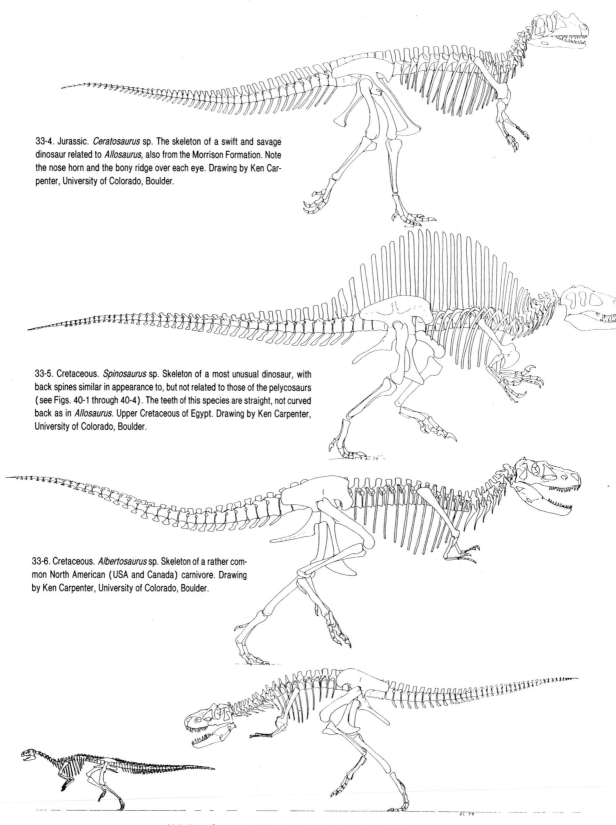

33-4. Jurassic. *Ceratosaurus* sp. The skeleton of a swift and savage dinosaur related to *Allosaurus,* also from the Morrison Formation. Note the nose horn and the bony ridge over each eye. Drawing by Ken Carpenter, University of Colorado, Boulder.

33-5. Cretaceous. *Spinosaurus* sp. Skeleton of a most unusual dinosaur, with back spines similar in appearance to, but not related to those of the pelycosaurs (see Figs. 40-1 through 40-4). The teeth of this species are straight, not curved back as in *Allosaurus.* Upper Cretaceous of Egypt. Drawing by Ken Carpenter, University of Colorado, Boulder.

33-6. Cretaceous. *Albertosaurus* sp. Skeleton of a rather common North American (USA and Canada) carnivore. Drawing by Ken Carpenter, University of Colorado, Boulder.

33-7. Upper Cretaceous. *Albertosaurus* chasing the ornithopod *Parksosaurus.* Both dinosaurs roamed what is now Alberta in western Canada, and down into Montana, Wyoming, Colorado, and Utah in the United States. Drawing by Ken Carpenter, University of Colorado, Boulder.

33-8. Upper Cretaceous. The carnivore *Albertosaurus* (*Gorgosaurus*) *libratus* (Lambe) standing over its fallen prey, the duckbilled *Lambeosaurus*. Both skeletons are reconstructed from bones found in the Oldman Formation of western Canada. The skeletons (above), as well as the model by Maidi Wiebe (below) are on exhibit at the Field Museum in Chicago. Courtesy: Field Museum of Natural History, Chicago.

33-9. Upper Cretaceous. *Albertosaurus (Gorgosaurus) libratus* (Lambe). Closeup of the skull of the carnivorous dinosaur in Fig. 33-8. Oldman Formation, Alberta, Canada. Courtesy: Field Museum of Natural History, Chicago.

trom of Yale University. This small carnivore was only about four to four and one-half feet tall and weighed approximately between 68 and 75 kilograms—a small, but rather fierce predator, and fast on his feet chasing his prey. There is no doubt that *Deinonychus* hunted in packs in order to bring down animals larger than itself. *Deinonychus* possessed three claws (with sharp recurved talons) on both its arms and feet.

A favorite food of the carnivores was the ubiquitous ornithischians, particularly the duck-billed, plant-eating dinosaurs (see Fig. 33-8).

Note: The questionable family **Poposauridae** (**= Rauisuchidae**) which Romer (1966, p. 369R) listed as part of the infraorder Carnosauria, has been reassigned to to the Order Thecodontia, Subclass Archosauria, Suborder Pseudosuchia (Chapter 31—Ruling reptiles).

33-10. Upper Cretaceous. *Albertosaurus* (*Gorgosaurus*) *libratus* (Lambe). Another view of the skeleton in Fig. 33-8, showing the enormous rib cage of this ferocious carnivore. Courtesy: Field Museum of Natural History, Chicago.

33-11. Upper Cretaceous. *Tyrannosaurus rex.* The largest of the carnivorous dinosaurs that roamed North America and Asia. Note the ridiculously small forelimbs for such a large animal. A life-sized model on exhibit in the Children's Museum, Indianapolis, Indiana, Courtesy: Richard Rush Studio, Chicago.

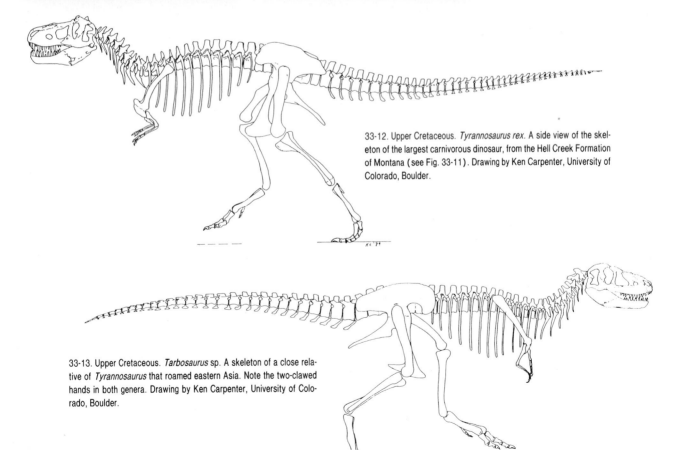

33-12. Upper Cretaceous. *Tyrannosaurus rex.* A side view of the skeleton of the largest carnivorous dinosaur, from the Hell Creek Formation of Montana (see Fig. 33-11). Drawing by Ken Carpenter, University of Colorado, Boulder.

33-13. Upper Cretaceous. *Tarbosaurus* sp. A skeleton of a close relative of *Tyrannosaurus* that roamed eastern Asia. Note the two-clawed hands in both genera. Drawing by Ken Carpenter, University of Colorado, Boulder.

34 DINOSAURIA (Saurischia-Sauropodomorpha and Sauropoda)

The sauropods were large amphibious and herbivorous quadruped dinosaurs which roamed lakes, savannahs, swamps, and marshy lands during the Jurassic and Cretaceous periods. These heavy-bodied dinosaurs, with their long necks and tails, lumbered around on elephantlike feet in search of succulent marsh grasses and plants. They include such forms as the titanosaur *Hypselosaurus* (see Fig. 34-5), and the giant sauropod *Brontosaurus* (see Figs. 34-6 through 34-10). Their Triassic and Early Jurassic ancestors, the prosauropods, were smaller and less completely quadrupedal; the best known is *Plateosaurus* (see Figs. 34-1 through 34-4).

Classification: The suborder Sauropodomorpha is composed of the two infraorders Prosauropoda and Sauropoda. In the Prosauropoda, there are three families: **Anchisauridae, Plateosauridae,** and **Melanorosauridae.** The family **Anchisauridae (Thecodontosauridae)** is the ancestral stock from which the sauropodomorphs evolved and includes such forms as *Anchisaurus (Yaleosaurus), Efraasia,* and *Thecodontosaurus.*

Genera: **Plateosauridae:** *Ammosaurus, Aristosaurus, Lufengosaurus, Massospondylus,* and *Plateosaurus* (see Figs. 34-1 through 34-4). **Melanorosauridae:** *Eucnemesaurus, Euskelosaurus, Melanorosaurus, Plateosauravus, Rioiasaurus,* and *Vulcanodon.*

The infraorder Sauropoda (Opisthocoelia or Cetosauria), has five families representing all forms from the Jurassic and Cretaceous Periods. The five families are: **Cetiosauridae (Brachiosauridae), Euhelopodidae, Camarasauridae, Diplodocidae,** and **Titanosauridae.**

Generic types for the family **Cetiosauridae (Brachiosauridae):** JURASSIC: *Amygdalodon, Apatosaurus, Astrodon (Pleurocoelus), Bothriospondylus, Brachiosaurus, Cetiosaurus (Cardiodon), Dystrophaeus, Elosaurus, Haplocanthosaurus (Haplocanthus), Omeisaurus, Pelorosaurus (Dinodocus), Rhoetosaurus,* and *Tienshanosaurus;* CRETACEOUS: *Astrodon (Pleurocoelus), Austrosaurus, Cetiosaurus (Cardiodon), Omeisaurus, Parrosaurus (Neosaurus), Pelorosaurus (Dinodocus),* and *Tienshanosaurus.* **Euhelopodidae:** CRETACEOUS: *Euhelopus (Helopus).*

Camarasauridae: JURASSIC: *Camarasaurus (Morosaurus)* (see Fig. 34-4); CRETACEOUS: *Camarasaurus (Morosaurus).* **Diplodocidae:** JURASSIC: *Diplodocus.* **Titanosauridae:** JURASSIC: *Amphicoelias, Barosaurus, Brontosaurus* (see Figs. 34-6 through 34-10), *Dicraeosaurus, Mamenchisaurus,* and *Tornieria (Gigantosaurus);* CRETACEOUS: *Aegyptosaurus, Aepisaurus, Alamosaurus, Algoasaurus, Antarctosaurus, Argyrosaurus, Asiatosaurus, Chiayusaurus, Hypselosaurus (Magyarosaurus)* (see Fig. 34-5), *Laplatasaurus, Macrurosaurus, Mongolosaurus, Rebbachisaurus, Succinodon,* and *Titanosaurus.*

Geographical locations for Triassic age prosauropods: Africa: *Massospondylus, Thecondontosaurus, Eucnemesaurus* (Van Hoepen Beds, South Africa), *Gryponyx* (Transvaal region of South Africa), and, from the Stormberg Beds of South Africa, *Aetonyx, Dromicosaurus, Gryponyx, Gryposaurus (Aristosaurus), Melanorosaurus,* and *Plateosauravus.* Asia: *Lufengosaurus, Massospondylus, Thecodontosaurus,* and *Yunanosaurus.* Europe: *Thecodontosaurus* and *Plateosaurus.* South America: *Spondylosoma.* North America: *Thecodontosaurus* and *Yaleosaurus* (Connecticut Valley). Arctic: *Arctosaurus.*

Geographical locations for Jurassic and Cretaceous sauropods: JURASSIC: Australia: *Rhoetosaurus* (Queensland). Africa: *Barosaurus, Bothriospondylus* (Madagascar), *Brachiosaurus, Cetiosaurus, Dicraeosaurus* (Tanganyika), and *Tornieria (Gigantosaurus).* Asia: *Mamenchisaurus, Omeisaurus* (China), and *Tienshanosaurus.* South America: *Amygdalodon.* Europe: *Apatosaurus, Astrodon* (Portugal and England), *Bothriospondylus, Brachiosaurus, Camarasaurus, Cetiosaurus* (England), and *Pelorosaurus* (Portugal and England). North America: *Amphicoelias* (Colorado), *Apatosaurus, Astrodon, Barosaurus* (South Dakota), *Brachyiosaurus, Camarasaurus, Diplodocus* (Colorado and Wyoming), *Dystrophaeus* (Utah), *Elosaurus,* and *Haplocanthosaurus (Haplocanthus)* (Colorado). CRETACEOUS: Australia: *Austrosaurus.* Africa: *Aegyptosaurus* (Egypt), *Algoasaurus* (South Africa), *Cetiosaurus (Cardiodon), Laplatasaurus* (Madagascar), and *Rebbachiasaurus.* Asia: *Antarctosaurus,*

34-1. Upper Triassic. *Plateosaurus trossingensis*. Skeletal mount of a prosauropod (sauropodomorph) dinosaur. Keuper Beds, Trossingen, Germany. Courtesy: Wurtemberg Natural History Collection, Stuttgart, Germany.

34-2. Upper Triassic. *Plateosaurus* sp. Drawing by Ken Carpenter, University of Colorado, Boulder.

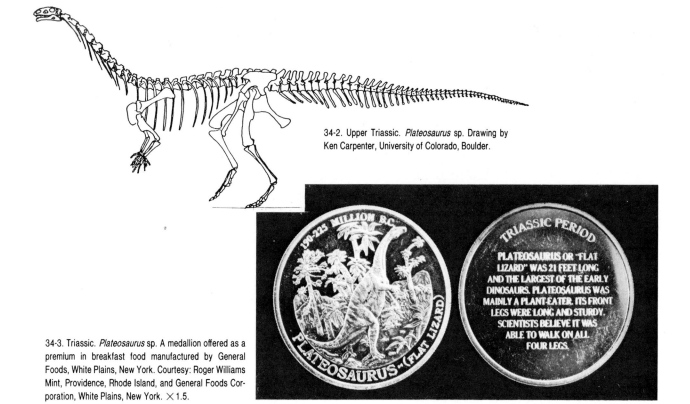

34-3. Triassic. *Plateosaurus* sp. A medallion offered as a premium in breakfast food manufactured by General Foods, White Plains, New York. Courtesy: Roger Williams Mint, Providence, Rhode Island, and General Foods Corporation, White Plains, New York. ✕1.5.

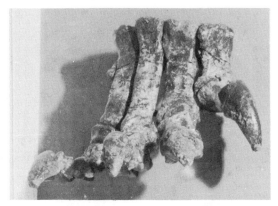

34-4. Triassic. *Plateosaurus* sp. Right hind foot from the Stormberg Beds, Transvaal, South Africa. Courtesy: South African Museum, Cape Town, Republic of South Africa. Reduced ¼ actual size.

34-5. Upper Cretaceous. Eggs atrributed to the titano-saurid sauropod *Hypselosaurus priscus* Mathéron. Danian (Garumnian) of Southern France. 20 cm, inset 15 cm. Courtesy: Geological Enterprises, Ardmore, Oklahoma.

34-6. Upper Cretaceous. *Apatosaurus (Brontosaurus) ajax* Marsh. Skeleton of a giant sauropod dinosaur on exhibit at the Field Museum of a giant sauropod dinosaur on exhibit at the Field Museum in Chicago, Illinois. The skeleton was reconstructed from bones recovered from the Morrison Formation, near Fruita, Colorado. Note: The skull shown is not part of the skeleton. Courtesy: Field Museum of Natural History, Chicago.

34-7. Upper Jurassic. *Camarasaurus* sp. An isolated tooth of another type of sauropod. This animal was a herbivore. The lack of a cutting edge with serrations on the above tooth indicate that this was a plant-eating dinosaur. Cleveland/Lloyd Dinosaur Quarry, Utah. ✕ 1.5. Courtesy: University of Utah.

34-8. Upper Jurassic. *"Brontosaurus"—Apatosaurus* is the correct scientific name for this giant sauropod that browsed in and around lakes and streams in what is now Utah, Wyoming and Colorado. Primitive crocodiles, *Goniopholis,* sun themselves on the banks of the lake. Painting by Charles R. Knight. Courtesy: Field Museum of Natural History, Chicago.

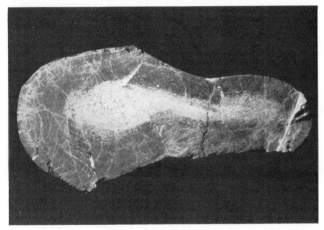

34-9. Upper Jurassic. A cross-section of a limb bone of a sauropod dinosaur showing the cellular structure. Morrison Formation, Colorado.

34-10. Upper Jurassic. *Apatosaurus (Brontosaurus) excelsus* (Marsh). Rear view of the type skeleton of the herbivorous sauropod on exhibit at the Peabody Museum of Yale University (from an old postcard from that institution). Courtesy: Yale University, New Haven, Connecticut.

Asiatosaurus (Mongolia), *Chiayüsaurus, Euhelopus (Helopus)* (China), *Laplatasaurus, Mongolosaurus* (Mongolia), *Omeisaurus* (China), *Tienshanosaurus* (China), and *Titanosaurus.* South America: *Antarctosaurus, Argyrosaurus, Laplatasaurus,* and *Titanosaurus.* Europe: *Aepisaurus, Astrodon* (Portugal and England), *Camarasaurus (Morosaurus), Cetiosaurus (Cardiodon)* (England), *Hypselosaurus (Magyarosaurus)* (Transylvania, Hungary, and France), *Macrurosaurus* (England), *Succinodon,* and *Titanosaurus.* North America: *Alamosaurus* (New Mexico), *Astrodon, Camarasaurus (Morosaurus), Chiayüsaurus* (Kansas), and *Parrosaurus (Neosaurus).*

35 DINOSAURIA (Ornithischia-Ornithopoda and Ankylosauria)

This chapter deals with the hadrosaurians (duckbilled) dinosaurs, as well as the pachycephalosaurs and ankylosaurians, all three of which are suborders of the order Ornithischia. The ornithopods that are commonly known as duckbilled dinosaurs are most curious reptiles. These herbivorous, aquatic reptiles dominated the wetlands, swamps, and estuaries of the Late (Upper) Cretaceous. They were quite abundant, but died out before the close of the Mesozoic leaving no known descendants. Varieties in duckbilled dinosaurs are evidenced by the skull structures. Some genera had head crests which were quite ornate, i.e., *Corythosaurus* (see Figs. 35-5 and 35-6) and *Lambeosaurus* (see Fig. 35-11) for example.

The "parrot-faced" dinosaur (see Fig. 35-18) forms a link between ornithopods and ceratopsians; some authorities classify it as the most primitive horned dinosaur (see Chapter 37). Single representatives of the pachycephalosaurs and Ankylosaurs are added to the end of this chapter.

Classification of the ornithischians: The order Ornithischia has five suborders (two of which will be discussed separately in the following two chapters): Ornithopoda, Stegosauria, Ankylosauria, Pachycephalosauria and Ceratopsia. The suborders Stegosauria and Ceratopsia will be presented in Chapters 36 and 37 respectively. The suborders Ornithopoda, Ankylosauria, and Pachycephalosauria are presented in this chapter.

There are six families representing the suborder Ornithopoda, they are as follows: **Heterodontosauridae, Hypsilophodontidae, Iguanodontidae, Hadrosauridae (Trachodontidae, Psittacosauridae, and Pachycephalosauridae ("Troödontidae").**

The following genera are representatives of the above families: **Heterodontosauridae:** TRIASSIC: *Heterodontosaurus.* **Hypsilophodontidae:** TRIASSIC: *Geranosaurus, Lycorhinus,* and *Tatisaurus;* JURASSIC: *Laosaurus (Dryosaurus), Nanosaurus,* and *Marcellognathus;* CRETACEOUS: *Hypsilophodon, Laosaurus (Dryosaurus), Parksosaurus* (see Figs. 33-7 and 35-1), *Stenopelix,* and *Thescelosaurus.* **Iguanodontidae:** JURASSIC: *Campto-*

saurus, Cryptodraco (Cryptosaurus), Dysalotosaurus, and *Iguanodon (Sphenospondylus);* CRETACEOUS: *Anoplosaurus, Camptosaurus* (see Fig. 35-3), *Craspedodon, Iguanodon* (see Fig. 35-2), *Kangnasaurus, Rhabdodon (Orithomerus),* and *Vectisaurus.* **Hadrosauridae (Trachodontidae):** *Anatosaurus, Bactrosaurus* (see Fig. 35-4), *Brachlyophosaurus, Cheneosaurus, Claorhynchus, Claosaurus, Corythosaurus* (see Figs. 35-5 and 35-6), *Dysganus, Edmontosaurus* (see Figs. 35-7 and 35-8), *Hadrosaurus* (see Fig. 35-10), *Hypacrosaurus, Hypsibema, Kritosaurus (Gryposaurus)* (see Fig. 35-9), *Lambeosaurus (Stephenosaurus)* (see Fig. 35-11), *Lophorhothon, Maiasaura* (see Fig. 35-14), *Mandschurosaurus (Nipponosaurus), Ornithotarsus, Orthomerus (Hecatasaurus), Parasaurolophus, Procheneosaurus (Tetragonosaurus), Prosaurolophus, Saurolophus* (see Fig. 35-12), *Tanius (Tsintaosaurus)* (see Fig. 35-13), and *Yaxartosaurus.* **Psittacosauridae:** CRETACEOUS: *Protiguanodon* and *Psittacosaurus* (see Fig. 35-18). **Pachycephalosauridae ("Troödontidae"):** CRETACEOUS: *Pachycephalosaurus, Polydontosaurus,* and *Stegoceras (Troödon)* (see Fig. 35-19).

Finally, the suborder Ankylosauria has two families: **Acanthopholidae** and **Nodosauridae (Ankylosauridae).** The generic listing for the above families is as follows: **Acanthopholidae:** CRETACEOUS: *Acantholis, Hylaeosaurus (Hylosaurus), Loricosaurus, Onychosaurus, Rhodanosaurus,* and *Struthiosaurus (Crataeomus).* **Nodosauridae (Ankylosauridae):** Anodontosaurus, Brachypodosaurus, Dyoplosaurus, Edmontonia, Euoplocephalus (Ankylosaurus) (see Fig. 35-20), *Heishansaurus, Hierosaurus, Hoplitosaurus, Lametasaurus, Nodasaurus, Palaeoscincus, Panoplosaurus, Paracanthodon, Peishansaurus, Pinacosaurus, Polacanthus, Sauroplites, Scolosaurus, Silvisaurus, Stegopelta, Stegosaurides, Syrmosaurus, Talarurus,* and *Viminicaudus.*

35-1. Upper Cretaceous. *Parksosaurus* sp. A fast running dinosaur which roamed the Northwestern portion of North America. The skeletal remains are found in Alberta, Canada, and in Montana and Wyoming. Drawing by Ken Carpenter, University of Colorado, Boulder.

35-2. Cretaceous. *Iguanodon,* a herbivorous dinosaur that lived in swamps. Skeletal remains have only been found in Europe. Above: Skeletal reconstruction of Iguanodons from Bernissart, Mons, Belgium. These particular skeletons are from the Lower Cretaceous (Wealden Beds), and are on exhibit at the Brussels Museum, Brussels, Belgium. The above picture is from an old postal card from that museum. See, left, an early attempt at a reconstruction of *Iguanodon*. Courtesy: Musée Royal d'Histoire Naturelle, Bruxelles, Belgium.

35-3. Upper Jurassic to Cretaceous. *Camptosaurus,* a small dinosaur probably descended from the thecodonts. Skeletal material of this dinosaur has been found in North America and Europe, but is far more abundant in the fossil beds of Wyoming. Drawings by Ken Carpenter, University of Colorado, Boulder.

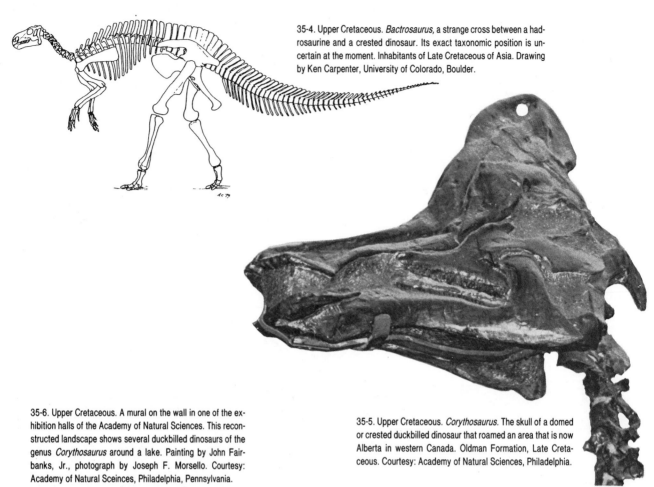

35-4. Upper Cretaceous. *Bactrosaurus,* a strange cross between a hadrosaurine and a crested dinosaur. Its exact taxonomic position is uncertain at the moment. Inhabitants of Late Cretaceous of Asia. Drawing by Ken Carpenter, University of Colorado, Boulder.

35-6. Upper Cretaceous. A mural on the wall in one of the exhibition halls of the Academy of Natural Sciences. This reconstructed landscape shows several duckbilled dinosaurs of the genus *Corythosaurus* around a lake. Painting by John Fairbanks, Jr., photograph by Joseph F. Morsello. Courtesy: Academy of Natural Sceinces, Philadelphia, Pennsylvania.

35-5. Upper Cretaceous. *Corythosaurus.* The skull of a domed or crested duckbilled dinosaur that roamed an area that is now Alberta in western Canada. Oldman Formation, Late Cretaceous. Courtesy: Academy of Natural Sciences, Philadelphia.

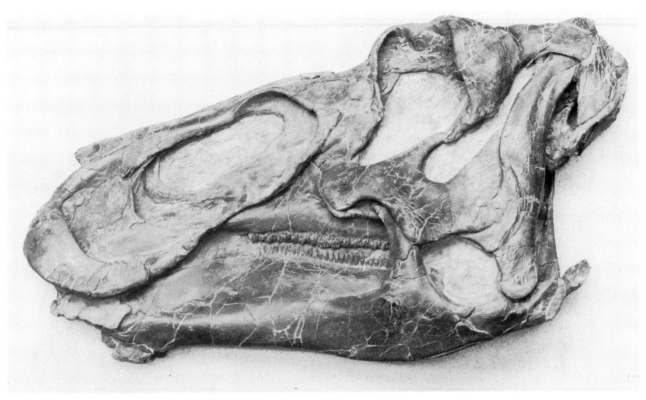

35-7. Upper Cretaceous. *Edmontosaurus regalis* Lambe. The skull of an ornithischian duckbilled dinosaur from the Oldman Formation of western Canada. Courtesy: Field Museum of Natural History, Chicago.

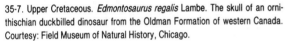

35-8. Upper Cretaceous. *Edmontosaurus* sp. The skeleton of a fairly common duckbilled dinosaur that ranged from western Canada to New Jersey. Drawing by Ken Carpenter, University of Colorado, Boulder.

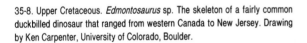

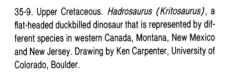

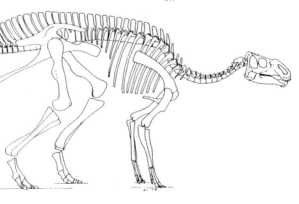

35-9. Upper Cretaceous. *Hadrosaurus (Kritosaurus)*, a flat-headed duckbilled dinosaur that is represented by different species in western Canada, Montana, New Mexico and New Jersey. Drawing by Ken Carpenter, University of Colorado, Boulder.

35-10. Cretaceous. An isolated tooth from a large battery of teeth of a duckbilled dinosaur. Foremost Formation, Medicine Hat, Alberta, Canada. ×1.5.

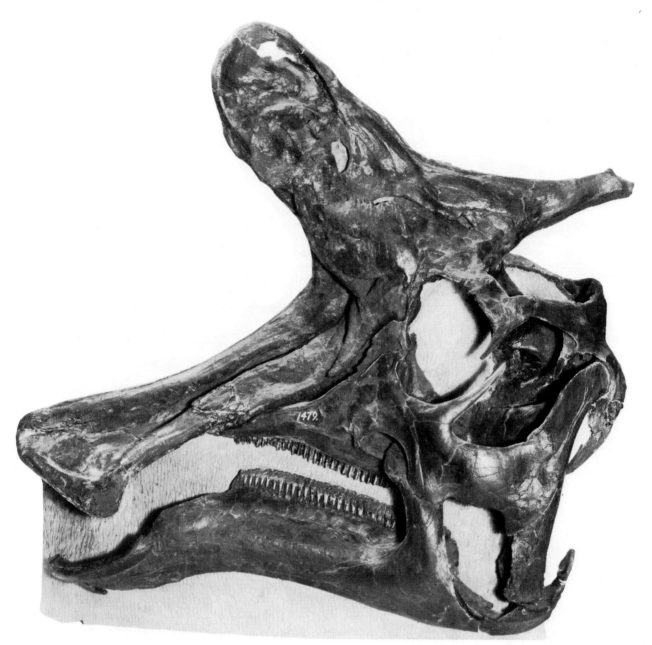

35-11. Upper Cretaceous. *Lambeosaurus lambei* Parks. The skull of a crested hadrosaur (duckbilled dinosaur), from the Oldman Formation (Campanian), western Canada. Courtesy: Field Museum of Natural History, Chicago.

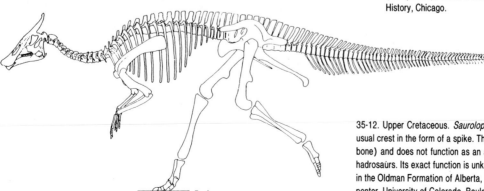

35-12. Upper Cretaceous. *Saurolophus* sp. A duckbilled dinosaur with an unusual crest in the form of a spike. This spike is solid (an extension of the nasal bone) and does not function as an air passage as in other crested species of hadrosaurs. Its exact function is unknown at the present time. Species is found in the Oldman Formation of Alberta, Canada, and in Asia. Drawing by Ken Carpenter, University of Colorado, Boulder.

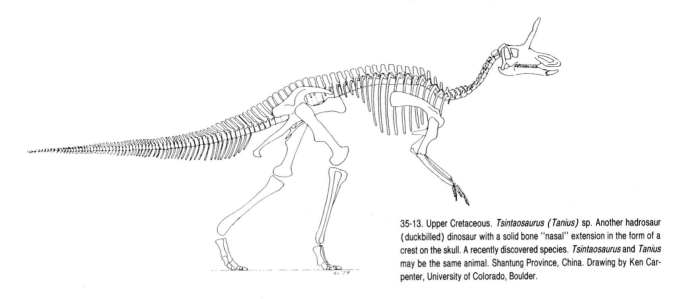

35-13. Upper Cretaceous. *Tsintaosaurus (Tanius)* sp. Another hadrosaur (duckbilled) dinosaur with a solid bone "nasal" extension in the form of a crest on the skull. A recently discovered species. *Tsintaosaurus* and *Tanius* may be the same animal. Shantung Province, China. Drawing by Ken Carpenter, University of Colorado, Boulder.

35-14. Upper Cretaceous. Jack Horner, a research assistant at the Geology Department of Princeton University, displaying his discovery of a new type of a baby hadrosaurine (duckbilled) dinosaur. The skeleton above is a cast made from a completed reconstruction by Mr. Horner from the bones he discovered in 1978 in the Two Medicine Formation near Choteau, Montana. This baby duckbill was found in a nest with 14 others just like it. Nearby the nest, Horner discovered a large skull (possibly the mother) of a hadrosaur that has been named *Maiasaura peeblesorum* Horner and Makela. Photo by J. I. Merritt, Princeton University, Princeton, New Jersey.

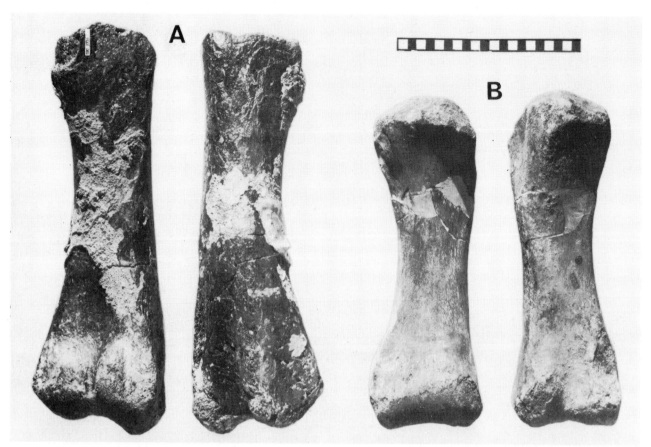

35-15. Upper Cretaceous. Hadrosaurinae indeterminate. A. distal·half of right humerus bone. B. right metatarsal bone. Bones from a duckbilled dinosaur from the Black Creek Formation (Campanian) of North Carolina. From Baird and Horner, 1979; "Cretaceous dinosaurs of North Carolina," Brimleyana. Courtesy: University of North Carolina, Chapel Hill.

35-16. Upper Cretaceous. Hadrosaur vertebra (species indeterminate). Navesink Formation (Middle Maestrichtian), Big Brook, Monmouth County, New Jersey. Approx. half size.

35-17. Upper Cretaceous. Hadrosaur toe bone (species indeterminate). Teapot Sandstone Member, Mesaverde Formation (Late Campanian), Big Horn Basin, Washakie County, Wyoming. ✕1.

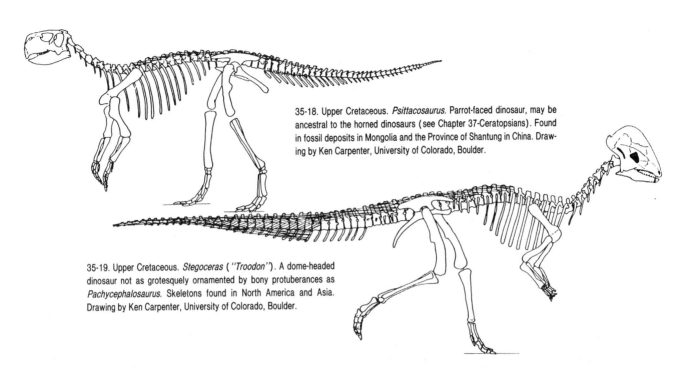

35-18. Upper Cretaceous. *Psittacosaurus*. Parrot-faced dinosaur, may be ancestral to the horned dinosaurs (see Chapter 37-Ceratopsians). Found in fossil deposits in Mongolia and the Province of Shantung in China. Drawing by Ken Carpenter, University of Colorado, Boulder.

35-19. Upper Cretaceous. *Stegoceras* (*"Troodon"*). A dome-headed dinosaur not as grotesquely ornamented by bony protuberances as *Pachycephalosaurus*. Skeletons found in North America and Asia. Drawing by Ken Carpenter, University of Colorado, Boulder.

35-20. Upper Cretaceous. *Ankylosaurus* (was called *Palaeoscincus*). This Ankylosaurian dinosaur is added here to the end of the chapter on ornithopods even though it is not part of this latter group. It is the only representative of the suborder Ankylosauria in this book. Painting by Charles R. Knight, Courtesy: Field Museum of Natural History, Chicago.

36 DINOSAURIA (Ornithischia-Stegosauria)

The stegosaurians were quadrupedal, armored ornithischians with small heads and rotund bodies (averaging some twenty feet in length) with bony plates protruding divergently from the back of the nape of the neck, along the entire back of the animal, and extending almost the full length of the tail, where long pointed spikes protruded from the termination of the tail. These armored stegosaurians differed from the later armored dinosaurs, the ankylosaurs (see preceding chapter), in their types of body armor. The ankylosaurians had a close-fitting armor plating and spikes sticking out sideways from the body and tail (compare Figures 35-20 and 36-2 to see the differences in armor arrangement and pattern). The ankylosaurs took the place of the stegosaurians in the Cretaceous, their range being Middle to Late (Upper) Cretaceous.

The known geological range for the stegosaurians is from Jurassic (Morrison Time) to the Early (Lower) Cretaceous. Their geographical range is as follows: Asia: *Chialingosaurus*. Africa: *Kentrosaurus (Doryphorosaurus)* (Tanganyika). Europe: *Craterosaurus* (England), *Lexovisaurus* (France), *Lusitanosaurus* (Portugal), *Priodontognathus (Omosaurus)* (Portugal and France), North America: *Stegosaurus (Diracodon)* (Colorado and Wyoming). *Stegosaurus* (see Figs. 36-1 and 36-2) is a good example of these curious, armored, plant-eating dinosaurs.

Classification of the stegosaurians: The suborder Stegosauria has only one family: **Stegosauridae.** Typical genera of the family **Stegosauridae:** JURASSIC: *Chialingosaurus, Kentrosaurus (Doryphorosaurus)* (see Fig. 36-3), *Lexovisaurus, Priodontosaurus (Omosaurus), Saurechinodon (Echinodon),* and *Stegosaurus (Diracodon)* (see Figs. 36-1 and 36-2); CRETACEOUS: *Craterosaurus* and *Priconodon.*

36-1. Jurassic. *Stegosaurus* sp. Skeleton on exhibit. Note the tall dorsal "spines" (bony plates) and the multiple spiked tail. Although you cannot see them in this illustration, the skeleton of *Stegosaurus* is standing next to an upright skeleton of the large marine turtle *Archelon* (you can see the arm and finger bones to the right) and the sauropod *Apatosaurus (Brontosaurus)*. (See Fig. 34-10 for this latter skeleton.) From an old post card. Courtesy: Yale Peabody Museum, New Haven, Conn.

36-2. Upper Jurassic. *Stegosaurus* sp. Reconstruction of a plate-back dinosaur. In this painting by Charles R. Knight, two individuals are seen rooting around for food near a lake, in an area which is now called southeastern Wyoming. Courtesy: Field Museum of Nautral History, Chicago.

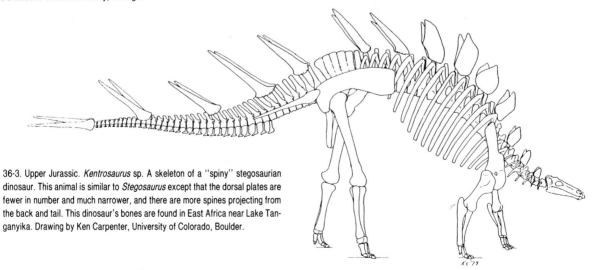

36-3. Upper Jurassic. *Kentrosaurus* sp. A skeleton of a "spiny" stegosaurian dinosaur. This animal is similar to *Stegosaurus* except that the dorsal plates are fewer in number and much narrower, and there are more spines projecting from the back and tail. This dinosaur's bones are found in East Africa near Lake Tanganyika. Drawing by Ken Carpenter, University of Colorado, Boulder.

37 DINOSAURIA (Ornithischia-Ceratopsia)

The ceratopsians, or horned dinosaurs as they are commonly called, were the last of the ornithischian dinosaurs. Their existence seems to have been restricted to the Late (Upper) Cretaceous.

Although the most famous discoveries of skeletons (and eggs) of ceratopsians are from Asia (primarily Outer Mongolia), the majority of the fossils of ceratopsians are from North America. There is one possible skeleton known from South America as well.

The ceratopsians were quadrupeds of not too large a size, although *Triceratops* (see Figs. 37-9 and 37-10) did range in size from 15 to 20 feet in body length.

The skull structure of the ceratopsians is of special interest. For one thing, the skulls are rather large for the body of the animal and there are peculiar bony growths and armament on the skulls of some species. Most species have oversized posterior skull crests of bone which form "collars" (see Figs. 37-9 and 37-10) with or without openings—see Figs. 37-11 and 37-12 for bone crests with openings. Another feature which separates the different species or groups of ceratopsians are the parrotlike beaks on some skulls (see Figs. 37-1 through 37-7), *Protoceratops* being a good example, as well as the "rhinolike" (or "bisonlike") horns above the nasal (nose) area on other species (see Figs. 37-9 through 37-11), *Monoclonius* and *Triceratops* being good examples. There is no doubt that these curious protuberances were defensive in nature. The horn of *Triceratops* must have been a formidable weapon against aggressors.

The bodies of the ceratopsians are almost rhinolike as well. They had short necks, a shortened spikelike pubis, platycoelous vertebra, and four hooflike toes on all four legs (the front legs being the shortest, as with most ornithischians).

The ceratopsian *Protoceratops* (see Figs. 37-2 through 37-7) of Mongolia, was discovered by the expeditions of Roy Chapman Andrews, a famous paleontologist of the 1920s. The name *Protoceratops* was given to these dinosaurs discovered by Andrews and his party even though these particular primitive forms had no horns on their skulls.

The most interesting discovery by Andrews and his assistant, Walter Granger, was the "nests" of dinosaur eggs of *Protoceratops* (see Figs. 37-2, 37-5, 37-7 and 37-8). This discovery of the eggs in 1923, in Mongolia, made the headlines in newspapers all over the world.

Classification of the ceratopsians:

The suborder Ceratopsia of the order Ornithischia, has three families: **Protoceratopsidae, Ceratopsidae,** and **Pachyrhinosauridae.**

The family **Protoceratopsidae** has the following genera (all Late (Upper) Cretaceous in age): *Leptoceratops* (see Fig. 37-1) and *Montanoceratops* from North America, and *Protoceratops* (see Figs. 37-2 through 37-7) from Mongolia (eastern Asia).

The genera comprising the family **Ceratopsidae** are as follows (all Late (Upper) Cretaceous in age): *Anchiceratops, Arrhinoceratops, Centrosaurus, Ceratops (Proceratops), Chasmosaurus (Protorosaurus), Diceratops, Eoceratops, Monoclonius (Brachyceratops)* (see Fig. 37-11), *Pentaceratops, Styracosaurus, Torosaurus,* and *Triceratops (Polyonax)* (see Figs. 37-9 and 37-10), also *Microceratops* from eastern Asia.

Finally, we have as the sole representative of the family **Pachyrhinosauridae,** *Pachyrhinosaurus* (See Fig. 37-12) of the Late (Upper) Cretaceous of North America.

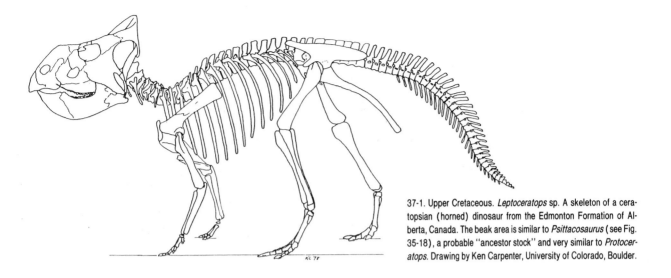

37-1. Upper Cretaceous. *Leptoceratops* sp. A skeleton of a ceratopsian (horned) dinosaur from the Edmonton Formation of Alberta, Canada. The beak area is similar to *Psittacosaurus* (see Fig. 35-18), a probable "ancestor stock" and very similar to *Protoceratops*. Drawing by Ken Carpenter, University of Colorado, Boulder.

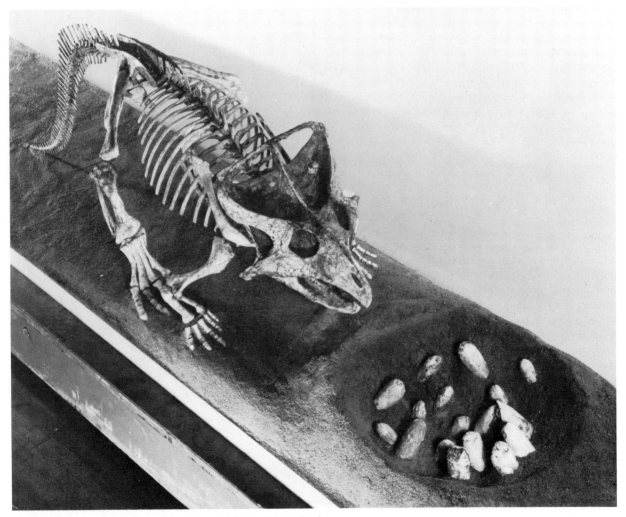

37-2. Upper Cretaceous. *Protoceratops andrewsi* Granger and Gregory. An anterodorsal view of the skeleton of a ceratopsian dinosaur (see Figs. 37-3 and 37-4 for other views of this same skeleton). Note the nest of eggs found in association with skeletal remains of this genus. Djadochta Beds, Shabarakh Usu, Gobi Desert, Mongolia. Courtesy: Field Museum of Natural History, Chicago.

37-3. Upper Cretaceous. *Protoceratops andrewsi* Granger and Gregory. A posterior view of the same skeleton in Figs. 37-2 and 37-4. Djadochta Beds, Shabarakh Usu, Gobi Desert, Mongolia. Courtesy: Field Museum of Natural History, Chigago.

37-4. Upper Cretaceous. *Protoceratops andrewsi* Granger and Gregory. A lateral view of the skeleton in Figs. 37-2 and 37-3. Djadochta Beds, Shabarakh Usu, Gobi Desert, Mongolia. Courtesy: Field Museum of Natural History, Chicago.

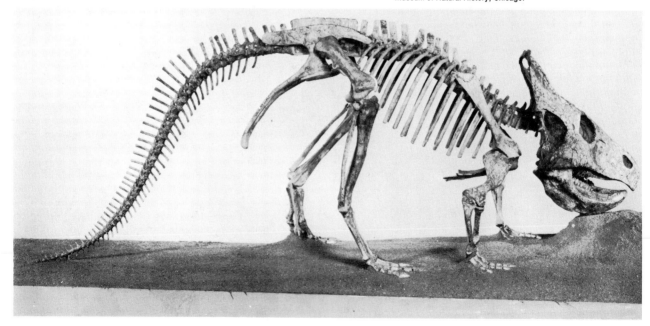

37-5. Upper Cretaceous. *Protoceratops andrewsi* Granger and
Gregory. A reconstruction of the Ceratopsian and nest of eggs figured
in Fig. 37-2, in an area of what is now the Gobi Desert in Mongolia,
Asia. Painting by Maidi Weibe. Courtesy: Field Museum of Natural History, Chicago.

37-6. Upper Cretaceous. *Protoceratops.* A reconstruction of a group of ceratopsians and a nest of eggs from the Late Cretaceous of Mongolia. Painting by Charles R. Knight. Courtesy: Field Museum of Natural History, Chicago.

37-7. Upper Cretaceous. *Protoceratops* sp. Four eggs found in association with the skeleton in Figs. 37-2, 3, and 4. (See also Fig. 37-8.) Djadochta Beds, Shabarakh Usu, Gobi Desert, Mongolia. Courtesy: Field Museum of Natural History, Chicago.

37-8. Upper Cretaceous. *Protoceratops* sp. A nest of eggs probably belonging to a ceratopsian dinosaur from the Late Cretaceous of the Gobi Desert. Djadochta Beds, Shabarakh Usu, Mongolia. Courtesy: Field Museum of Natural History, Chicago.

37-9. Upper Cretaceous. *Triceratops calicornis.* The skull (minus the lower jaw) of a ceratopsian or "horned" dinosaur from the Late Cretaceous of Chalk Butte, Carter County, Montana. Courtesy: Field Museum of Natural History, Chicago.

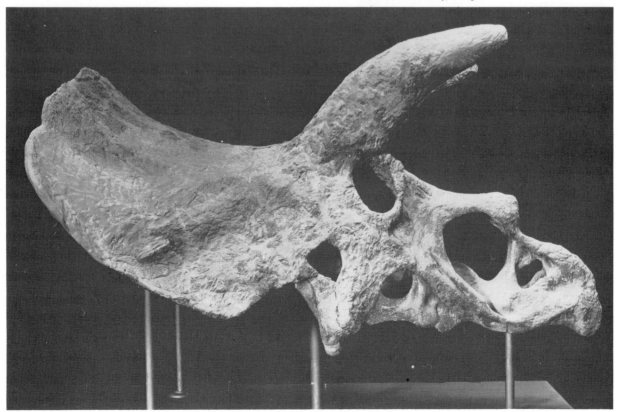

37-10. Upper Cretaceous. *Triceratops* sp. Skeleton of the largest of the horned (ceratopsian) dinosaurs. This genus had 3 horns on its head—1 on its snout, and 2 even longer ones projecting from either side of the top of its skull above the eye area. Habitat: North America. Drawing by Ken Carpenter, University of Colorado, Boulder.

37-11. Upper Cretaceous. *Monoclonius* (*Brachyceratops*). A skeleton of the short-frilled ceratopsian dinosaur from the Judith River Formation of Montana and the Oldman Formation of Western Canada. This horned dinosaur had a large ''rhinoceros'' type of horn above its nose, two smaller horns above its eyes, and a parrotlike beak. (See Fig. 35-18, *Psittacosaurus;* and Fig. 37-1, *Leptoceratops* for comparison). Drawing by Ken Carpenter, University of Colorado, Boulder.

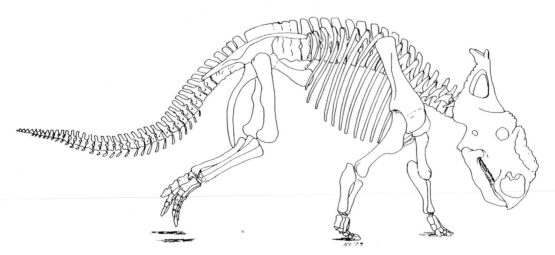

37-12. Upper Cretaceous. *Pachyrhinosaurus* sp. The skeleton of a divergent ceratopsian. Note the ''boss'' or rough bony pad (mass) on the top of the skull near the snout, instead of horns as with the typical ceratopsians. From the Edmonton Beds of Western Canada. Drawing by Ken Carpenter, University of Colorado, Boulder.

38 DINOSAURIA (Ichnites—"tracks")

The dinosaur trackways of the Connecticut Valley area of the eastern United States are given separate names from the dinosaurs that made them, as dinosaur bones and trackways are almost never found in association with each other. In the Newark Supergroup, a band of reddish arkosic sandstones and red and gray shales that runs from Nova Scotia to North Carolina, only half-a-dozen dinosaur skeletons have been found, while the footprints number in the scores of thousands. Thus, the trackways fill a conspicuous gap in our knowledge.

The areas most productive of footprints are the road cuts and quarries of the Connecticut Valley and north-central New Jersey. Some of the more recent finds of bedrock surfaces crisscrossed by trackways, such as those at Rocky Hill Dinosaur State Park in Connecticut and at Roseland and Lincoln Park in New Jersey, have yielded much new information regarding the herding habits and the gait or style of walking, running, and even kick-swimming, as recorded in the soft mud of alluvial plains and tidal flats.

The most common footprints (ichnites) preserved in the shales and sandstones of the Newark Supergroup are various species of *Anchisauripus* and *Grallator* (see Figs. 38-1 and 38-2). These are three-toed tracks made by dinosaurs that walked exclusively on their hind legs, and ranged from turkey-sized to more than seven meters long. Analysis of the foot structure revealed in the imprints indicates that the trackmakers were coelurosaurian theropod dinosaurs (see Chapter 32), which must have preyed upon each other as well as on ornithischian dinosaurs and a variety of small reptiles that inhabited the same area.

The broad, four-toed hind foot prints of *Anomoepus* (see Figs. 38-3 through 38-5) are sometimes accompanied by smaller forefoot impressions, indicating that the dinosaurs could walk either bipedally or quadrupedally. Their foot structure strongly suggests that *Anomoepus* tracks were made by primitive ornithischian dinosaurs that were similar in body build to *Parksosaurus* and *Camptosaurus* (see Figs. 35-1 and 35-3) although, of course, much earlier in time. As skeletal remains of ornithischian dinosaurs are extremely scarce in rocks of the Triassic and earliest Jurassic periods, these footprints provide valuable supple-

mentary information about dinosaur evolution and about predator–prey relationships in dinosaur communities.

For more than a century the Newark rocks were thought to be Triassic in age, but recent investigation has

38-1. Jurassic. Top: *Anchisauripus* sp. Bottom: Smaller print (top left), *Grallator* sp.; larger print, *Anchisauripus*. Ichnites or tracks are in red sandstone from the Connecticut Valley and north-central New Jersey. Bottom illustration courtesy: Brooklyn Children's Museum, Brooklyn, New York. Both approx. half size.

409

38-2. Jurassic. Large slabs of tracks (trackways, if more than one track made by the same species is found on the slab) showing *Anchisauripus, Grallator,* and other theropod dinosaur tracks. Towaco Sandstone, Towaco, Morris County, New Jersey. Courtesy: University Museum, Rutgers University, New Brunswick, New Jersey.

shown that all the Connecticut Valley tracks and most of those from New Jersey are actually Early Jurassic. Active research on these tracks is still underway.

38-3. Jurassic. *Anomoepus crassus* Hitchcock. A very fine footprint of a four-toed dinosaur, possibly an ornithischian. One of the most common tracks found in the arkosic red sandstone beds of New Jersey. Towaco Sandstone, Roseland, Essex County, New Jersey. Note the skin impression in the heel area. ×1.

38-4. *Anomoepus* sp. Another example, superimposed over ripplemarks (see also Fig. 21-2). Roseland, Essex County, New Jersey. Slightly less than actual size.

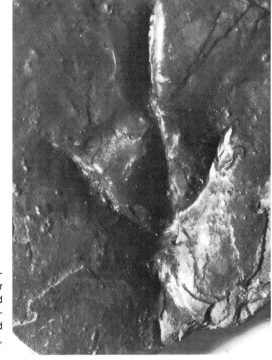

38-5. *Anomoepus* sp. An isolated track cast of the dinosaur common to the arkosic red sandstone of the eastern portion of the United States. Old Saybrook, Middlesex County, Connecticut. ×1 (approx.).

39 AVES (Birds)

The origin of birds goes back to the archosaurians, although their development was entirely separate from that of the flying reptiles or Pterosauria (see Chapter 31).

Pterosaurs and primitive birds shared the skies in the Jurassic and Cretaceous Periods, presumably competing for food; and it appears that the birds eventually developed such a superiority that the flying reptiles were driven to extinction.

The earliest known bird, *Archaeopteryx* (see Figs. 39-1 and 39-2), dates from the Late Jurassic and is known from half-a-dozen skeletons entombed in the lithographic limestones of Bavaria in southern Germany. These fine-grained sediments preserved the imprints of the wing and tail feathers, the claws, and even the form of the brain. The asymmetrical vanes of its wing feathers prove that *Archaeopteryx* could fly (not merely glide), though it was probably not a strong flyer. Unlike later birds it had sharp teeth in both upper and lower jaws as well as a long tail fringed with feathers.

Except for its feathers, the skeleton of *Archaeopteryx* could easily be mistaken for that of a small coelurosaurian dinosaur (see Chapter 32). The multitude of similarities have convinced most scientists that birds evolved from small dinosaurs (possibly already warm-blooded) in which the body scales had evolved into feathers—first as down for insulation, later as stiff pinnules suitable for flight. Birds have dinosaurlike hind limbs and feet, allowing them to walk in a semierect position. The hollow bones they inherited from dinosaurs became further lightened and filled with air sacs that formed as outgrowths from the lungs. The lengthened ilium became firmly attached to additional sacral vertebrae, forming a synsacrum. The pubis migrated to a downward-and-backward position, as it did (independently) in ornithischian dinosaurs, while the acetabulum remained perforated, as in dinosaurs in general. The skull is basically archosaurian in style, lacking a pineal opening, but with enlarged orbits to accomodate the powerful eyes and a relatively huge cranium to house the highly developed brain. The openings in front of the orbits and in the temporal region became further enlarged to eliminate excess weight. Teeth were retained in a few Cretaceous birds, but by Tertiary time the teeth had disappeared and the upper and lower jaws were sheathed with horny bills like those seen in turtles and some dinosaurs.

As powerful muscles are needed to provide a downward-and-backward thrust for the wings, the shoulder girdle became strong and deep with a keel on the sternum and another at the fused ends of the clavicle (the wishbone). Although *Archaeopteryx* retained three clawed fingers, the hand is vestigial in most later birds.

A listing (classification) of birds in the fossil record:

In the class Aves (birds) there are two basic subclasses: Archaeornithes (Sauriurae) (ancient birds) and Neornithes (new or "modern" birds).

In the subclass Archaeornithes (Sauriurae) there is only one order Archaeopterygiformes, with one family, **Archaeopterygidae.** *Archaeopteryx* (see Figs. 39-1 and 39-2) from the Jurassic of Europe is the only genus so far discovered.

In the subclass Neornithes there are three superorders: Odontognathae (Odontoholcae), Palaeognathae, and Neognathae.

In the superorder Odontognathae (Odontoholcae) there is only one order, Hesperornithiformes, with one family, **Hesperornithidae,** with two known genera: *Coniornis,* and *Hesperornis* from the Cretaceous of North America. These are the last of the toothed diving birds.

In the superorder Palaeognathae there are seven orders: Tinamiformes, Struthioniformes, Rheiformes, Casuariformes, Aepyornithiformes, Dinornithiformes, and Apterygiformes.

The following is a condensed listing of the above orders along with their families and typical generic types: Tinamiformes: family **Tinamidae:** PLIOCENE: *Cayetanornis,* and *Tinamisornis;* PLEISTOCENE: *Nothoprocta, Nothura,* and *Rhynchotus.* Struthioniformes: family **Eleuthernithidae:** EOCENE: *Eleutherornis.* PLIOCENE/PLEISTOCENE: *Struthio.* Rheiformes: families **Opisthodactylidae:** MIOCENE: *Opisthodactylus.* **Rheidae:** PLIOCENE: *Heterorhea;* PLEISTOCENE: *Rhea.* Casuariformes: families **Casuariidae:** PLEISTOCENE: *Casuarius.* **Dromaiidae:** PLIOCENE: *Dromaius.* **Dro-**

39-1. Upper Jurassic. *Archaeopteryx lithographica* Meyer. A restoration of the early "toothed bird" by Dr. Hayward of the British Museum. The original fossil was recovered from Lithographic Limestone in Bavaria, southern Germany. Courtesy: Field Museum of Natural History, Chicago.

mornithidae: PLEISTOCENE: *Dromornis*. Aepyorni- thiformes: family **Aepyornithidae**: PLEISTOCENE: *Aepyornis*. Dinornithiformes: families: **Emeidae**: PLEISTOCENE: *Palapteryx (Emeus)* (see Figs. 39-17 and 39-18). **Dinornithidae**: PLEISTOCENE: *Dinornis (Moa)* (see Figs. 39-19 through 39-21). Apterygiformes: family **Apterygidae**: PLEISTOCENE: *Apteryx*.

In the superorder Neognathae, there are twenty-four orders: Gaviiformes, Podicipediformes, Procellariiformes (Tubinares), Sphenisciformes, Pelecaniformes (Stegano- podes), Ciconiiformes (Ardeiformes Gressores), Anseri- formes, Falconiformes (Accipitriformes), Galliformes (Galli): family **Cracidae**: EOCENE: *Gallinuloides* (see Figs. 39-3, 39-4, 39-7, and 39-8), Ralliformes (Grui- formes), Diatrymiformes, Ichthyornithiformes, Charadri- iformes/suborder Charadrii (Grallae): family **Scolopaci-**

39-2. Upper Jurassic. *Archaeopteryx* sp. An old engraving showing another interpretation of the ancient bird found in southern Germany (see Fig. 39-1). Note the claws attached to the forward edge of the wings, much like in the Chiroptera (bats). Malm Zeta, Solnhofen, Bavaria, Germany.

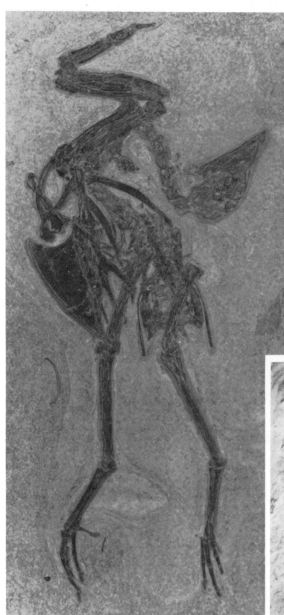

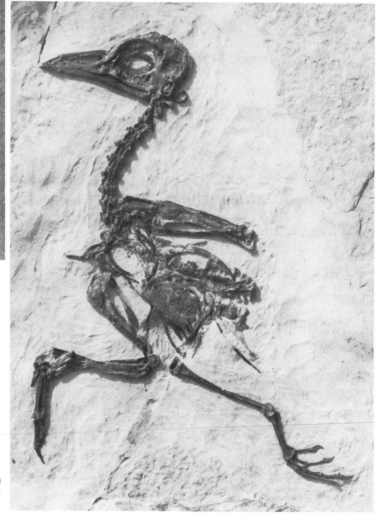

dae: EOCENE: *Rhynchaeytes* (see Figs. 39-5 and 39-6); MIOCENE: *Pelecyornis* (see Fig. 39-13); PLIOCENE: *Mesembriornis* (see Figs. 39-14 through 39-16), Columbiformes, Psittaciformes, Cuculiformes, Strigiformes, Caprimulgiformes, Apodiformes, Coliiformes, Trogoniformes, Coraciiformes, Piciformes, and Passeriformes.

Along with the above twenty-four orders, there are the following suborders: Pitaethontes, Odontopterygia, Pelecani, Fregatae, Cladornithes, Phoenicopteri, Plataleae, Ardeae, Ciconiae, Anseres, Anhimia, Sarcoramphi (Cathartides), Accipitres (Falcones), Ralli, Mesitornithides, Turnices (Hemipodes), Grues, Rhynocheti, Euryptygae, Cariamae, Otides, Charadrii (Grallae), Lari, Alcae, Pterocletes, Columbae, Steatornithes, Apodi, Trochili, Halcyones, Bucerotes, Coraciae, Meropes, Pici, Galbulae, Capitones, Ramphastides, Eurylaimi, Menurae, and Passeres.

The following families comprise the above twenty-four orders and forty suborders of the class Aves (with the exception of extant (recent) families which are omitted): **En-**

39-3. Eocene. *Gallinuloides wyomingensis* Eastman. The skeleton of a primitive pheasant (galliform) in the suborder Gallinae. Green River Formation, from Fossil, Lincoln County, Wyoming. Type specimen (MCZ1598), Courtesy: Museum of Comparative Zoology, Harvard University, Cambridge, Massachusetts. ✕1.

39-4. Eocene. Another fine skeleton of a bird in lateral repose. Green River Formation, Kemmerer, Lincoln County, Wyoming. ✕1. Courtesy: Siber and Siber Ltd., Aathal, Switzerland.

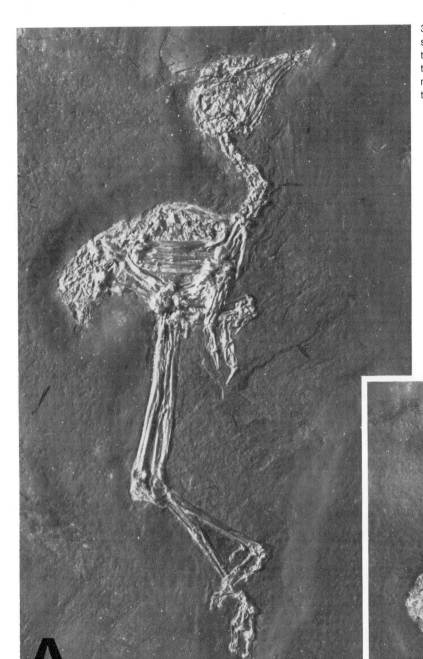

39-5. Eocene. *Rhynchaeytes messelensis.* A skeleton of a fossil snipe in the suborder Charadrii. A. Upper left corner lighting to highlight bones. B. Flat lighting to show other details of interest on the specimen. Specimen from the Oil-shale mines near Messel (Darmstadt), Hessen, Germany. 19 cm. Courtesy: Siber and Siber Ltd., Aathal, Switzerland.

aliornithidae, Lonchodytidae, Gaviidae (Colymbidae), Baptornithidae, Podicipedidae, Diomedeidae, Procellariidae, Oceanitidae, Pelecanoididae, Spheniscidae, Elopterygidae, Phalacrocoracidae, Anhingidae, Sulidae, Phaethontidae, Odontopterygidae, Pseudodontornithidae, Cyphornithidae, Pelecanidae, Cladornithidae, Torotigidae, Scaniornithidae, Telmabatidae, Agnopteridae, Phoenicopteridae, Palaelodidae, Plegadornithidae, Plataleidae (Threskiornithidae), Ardeidae, Ciconiidae, Parancyrocidae, Anatidae, Vulturidae, Neocathartidae, Sagittariidae, Pandionidae, Accipitridae, Falconidae,

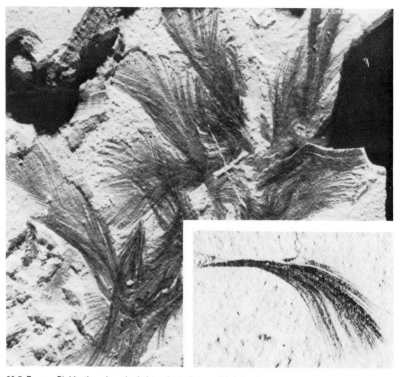

39-7. Eocene. Bird feathers (species indeterminate, but possibly belonging to the pheasant *Gallinuloides* (see Fig. 39-3). Inset shows a single feather. Green River Formation, Garfield County, Colorado. ×1. Courtesy: Geological Enterprises, Ardmore, Oklahoma.

39-6. Eocene. *Rhynchaeites messelensis*. An epoxy cast of another specimen of the fossil snipe (see Fig. 39-5A and B). Oil-shale mines, Messel near Darmstadt, Hessen, Germany. Courtesy: Geological Enterprises, Ardmore, Oklahoma.

39-8. Eocene. Bird tracks (species indeterminate, but evidently a shorebird). Green River Formation, Garfield County, Colorado. ×1.25.

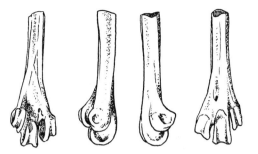

39-9. Miocene. One terminal end (articulated facette) of a unidentified marine bird, possibly a tern. Four views of the articulation. ×1. Pungo River Marl Formation (Helvetian), Aurora, Beaufort County, North Carolina.

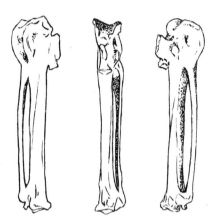

39-10. Miocene. A complete tarsometatarsus bone of a marine bird, possibly a gull or tern. Three views of a complete bone. ×1.5.Pungo River Marl Formation (Helvetian), Aurora, Beaufort County, North Carolina.

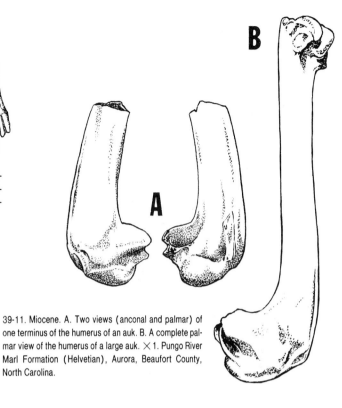

39-11. Miocene. A. Two views (anconal and palmar) of one terminus of the humerus of an auk. B. A complete palmar view of the humerus of a large auk. ×1. Pungo River Marl Formation (Helvetian), Aurora, Beaufort County, North Carolina.

39-12. Miocene. Top row: various bones of grebes, terns and gulls. On the extreme left-hand side, a coracoid of a grebe (duck). Bottom row: Various bones of auks. The bone on the extreme left-hand side is a palmar view of one facet of an auk's humerus bone (see Fig. 39-11B), while the bone on the extreme right-hand side of the picture is a neck vertebra. The various bones represent: humeri, tarsometatarsi, rami, coracoids, and tibia and fibia of various species of marine birds. ×1. Pungo River Marl Formation (Helvetian), Aurora, Beaufort County, North Carolina.

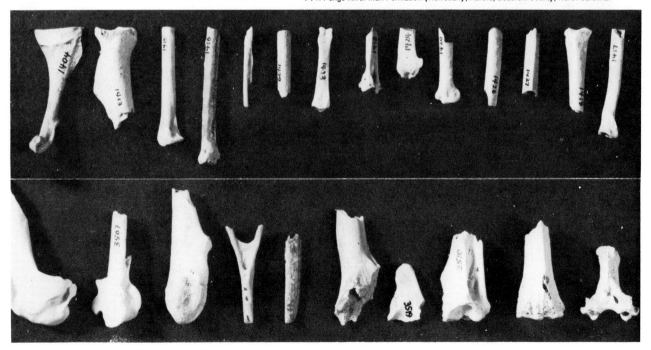

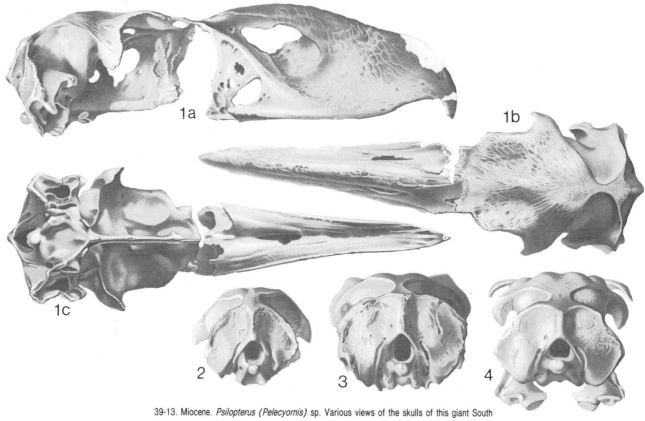

39-13. Miocene. *Psilopterus (Pelecyornis)* sp. Various views of the skulls of this giant South American wading bird from the order Ralliformes (Family Cariamidae). Santa Cruz Formation of Patagonia. la. Lateral view of skull (minus lower jaw). lb. Dorsal view of skull roof of same skull. lc. Ethmoidal or ventral view of same skull. 2. Posterior view of another skull showing the occipital condyle. 3. Posterior view of skull in la,b,c. 4. Posterior view of yet another skull. Drawings by Bruce Horsfall. Courtesy: Princeton University, Princeton, New Jersey.

39-14. Miocene. *Phorusrhacos (Mesembriornis)* sp. Reconstruction of a flightless bird from Argentina. Courtesy: Field Museum of Natural History, Chicago.

39-15. Miocene. *Phorusrhacos (Phororhacos) longissimus*
Ameghino. The skeleton of a giant flightless bird (see Fig. 39-14).
Santa Cruz Formation of Argentina. Courtesy: Field Museum of
Natural History, Chicago.

39-16. Miocene. *Phorusrhacos (Phororhacos) longissimus* Ameghino. An illustration of what the Patagonian flightless bird might have looked like in life. Painting by J. C. Hanson. Courtesy: Field Museum of Natural History, Chicago.

Cracidae, Ophisthocomidae, Megapodidae, Numididae, Phasianidae, Rallidae, Idiornithidae, Turnicidae, Gruidae, Aramidae, Phororhachidae, Brontornithidae, Dryornithidae, Cunampaiidae, Bathornithidae, Cariamidae, Otididae, Diatrymatidae, Gastornithidae, Ichthyornithidae, Apatornithidae, Cimolopterygidae, Scolopacidae, Recurvirostridae, Jacanidae, Burhinidae, Laridae, Stercorariidae, Alcidae, Pteroclidae, Columbidae, Raphidae, Psittacidae, Cuculidae, Protostrigidae, Strigidae, Tytonidae, Nyctibiidae, Caprimulgidae, Aegialornithidae, Apodidae, Trochilidae, Trogonidae, Momotidae, Halcyonidae, Bucerotidae, Coracidae, Phoeniculidae, Upupidae, Meropidae, Picidae, Bucconidae, Ramphastidae, Syctalopidae (Rhinocryptidae/Pteroptochidae), Furnariidae, Formicariidae, Tyrannidae, Alaudidae, Palaeospizidae, Hiruninidae, Dicruridae, Oriolidae, Corvidae, Callaeatidae, Paridae, Sittidae, Certhiidae, Chamaeidae, Timaliidae, Pycnonotidae, Palaeoscinidae, Cinclidae, Troglodytidae, Mimidae, Turdidae, Muscicapidae, Sylvidae, Prunellidae, Motacillidae, Bombycillidae, Laniidae, Sturnidae, Meliphagidae, Vireonidae, Coerebidae, Parulidae, Icteridae, Ploceidae, Tangridae, and Fringillidae.

39-17. Pleistocene. *Emeus crassus* (Owen). Skeleton of a moa or flightless bird from the Pleistocene of New Zealand. Courtesy: Field Museum of Natural History, Chicago.

39-18. Pleistocene. *Emeus crassus* (Owen). Reconstruction of the flightless moa in Fig. 39-17. Painting by J. C. Hanson. Courtesy: Field Museum of Natural History, Chicago.

39-19. Pleistocene. *Dinornis giganteus* Owen, the extinct giant flightless bird. This painting by Charles R. Knight shows a gathering of moas at a water hole in the uplands of post-Pleistocene New Zealand. Courtesy: Field Museum of Natural History, Chicago.

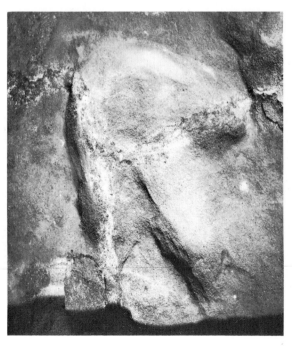

39-20. Pleistocene. A track or footprint of a giant moa bird from New Zealand. Courtesy: New Zealand National Museum, Wellington, New Zealand. Approx. ⅓ actual size.

39-21. Pleistocene. Sculptured reconstruction of a moa bird (now extinct) on exhibit at the National Museum of New Zealand. Courtesy: New Zealand National Museum, Wellington, New Zealand.

40 MAMMALLIKE REPTILES

The synapsids, reptiles which ultimately gave rise to the primitive and earliest forms of mammals, arrived early on the geological scene. Primitive forms of the subclass Synapsida made their appearance as far back as the Early Pennsylvanian. The synapsids persisted until the Triassic Period, and were ultimately replaced by the archosaurians.

The Permian and Triassic Periods brought forth a multitude of mammallike reptile forms, including the pelycosaurs, therapsids, theriodonts (including the therocephalians), anomodonts, dicynodonts, and tritylodonts, among others.

The mammals themselves evolved from the ancestral synapsid stock in the Late Triassic Period. By the end of the Cretaceous they had diversified considerably although remaining small in size.

Classification of mammallike reptiles: The subclass Synapsida contains two basic orders: Pelycosauria (Thero-morpha) and Therapsida.

There are three suborders in the order Pelycosauria (Theromorpha): Ophiacodontia, Sphenacodontoidea, and Edaphosauria. The suborder Ophiacodontia in turn contains two families: **Ophiacodontidae** and **Eothyrididae.** Typical generic examples of the two families above are as follows: **Ophiacodontidae:** PENNSYLVANIAN: *Clepsydrops (Archaeobelus),* and *Protoclepsydrops;* PERMIAN: *Basicranion, Ophiacodon (Arribasaurus)* (see Figs. 40-5 and 40-6), and *Varanosaurus (Poecilospondylus)* (see Fig. 40-3). **Eothyrididae:** PERMIAN: *Baldwinonus, Bayloria, Eothyris, Stereophallodon, Stereorhachis,* and *Tetraceratops.*

The suborder Sphenacodontoidea also has two families. Generic examples: **Varanopsidae:** PERMIAN: *Aerosaurus, Anningia (Galesphyrus), Homodontosaurus, Scoliomus,* and *Varanops (Varanoops).* **Sphenacodontidae:**

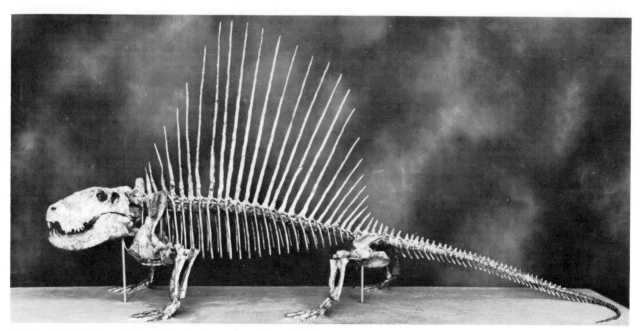

40-1. Permian. *Dimetrodon grandis.* The skeleton of a pelycosaurian reptile from the Early Permian red beds of northwestern Texas. Courtesy: Field Museum of Natural History, Chicago.

40-2. Permian. *Edaphosaurus pogonias* Cope. Reconstruction of an Early Permian reptile from Texas. Painting by Maidi Wiebe. Courtesy: Field Museum of Natural History, Chicago.

40-3. Permian. *Edaphosaurus* and *Dimetrodon,* mammallike reptiles, with *Varanosaurus,* a primitive pelycosaur (left), and the nectridean amphibian *Diplocaulus* (right-lower, in pool), in a Permian setting. Painting by Charles R. Knight. Courtesy: Field Museum of Natural History, Chicago.

PENNSYLVANIAN: *Macromerion;* PERMIAN: *Bathygnathus, Ctenospondylus, Dimetrodon (Bathyglyptus)* (see Figs. 40-1 and 40-3), *Haptodus (Pantelosaurus), Neosaurus, Oxyodon, Secodontosaurus, Sphenacodon (Elcabrosaurus)* (see Figs. 40-7 and 40-8), *Steppesaurus,* and *Thrausmosaurus;* TRIASSIC: *Ctenosaurus.*

The suborder Edaphosauria has four families: **Nitosauridae:** PENNSYLVANIAN: *Petrolacosaurus (Podargosaurus);* PERMIAN: *Colobmycter, Mycterosaurus (Eu-* *matthevia),* and *Nitosaurus.* **Lupeosauridae:** PERMIAN: *Lupeosaurus.* **Edaphosauridae:** PENNSYLVANIAN/ PERMIAN: *Edaphosaurus (Brachycnemius)* (see Figs. 40-2 through 40-4). **Caseidae:** PENNSYLVANIAN: *Caseopsis;* PERMIAN: *Angeolosaurus, Casea, Caseoides, Cotylorhynchus,* and *Ennatosaurus.*

The order Therapsida is composed of three suborders: Phthinosuchia (Eotitanosuchia), Theriodontia, and Anomodontia.

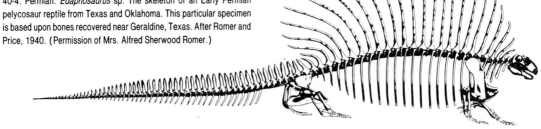

40-4. Permian. *Edaphosaurus* sp. The skeleton of an Early Permian pelycosaur reptile from Texas and Oklahoma. This particular specimen is based upon bones recovered near Geraldine, Texas. After Romer and Price, 1940. (Permission of Mrs. Alfred Sherwood Romer.)

40-5. Permian. *Ophiacodon mirus* Marsh. Reconstruction of the theromorph reptile from the Early Permian Abo Formation of New Mexico. See Fig. 40-6 for the skeleton of this specimen. Painting by Maidi Wiebe. Courtesy: Field Museum of Natural History, Chicago.

40-6. Permian. *Ophiacodon mirus* Marsh. The skeleton of a theromorph reptile from the Early Permian Abo Formation of New Mexico. See Fig. 40-5 for a reconstruction. Courtesy: Field Museum of Natural History, Chicago.

The suborder Phthinosuchia (Eotitanosuchia) has only one family: **Phthinosuchidae:** PERMIAN: *Biarmosaurus, Biarmosuchus, Eotitanosuchus, Gorgodon, Knoxosaurus, Phthinosaurus,* and *Phthinosuchus.*

The suborder Theriodontia has in turn six infraorders: Gorgonopsia, Cynodontia, Tritylodontoidea, Therocephalia, Bauriamorpha, and Ictidosauria.

40-7. Permian. *Sphenacodon ferox* Marsh. Reconstruction of another theromorphic reptile from the Early Permian Abo Formation of New Mexico. See Fig. 40-8 for a skeleton of this species. Painting by Maidi Wiebe. Courtesy: Field Museum of Natural History, Chicago.

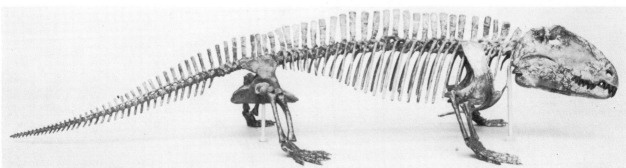

40-8. Permian. *Sphenacodon ferox* Marsh. The skeleton of a theromorph reptile from the Early Permian Abo Formation of New Mexico. (See Fig. 40-7.) Courtesy: Field Museum of Natural History, Chicago.

40-9. Permian. *Aulacocephalodon peavoti*. The skeleton of a large dicynodontid reptile from the Karroo Series, Lower Beaufort beds of South Africa. Courtesy: Field Museum of Natural History, Chicago.

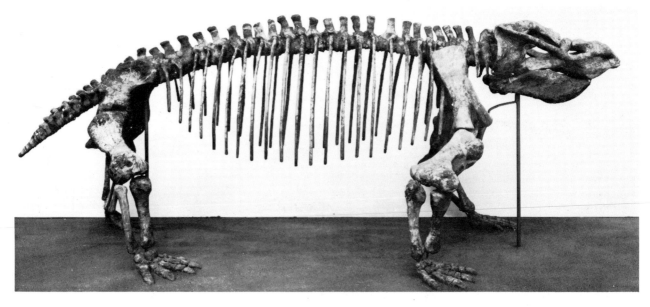

40-10. Permian. *Inostrancevia (?Amalitzkia) alexandri* Amalitzky. The skull of a gorgonopsid (therapsid) from the Soviet Union. Courtesy: Soviet Academy of Sciences, U.S.S.R.

40-11. Triassic. *Cynognathus* (top and left), a cynodont, and *Kannemeyeria* (lower right), a dicynodont, both mammallike reptiles from the Triassic beds of South Africa. Painting by Charles R. Knight. Courtesy: Field Museum of Natural History, Chicago.

40-12. Upper Triassic. *Placerias gigas* Camp and Welles. Sculptured restoration of the last surviving dicynodont (therapsid), found in Arizona and North Carolina. Courtesy: Richard Rush Studio, Chicago.

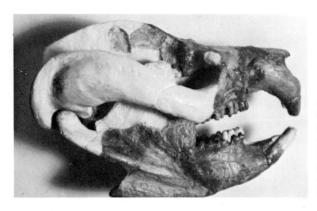

40-13. Early Jurassic. *Bienotherium yunnanense* Young. The skull of a tritylodont (therapsid) from the Lufeng Series (Lower Jurassic) of Yunnan Province, China. Note that the posterior portion of the skull and lower jaw has been restored in plaster of paris. This is a most common procedure in paleontology. The darker material represents the actual fossilized bone parts recovered from this particular animal. Courtesy: Geological Survey of the People's Republic of China. Photo by D. Baird.

The infraorder Gorgonopsia has seventeen definitive families and one questionable family. They are as follows (with generic examples, all Permian in age): **Galesuchidae:** *Galesuchus.* **Hipposauridae (Ictidorhinidae):** *Hipposaurus.* **Cynariopsidae:** *Cynariops.* **Rubidgeidae:** *Rubidgea.* **Gorgonopsidae:** *Gorgonopsis.* **Scymnognathidae:** *Scymnognathus (Scymnosuchus).* **Aelurosauridae:** *Aelurosaurus.* **Galerhinidae:** *Galerhinus.* **Gorgonognathidae:** *Gorgonognathus.* **Arctognathoidae:** *Arctognathoides.* **Scylacopsidae:** *Scylacops.* **Sycosauridae:** *Sycosaurus.* **Arctognathidae:** *Arctognathus.* **Aelurosauropsidae:** *Aeluropsis.* **Scylacocephalidae:** *Scylacocephalus.* **Broomisauridae:** *Broomisaurus (Scymnorhinus).* **Inostranceviidae:** *Inostrancevia* (see Fig. 40-10). **?Burnetiidae (Burnetamorpha):** *Burnetia* and *Styracocephalus.*

The infraorder Cynodontia has five families: **Procynosuchidae:** PERMIAN: *Procynosuchus.* **Thrinaxodontidae (Galesauridae):** PERMIAN: *Galesaurus;* TRIASSIC: *Thrinaxodon (Ictidopsis).* **Cynognathidae:** TRIASSIC: *Cynognathus* (see Fig. 40-11). **Diademodontidae:** TRIASSIC: *Diademodon (Cyclogomphodon).* **Traversodontidae:** TRIASSIC: *Theropsis* and *Traversodon.*

The infraorder Tritylodontoidea has only one family: **Tritylodontidae:** TRIASSIC: *Tritylodon;* JURASSIC: *Bienotherium* (see Fig. 40-13). *Oligokyphus (Mucrotherium).*

The infraorder Therocephalia has five families (all Permian in age): **Pristerognathidae:** *Pristerognathus.* **Alopecodontidae:** *Alopecodon.* **Trochosauridae (Lycosuchidae):** *Lycosuchus* and *Trochosaurus.* **Whaitsiidae:** *Whaitsia.* **Euchambersiidae:** *Euchambersia.*

The infraorder Bauriamorpha has seven definitive families and one questionable family: **Lycideopsidae:** PERMIAN: *Lycideops.* **Ictidosuchidae:** PERMIAN: *Ictidosuchus.* **Nanictidopsidae:** PERMIAN: *Nanictidops.* **Silpholestidae:** PERMIAN: *Silpholestes.* **Scaloposauridae:** PERMIAN: *Scalposaurus.* **Ericiolacertidae:** TRIASSIC: *Ericiolacerta.* **Bauriidae:** TRIASSIC: *Bauria (Bauriodes).* **?Rubidginidae:** PERMIAN: *Rubidgina.*

The infraorder Ictidosauria has two families: **Diarthrognathidae:** TRIASSIC: *Diarthrognathus,* and the questionable family **Haramiyidae (Microcleptidae):** TRIASSIC/JURASSIC: *Haramiya (Microcleptes).*

The suborder Anomodontia has four infraorders: Dinocephalia, Venyukoviamorpha, Dromasauria, and Ducynodontia.

The infraorder Dinocephalia is composed of two superfamilies: Titanosuchoidea and Tapinocephaloidea. The superfamily Titanosuchoidea has in turn four families (all Permian): **Brithopodidae (Titanophoneidae):** *Brithopus* and *Titanophoneus.* **Estemmenosuchidae:** *Estemmenosuchus.* **Anteosauridae:** *Anteosaurus (Dinosuchus).* **Titanosuchidae (Jonkeriidae):** *Jonkeria* and *Titanosuchus.* The superfamily Tapinocephaloidea has two definite families: **Deuterosauridae:** PERMIAN: *Deuterosaurus,* and **Tapinocephalidae:** PERMIAN: *Tapinocephalus,* as well as three questionable families (all Permian in age): **Driveriidae:** *Driveria.* **Mastersoniidae:** *Mastersonia.* **Tappenosauridae:** *Tappenosaurus.*

The infraorder Venyukoviamorpha has only one family: **Venyukoviidae:** PERMIAN: *Venyukovia (Venjukovia).*

The infraorder Dromasauria also has only one family: **Galeopsidae:** PERMIAN: *Galeops.*

Finally, the infraorder Dicynodontia has six families: **Endothiodontidae:** PERMIAN: *Brachyuraniscus (Brachyprosopus), Broilius, Chelyposaurus, Composodon, Emydops, Endothiodon, Koupia, Newtonella,* and *Robertia.* TRIASSIC: *Myosaurus.* **Dicynodontidae:** PERMIAN: *Aulacephalodon* (see Fig. 40-9), *Dicynodon, Dicynodontoides, Eosimops, Geikia, Gordonia, Haughtoniana, Kingoria, Oudenodon (Udenodon), Placerias* (see Fig. 40-12), and *Rhachiocephalus (Eocyclops).* TRIASSIC: *Jimusaria* and *Rhadiodromus.* **Lystrosauridae:** TRIASSIC: *Lystrosaurus (Mochlorhinus).* **Kannemeyeriidae:** TRIASSIC: *Barysoma, Kannemeyeria* (see Fig. 40-11), and *Parakannemeyeria.* **Stahleckeriidae:** TRIASSIC: *Stahleckeria.* **Shansiodontidae:** TRIASSIC: *Shansiodon* and *Tetragonias.*

Note: Some of this classification may be a bit out-of-date. Only the cynodonts have been updated by Hopson and Kitching, and at the present time it is impossible to revise the listing.

41 MAMMALIA (Primitive types)

The primitive or early mammals that evolved out of reptilian stock (see preceding chapter) survived and prospered during the long reign of the dinosaurs. These small shrewlike mammals were the ancestral mammalian stock that gave rise to many varieties of forms (triconodonts, docodonts, monotremes, multituberculates, and therians, including marsupials) during the Late Mesozoic and into the Cenozoic era.

In the class Mammalia there are three subclasses: Prototheria, Allotheria, and Theria.

In the subclass Prototheria there is the order Monotremata (monotremes) with two families: **Ornithorhynchidae** and **Tachyglossidae (Echidnidae)**. Typical genera of the above families: **Ornithorhynchidae**: PLEISTOCENE: *Ornithorhynchus (Platypus)*. **Tachyglossidae (Echidnidae)**: PLEISTOCENE: *Tachyglossus (Echidna)* and *Zaglossus*.

In the subclass Allotheria, there is the order Multituberculata, with its three suborders: Plagiaulacoidea, Ptilodontoidea, and Taeniolaboidea.

The suborder Plagiaulacoidea has only one family, **Plagiaulacidae**. Typical genera: JURASSIC: *Ctenacodon (Allodon)* and *Psalodon*. CRETACEOUS: *Loxaulax*.

The suborder Ptilodontoidea has three families: **Ectypodidae, Cimolodontidae,** and **Ptilodontidae**. Typical genera of the above families: **Ectypodidae**: CRETACEOUS: *Cimexomys*, and *Mesodma (Parectypodus)* (see Figs. 41-1 and 41-2). PALEOCENE: *Ectypodus, Mesodma*, and *Neoplagiaulax*. **Cimolodontidae**: CRETACEOUS: *Cimolodon*; PALEOCENE: *Anconodon*. **Ptilodontidae**: PALEOCENE: *Kimbetohia, Prochetodon*, and *Ptilodus*.

There are three families in the suborder Taeniolaboidea: **Cimolomyidae, Eucosmodontidae,** and **Taeniolabidae**. Typical genera of these families: **Cimolomyidae**: CRETACEOUS: *Cimolomys*; PALEOCENE: *Sphenopsalis*. **Eucosmodontidae**: CRETACEOUS: *Eucosmodon*, PALEOCENE: *Microcosmodon*. **Taeniolabidae**: CRETACEOUS: *Catopsalis* (see Fig. 41-3); PALEOCENE: *Taeniolabis (Polymastodon)*.

In the subclass Theria, the infraclass Trituberculata has two orders: Symmetrodonta and Pantotheria. In turn the order Symmetrodonta has two families: **Spalacotheriidae** and **Amphodontidae**. Typical genera of the above: **Spalacotheriidae**: JURASSIC: *Eurylambda, Spalacotherium*, and *Tinodon;* CRETACEOUS: *Spalacotheroides*. **Amphodontidae**: JURASSIC: *Amphodon;* CRETACEOUS: *Manchurodon*.

There are three families in the order Pantotheria: **Amphitheriidae, Paurodontidae,** and **Dryolestidae**. Typical genera: **Amphitheriidae**: JURASSIC: *Amphitherium*. **Paurodontidae**: JURASSIC: *Araeodon, Archaeotrigon, Brancatherulum, Paurodon,* and *Tathiodon*. **Dryolestidae**: JURASSIC: *Amblotherium (Stylodon), Dryolestes, Kurtodon, Melanodon,* and *Phascolestes*.

In the infraclass Metatheria, there is the order Marsupialia (marsupials) with its four suborders: Polyprotodonta, Peramelida, Caenolestoidia, and Diprotodonta.

In the suborder Polyprotodonta, there are seven families: **Didelphidae, Caroloameghiniidae, Borhyaenidae, Necrolestidae, ?Microtragulidae, Dasyuridae,** and **Notoryctidae**. Typical genera of the above families: **Didelphidae**: CRETACEOUS: *Alphodon, Boreodon, Campodus, Delphonodon, Diaphorodon, Didelphodon, Ectonodon, Eodelphis, Pediomys* and *Thlaeodon;* PALEOCENE: *Derorhynchus, Didelphopsis, Gaylordia, Guggenheimia, Ischyrodelphis, Mirandatherium (Mirandaia), Peradectes, Protodidelphis, Schaefferia,* and *Thylacodon;* EOCENE: *Coöna, Nanodelphys,* and *Peratherium (Herpetotherium);* OLIGOCENE: *Didectodelphis, Didelphidectes,* and *Microbiotherium (Hadrorhynchus);* MIOCENE: *Microbiotherium* and *Peratherium;* PLIOCENE: *Cladodidelphis, Lutreolina, Marmosa (Marmosops), Paradidelphys, Perazoyphium, Sparassocynus, Thylatheridium,* and *Thylophorops;* PLEISTOCENE: *Caluromys, Chironectes; Didelphis, Marmosa, Metachirus, Peramys,* and *Philander*. **Caroloameghiniidae**: EOCENE: *Caroloameghinia*. **Borhyaenidae**: PALEOCENE: *Eobrasilia, Palaeocladosictis,* and *Patene;* EOCENE: *Aminiheringia (Dilestes), Argyrolestes, Nemolestes,* and *Patene;* OLIGOCENE: *Cladosictis (Hathlyacynus), Notogale, Phar-*

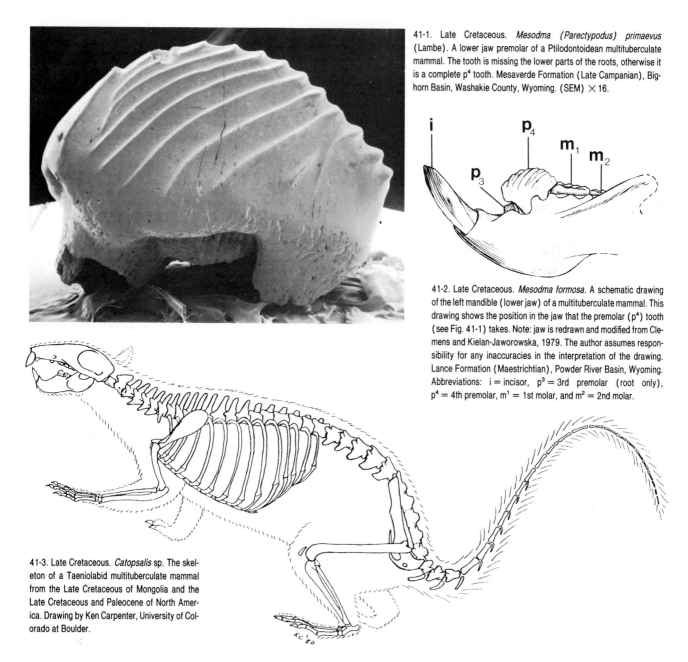

41-1. Late Cretaceous. *Mesodma (Parectypodus) primaevus* (Lambe). A lower jaw premolar of a Ptilodontoidean multituberculate mammal. The tooth is missing the lower parts of the roots, otherwise it is a complete p⁴ tooth. Mesaverde Formation (Late Campanian), Big-horn Basin, Washakie County, Wyoming. (SEM) × 16.

41-2. Late Cretaceous. *Mesodma formosa*. A schematic drawing of the left mandible (lower jaw) of a multituberculate mammal. This drawing shows the position in the jaw that the premolar (p⁴) tooth (see Fig. 41-1) takes. Note: jaw is redrawn and modified from Clemens and Kielan-Jaworowska, 1979. The author assumes responsibility for any inaccuracies in the interpretation of the drawing. Lance Formation (Maestrichtian), Powder River Basin, Wyoming. Abbreviations: i = incisor, p³ = 3rd premolar (root only), p⁴ = 4th premolar, m¹ = 1st molar, and m² = 2nd molar.

41-3. Late Cretaceous. *Catopsalis* sp. The skeleton of a Taeniolabid multituberculate mammal from the Late Cretaceous of Mongolia and the Late Cretaceous and Paleocene of North America. Drawing by Ken Carpenter, University of Colorado at Boulder.

sophorus (Pleisofelis), Proborhyaena, and *Pseudoborhyaena.* MIOCENE: *Acrocyon, Agustylus, Apera, Borhyaena* (see Fig. 41-5), *Cladosictis (Hathlyacynus)* (see Fig. 41-4), *Conodonictis, Ictioborus, Lycopsis, Napodonictis, Perathereutes, Prothylacynus, Sipalocyon,* and *Thylacodictis (Amphiproviverra);* PLIOCENE: *Achlysictis, Acrohyaenodon, Borhyaenidium, Chasicostylus, Hyaenodonops, Notosmilus, Parahyaenodon, Stylocinus,* and *Thylacosmilus.* **Necrolestidae:** MIOCENE: *Necrolestes.* **?Microtragulidae:** PLIOCENE/PLEISTOCENE: *Microtragulus (Argyrolagus).* **Dasyuridae:** PLEISTOCENE: *Dasyurus, Glaucodon, Phascogale (Antechinus), Sarcophilus,* and *Thylacinus.* **Notoryctidae** is a recent family of marsupials.

There is only one family in the suborder Peramelida, **Peramelidae,** with a Pleistocene representative, *Perameles.* There are three families in the suborder Caenolestoidia: **Caenolestidae,** with a Pliocene representative, *Pliolestes* (among others); **Groeberiidae,** with its Oligocene representative, *Groeberia;* and **Polydolopidae,** with a typical representative, *Polydolops,* from the Paleocene/Eocene.

The suborder Diprotodonta has five families: **Phalangeridae,** with a typical genera, *Phalanger (Cuscus)* from the Pliocene; **Thylacoleonidae,** with *Thylacoleo* from the Pleistocene; **Phascolomidae (Phascolomyidae),** with *Phascolomis (Phascolomys)* from the Pleistocene; **Macropodidae** with *Macropus (Halmaturus)* from the Pliocene and *Palorchestes* from the Pleistocene (see Fig. 41-6); and **Diprotodontidae,** with *Diprotodon* as a representative from the Pleistocene.

In an uncertain subclass, there are two orders: Docodontidae and Triconodonta. Each order has two families.

They are, respectively: **Docodontidae** with *Docodon (Dicrocynodon)* from the Jurassic; **?Morganucodontidae** with the Triassic representative, *Morganucodon;* **Amphilestidae** with *Amphlestes* from the Jurassic; and **Triconodontidae** with its Late Jurassic representative, *Triconodon.*

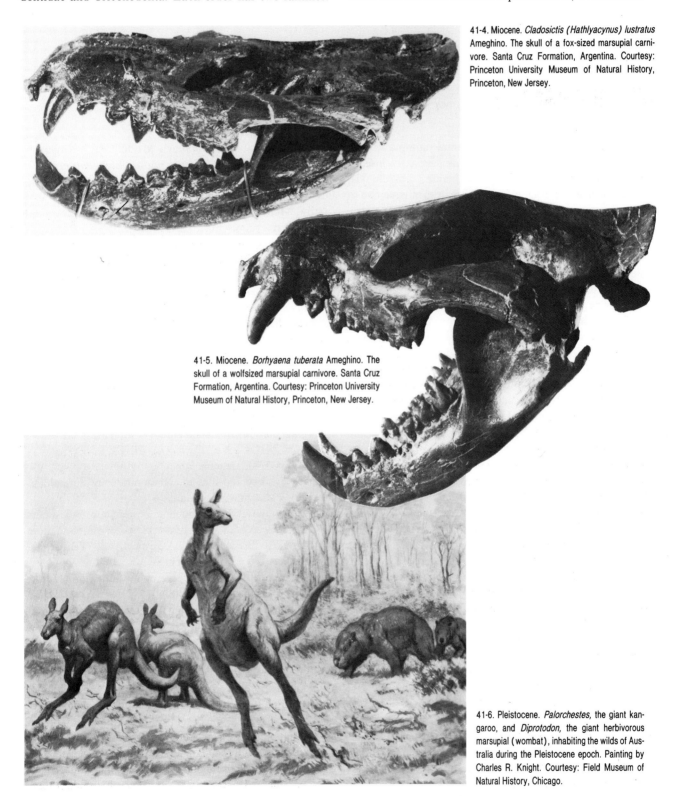

41-4. Miocene. *Cladosictis (Hathlyacynus) lustratus* Ameghino. The skull of a fox-sized marsupial carnivore. Santa Cruz Formation, Argentina. Courtesy: Princeton University Museum of Natural History, Princeton, New Jersey.

41-5. Miocene. *Borhyaena tuberata* Ameghino. The skull of a wolfsized marsupial carnivore. Santa Cruz Formation, Argentina. Courtesy: Princeton University Museum of Natural History, Princeton, New Jersey.

41-6. Pleistocene. *Palorchestes*, the giant kangaroo, and *Diprotodon*, the giant herbivorous marsupial (wombat), inhabiting the wilds of Australia during the Pleistocene epoch. Painting by Charles R. Knight. Courtesy: Field Museum of Natural History, Chicago.

42 MAMMALIA (Insectivores)

The eutherians, or primitive placental mammals, gave rise to an interesting group of animals, the early insectivores, in which we find the proteutherians, apatemyids, and lipotyphlans, among others. These in turn gave rise to our modern day insectivores, e.g., hedgehogs, moles, shrews, elephant shrews, "flying lemurs," zalambdodonts, and bats.

The infraclass Eutheria has four orders: Insectivora, Tillodontia, Taeniodontia, and Chiroptera.

In the order Insectivora, there are in turn five suborders: Proteutheria, Macroscelidea, Dermoptera, Lipotyphla, and Zalambdodonta.

In the suborder Proteutheria, there are two superfamilies: Leptictoidea and Apatemyoidea. There are eight families in the Leptictoidea: **Leptictidae, Zalambdalestidae, Anagalidae, Paroxyclaenidae, Tupaiidae, Pantolestidae, Ptolemaiidae,** and **Pentacodontidae.** Representative genera of the above families: **Leptictidae:** CRETACEOUS: *Procerberus;* PALEOCENE: *Adunator, Diacodon, Diaphyodectes, Myrmecoboides* (see Fig. 42-1), and *Prodiacodon (Paleolestes).* **Zalambdalestidae:** CRETACEOUS: *Zalambdalestes.* **Anagalidae:** OLIGOCENE: *Anagale.* **Paroxyclaenidae:** EOCENE: *Dulcidon, Kopiodon, Paroxyclaenus, Pugiodens (Vulpavoides),* and *Russellites;* OLIGOCENE: *Kochictis.* **Tupaiidae:** PALEOCENE: *?Adapisoriculus.* **Pantolestidae:** PALEOCENE: *Apheliscus, Pagonomus,* and *Propalaeosinopa* (Bessoecetor); EOCENE: *Galethylax, Opsiclaenodon,* and *Palaeosinopa;* OLIGOCENE: *Chadronia* and *Dyspterna.* **Ptolemaiidae:** OLIGOCENE: *Ptolemaia.* **Pentacodontidae:** PALEOCENE: *Aphronorus, Bisonalveus, Coriphagus (Mixoclaenus), Pentacodon,* and *Protentomodon,* EOCENE: *Amaramnis.*

There is one family in the superfamily Apatemyoidea, **Apatemyidae:** PALEOCENE: *Apatemys (Teilhardella), Jepsenella, Labidolemur,* and *Unuchinia (Apator),* EOCENE: *Eochiromys, Heterohyus (Amphichiromys),* and *Stehlinella (Stehlinius),* OLIGOCENE: *Sinclairella* (See Figs. 42-2 and 42-3).

The suborder Macroscelidea has one family: **Macroscelididae** with the following generic representatives: OLIGOCENE: *Metoldobotes;* MIOCENE: *Rhynchocyon,* PLEISTOCENE: *Elephantulus (Elephantomys).*

In the suborder Dermoptera, there are two superfamilies: Mixodectoidea and Plagiomenoidea. There is one family, **Mixodectidae** in the superfamily Mixodectoidea, with generic representatives: PALEOCENE: *Dracontolestes* and *Mixodectes (Indrodon).* Two families in the Plagiomenoidea: **Plagiomenidae:** EOCENE: *Plagiomene,* PALEOCENE: *Planetetherium;* and **Galeopithecidae** (with no fossil representatives).

In the suborder Lipotyphla, there are two superfamilies: Erinaceoidea and Soricoidea. In the superfamily Erinaceoidea there are four families: **Adapsoricidae:** PALEOCENE: *Adapisorex;* EOCENE: *Macrocranion (Aculeodens)* (see Fig. 42-4). **Dimylidae:** OLIGOCENE: *Cordylodon;* MIOCENE: *Dimylus.* **Talpidae:** EOCENE: *Geotrypus;* OLIGOCENE: *Cryporyctes;* MIOCENE: *Mydecodon;* PLIOCENE: *Hydroscapheus;* PLEISTOCENE: *Condylura.* **Erinaceidae:** OLIGOCENE: *Amphechinus;* MIOCENE: *Erinaceus;* PLIOCENE: *Galerix;* PLEISTOCENE: *Hemiechinus.*

There are two families in the superfamily Soricoidea: **Plesiosoricidae** and **Soricidae.**

In the suborder Zalambdodonta there are two superfamilies: Tenrecoidea and Chrysochloroidea. Each of these superfamilies has one family, respectively: **Tenrecidae (Centetidae)** and **Chrysochloridae.**

In the order Tillodontia there is one family, **Esthonychidae.** In the order Taeniodontia there is one family, **Stylinodontidae.**

In the order Chiroptera, there are two suborders: Megachiroptera ("fruit-eating" bats) and Microchiroptera. There is one family in the suborder Megachiroptera, **Pteropodidae.** In the suborder Microchiroptera, there are four superfamilies: Emballonuroidea, Rhinolophoidea, Phyllostomatoidea, and Vespertilionoidea.

The following families appear in the superfamilies above: Emballonuroidea: **Rhinopomatidae, Emballonuridae,** and **Noctilionidae.** Rhinolophoidea: **Nycteridae, Megadermatidae, Rhinolophidae,** and **Hipposideridae.** Phyllostomatoidea: **Phyllostomatidae** and **Desmodontidae.**

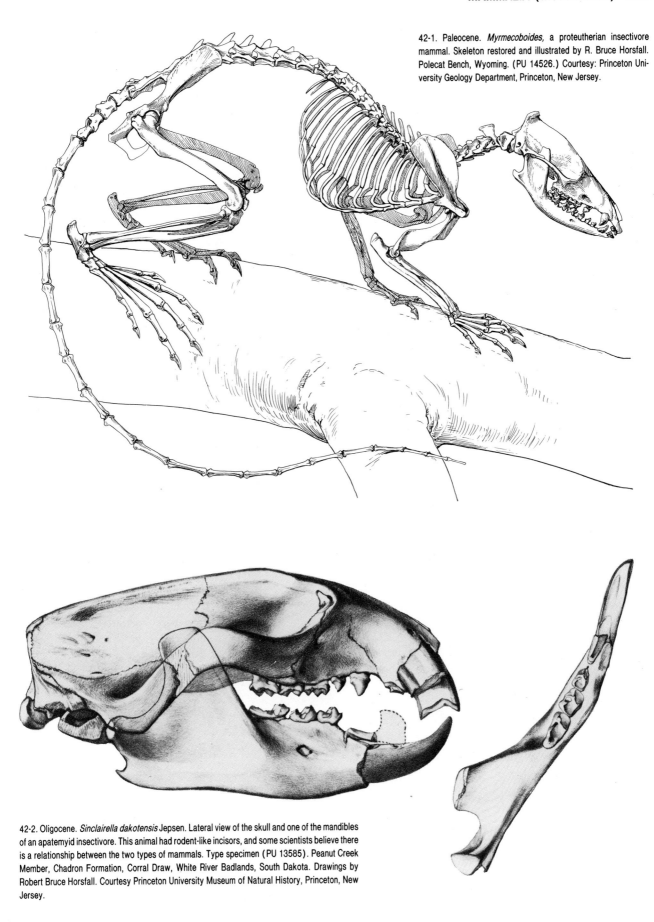

42-1. Paleocene. *Myrmecoboides,* a proteutherian insectivore mammal. Skeleton restored and illustrated by R. Bruce Horsfall. Polecat Bench, Wyoming. (PU 14526.) Courtesy: Princeton University Geology Department, Princeton, New Jersey.

42-2. Oligocene. *Sinclairella dakotensis* Jepsen. Lateral view of the skull and one of the mandibles of an apatemyid insectivore. This animal had rodent-like incisors, and some scientists believe there is a relationship between the two types of mammals. Type specimen (PU 13585). Peanut Creek Member, Chadron Formation, Corral Draw, White River Badlands, South Dakota. Drawings by Robert Bruce Horsfall. Courtesy Princeton University Museum of Natural History, Princeton, New Jersey.

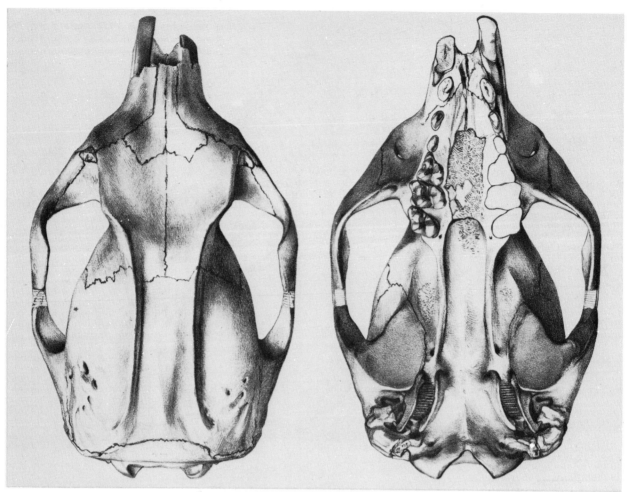

42-3. Oligocene. *Sinclairella dakotensis* Jepsen. Left: dorsal view of the skull, right: palatal view of the same skull. (See Fig. 42-2 for the lateral view of the same skull and the isolated right mandible of the same animal.) Type specimen (PU 13585). Peanut Creek Member, Chadron Formation, Corral Draw, White River Badlands, South Dakota. Drawings by R. Bruce Horsfall. Courtesy: Princeton University Museum of Natural History, Princeton, New Jersey.

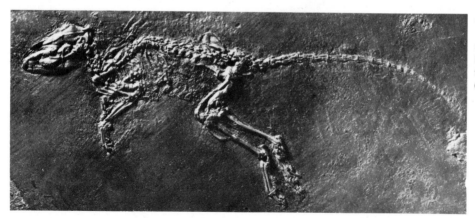

42-4. Eocene. *Macrocranion (Aculeodens) tupiadon.* An epoxy cast of a lipotyphlan insectivore. Oil-Schist Mines, Messel bei Darmstadt, Hessen, West Germany. Courtesy: Geological Enterprises, Ardmore, Oklahoma.

Vespertilionoidea: **Natalidae, Furipteridae, Thyropteridae, Myzopodidae, Vespertilionidae, Mystacinidae,** and **Molossidae.**

In an uncertain superfamily, there are three families: **Archaeonycteridae, Icaronycteridae,** and **Palaeochirop-** **terygidae.** Typical genera of the above families: **Archaeonycteridae**: EOCENE: *Archaeonycteris.* **Icaronycteridae:** EOCENE: *Icaronycteris* (see Figs. 42-6 through 42-8). **Palaeochiropterygidae:** EOCENE: *Palaeochiropteryx* (see Fig. 42-5).

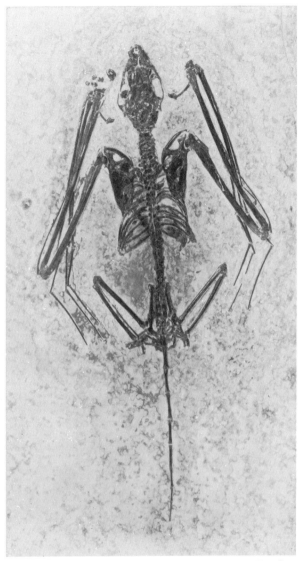

42-5. Eocene. *Palaeochiropteryx tupiadon.* The skeleton of a fossil bat. Romer (1966) has it in a Superfamily "uncertain." Oilschist Mines, Messel bei Darmstadt, Hessen, West Germany. 6.5 cm. Courtesy: Siber and Siber Ltd., Aathal, Switzerland.

42-6. Eocene. *Icaronycteris index* Jepsen. Dorsoventral view of a fossil bat. Type specimen (PU 18150). Green River Formation, Fossil Lake, Sweetwater County, Wyoming. Courtesy: Princeton University Museum of Natural History, Princeton, New Jersey.

42-7. Eocene. *Icaronycteris index* Jepsen. A close-up of the ventral (palatal veiw) side of the skull of the bat figured in Fig. 42-6. Note the fairly complete dentition. Type specimen (PU 18150). Green River Formation, Fossil Lake, Sweetwater County, Wyoming. Courtesy: Princeton University Museum of Natural History, Princeton, New Jersey.

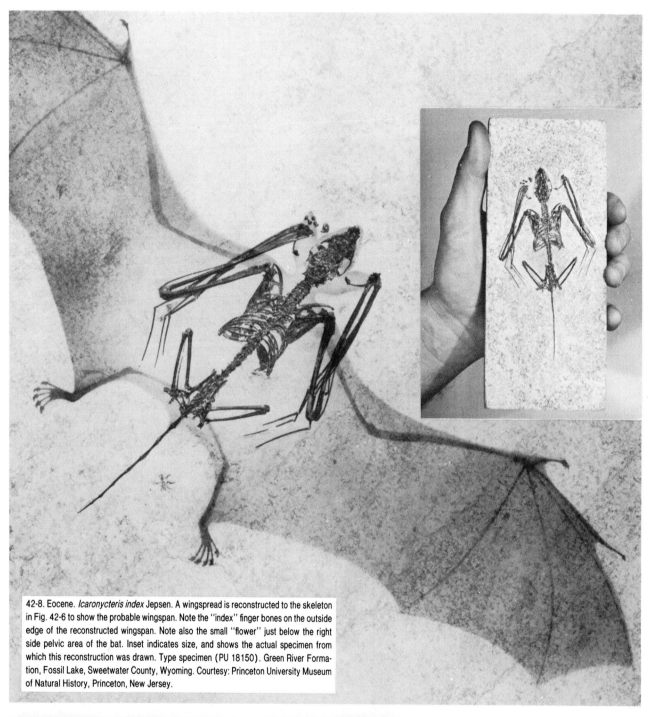

42-8. Eocene. *Icaronycteris index* Jepsen. A wingspread is reconstructed to the skeleton in Fig. 42-6 to show the probable wingspan. Note the "index" finger bones on the outside edge of the reconstructed wingspan. Note also the small "flower" just below the right side pelvic area of the bat. Inset indicates size, and shows the actual specimen from which this reconstruction was drawn. Type specimen (PU 18150). Green River Formation, Fossil Lake, Sweetwater County, Wyoming. Courtesy: Princeton University Museum of Natural History, Princeton, New Jersey.

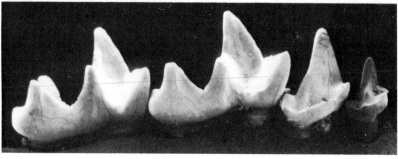

42-9. Pleistocene. Partial dentition (mandible) of a cave bat (species indeterminate). Onyx Cave, Hagerstown, Washington County, Maryland. (SEM) ×24.

43 MAMMALIA (Primates)

The order Primates includes humans as well as lemurs, apes, and monkeys, and is, therefore, of particular interest to us all. Many primates are forest dwellers. Since we find so very few fossilized Tertiary forest deposits, we find relatively few fossil remains of primates.

The order Primates has five suborders: Plesiadapoidea (primitive and aberrant types), Lemuroidea (small squirrellike lemurs), Tarsioidei (small, peculiar "hopping" animals), Platyrrhini (marmosets and monkeys from South America), and Catarrhini (advanced types—apes and humans).

In the suborder Plesiadapoidea, there are three families: **Phenacolemuridae (Paromomyidae), Carpolestidae,** and **Plesiadapidae.** Typical genera of the above families: **Phenacolemuridae (Paromomyidae):** PALEOCENE: *Palaechthon, Palenochtha, Paromomys, Plesiolestes,* and *Phenacolemur,* EOCENE: *Phenacolemur.* **Carpolestidae:** PALEOCENE: *Carpodaptes, Carpolestes (Litotherium), Elphidotarsius,* and *Saxonella,* EOCENE: *Carpolestes (Litotherium).* **Plesiadapidae:** PALEOCENE: *Chiromyoides, Plesiadapis (Nothodectes)* (see Fig. 43-1), and *Pronothodectes;* EOCENE: *Platychoerops.* There are four families in the suborder Lemuroidea: **Adapidae (Notharctidae), Lemuridae, Daubentoniidae (Cheiromyidae),** and **Lorisidae.** Typical genera of the above families: **Adapidae (Notharctidae):** EOCENE: *Adapis (Palaeolemur), Anchomomys, Caenopithicus, Gesneropithex, Lantianius, Notharctus (Hipposyus), Pelycodus, Pronycticebus, Protoadapis (Europolemur),* and *Smilodectes (Aphanolemur).* **Lemuridae:** PLEISTOCENE: *Archaeoindris (Bradylemur), Cheirogaleus, Hadropithecus, Indri (Indris), Lemur, Lepilemur, Lichanotus (Avahi), Megaladapis, Megalindris, Mesopropithecus, Neopropithecus, Palaeopropithecus (Bradytherium)* (see Fig. 43-2), *Peloriadapis, Prohapalemur,* and *Propithecus.* **Daubentoniidae (Cheiromyidae):** PLEISTOCENE: *Daubentonia (Cheiromys).* **Lorisidae:** MIOCENE: *Galago (Galagoides,* and *Progalago;* PLIOCENE: *Indraloris.*

There are four families in the suborder Tarsioidei: **Anaptomorphidae, Omomyidae, Tarsiidae,** and **Microsyop-**sidae. Typical genera in the above: **Anaptomorphidae:** PALEOCENE: *Berruvius;* EOCENE: *Absarokius, Anaptomorphus, Anemorhysis, Tetonius (Paratetonius), Tetonoides, Trogolemur, Uintalacus, Uintanius,* and *Uintasorex.* **Omomyidae:** PALEOCENE: *Navajovicus;* EOCENE: *Cantius, Chlororhysis, Chumashius, Dyseolemur, Hemiacodon, Loveina, Lushius, Macrotarsius, Niptomomys, Omomys (Euryacodon, Ourayia, Periconodon, Shoshonius, Stockia, Teilhardinia, Utahia,* and *Washakius (Yumanius);* OLIGOCENE: *Rooneyia;* MIOCENE: *Ekgmowechashala.* **Tarsiidae:** EOCENE: *Microchoerus, Nannopithex, Neorolemur,* and *Pseudoloris.* **Microsyopsidae:** EOCENE: *Alsaticopithecus, Craseops, Cynodontomys,* and *Microsyops.*

In the suborder Platyrrhini, there are two families: **Callithricidae (Hapalidae)** and **Cebidae.** Typical genera of the above: **Callithricidae (Hapalidae):** OLIGOCENE: *Dolichocebus;* PLEISTOCENE: *Callithrix.* **Cebidae:** MIOCENE: *Cebupithecia, Homunculus, Neosaimiri,* and *Pitheculus;* PLEISTOCENE: *Alouatta (Mycetes), Brachyteles (Eriodes), Callicebus, Cebus,* and *Xenothrix.*

In the suborder Catarrhini, there are three superfamilies: Parapithecoidea, Cereopithecoidea, and Hominoidea.

There is only one family in the superfamily Parapithecoidea: **Parapithecidae,** with the following genera from the Oligocene: *Apidium* and *Parapithecus.*

There is only one family in the superfamily Cercopithecoidea: **Cercopithecidae.** Typical genera: MIOCENE: *Macaca (Rhesus)* and *Prohylobates (Victoriapithecus);* PLIOCENE: *Cercopithecus, Dolichopithecus, Libypithecus, Macaca (Rhesus), Mesopithecus, Papio (Brachygnathopithecus),* and *Pygathrix.* PLEISTOCENE: *Cercocebus, Cercopithecoides, Dolichopithecus, Macaca (Rhesus), Papio (Brachygnathopithecus, Presbytis, (Semnopithecus), Procynocephalus, Rhinopithecus, Szechuanopithecus,* and *Trachypithecus.*

There are three families representing the superfamily Hominoidea: **Oreopithecidae, Pongidae (Simiidae),** and **Hominidae.** Typical genera of the above families: **Oreopithecidae:** MIOCENE: *Oreopithecus;* PLIOCENE: *Or-*

43-1. Paleocene. *Plesiadapis clarkii.* The upper jaw teeth of a primitive Late Paleocene primate from Colorado. Courtesy: Field Museum of Natural History, Chicago.

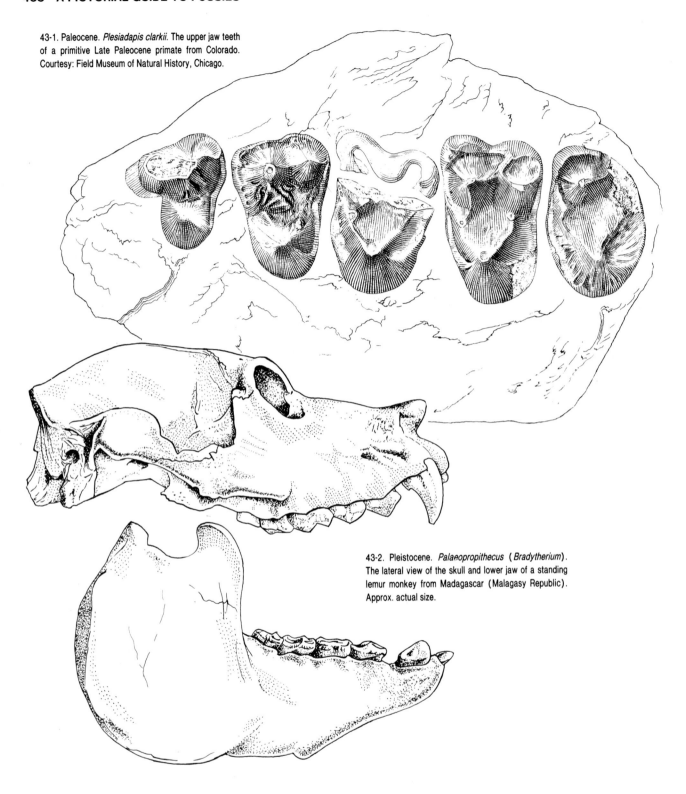

43-2. Pleistocene. *Palaeopropithecus* (*Bradytherium*). The lateral view of the skull and lower jaw of a standing lemur monkey from Madagascar (Malagasy Republic). Approx. actual size.

eopithecus. **Pongidae (Simiidae):** EOCENE: *Amphipithecus* and *Pondaungia;* OLIGOCENE: *Aegyptopithecus, Aeolopithecus, Moeripithecus, Oligopithecus,* and *Propliopithecus;* MIOCENE: *Dryopithecus (Proconsul)* and *Piopithecus (Plesiopithecus);* PLIOCENE: *Dryopithecus* *(Proconsul);* PLEISTOCENE: *Gigantopithecus (Gigantanthropus), Hylobates, Pan (Gorilla),* and *Pongo (Simia).* **Hominidae:** MIOCENE: *Ramapithecus (Bramapithecus);* PLIOCENE: *Ramapithecus (Bramapithecus);* PLEISTOCENE: *Australopithecus (Zijanthropus),* and *Homo (Sinanthropus).*

44 MAMMALIA (Carnivorous types)

The creodonts and the carnivores make up the two basic types of carnivorous mammals.

The creodonts (order Creodonta, primitive flesh-eaters) are placental animals that are allied, but not closely related, to the true carnivores. The creodonts may have evolved from an insectivore of the Late Cretaceous, such as *Deltatheridium*.

The hyaenodonts are typical of the carnivorous creodonts that roamed what are now the Badlands of western Nebraska and South Dakota, during the Late Eocene and Early to Middle Oligocene. These strange looking animals were about the size of an average sheep dog, and were most likely scavengers as well as hunters.

The carnivores (order Carnivora, advanced flesh-eaters) did not evolve from the creodonts, but rather from a distinctive lineage of the insectivores, most probably through the family **Miacidae**. Some genera, e.g., *Didymictis, Miacis*, and *Vulpavus*, surely gave rise to the carnivores. In summary, the creodonts and the carnivores had as ancestors distinct and divergent lineages of the earlier insectivores.

The order Creodonta has two suborders: Deltatheridia and Hyaenodontia.

In the suborder Deltatheridia, there are three families: **Deltatheridiidae (Palaeoryctidae), Didymoconidae (Tschelkariidae),** and **Micropternodontidae.** Typical genera of the above families: **Deltatheridiidae (Palaeoryctidae):** CRETACEOUS: *Cimolestes, Deltatheridium, Deltatheriodes, Hyotheridium,* and *Sarcodon;* PALEOCENE: *Aboletylestes, Acmeodon, Avunculus, Cimolestes (Nyssodon), Gelastops (Emperodon), Palaeoryctes,* and *Pararyctes;* EOCENE: *Didelphodus (Phanacops).* **Didymoconidae (Tschelkariidae):** EOCENE: *Mongoloryctes;* OLIGOCENE: *Ardynictis* and *Didymoconus (Tschelkaria).* **Micropternodontidae:** EOCENE/OLIGOCENE: *Micropternodus (Kentrogomphus).*

There are two families in the suborder Hyaenodontia: **Hyaenodontidae** and **Oxyaenidae.** Typical genera of above: **Hyaenodontidae:** EOCENE: *Apataelurus, Arfia, Cynohyaenodon, Galethylax, Hyaenodon (Taxotherium),*

Imperatoria (Prodissopsalis), Limnocyon (Telmatocyon), Machaeroides, Metasinopa, Oxyaenodon, Paracynohyaenodon, Propterodon, Protoproviverra, Prototomus (Prolimnocyon), Quercytherium, Sinopa, Thereutherium, and *Tritemnodon;* OLIGOCENE: *Hemipsalodon, Hyaenodon* (see Figs. 44-1 through 44-3), *Ischnognathus, Metasinopa,* and *Tylodon;* PLIOCENE: *Dissopsalis.* **Oxyaenidae:** PALEOCENE: *Dipsalodon* and *Oxyaena (Dipsalidictis);* EOCENE: *Ambloctonus (Amblyctonus), Argillotherium, Dipsalidictides, Oxyaena, Palaeonictis, Paroxyaena, Patriofelis (Orecyon), Protopsalis,* and *Sarkastodon.*

In the order Carnivora, there are two suborders: Fissipedia and Pinnipedia. The latter suborder is not discussed here; but, rather, in Chapter 51 (marine mammals).

There are three infraorders in the suborder Fissipedia: Miacoidea, Aeluroidea, and Arctoidea. In the infraorder Miacoidea, there is only one family, **Miacidae.** Typical genera of this family: PALEOCENE: *Didymictis, Ictidopappus,* and *Viverravus;* EOCENE *Didymictis, Miacis (Mimocyon) Oödectes, Palaearctonyx, Petersonella (Pleurocyon), Plesiomiacis, Tapocyon, Uintacyon, Vassocyon, Viverravus,* and *Vulpavus.*

There are three families in the infraorder Aeluroidea: **Viverridae, Hyaenidae,** and **Felidae.** Typical genera of the above families: **Viverridae:** EOCENE: *Palaeoprionodon,* OLIGOCENE: *Amphicticeps, Haplogale, Herpestes (Calogale), Palaeoprionodon,* and *Stenoplesictis;* MIOCENE: *Herpestes (Calogale), Jourdanictis, Leptoplesictis, Semigenetta,* and *Viverra (Anictis);* PLIOCENE: *Vishnuictis* and *Viverra (Anictis);* PLEISTOCENE: *Atilax, Crossarchus, Cryptoprocta, Cynictis, Fossa, Genetta, Leecyaena, Macrogalidia, Mungos, Paguma, Paradoxurus, Suricata, Vishnuictis,* and *Viverra (Anictis).* **Hyaenidae:** MIOCENE: *Crocuta (Crocotta), Hyaenictis, Ictitherium (Galeotherium), Lycyaena, Progenetta (Miohyaena),* and *Tungurictis;* PLIOCENE: *Crocuta (Crocotta), Hyaena,* (see Fig. 44-15), *Hyaenictis, Ictitherium (Galeotherium), Lycyaena,* and *Progenetta (Miohyaena);* PLEISTOCENE: *Chasmaporthetes (Ailurena), Crocuta (Crocotta), Hyaena, Hyaenictis,* and *Lycyaenops (Euryboas).* **Felidae:**

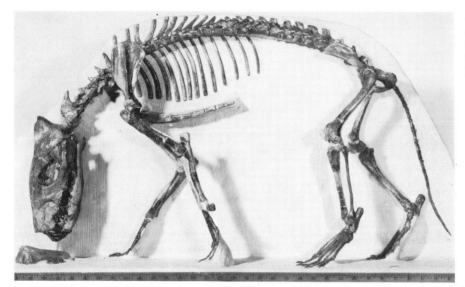

44-1. Oligocene. *Hyaenodon cruentus* Leidy. A skeleton of a creodont carnivore from the White River Badlands of South Dakota. Courtesy: Princeton University Museum of Natural History, Princeton, New Jersey.

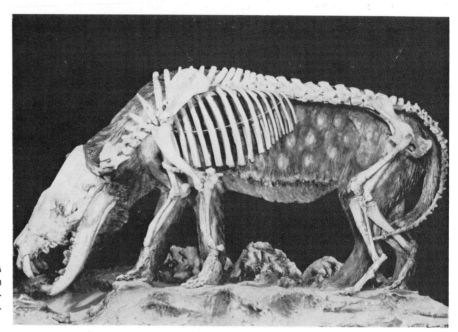

44-2. Oligocene. *Hyaenodon cruentus* Leidy. A skeletal mount with a partial skin restoration (on the right side) of a creodont carnivore. Chadronian, Sioux County, Nebraska. Courtesy: Nebraska State Museum, Lincoln, Nebraska.

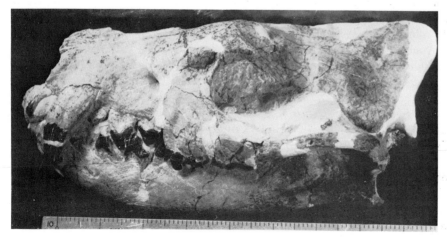

44-3. Oligocene. *Hyaenodon cruentus* Leidy. The skull of a creodont carnivore. Brulé Formation, White River Badlands, Shannon County, South Dakota. Courtesy: Black Hills Institute for Paleontological Research, Hill City, South Dakota.

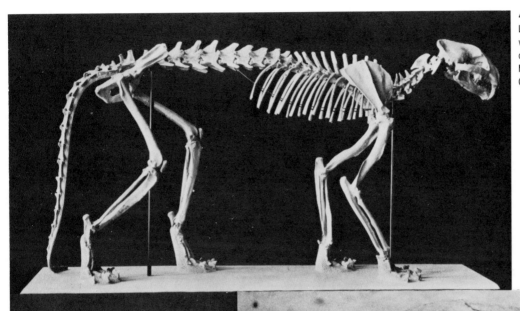

44-4. Oligocene. *Dinictis felina* Leidy. A skeleton of a felid carnivore from the White River Badlands of South Dakota. Courtesy: Field Museum of Natural History, Chicago.

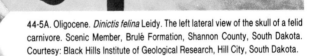

44-5A. Oligocene. *Dinictis felina* Leidy. The left lateral view of the skull of a felid carnivore. Scenic Member, Brulé Formation, Shannon County, South Dakota. Courtesy: Black Hills Institute of Geological Research, Hill City, South Dakota.

44-5B. Oligocene. *Dinictis felina* Leidy. The right lateral view of another skull of a felid carnivore. Note the sharp incisor teeth (or "tusks"). Scenic Member, Brulé Formation, Shannon County, South Dakota. Courtesy: Black Hills Institute of Geological Research, Hill City, South Dakota.

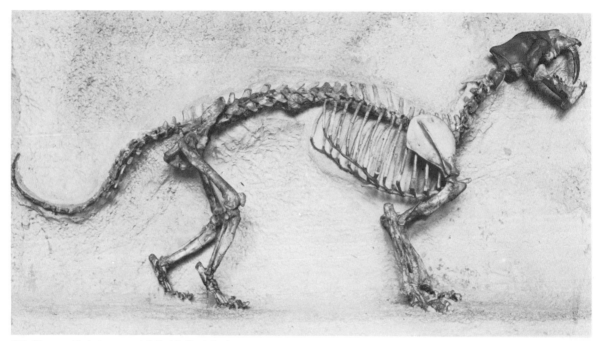

44-6. Oligocene. *Hoplophoneus mentalis* Sinclair. The skeletal mount of a felid carnivore which competed with *Dinictis* (see Fig. 44-4). Chadron Formation, White River Badlands, South Dakota. Courtesy: Princeton University Museum of Natural History, Princeton, New Jersey.

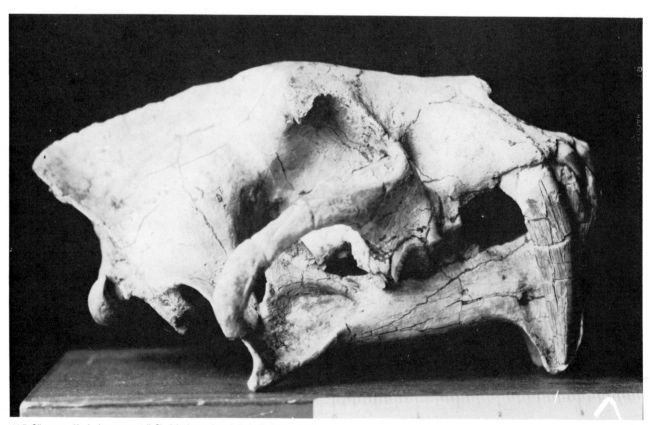

44-7. Oligocene. *Hoplophoneus mentalis* Sinclair. A complete skull of a "saber-toothed" felid carnivore (see Fig. 44-6). Chadron Formation, White River Badlands, Scenic, Pennington County, South Dakota. Courtesy: Princeton University Museum of Natural History, Princeton, New Jersey.

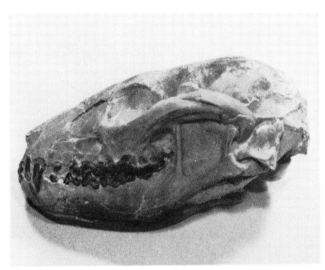

44-8. Oligocene. *Hesperocyon* sp. The skull of a canid carnivore. Brulé Formation (Scenic Member), Shannon County, South Dakota. Courtesy: Black Hills Institute of Geological Research, Hill City, South Dakota.

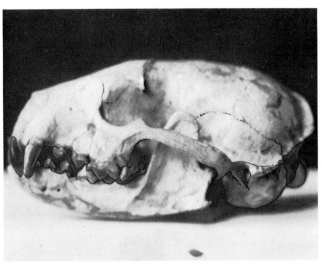

44-9. Miocene. *Promartes olcotti* Riggs. The skull of a mustelid carnivore (arctoid). Harrison Beds, Raw Hide Creek. Goshen County, Wyoming. Specimen recovered on a Field Museum expedition in 1906. Courtesy: Field Museum of Natural History, Chicago.

44-10. Miocene. *Temnocyon* sp. A doglike carnivore skeleton. Gering Formation, Redington Gap, Redington, Morrill County, Nebraska. Courtesy: University of Nebraska State Museum, Lincoln.

EOCENE: *Aelurogale, Eofelis,* and *Eusmilus;* OLIGO-CENE: *Dinailurictis, Dinictis* (see Figs. 44-4 and 44-5), *Hoplophoneus* (see Figs. 44-6 and 44-7), *Nimravus,* and *Proailurus;* MIOCENE: *Archaelurus, Dinaelurus, Ekmoiteptecela, Hyaenaelurus, Machairodus, Pogonodon, Pseudaelurus,* and *Sansanosmilus;* PLIOCENE: *Felis,* *Homotherium, Ischyrosmilus, Machairodus, Meganterion, Mellivorodon, Nimravides, Paramachaerodus, Pseudaelurus, Smilodon, Vinayakia,* and *Vishnufelis;* PLEISTOCENE: *Acinonyx, Dinobastis, Felis (Panthera)* (see Fig. 44-14), and *Smilodon* (see Figs. 44-21 through 44-23).

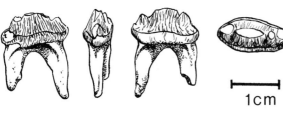

1cm

44-11. Miocene. An unidentified canid molar tooth in four views. From left to right: Outer or labial face; lateral (side) view; inner or lingual face; and top or occlusal face. Pungo River Marl Formation (Helvetian), Aurora, Beaufort County, North Carolina.

44-12. Pleistocene. *Canis* sp. A molar tooth of a dire wolf. Lake Monroe, Enterprise, Volusia County, Florida. × 1.5.

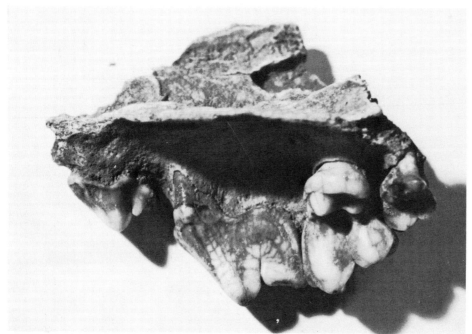

44-13. Pleistocene. *Canis lupus* Linnaeus. Fragment of the upper jaw of a wolf. Note the robust molars (teeth). Eemien Stage, Loveral, Belgium. × 1.5.

44-14. Pleistocene. *Felis led*. Fragment of the lower jaw of a tiger. Eemien Stage, Loveral, Belgium. × 1.5.

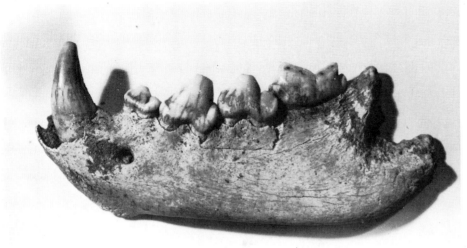

44-15. Pleistocene. *Hyaena spelaea*. An almost complete left mandible (lower jaw) of a large hyaena carnivore. Eeimien Stage. Loveral, Belgium. ×1.

44-16A. Pleistocene. *Ursus spelaeus* Rosenmüller and Heinroth. A mounted skeleton of a cave bear. On display at the Fossilien Museum, Holland. Courtesy: Fossilien Museum, Walchern, The Netherlands.

44-16B. Pleistocene. *Ursus spelaeus* Rosenmüller and Heinroth. Close-up of the skull of the European cave bear in Fig. 44-16A. Courtesy: Fossilien Museum, Walchern, Holland.

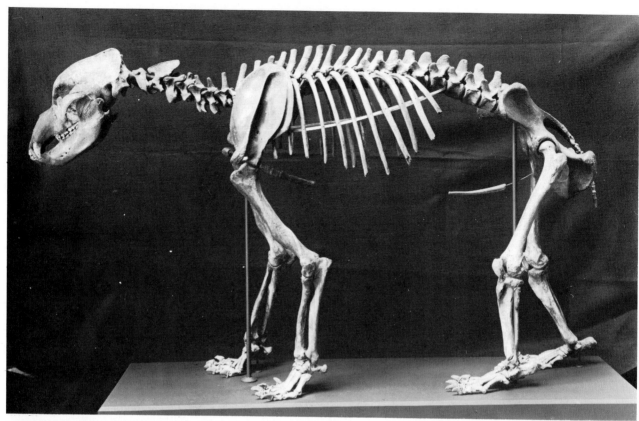

44-17. Pleistocene. *Ursus spelaeus* Rosenmüller and Heinroth. Mounted skeleton of a giant European cave bear from Trieste, Italy. Courtesy: Field Museum of Natural History, Chicago.

44-18. Pleistocene. *Ursus spelaeus* Rosenmüller and Heinroth. Two views of an isolated molar tooth of the old world (European) giant cave bear. From left to right: Occlusal (top) view and outer face (one root broken off). Höhle im Gesäuse (Mixnix), Austria. ×3.

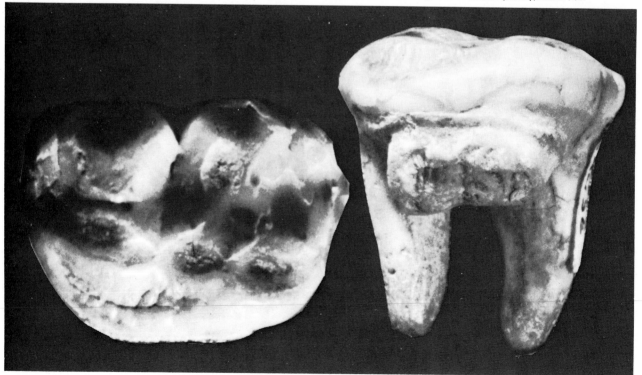

44-19. Pleistocene. Diggings at Rancho La Brea (La Brea tar pits), on Wilshire Boulevard in the heart of Los Angeles. This is the initial dig to uncover the many animals trapped in the pools of tar seeping up from beneath the ground under the very streets of the city. See Fig. 44-20 for the area before this hole was dug, and Fig. 44-21 for a reconstruction of the area during prehistoric times. Courtesy: Los Angeles County Museum, Los Angeles, California.

44-20. The famous La Brea tar pits before digging began to discover the hundreds of thousands of bones of trapped animals in the ancient tar pools lying beneath this spot. See Fig. 44-19 for the spot where the man is standing after digging. Note the oil fields in the background. This site is now known as Hancock Park and is situated on Wilshire Boulevard in the heart of Los Angeles. Courtesy: Field Museum of Natural History, Chicago.

There are four families, **Mustelidae, Canidae, Procyonidae,** and **Ursidae,** in the infraorder Arctoidea. **Mustelidae:** EOCENE: *Amphictis;* OLIGOCENE: *Amphictis, Drassonax, Palaeogale, Plesiogale,* and *Potamotherium;* MIOCENE: *Aelurocyon, Brachypsalis, Broiliana, Craterogale, Dinogale, Hypsoparia, Laphyctis, Leptarctus, Limnonyx, Martes, Megalictis, Melidellavus, Melodon, Miomephitis, Miomustela, Mionictis, Mustela, Oligobunis, Palaeomeles, Paralutra, Paraligobunis, Plesiogula, Plesiomeles, Plionictis, Promartes* (see Fig. 44-9), *Proputorius, Pseudictis, Sthenictis, Stromeriella, Taxodon, Trocharion, Trochictis,* and *Trochotherium;* PLIOCENE: *Arctomeles, Baranogale, Brachyopsigale, Buisnictis, Cernictis, Conepatus, Enhydrictis, Enhydriodon, Eomellivora, Hadrictis, Lutra, Lutravus, Martinogale, Meles, Mellivora, Parataxidea, Pliotaxidea, Promeles, Sa-* *badellictis, Semantor, Sinictis, Sivalictis, Sivanonyx, Taxidea, Vishnuonyx,* and *Vormela;* PLEISTOCENE: *Aonyx, Arctonyx, Brachyprotoma, Canimartes, Charronia, Cyrnaonyx, Enhydra, Galera, Grison, Grisonella, Gulo, Lyncodon, Mephitis, Nesolutra, Osmotherium, Parameles, Promellivora, Rhabdogale, Spilogale,* and *Stipanicicia.* **Canidae:** EOCENE: *Amphicyonodon, Cynodictis, Procynodictis, Pseudamphicyon,* and *Simamphicyon;* OLIGOCENE: *Amphicyon, Brachyrhynchocyon, Campylocynodon, Daphoenocyon, Daphoenus, Haplocyon, Hesperocyon* (see Fig. 44-8), *Mesocyon, Nothocyon, Oxetocyon, Pachycynodon, Parictis, Proamphicyon, Protemnocyon,* and *Pseudocyon;* MIOCENE: *Absonodaphoneus, Actiocyon, Aelurodon, Agnotherium, Aletocyon, Alopecocyon, Amphicyonopsis, Borocyon, Cynarctoides, Cynarctus, Cynodesmus, Daphoenodon, Enhydrocyon, Euoplo-*

44-21. Pleistocene. The death trap of Rancho La Brea. In this famous painting by Charles R. Knight, we see a reconstruction of a tar pool (asphalt) during the Pleistocene Epoch (some 200,000 years ago). The animals who were unfortunate enough to become mired in these pools were soon joined by their predators, who also became entrapped. The painting shows *Smilodon* (the saber-toothed cat at the right) and a dire wolf (left). Vultures *(Teratornis)* are shown to the left and center of the mural. This area is situated in the very heart of Los Angeles. Courtesy: Field Museum of Natural History, Chicago.

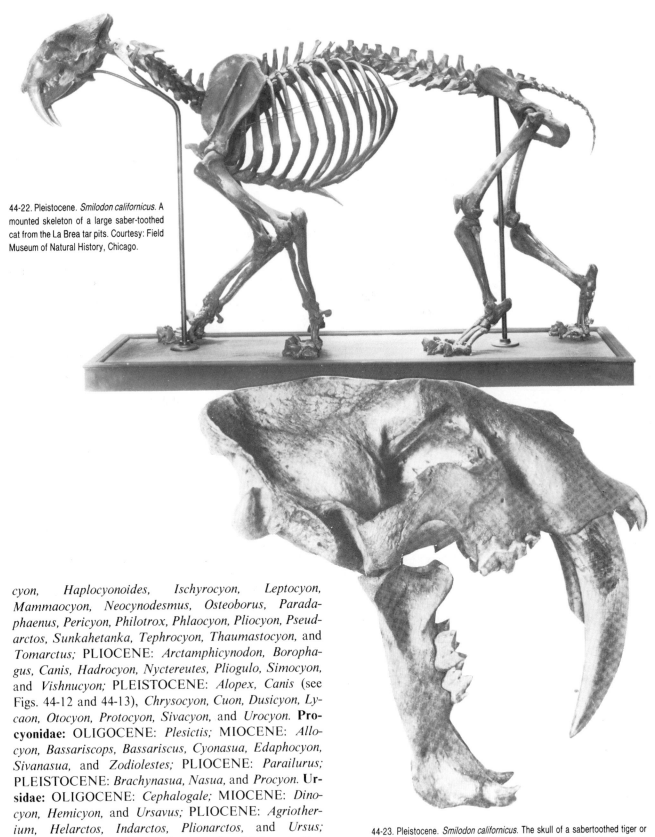

44-22. Pleistocene. *Smilodon californicus*. A mounted skeleton of a large saber-toothed cat from the La Brea tar pits. Courtesy: Field Museum of Natural History, Chicago.

cyon, Haplocyonoides, Ischyrocyon, Leptocyon, Mammaocyon, Neocynodesmus, Osteoborus, Paradaphaenus, Pericyon, Philotrox, Phlaocyon, Pliocyon, Pseudarctos, Sunkahetanka, Tephrocyon, Thaumastocyon, and *Tomarctus;* PLIOCENE: *Arctamphicynodon, Borophagus, Canis, Hadrocyon, Nyctereutes, Pliogulo, Simocyon,* and *Vishnucyon;* PLEISTOCENE: *Alopex, Canis* (see Figs. 44-12 and 44-13), *Chrysocyon, Cuon, Dusicyon, Lycaon, Otocyon, Protocyon, Sivacyon,* and *Urocyon.* **Procyonidae:** OLIGOCENE: *Plesictis;* MIOCENE: *Allocyon, Bassariscops, Bassariscus, Cyonasua, Edaphocyon, Sivanasua,* and *Zodiolestes;* PLIOCENE: *Parailurus;* PLEISTOCENE: *Brachynasua, Nasua,* and *Procyon.* **Ursidae:** OLIGOCENE: *Cephalogale;* MIOCENE: *Dinocyon, Hemicyon,* and *Ursavus;* PLIOCENE: *Agriotherium, Helarctos, Indarctos, Plionarctos,* and *Ursus;* PLEISTOCENE: *Ailuropoda, Arctodus, Melursus, Tremarctos,* and *Ursus* (see Figs. 44-16 through 44-18).

44-23. Pleistocene. *Smilodon californicus.* The skull of a sabertoothed tiger or cat from the La Brea tar pits. Courtesy: Field Museum of Natural History, Chicago.

45 MAMMALIA (Archaic Ungulates)

This chapter deals with the archaic ungulates (hoofed mammals) rather than the typical hoofed cows, horses, etc. (see Chapter 48). The primitive ungulates include the condylarths and pantodonts.

These animals are extinct, but at least one specialized order, the condylarths, may have given rise to some later ungulate forms. The origins for the condylarths are traced back to the insectivores (as with the primates and carnivores of the preceding chapters). The condylarths are known only from rocks of early Tertiary age (particularly the Paleocene). A typical example of a condylarth is *Phenacodus* (see Fig. 45-2). There is a distinct possibility that the South American ungulates, the litopterns (see Chapter 47), may be related (in descent) to the condylarths.

The pantodonts, another primitive ungulate group, grew to massive proportions, and were a suborder, along with the dinoceratans, xenungulates, and pyrotherians, in the order Amblypoda. It is possible that all had a common origin.

A typical Late Paleocene pantodont was *Barylambda* (see Fig. 45-1), while the most common pantodont of the Early Eocene, was the ubiquitous *Coryphodon (Loxolophodon)* (see Fig. 45-3), found especially in the Willwood Formation (Wasatchian) of Wyoming. An outstanding feature of the skull of *Coryphodon* is the flared occipital area. The incisor teeth of *Coryphodon* are quite distinctive (see Fig. 45-3B and 45-3c). The cheek teeth (premolars and molars) of *Coryphodon* are lophodontid in design (see Fig. 45-3A, 45-3D, and 45-3E).

The uintatheres were another amblypod group which attained enormous rhino-like proportions and had some similarities to their cousins the pantodonts. In the case of the uintatheres, besides the enlarged occipital regions, enormous canines, etc., there is a peculiar dinocerate characteristic of an enlargement and flaring-out of the parietal region of the skull to form hornlike crests, divergent on the cranium.

The xenungulates were a distinctive South American form with some parallel features of the uintatheres as well

as the pantodonts. They were confined to South America during the Late Paleocene.

The pyrotheres are also restricted to the South American continent during the Oligocene. Many features of these mammals demonstrate parallelism with the proboscideans.

The order Condylartha has seven families: **Arctocyonidae (Oxyclaenidae), Mesonychidae, Hyopsodontidae, Meniscotheriidae, Periptychidae, Phenacodontidae,** and **Didolodontidae.** Typical genera of the above families: **Arctocyonidae (Oxyclaeindae):** CRETACEOUS: *Protungulatum;* PALEOCENE: *Arctocyon, Arctocyonides, Baioconodon, Carcinodon, Chriacus, Claenodon, Colpoclaenus, Deltatherium, Deuterogonodon, Elphidophorus, Eoconodon, Goniacodon, Loxolophus, Mentoclaenodon, Metachriacus, Mimotricentes, Oxyclaenus, Paradoxodonta, Prothryptacodon, Protogonodon, Protungulatum, Spanoxyodon, Thryptacodon, Tricentes,* and *Triisodon;* EOCENE: *Paratriisodon* and *Thryptacodon.* **Mesonychidae:** PALEOCENE: *Dissacus* and *Microclaenodon;* EOCENE: *Andrewsarchus, Disarchus, Gandakasia, Hapalodectes, Harpogolestes, Hessolestes, Ichthyolestes, Mesonyx, Pachyaena,* and *Synoplotherium.* OLIGOCENE: *Apterodon* and *Mesonyx.* **Hyopsodontidae:** PALEOCENE: *Apheliscus, Asmithwoodwardia, Dracoclaenus, Jepsenia, Louisina, Oxyacodon, Tiznatzinia,* and *Tricuspiodon;* EOCENE; *Hyopsodus* and *Kopidodon;* OLIGOCENE: *Epapheliscus.* **Meniscotheriidae:** *Meniscotherium, Orthaspidotherium,* and *Pleuraspidotherium.* **Periptychidae:** PALEOCENE: *Anisonchus, Carsioptychus, Conacodon, Ectoconus, Haploconus, Hemithlaeus,* and *Periptychus.* **Phenacodontidae:** PALEOCENE: *Desmatoclaenus, Ectocion, Phenacodus* (see Fig. 45-2), and *Tetraclaenodon;* EOCENE: *Almogaver.* **Didolodontidae:** PALEOCENE: *Ernestokokenia* and *Lamegoia;* EOCENE: *Didolodus* and *Enneoconus;* OLIGOCENE: *Lophiodolodus;* MIOCENE: *Megadolodus.*

In the order Amblypoda, there are four suborders: Pantodonta, Dinocerta, Xenungulata, and Pyrotheria.

In the suborder Pantodonta, there are five families:

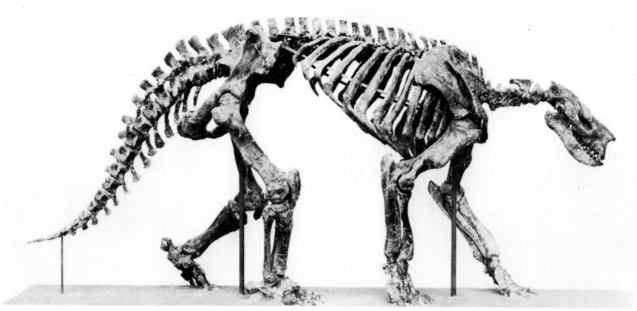

45-1. Paleocene. *Barylambda faberi* Patterson. Skeleton of the Late Paleocene pantodont mammal. Rifle Formation, Mesa County, Colorado. Courtesy: Field Museum of Natural History, Chicago.

45-2. Eocene. *Phenacodus* sp. A skull of a condylarth ungulate from the Wasatchian of Wyoming. Willwood Formation, Basin, Big Horn County, Wyoming. Courtesy: South Dakota School of Mines and Technology, Rapid City.

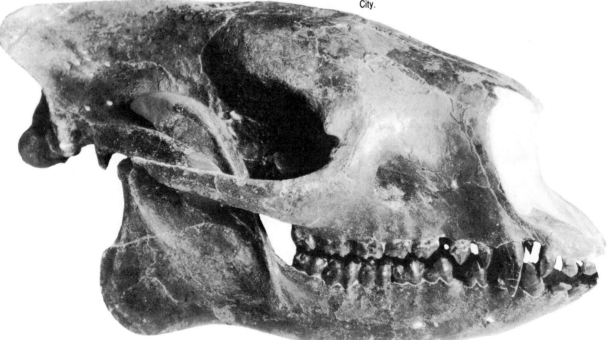

Pantolambdidae: PALEOCENE: *Pantolambda:* **Barylambdidae:** PALEOCENE: *Barylambda* (see Fig. 45-1): **Titanoideidae:** PALEOCENE: *Titanoides:* **Coryphodontidae:** PALEOCENE/EOCENE: *Coryphodon (Loxolophodon)* (see Fig. 45-3). **Pantolambdodontidae:** EOCENE: *Pantolambdodon.*

In the suborder Dinocerata, there are three families: **Prodinocerotidae:** PALEOCENE: *Prodinoceras:* **Uintatheriidae:** EOCENE: *Uintatherium (Dinoceras):* **Gobiatheriidae:** *EOCENE:* **Gobiatherium.**

In each of the suborders Xenungulata and Pyrotheria, there is only one family. Xenungulata, **Carodniidae,** with a typical genus *Carodnia* from the Paleocene. Pyrotheria: **Pyrotheriidae,** with a typical genus *Pyrotherium* from the Oligocene.

45-3. Eocene. *Coryphodon (Loxolophodon)*. Isolated teeth from an archaic amblypod ungulate especially common in the Early Eocene of Wyoming. A. Three views of a lower jaw molar tooth. B. Upper jaw incisor tooth (3 views). C. Lower jaw incisor tooth (3 views). D. Three views of an upper jaw cheek tooth. E. Lower jaw molar tooth (3 views). Upper Willwood Formation (Wasatchian), 15 Mile Creek, West of Worland, Washakie County, Wyoming. ×1.

46 MAMMALIA (Subungulates)

A variety of animals, seemingly unrelated in appearance, are classed together as subungulates. They include a) the old world hoofed conies or hyraces; b) *Arsinoitherium* (see Fig. 46-1), a large horned animal, probably a marsh dweller; c) the proboscideans, including elephants and their primitive (extinct) relatives such as the mastodonts and mammoths of the Pleistocene; d) the sirenians (sea-cows); and e) the desmostylians, a peculiar extinct amphibious type of mammal. These five groups of subungulates represent both land-dwelling and sea-going varieties.

The first group, the hyracoids or conies (rodent-like nail-bearing hoofed animals) are represented today on the continent of Africa (some species are also found in the Levant, particularly Syria) with the predominant generic type being *Hyrax* (Procavia). The fossil record for conies goes back as far as the Early Oligocene, where skeletal remains are not uncommon to fossil deposits in Egypt. These show primitive forms of hyracoids differing greatly from the few existing forms of today. No examples of this group are in this chapter.

The second group, the order Embrithopoda, includes the rhinoceros-like, horned, marsh-dwelling(?) animals found in fossil beds of Oligocene age, also in Egypt. Their distinguishing feature (besides the large horns is their 5-digit, splayed-out "paws," possibly webbed between the digits, which allowed the heavy animal to walk on mud flats such as in swampy or marshy areas. Nothing is known of the ancestors or any descendants of this group. Its only fossil genus is *Arsinoitherium,* and it seems to be restricted to the Oligocene of Egypt.

The third and largest group is the Proboscideans: the elephants, mastodons, and other long-nosed, tusk-bearing related forms. The origin for all proboscideans was on the continent of Africa.

The order Proboscidea is made up of three suborders: Moeritherioidea, Euelephantoidea, and Deinotheroidea. In the suborder Moeritherioidea, there is only one family, **Moeritheriidae,** with one generic type, *Moeritherium.* Its geological range was Late Eocene to Early Oligocene on the African continent.

The suborder Euelephantoidea has three families: **Gomphotheriidae (Trilophodontidae), Mastodontidae (Mammutidae),** and **Elephantidae.** The following generic types comprise the family **Gomphotheriidae (Trilophodontidae):** OLIGOCENE: *Palaeomastodon* and *Phiomia;* MIOCENE: *Gomphotherium (Megabelodon), Platybelodon (Torynobelodon), Rhynchotherium (Aybelodon), Serridentinus (Hemimastodon), Synconolophus,* and *Tetralophodon (Lydekkeria);* PLIOCENE: *Amebelodon, Anancus (Dibunodon), Eubelodon, Gnathabelodon, Gomphotherium (Megabelodon)* (see Fig. 46-2), *Serridentinus (Hemimastodon), Stegomastodon (Aleamastodon), Synconolophus,* and *Tetralophodon (Lydekkeria);* PLEISTOCENE. *Anancus (Dibunodon), Cuvieronius (Cordillerion), Notiomastodon,* and *Stegomastodon (Aleamastodon).* Family **Mastodontidae (Mammutidae):** MIOCENE: *Mammut (Mastodon),* PLIOCENE: *Mammut (Mastodon);* PLEISTOCENE: *Mammut (Mastodon)* (see Figs. 46-3 through 46-7, and 46-17). Family **Elephantidae:** MIOCENE: *Stegolophodon;* PLIOCENE: *Stegodon (Parastegodon),* and *Stegolophodon;* PLEISTOCENE: *Elephas (Hypselephas)* (see Fig. 46-9), *Loxodonta (Hesperoloxodon), Mammuthus (Archidiskodon)* (see Figs. 46-8 through 46-13, also 46-15 and 46-18), *Stegodon (Parastegodon),* and *Stegolophodon.*

The fourth group of subungulates is the water-dwelling sirenians (sea-cows), the "mermaids" of mythology and folklore. The order Sirenia has two families: **Dugongidae (Halicoridae)** (the dugongs), and **Manatidae (Trichechidae)** (the manatees). The dugongs go back as far as the Oligocene with *Anotherium.* Their fossil rib bones are most common in the Florida phosphate deposits (see Fig. 46-19). Fossil manatee material has also been recovered in the phosphate pits of Florida, in Pleistocene deposits, while earlier (Miocene-Pliocene) material is found in South American fossil deposits.

The fifth and last group of subungulates is the desmostylians, peculiar amphibious mammals, hippopotamus-like in body structure, with small cylindrical-cusped molar teeth, reminiscent of worn down "mastodon" teeth (see Fig. 46-21). These animals no doubt favored life along

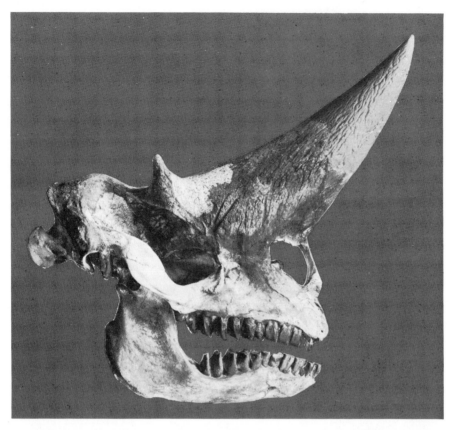

46-1. Oligocene. *Arsinoitherium zitteli* Beadmell. A large horned subungulate mammal from Egypt. A possible marsh dweller in the order Embrithopoda. After Andrews. Courtesy: Field Museum of Natural History, Chicago.

46-2. Pliocene. *Gomphotherium (Megabelodon?)* lulli. A long-tusked mastodontid elephant. A complete skull and tusks. Cherry County, Nebraska. Courtesy: University of Nebraska State Museum, Lincoln.

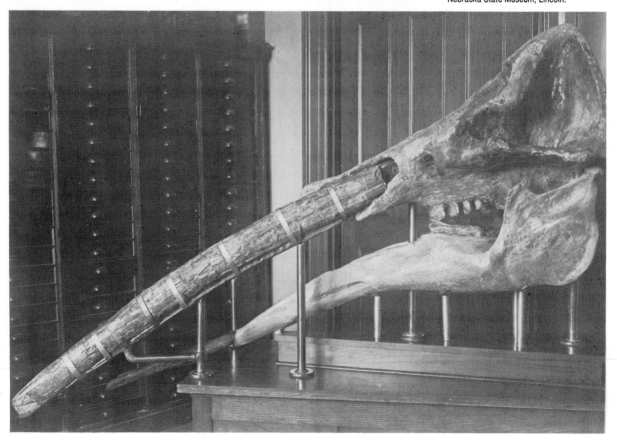

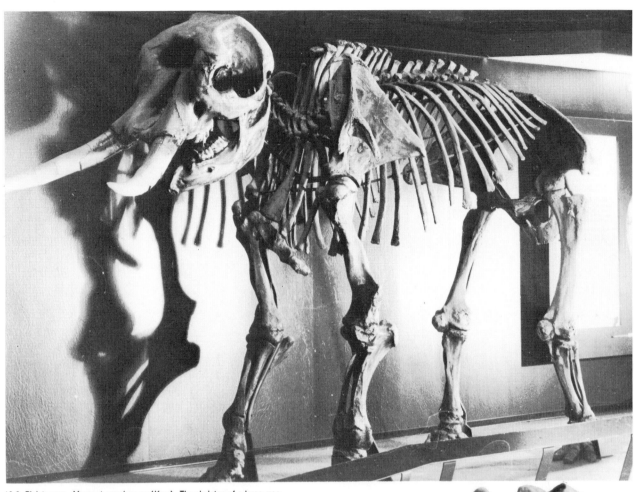

46-3. Pleistocene. *Mammut americanum* (Kerr). The skeleton of a large mastodon collected in New Jersey. Courtesy: New Jersey State Museum, Trenton.

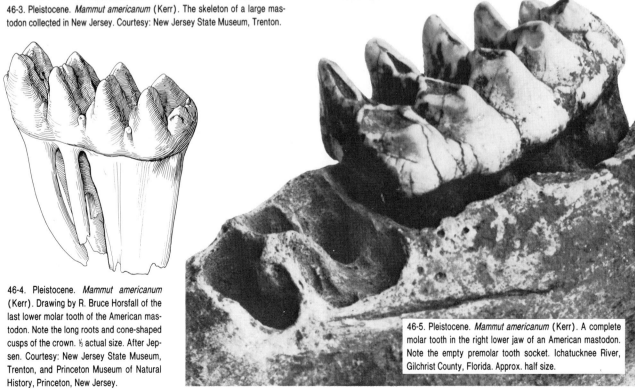

46-4. Pleistocene. *Mammut americanum* (Kerr). Drawing by R. Bruce Horsfall of the last lower molar tooth of the American mastodon. Note the long roots and cone-shaped cusps of the crown. ⅓ actual size. After Jepsen. Courtesy: New Jersey State Museum, Trenton, and Princeton Museum of Natural History, Princeton, New Jersey.

46-5. Pleistocene. *Mammut americanum* (Kerr). A complete molar tooth in the right lower jaw of an American mastodon. Note the empty premolar tooth socket. Ichatucknee River, Gilchrist County, Florida. Approx. half size.

46-6. Pleistocene. *Mammut* sp. Fragment of the upper jaws of a mastodon with two complete molar teeth in place. Ichatucknee River, Gilchrist County, Florida. ⅓ actual size.

coastlines, and perhaps swamps or marshes near the oceanside. Some species, such as *Desmostylus,* had small vestigial tusklike incisor teeth (uppers and lowers). The upper incisors were not as pronounced in *Paleoparadoxia* (see Fig. 46-20). *Desmostylus* (see Fig. 46-21) fossil remains are found along the western coast of North America in fossil deposits of Miocene age, and are particularly

46-7. Pleistocene. Mammut sp. An isolated vertebral centrum of a mastodon elephant. Blue Springs, Orange City, Volusia County, Florida. Approx. half size.

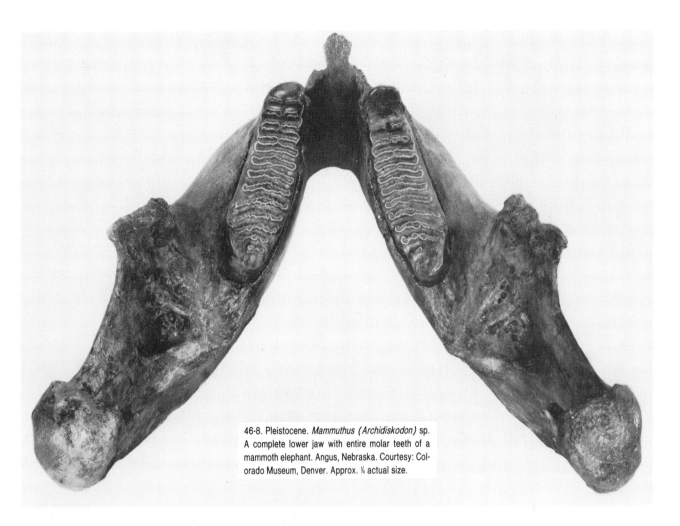

46-8. Pleistocene. *Mammuthus (Archidiskodon)* sp. A complete lower jaw with entire molar teeth of a mammoth elephant. Angus, Nebraska. Courtesy: Colorado Museum, Denver. Approx. ¼ actual size.

46-9. Pleistocene. *Mammuthus (Archidiskodon) maibeni.* Large skeleton of an Imperial mammoth elephant. The little elephant in front of *Mammuthus* is *Elephas falconeri,* the dwarf or pygmy elephant from Sicily, Italy. Courtesy: University of Nebraska State Museum, Lincoln.

46-10. Pleistocene. *Mammuthus meridionalis* (Nesti). Partial jaw section with one complete tooth of a mammoth elephant. Dredged from off the continental shelf of Europe (location: 10 miles off the coast of Schouw, Zeeland, North Sea, The Netherlands). Dredged by fishnets. Courtesy: Fossilien Museum, Walchern, Holland.

46-11. Pleistocene. *Mammuthus* sp. Two baby mammoth molar teeth. Collected at the bottom of a spring in the Ichatucknee River on the border of Columbia and Gilchrist Counties, Northcentral Florida. ½ actual size.

46-12. Pleistocene. *Mammuthus* sp. Isolated mammoth elephant tooth. Ichatucknee River, Gilchrist County, Florida. ⅓ actual size.

46-13. Pleistocene. *Mammuthus primigenius* Blümenbach. Complete mandibles (lower jaws), with the exception of a small part of the jaw articulation facette (left side), of a European mammoth elephant. All teeth are present and accounted for. Termonde, Belgium. ¼ actual size.

46-15. Pleistocene. *Mammuthus* sp. An isolated molar tooth of a mammoth elephant. Ichatucknee River, Gilchrist County, Florida. Approx. half size.

46-14. Pleistocene. Cross-section of a fragment of an elephant tusk showing "cross-hatching" effect, which indicates true "ivory." Bone Valley Phosphate Pits, Bartow, Polk County, Florida. ×1.5.

46-16. Pleistocene. An "elephant" dig along the Peace River above Wachulla Springs, Florida. The Boy Scout troop of Wachulla Springs, with their scoutmaster, Mitchell Hope, excavated the mammoth. Courtesy: Mitchell Hope, Wachulla, Florida.

46-17. Pleistocene. *Mammut,* the mastodon elephant, roaming the fields and streams during the Pleistocene epoch. Painting by Charles R. Knight. Courtesy: Field Museum of Natural History, Chicago.

46-18. Pleistocene. Mammoths. A herd of ''wooly mammoths'' (elephants) making their way through snow covered terrain in search of food. Painting by Charles R. Knight. Courtesy: Field Museum of Natural History, Chicago.

abundant in the Temblor Formation (Middle Miocene-Helvetian) at Shark Tooth Hill, near Bakersfield, Kern County, California. *Desmostylus* and *Paleoparadoxia* are also found in the eastern parts of Asia, particularly in Miocene fossil deposits in Japan (see Fig. 46-20).

The most abundant fossil remains of subungulates seem to be those of early relatives of elephants—the mastodonts

and mammoths. Although these names do become confused, the primary difference, at least in the dentition (tooth structure), between the two types is as follows: In the mastodons, the teeth are enameled and multicusped (the molars), see Figs. 46-2 through 46-6, for example. Figs. 46-4 and 46-5 are very fine examples showing the multicusped molar teeth. The mammoths, on the other

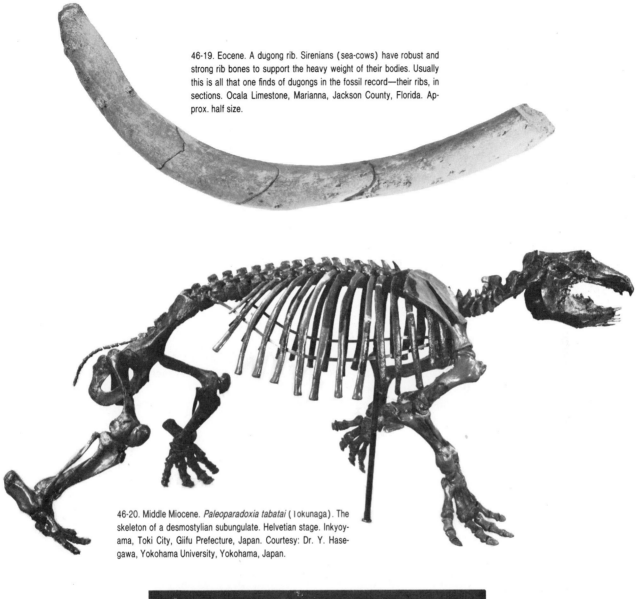

46-19. Eocene. A dugong rib. Sirenians (sea-cows) have robust and strong rib bones to support the heavy weight of their bodies. Usually this is all that one finds of dugongs in the fossil record—their ribs, in sections. Ocala Limestone, Marianna, Jackson County, Florida. Approx. half size.

46-20. Middle Miocene. *Paleoparadoxia tabatai* (Tokunaga). The skeleton of a desmostylian subungulate. Helvetian stage. Inkyoyama, Toki City, Giifu Prefecture, Japan. Courtesy: Dr. Y. Hasegawa, Yokohama University, Yokohama, Japan.

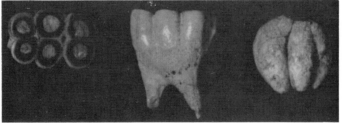

46-21. Miocene. *Desmostylus* sp. Three cheek teeth of a desmostylian subungulate. Temblor Formation (Helvetian), Shark Tooth Hill, Oildale, Kern County, California. Approx. half size.

hand, have long, "breadlike" molar teeth (see Figs. 46-8 through 46-13, and 46-15). These molar teeth differ greatly from the preceding types, particularly in that they have little if any enamel, except on the laminar ridges of the occlusal (biting) surface. The teeth are composed mostly of laminated osteodentinal material. The comparisons of the two types of dentition are dramatic and well separate the two families **Mastodontidae** and **Elephantidae.**

47 MAMMALIA (South American Ungulates)

South America was separated from other continents during part of the Cenozoic. Two orders of ungulates, the perissodactyls (see following chapter) and the artiodactyls (see Chapter 49), are not well represented in the fossil record of South America until the end of the Tertiary, by which time they had arrived, via a land bridge, from North America. Distinct species of ungulates were native to the region, notably three orders: Notoungulata, Astrapotheria, and Litopterna. Representatives in South America of the Notoungulata were (all Paleocene to Eocene). Suborder Notioprogonia: family **Arctostylopidae**—no representatives in South America. Family **Henricosborniidae:** *Henricosbornia (Hemistylops), Othnielmarshia (Postpithecus);* and *Peripantostylops.* Family **Notostylopidae:** *Edvardotrouessartia, Homalostylops (Acrostylops), Notostylops (Anastylops), Otronia,* and *?Seudenius.* In Notoungulata

Incertae Sedis: *Acamana.*

In the order Astrapotheria: Family **Trigonostylopidae:** *?Albertogaudrya (Scabellia), ?Shecenia,* and *Trigonostylops (Chiodon).* Family **Astrapotheriidae:** *Astraponotus (Notamynus), Astrapothericulus, Astrapotherium* (see Fig. 47-1), *Parastrapotherium, Proastrapotherium, Scaglia,* and *Kenestrapotherium.*

Finally, in the order Litopterna, there are three families: **Proterotheriidae, Macraucheniidae,** and **Adianthidae.** Typical genera of the family **Proterotheriidae** are: *Brachytherium, Polyacrodon (Decaconus), Proterotherium,* and *Xesmodon (Glyphodon).* Family **Macraucheniidae:** *Craumauchenia, Macrauchenia, Paranauchenia,* and *Victorlemoinea.* Family **Adianthidae:** *Adianthus (Adiatus), Adiantoides, Prodiantus,* and *Proheptoconus.*

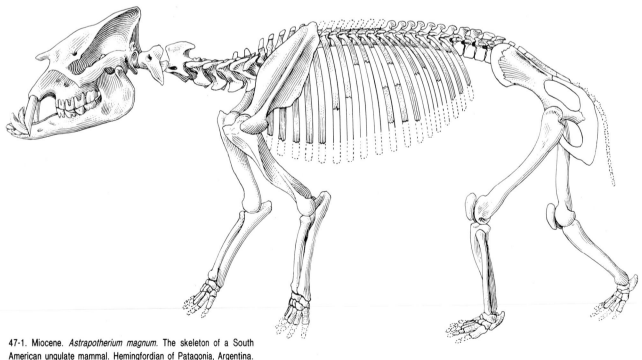

47-1. Miocene. *Astrapotherium magnum.* The skeleton of a South American ungulate mammal. Hemingfordian of Patagonia, Argentina. Courtesy: Field Museum of Natural History, Chicago.

48 MAMMALIA (Perissodactyls)

The perissodactyls, the "odd-toe" mammals, are a prolific and persistent order of ungulates, a diversified group of mammals that includes horses, palaeotheres, titanotheres, chalicotheres, tapirs, and rhinoceri. They are animals with grinding molars and premolars, predominantly grass eaters, similar in some respects to their ruminant cousins in the "even-toe" group—the artiodactyls (see following chapter)—except that perissodactyls do not chew their cud as do the ruminants.

The order Perissodactyla is comprised primarily of three suborders: Hippomorpha, Ancylopoda, and Ceratomorpha. The suborder Hippomorpha in turn has two superfamilies: Equoidea and Brontotherioidea. The two families in the Equoidea are **Equidae** and **Palaeotheriidae.** The latter is not as well known and is only represented in the fossil record by *Palaeotherium* and *Plagiolophus* of the Late Eocene to Early Oligocene of Europe. **Equidae** is represented in the fossil record by: EOCENE: *Anchilophus, Epihippus (Duchesnehippus), Haplohippus, Hyracotherium (Eohippus)* (see Figs. 48-2 and 48-8), *Lophiotherium, Orohippus, Pachynolophus,* and *Propalaeotherium.* OLIGOCENE: *Mesohippus* (see Fig. 48-1), *Miohippus,* and *Palaeotherium.* MIOCENE: *Anchitherium, Archaeohippus, Hipparion* (see Figs. 48-3 and 48-4), *Hypohippus, Merychippus, Miohippus,* and *Parahippus.* PLIOCENE: *Hipparion, Hypohippus, Merychippus* (See Figs. 48-3 and 48-4), *Nannippus* (see Figs. 48-3, through 48-5), *Neohipparion* (see Fig. 48-3), and *Pliohippus.* PLEISTOCENE: *Equus* (see Figs. 48-6 and 48-7), *Hipparion, Hippidion, Nannippus, Neohipparion, Onohippidium,* and *Parahipparion.*

The superfamily Brontotherioidea has one family **Brontotheriidae (Titanotheriidae).** Examples of this group are: *Brontops (Diploclonus)* (see Fig. 48-8), *Brontotherium, Manteoceras, Teleodus,* and *Titanodectes.* The family ranges from Early geological age.

The suborder Ancylopoda has two families: **Eomoropidae** and **Chalicotheriidae.** Neither family is included in this chapter. *Moropus,* **(Chalicotheriidae),** however, is represented by isolated bones in Fig. 49-3.

In the final suborder, Ceratomorpha, there are two superfamilies: Tapiroidea and Rhinocerotoidea (the tapirs and rhinoceri).

The superfamily Tapiroidea has, in turn, six families: **Isectolophidae, Helaletidae (Hyrachiidae), Lophialetidae, Depretellidae, Lophodontidae,** and **Tapiriidae.** We are only concerned in this chapter with **Tapiridae:** OLIGOCENE: *Protapirus.* MIOCENE: *Miotapirus, Palaeotapirus,* and *Tapiravus.* PLIOCENE: *Tapirus.* PLEISTOCENE: *Tapirus* (see Figs. 48-9 and 48-10).

The superfamily Rhinocerotoidea has three families: **Hyracodonidae, Amynodontidae,** and **Rhinocerotidae.** Faunal listing for **Hyracodonidae:** EOCENE: *Ardynia (Ergilia), Caenolophus, Teilhardia,* and *Trilopus,.* OLIGOCENE: *Ardynia (Ergilia), Hyracodon* (see Figs. 48-11 through 48-13), and *Parahyracodon.*

Examples of the family **Amynodontidae:** EOCENE: *Amynodon, Megalamynodon,* and *Sianodon.* OLIGOCENE: *Cadurcodon* and *Hysamynodon.*

The final family, **Rhinocerotidae:** EOCENE: *Eotrigonias, Epitriplophus, Fostercooperia (Cooperia), Ilianodon, Juxia, Pappaceras,* and *Prohyracodon.* OLIGOCENE: *Aceratherium, Amphicaenopus, Benaritherium, Brachypotherium, Caenopus* (see Fig. 48-14), *Diceratherium, Di-*

48-1. Oligocene. *Mesohippus bairdi* Leidy. A complete skull of the 3-toed horse of the White River Badlands. Brulé Formation, Pennington County, South Dakota.

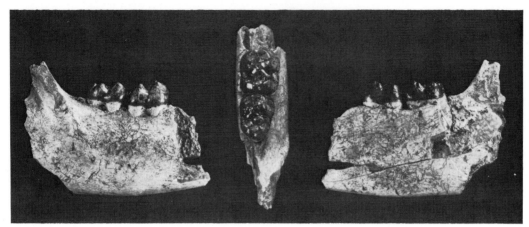

48-2. Eocene. *Hyracotherium (Eohippus)*. Three views of a lower jaw (mandible) fragment showing two complete cheek teeth of a "dawn-horse." Upper Willwood Formation (Wasatchian), 15 Mile Creek, West of Worland, Washakie County, Wyoming. ×1.

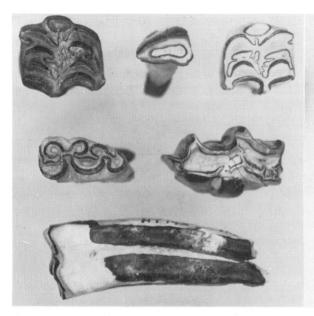

48-3. Miocene to Pliocene. Teeth of the horse species: *Merychippus, Nannippus, Hipparion,* and *Neohipparion*. All recovered from the Bone Valley Sediments (Phosphate pits), Polk County, Florida. Slightly larger than actual size.

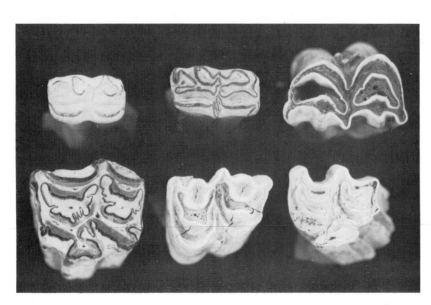

48-5. Pliocene. Cannon leg bone of *Nannippus,* the three-toed horse. Bone Valley Sediments (Phosphate pits), Polk County, Florida. Half size.

48-4. Miocene to Pliocene. Occlusal view (top) of the horse species: *Hipparion, Nannippus,* and *Merychippus*. Bone Valley Sediments (Phosphate pits), Polk County, Florida. ×1.

48-6. Pleistocene. *Equus* sp. The skeleton of our modern horse. Sheridan County, Nebraska. Courtesy: University of Nebraska State Museum, Lincoln.

48-7. Pleistocene. *Equus complicatus* Leidy. A series of cheek teeth of a modern-type horse. Santa Fe Springs, Florida. Slightly less than actual size.

cerorhinus, *Eggysodon, Epiaceratherium, (Allocerops), Indricotherium, Meninatherium, Paraceratherium (Aralotherium), Pleuroceros, Preaceratherium, Protaceratherium, Ronzotherium, Subhyracodon, Symphysorhachis, Tongriceros, Trigonias,* and *Urtinotherium.* MIOCENE: *Chilotherium, Diceratherium (Metacaenopus)* (see Fig. 48-15), *Dromaceratherium, Floridaceras, Gobitherium, Hispanotherium, Peraceras, Plesiaceratherium,* and *Teleoceras* (see Fig. 48-16). PLIOCENE: *Diceros, Gaindatherium, Gobitherium, Iranotherium, Peraceras, Sinotherium,* and *Teleoceras.* PLEISTOCENE: *Coelodonta* (see Figs. 48-17 and 48-18), *Elasmotherium, Ortho-*

48-8. Eocene. *Brontops,* a titanothere; and *Hyracotherium (Eohippus)* the dawn horse (dog-sized), at a river's edge. Painting by Charles R. Knight. Courtesy: Field Museum of Natural History, Chicago.

48-9. Pleistocene. *Tapirus americanus* Leidy. The occlusal (top) view of an isolated tapir cheek tooth. Peace River, Polk County, Florida. ×1.

48-10. Pleistocene. *Tapirus americanus* Leidy. Left upper jaw section of a tapir. Santa Fe Springs, Florida. Slightly less than actual size.

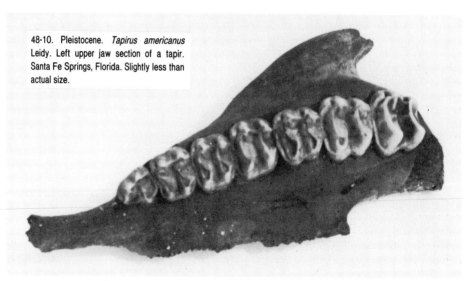

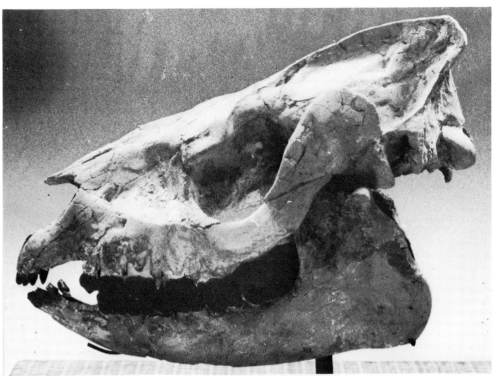

48-11. Oligocene. *Hyracodon nebrascensis* Leidy. A complete skull of the "running rhinoceros." Scenic Member, Brulé Formation, Shannon County, South Dakota. Courtesy: Black Hill Institute of Geological Research, Hill City, South Dakota. ⅓.

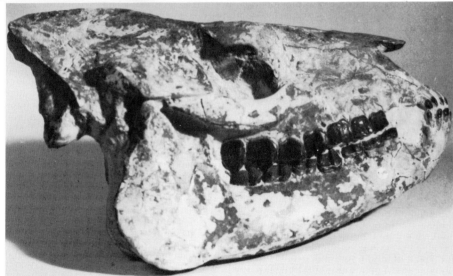

48-12. Oligocene. *Hyracodon nebrascensis* Leidy. Another complete skull of the running rhinoceros of the White River Badlands. Wall, Pennington County, South Dakota. Courtesy: Geological Enterprises, Ardmore, Oklahoma. ½ actual size.

48-13. Oligocene. *Hyracodon nebrascensis* Leidy. A fragment of the tooth enamel of a running rhinoceros molar. Brulé Formation, Scenic, Pennington County, South Dakota. ×1.

gonoceras, and *Rhinoceros.*

Three ages are represented in the fossil horse teeth in Florida's phosphate pits. These are: Late Miocene, Pliocene, and Pleistocene. It is possible to find fossils from all three ages mixed together in the tailings of mining operations. The teeth (mostly molars and premolars, sometimes incisors) of the following horses are found in these tailings: *Hipparion, Merychippus, Nannippus, Neohipparion,* and *Equus* (both species: *E. complicatus* and *E. leidyi*).

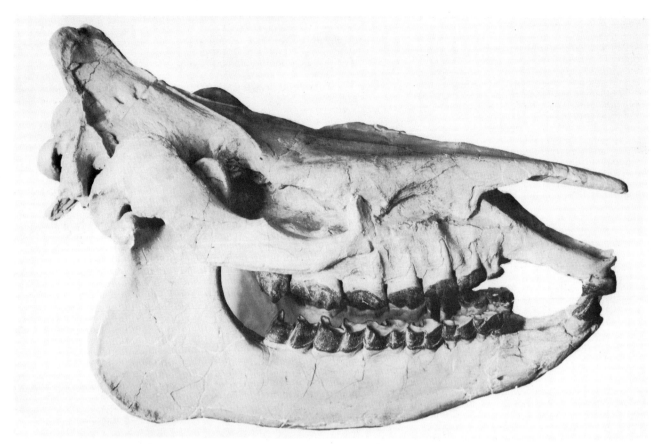

48-14. Oligocene. *Caenopus tridactylus* Osborn. The complete skull of a "hornless rhinoceros" from the White River Badlands. Poleslide Member, Brulé Formation, Scenic, Pennington County, South Dakota. Courtesy: South Dakota School of Mines and Technology, Rapid City.

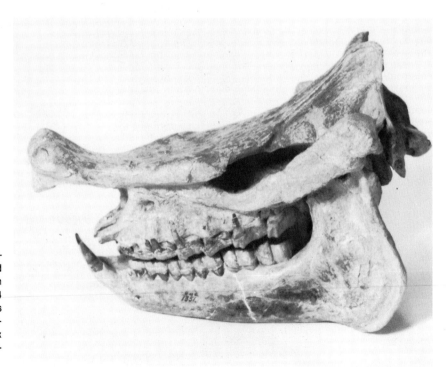

48-15. Miocene. *Diceratherium (Metacaenopus)* sp. A complete skull of the small-horned running rhinoceros (male), descended from *Caenopus* stock in the Oligocene. See Fig. 48-14 for comparison. Arikaree Group, Agate Springs Bone Bed (see Fig. 49-3 for a typical bone assemblage from this locality), Agate, Sioux County, Nebraska. Courtesy: Geological Enterprises, Ardmore, Oklahoma.

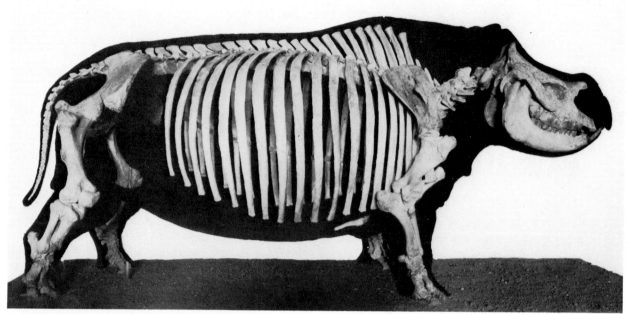

48-16. Miocene. *Teleoceras hicksi*. Skeletal mount of a short-legged rhinoceros with a recreated (silhouette) body outline. Type specimen. Arikareean, near Beecher Island (Arikaree River), Yuma County, Colorado. Courtesy: Colorado Museum, Denver.

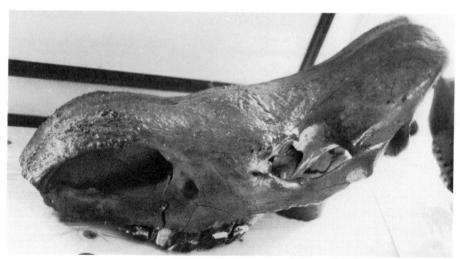

48-17. Pleistocene. *Coelodonta antiquitatis* Blumenbach. The skull of the European rhinoceros (see Fig. 48-18 for mandibles). Discovered in an excavation pit for a foundation at Nieuwpoort, Rotterdam, The Netherlands. Approx. ¼ actual size.

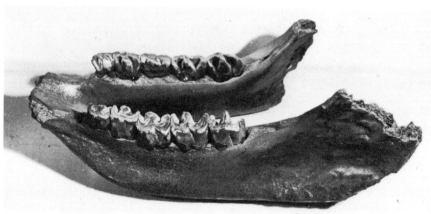

48-18. Pleistocene. *Coelodonta antiquitatis* Blumenbach. The mandibles or lower jaws of a European rhinoceros (see Fig. 48-17 for a skull of the same species). Termonde, Belgium. Approx. ½ actual size.

49 MAMMALIA (Artiodactyls)

The artiodactyls, the "even-toed" or cloven hoofed, are an even larger group of ungulates than the perissodactyls of the preceding chapter. They include pigs, peccaries, hippopotami, cows, sheep, goats, camels, deer, antelopes, and extinct related forms. The order Artiodactyla has three suborders: Palaeodonta, Suina (swine pigs, peccaries, and hippopotami—all animals with canine "tusks" and bunodont molar teeth), and Ruminantia (cud chewers).

The suborder Palaeodonta has five families: **Diacodectidae, Leptochoeridae, Homacodontidae, Dichobunidae,** and **Achaenodontidae.** No examples of this suborder are included in this chapter.

The suborder Suina has three superfamilies: Entelodontoidea, Suoidea, and Hippopotamoidea. The superfamily Entelodontoidea in in turn has three families: **Choeropotamidae, Cebochoeridae,** and **Entelodontidae (Elotheridae).** We will only be concerned with the last family.

Faunal listing for the family **Entelodontidae (Elotheridae):** EOCENE: *Eoentelodon.* OLIGOCENE: *Archaeotherium (Megachoerus), Entelodon (Elodon),* and *Ergilobia (Brachyodon).* MIOCENE: *Dinohyus* (see Figs. 49-3 and 49-4).

The superfamily Suoidea has two families: **Suidae** and **Tayassuidae (Dicotylidae).** The faunal listing for the family **Suidae** is as follows: EOCENE: *Paradoxodonides (Paradoxodon).* OLIGOCENE: *Palaeochoerus* and *Propalaeochoerus.* MIOCENE: *Conohyus, Diamantohyus, Hyotherium, Kubanochoerus, Libycochoerus, Listriodon, Palaeochoerus,* and *Xenochoerus.* PLIOCENE: *Chleuastochoerus, Conohyus, Dicoryphochoerus, Hippohyus (Hyosus), Listriodon, Lophochoerus, Palaeochoerus, Potamochoerus (Choiropotamus), Sanitherium, Schizochoerus, Sivachoerus, Sivahyus, Sus,* and *Tetraconodon.* PLEISTOCENE: *Babyroussa (Babirussa), Celebochoerus, Dicoryphochoerus, Hippohyus (Hyosus), Hylochoerus, Nyanzachoerus, Omochoerus (Mesochoerus), Orthostonyx, Phacochoerus (Afrochoerus), Potamochoerus (Choiropotamus), Sivachoerus, Sus* (to Recent), and *Tetraconodon.*

Faunal listing for the family **Tayassuidae (Dicotylidae):** OLIGOCENE: *Doliochoerus* and *Perchoerus (Both-*

rolabis) (see Fig. 49-5). MIOCENE: *Choeromurus (Choerotherium), Dyseohyus, Floridachoerus, Hesperhys (Desmathys), Perchoerus,* and *Prosthenops.* PLIOCENE: *Pecarichoerus, Prosthenops,* and *Selengonus.* PLEISTOCENE: *Catagonus, Leptotherium, Mylohyus, Platygonus (Parachoerus)* (see Figs. 49-6 and 49-7), and *Tayassu (Dicotyles).*

There are two families in the superfamily Hippopotamoidea: **Anthracotheriidae** and **Hippopotamidae.**

The faunal listing for the family **Anthracotheriidae** is: EOCENE: *?Anthracobune, Anthracohyus, Anthracokeryx, Anthracosenex, Anthracothema, Anthracotherium, Bothriodon (Aepinacodon), Gobiotypus, Haplobunodon, Lophiobunodon, Probrachyodus, Prominatherium, Rhagatherium (Amphirhagatherium),* and *Thaumastognathus.* OLIGOCENE: *Anthracochoerus, Bothriodon, Bothriogenys, Brachyodus, Bunobrachyodus, Elomeryx, Galasmodon, Hemimeryx, Heptacodon, Hyoboöps (Merycops), Microbunodon (Microselenodon), Octacodon,* and *Rhagatherium.* MIOCENE: *Anthracotherium, Arretotherium, Brachyodus, Elomeryx, Hyoboöps, Kukusepasutauka, Parabrachyodus,* and *Telmatodon (Gonotelma).* PLIOCENE: *Choeromeryx, Hemimeryx, Hyoboöps, Merycopotamus,* and *Telmatodon.*

Faunal listing for the family **Hippopotamidae:** PLIOCENE: *Hippopotamus (Hexaprotodon)* (to Recent). PLEISTOCENE: *Choeropsis.*

The suborder Ruminantia ("cud-chewers") has two infraorders: Tylopoda and Pecora. The infraorder Tylopoda has four superfamilies: Cainotheroidea, Anoplotheroidea, Merycoidodontoidea, and Cameloidea.

In the superfamily Cainotheroidea there is only one family: **Cainotheriidae (Caenotheriidae).** Some genera (all Oligocene in age): *Caenomeryx, Cainotherium (Caenotherium), Oxacron, Paroxacron,* and *Plesiomeryx.*

In the superfamily Anoplotheroidea there are three families: **Anoplotheriidae, Xiphodontidae,** and **Amphimerycidae.**

The faunal listing for the family **Anoplotheriidae** is: EOCENE: *Anoplotherium, Catodontherium, Dacryther-*

49-1. Oligocene. A view of a small part of the badlands of western South Dakota. The Brulé Formation (with its two members: Poleslide (upper) and Scenic (lower) sits upon the Chadron Formation, which in turn rests upon the late Cretaceous marine Pierre Shales. This view was taken directly south of Wall, Pennington County, South Dakota.

ium, Diplobune, and *Tapirulus.* OLIGOCENE: *Anoplotherium, Diplobune, Ephelcomenus (Hyracodontherium),* and *Tapirulus.*

Faunal listing for the family **Xiphodontidae:** EOCENE: *Dichodon (Tetraselenodon), Haplomeryx,* and *Xiphodon.* OLIGOCENE: *Dichodon* and *Xiphodon.*

Faunal listing for the family **Amphimerycidae:** EOCENE: *Amphimeryx* and *Pseudamphimeryx.* OLIGOCENE: *Amphimeryx.*

There are two families in the superfamily Merycoidodontoidea: **Agriochoeridae** and **Merycoidondontidae (Oreodontidae).** Faunal listing for the family **Agriochoeridae:** EOCENE: *Diplobunops* and *Protoreodon (Agriotherium)* (see Figs. 49-18 and 49-19). OLIGOCENE: *Agriochoerus.* MIOCENE: *Agriochoerus.*

Faunal listing (abbreviated) for the family **Merycoidodontidae (Oreodontidae):** OLIGOCENE: *Bathygenys, Eporeodon, Leptauchenia* (see Figs. 49-12 and 49-13), *Merycoidodon (Oreodon)* (see Figs. 49-14 through 49-17), *Oreontes, Parastenopsochoerus, Platyochoerus, Promes-*

49-2. Statue of Joseph Leidy, one of the pioneer paleontologists who worked with the varied mammalian fauna of the White River Badlands. Statue is situated in front of the Academy of Natural Sciences building in Philadelphia, Pennsylvania.

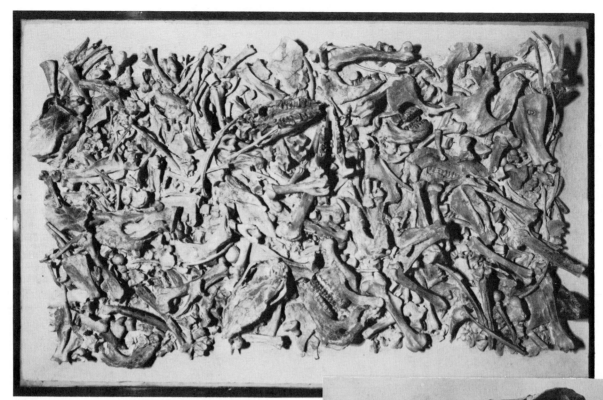

49-3. Miocene. Typical "bone bed" mammal skeletal material assemblage. In this mass burial are the skulls and other bones of *Dinohyus (Elotherium)*, a pig-like animal, along with the bones of *Diceratherium* (a rhino), and *Moropus* (a large horse-like herbivore). Agate Springs Bone Bed (Arikareean), Agate, Sioux County, Nebraska. Courtesy: Field Museum of Natural History, Chicago.

49-4. Miocene. *Dinohyus (Elotherium)*. The complete skull of a pig- like mammal. John Day Formation, Bridge Creek, Oregon. Courtesy: Princeton Museum of Natural History, Princeton, N.J.

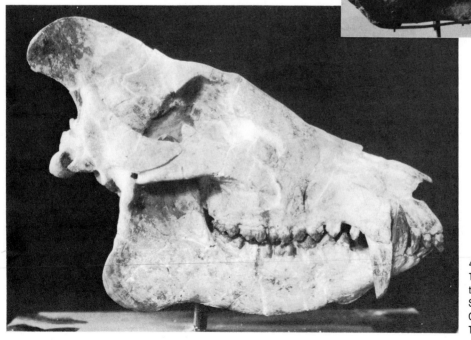

49-5. Oligocene. *Perchoerus probus* Leidy. The skull of an ancestral peccary (dicotylid). Scenic Member, Brulé Formation, Scenic, Pennington County, South Dakota. Courtesy: South Dakota School of Mines and Technology, Rapid City.

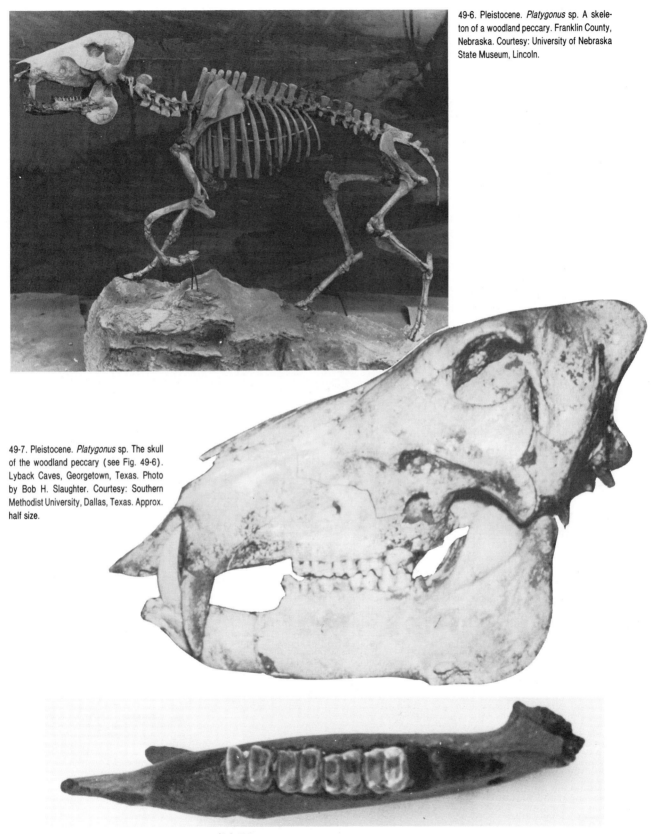

49-6. Pleistocene. *Platygonus* sp. A skeleton of a woodland peccary. Franklin County, Nebraska. Courtesy: University of Nebraska State Museum, Lincoln.

49-7. Pleistocene. *Platygonus* sp. The skull of the woodland peccary (see Fig. 49-6). Lyback Caves, Georgetown, Texas. Photo by Bob H. Slaughter. Courtesy: Southern Methodist University, Dallas, Texas. Approx. half size.

49-8. Pleistocene. Tapir (right lower jaw). Species indeterminate. Santa Fe Springs, Florida. Slightly less than actual size.

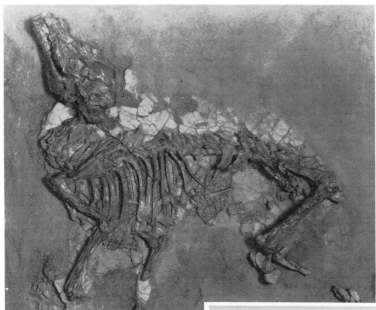

49-9. Oligocene. *Proceras (Pseudoproceras)* sp. (A pecoran.) The skeleton (a female) has an unborn calf in its pelvic region. Chadron Formation, Sioux County, Nebraska. Courtesy: University of Nebraska State Museum, Lincoln.

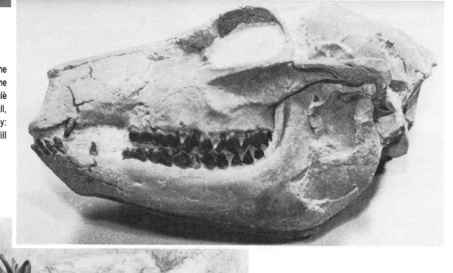

49-10. Oligocene. *Leptomeryx evansi* Leidy. The skull of a fossil chevrotain (a cud chewer) in the family Hypertragulidae. Scenic Member, Brulé Formation, White River Badlands, South of Wall, Pennington County, South Dakota. Courtesy: Black Hills Institute of Geological Research, Hill City, South Dakota.

49-11. Miocene. *Syndyoceras* sp. A group of pecorans in a museum display (one is a partial skeletal display). Sioux County, Nebraska. Courtesy: University of Nebraska State Museum, Lincoln.

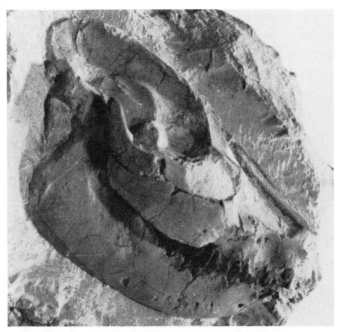

49-12. Oligocene. *Leptauchenia nitida* Leidy. The skull of a merycoidodont (oreodont) mammal. Poleslide Member, Brulé Formation, Shannon County, South Dakota. Courtesy: Black Hills Insititute of Geological Research, Hill City, South Dakota.

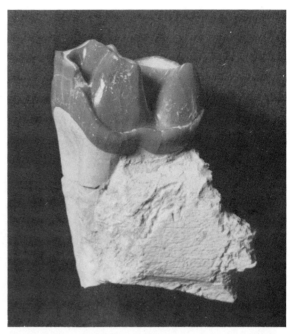

49-13. Oligocene. *Leptauchenia* sp. An isolated cheek tooth of an oreodont. Tooth is in a jaw bone fragment. Poleslide Member, Brulé Formation, South of Wall, Pennington County, South Dakota.

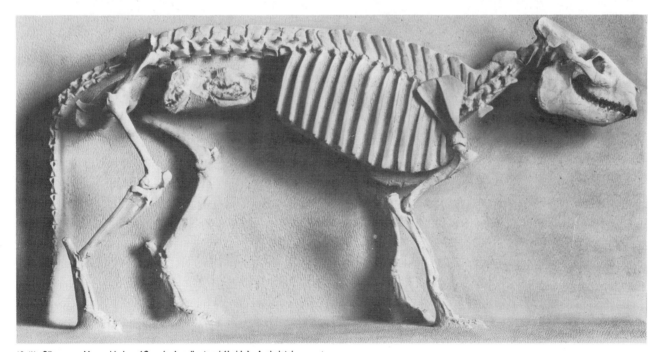

49-14. Oligocene. Merycoidodon *(Oreodon) culbertsoni* (Leidy). A skeletal mount of a pregnant female. Scenic Member, Brulé Formation (Oreodon Zone), Pennington County, South Dakota. Courtesy: South Dakota School of Mines and Technology, Rapid City.

oreodon, *Stenopsochoerus,* and *Trigenicus.* MIOCENE: *Brachycrus, Cyclopidius, Eporeodon, Hypsilops, Megoreodon, Merycoides, Mesoreodon, Oreodontoides, Phenacocoelus, Submerychoerus, Tichloleptus,* and *Ustatocho-*

erus. PLIOCENE: *Merychylus* and *Ustatochoerus.*

There are two families in the superfamily Cameloidea: **Oromerycidae** and **Camelidae.** A typical genus of the Oromerycidae is *Camelodon* from the Eocene.

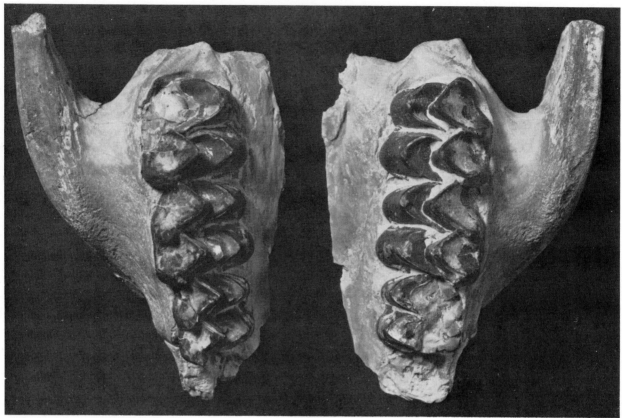

49-15. Oligocene. *Merycoidodon (Oreodon) culbertsoni* (Leidy). Two upper jaw sections showing the cheek teeth. Scenic Member, Brulé Formation, South of Wall, Pennington County, South Dakota. Slightly larger than actual size.

49-16. Oligocene. A typical coprolite (fecal pellet) found in association with *Merycoidodon* in the oreodont zone. Scenic Member, Brulé Formation, Pennington County, South Dakota. ×1. Note: Due to the phosphatic composition, most coprolites of the Brulé Formation are believed to have been produced by carnivores, not herbivores.

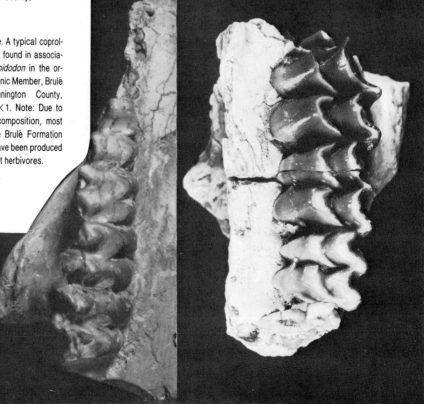

49-17. Oligocene. Two additional jaw sections of *Merycoidodon (Oreodon) culbertsoni* (Leidy). The specimen on the right is from a juvenile as seen from the fact that the teeth are not worn down. Brulé Formation, Pennington County, South Dakota. ×1.

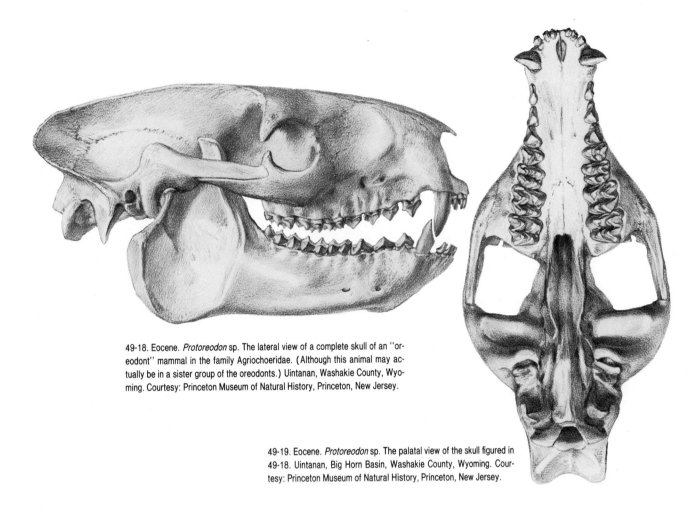

49-18. Eocene. *Protoreodon* sp. The lateral view of a complete skull of an "oreodont" mammal in the family Agriochoeridae. (Although this animal may actually be in a sister group of the oreodonts.) Uintanan, Washakie County, Wyoming. Courtesy: Princeton Museum of Natural History, Princeton, New Jersey.

49-19. Eocene. *Protoreodon* sp. The palatal view of the skull figured in 49-18. Uintanan, Big Horn Basin, Washakie County, Wyoming. Courtesy: Princeton Museum of Natural History, Princeton, New Jersey.

49-20. Oligocene. *Poëbrotherium wilsoni* Leidy. Complete skull of a camelid. Scenic Member, Brulé Formation, Shannon County, South Dakota. Courtesy: Geological Enterprises, Ardmore, Oklahoma.

49-21. Pleistocene. Isolated camel tooth (species indeterminate). Lake Monroe, Enterprise, Volusia County, Florida. ×1.5.

49-22. Pleistocene. *Gigantocamelus fricki* Barbour. The skeletal mount of a large ruminant camelid. (Photo taken May 3, 1939.) Courtesy: University of Nebraska State Museum, Lincoln.

An abbreviated faunal listing for the family **Camelidae:** OLIGOCENE: *Poebrotherium* (see Fig. 49-20). MIOCENE: *Alticamelus* and *Procamelus*. PLIOCENE: *Alticamelus, Camelus,* and *Procamelus*. PLEISTOCENE: *Camelops, Camelus, Gigantocamelus* (see Fig. 49-22), and *Megacamelus.*

There are three superfamilies in the infraorder Pecora: Traguloidea, Cervoidea, and Bovoidea. The superfamily Traguloidea is comprised of four families: **Hypertragulidae, Protoceratidae, Gelocidae,** and **Tragulidae.** Some typical generic examples from each family: **Hypertragulidae:** OLIGOCENE: *Leptomeryx* (see Fig. 49-10). **Protoceratidae:** OLIGOCENE: *Protoceras (Pseudoproceras)* (see Fig. 49-9). MIOCENE: *Syndyoceras* (see Fig. 49-11). **Gelocidae:** EOCENE: *Gelocus.* **Tragulidae:** PLEISTOCENE: *Tragulus.*

In the superfamily Cervoidea there are three families: **Palaeomerycidae, Cervidae,** and **Giraffidae.** Some typical generic examples of each family: **Palaeomerycidae:** MIOCENE: *Palaeomeryx.* **Cervidae:** PLEISTOCENE: *Cervus* and *Megaloceros* (see Fig. 49-23). **Giraffidae:** PLIOCENE TO RECENT: *Giraffa.*

In the superfamily Bovoidea there are two families: **Antilocapridae** and **Bovidae.** Some typical generic examples: **Antilocapridae:** PLEISTOCENE: *Antilocapra.* **Bovidae:** PLEISTOCENE: *Bison* (see Figs. 49-25 and 49-26).

49-23. Pleistocene. *Megaloceros giganteus* (Blumenbach). The giant deer or "Irish elk" roamed the hills and fields of Tipperary, Ireland and parts of the European continent during the interglacial periods of the Pleistocene epoch. Painting by Charles R. Knight. Courtesy: Field Museum of Natural History, Chicago.

49-25. Pleistocene. *Bison priscus* (Bajanns). The skull roof (frontal area and jaws missing) and horns of a large European buffalo. Dredged from the Rhine River at Rhenen, Holland. Courtesy: Fossilien Museum, Walchern, The Netherlands.

49-24. Pleistocene. An isolated deer tooth (species indeterminate). Lake Monroe, Enterprise, Volusia County, Florida. ✕ 1.5.

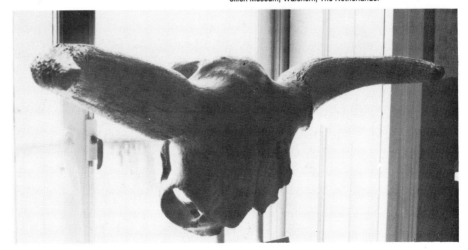

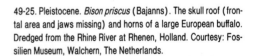

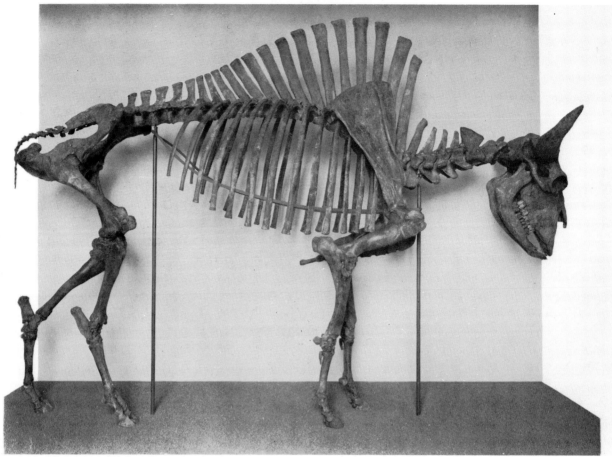

49-26. Pleistocene. *Bison bison antiquus* Leidy. Skeletal mount of a large buffalo. Folsom, New Mexico. Courtesy: Colorado Museum, Denver.

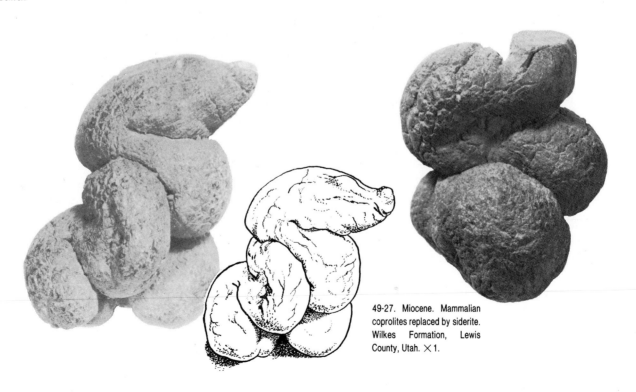

49-27. Miocene. Mammalian coprolites replaced by siderite. Wilkes Formation, Lewis County, Utah. ×1.

50 MAMMALIA (Edentates)

Edentates ("toothless" animals) include the ground sloths, glyptodons, and armadillos of the fossil record. The origin for these specialized forms was in South America, but it wasn't until the Late Tertiary that their descendants arrived in the north (particularly Florida), where their fossilized remains are abundant in some formations. Archaic forms that are possibly related are found in North American Paleocene, Eocene, and Oligocene deposits.

The order Edentata has two suborders: Palaenodonta and Xenarthra. The suborder Palaeanodonta has two families: **Metacheiromyidae** and **Epoicotheriidae.** In the suborder Xenarthra there are three infraorders—Loricata (Cingulata), Pilosa, and Vermilingua—with 3 superfamilies—Dasypodoidea, Palaeopectoidea, and Glyptodontoidea—in the infraorder Loricata (Cingulata), and two superfamilies—Megalonychoidea and Mylodontoidea—in the infraorder Pilosa. The infraorder Vermilingua has no superfamilies, just one family, **Myrmecophagidae.**

The three superfamilies we will be concerned with in this chapter are: Dasypodoidea, Glyptodontoidea, and Megalonychoidea. In the superfamily Dasypodoidea, we have the family **Dasypodidae** (armadillos). A good example or representative of this family would be *Chlamytherium* of the Pleistocene (see Fig. 50-1). In the superfamily Glyptodontoidea, we have the family **Glyptodontidae** (glyptodons), which includes a representative of the Pleistocene, *Boreostracon* (see Fig. 50-1). Finally, in the superfamily Megalonychoidea, we have two families, **Megalonychidae** and **Megatheridae** (the ground sloths), represented by the genera *Megalonyx* (see Figs. 50-3 through 50-5), and *Megatherium* (see Figs. 50-2 and 50-6), respectively. Both genera are of Pleistocene age.

50-I. Pleistocene. Giant armadillo body armor (buckle scutes). A. *Chlamytherium septentrionalis* Leidy. B. *Boreostracon floridanus* Simpson. Peace River, Wachulla, Polk County, Florida. ×1.5.

50-2. Pleistocene. *Megatherium* sp. Three views of a cheek tooth of a giant ground sloth. Lake Monroe, Enterprise, Volusia County, Florida. ⅔ actual size.

50-3. Pleistocene. *Megalonyx* sp. A selection of the cheek teeth of another type of giant ground sloth. Top row shows the occlusal (biting) surfaces of the teeth. Ichatucknee River, Gilchrist County, Florida. Approx. ⅓ actual size.

50-4. Pleistocene. *Megalonyx* sp. A skeletal mount of a large ground sloth. Garden County, Nebraska. Courtesy: University of Nebraska State Museum, Lincoln.

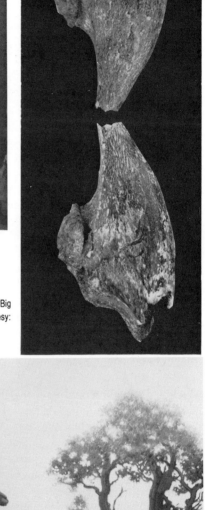

50-5. Pleistocene. *Megalonyx* sp. Two views of a ground sloth's claw. Big Brook, Marlboro Township, Monmouth County, New Jersey. Half size. Courtesy: New Jersey State Museum, Trenton.

50-6. *Great ground sloths.* Pictured are three giant ground sloths, *Megatherium*, and two glyptodons foraging for food during an interglacial period of the Pleistocene epoch. Painting by Charles R. Knight. Courtesy: Field Museum of Natural History, Chicago.

51 MAMMALIA (Marine forms/Whales, Porpoises, Etc.)

Marine mammals of the order Cetacea, are mammals adapted for an aquatic form of life. They originated on land, but their preference was for life in a watery environment. The whales (including their close smaller relatives, the porpoises) are the predominant form of mammal life in the sea. They inhabit all of the world's oceans.

The primitive whales (archaeocetes) are represented in the fossil record by the zeuglodonts or yoke-toothed whales. This group of early whales probably originated as terrestrial carnivores, but, due to their predilection for eating fish, soon took to the sea.

The original archaeocetes had a long period of dominance in the Eocene. The Middle Eocene had such forms as *Protocetus* and *Prozeuglodon* as principal types. The cetaceans were about the size of our present day porpoises, with lengthy snouts and carnivorous cheek teeth, along with conical or peglike anterior (front) teeth. The Late Eocene brought forth a most huge and grotesque type of archaeocete, *Basilosaurus (Zeuglodon)* (see Fig. 51-1), which dominated the seas until the end of the Eocene. It was replaced in the Oligocene by a shorter form of archaeocete, the **Dorudontidae,** which endured into the Early Miocene.

The archaeocetes were replaced by the odontocetes ("toothed whales"), many species of which gave rise to a large portion of the present-day cetaceans, such as porpoises, dolphins, and even the large sperm whales. The most primitive forms of the odontocetes were the squalodonts, who started to appear in the Late Oligocene. In some ways, the squalodonts resembled the earlier archaeocetes, with similar yoke or sharklike cheek teeth. The squalodonts had such early forms as *Prosqualodon* (Oligocene) and *Squalodon* (Early to Late Miocene) (see Figs. 51-2 and 51-3), as well as other allied forms (see lists to follow).

The remainder of the whales and porpoises that comprise the suborders Odontoceti and Mysticeti are featured in this chapter along with a discussion of the suborder Pinnipedia of the order Carnivora (not included in Chapter 44). This last group includes seals, sea lions, and walruses. Because they are marine mammals—even though in the order Carnivora—the author thought it best to include them in this particular chapter.

Faunal listing of the order Cetacea: suborder Archaeoceti: family **Protocetidae:** EOCENE: *Eocetus (Mesocetus),* and *Procetus.* Family **Dorudontidae:** EOCENE: *Dorudon* and *Zygorhiza;* MIOCENE: *Kekenodon* and *Phococetus.* Family **Basilosauridae (Zeuglodontidae):** EOCENE: *Basilosaurus (Zeuglodon)* (see Fig. 51-1), *Mammalodon,* and *Prozeuglodon;* OLIGOCENE: *Platyosphys.*

In the suborder Odontoceti: family **Agorophiidae:** EOCENE: *Agorophis* and *Xenorophis.* Family **Squalodontidae:** OLIGOCENE: *Microcetus* and *Protosqualodon;* MIOCENE: *Colophonodon, Metasqualodon, Microsqualodon, Neosqualodon, Parasqualodon, Phoberodon, Phocogenius, Protosqualodon, Saurocetus, Squalodon (Phocodon)* (see Figs. 51-2 and 51-3), *Tangarasaurus,* and *Trirhizodon;* PLIOCENE: *Prionodelphis.* Family **Platanisidae:** MIOCENE: *Hesperoinia, Proinia,* and *Zarhachis;* PLIOCENE: *Anisodelphis, Goniodelphis, Hesperocetus, Ischyrorhynchus,* and *Saurodelphis (Saurocetes).* Family **Ziphiidae:** MIOCENE: *Anoplonassa, Belemnoziphius, Cetorhynchus, Choneziphius, Eboroziphius, Incacetus, Mesoplodon, Notocetus (Diochotichius), Palaeoziphius, Proroziphius, Squalodelphis, Ziphioides,* and *Ziphirostrum (Miozyphius);* PLIOCENE: *Berardiopsis, Choneziphius,* and *Mesoplodon.* Family **Delphinidae:** MIOCENE: *Agabelus, Allodelphis, Araeodelphis, Belosphys, Ceterhinops, Delphinavus, Delphinodon, Delphinopsis, Doliodelphis, Grypolithax, Iniopsis, Ixacanthus, Kentriodon* (see Figs. 51-10 and 51-15), *Lamprolithax, Leptodelphis, Liolithax, Loxolithax, Macrochirifer, Macrodelphis, Megadelphis, Miodelphis, Nannolithax, Oedolithax, Pelodelphis, Pithanodelphis, Platylithax, Priscodelphinus, Sinanodelphinus, Sormatodelphis, Stereodelphis,* and *Tretosphrys;* PLIOCENE: *Delphinus, Lonchodelphis, Miodelphis, Orcinus, Pontistes, Pontivaga,* and *Steno;* PLEISTOCENE: *Delphinus, Globicephala, Orcinus, Pseudorca, Stenodelphis,* and *Tursiops (Tursio).* Family **Eurhinodelphidae:** MIOCENE: *Eurhinodelphis* (see Figs. 51-12 and 51-13), and *Ziphiodelphis.* Family

51-1. Eocene. *Zeuglodon (Basilosaurus).* A group of primitive "shark-toothed" whales. Painting by Charles R. Knight. Courtesy: Field Museum of Natural History, Chicago.

51-2. Miocene. *Squalodon* cf. *tiedemani.* The fused upper and lower jaws of a squalodont whale or "shark-tooth" whale. Pungo River Marl Formation (Helvetian), Aurora, Beaufort County, North Carolina. ⅛ actual size.

Hemisyntrachelidae: MIOCENE: *Lophocetus;* PLIOCENE: *Hemisyntrachelus.* Family **Acrodelphidae:** MIOCENE: *Acrodelphis, Champsodelphis, Eoplatanista, Heterodelphis, Pomatodelphis,* and *Schizodelphis* *(Cyrtodelphis);* PLIOCENE: *Pomatodelphis,* and *Schizodelphis.* Family **Monodontidae (Delphinapteridae):** PLEISTOCENE: *Delphinapterus* and *Monodon.* Family **Phocaenidae:** MIOCENE: *Palaeophocaena* and *Proto-*

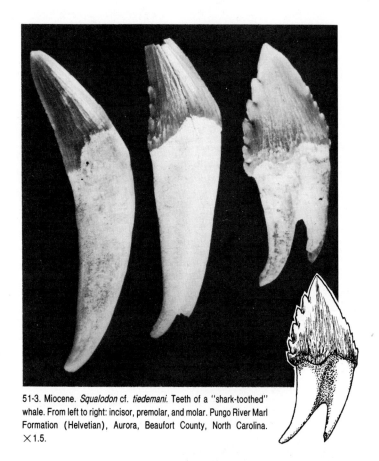

51-3. Miocene. *Squalodon* cf. *tiedemani*. Teeth of a "shark-toothed" whale. From left to right: incisor, premolar, and molar. Pungo River Marl Formation (Helvetian), Aurora, Beaufort County, North Carolina. ×1.5.

51-4. Miocene. A tympanic "earbone" of a whale (species indeterminate). Bone Valley Sediments, Brewster, Polk County, Florida. Slightly less than actual size.

51-5. Miocene. Three views of a tympanic "earbone" of a whale (species indeterminate). Pungo River Marl Formation (Helvetian), Aurora, Beaufort County, North Carolina. Less than half size.

phocaena; PLEISTOCENE: *Phocaena (Phocaenoides)* and *Phocaenopsis.* Family **Physeteridae:** MIOCENE: *Apenophyseter, Aulophyseter, Diaphorocetus, Dinoziphius, Hoplocetus, Idiophyseter, Idiorophius, Ontocetus, Orycterocetus, Physeter, Physeterula, Placoziphius, Prophyseter, Scaldicetus, Scaptodon* and *Thalassocetus;* PLIOCENE: *Balaenodon, Kogia, Kogiopsis, Physeter, Physetodon,* and *Priscophyseter;* PLEISTOCENE:

Physeter.

Suborder Mysticeti, family **Patriocetidae:** EOCENE: *Archaeodelphis;* OLIGOCENE: *Patriocetus.* Family **Cetotheriidae:** OLIGOCENE: *Cetotheriopsis* and *Pachycetus;* MIOCENE: *Aglaocetus, Cephalotropis, Cetotherium, Herpetocetus, Mesocetus, Palaeobalaena, Plesiocetopsis, Rhegnopsis, Siphonocetus,* and *Tretulias,* PLIOCENE: *Amphecetus, Nannocetus,* and *Plesiocetopsis.* Family **Es-**

51-6. Miocene. Whale vertebra (species indeterminate) with outer face or epiphyses in place. Pungo River Marl Formation (Helvetian), Aurora, Beaufort County, North Carolina. Slightly less than actual size.

51-7. Miocene. Three views of a small whale (species indeterminate) vertebra. Note that the epiphyses are intact (on edge of vertebra, see bottom right side). Pungo River Marl Formation (Helvetian), Aurora, Beaufort County, North Carolina. Approx. half size.

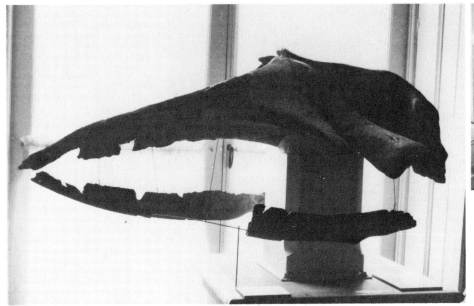

51-8. Pliocene. *Balaenula balaenopsis.* The skull of a baleen whale, Kallo Harbour Works, Contact of the Kattendijk and Luchtal Sands, Antwerpen, Belgium. Courtesy: Fossilien Museum, Walchern, The Netherlands.

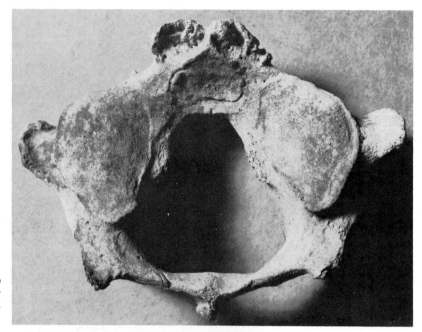

51-9. Miocene. *Rhabdosteus* sp. A complete atlas vertebra from a porpoise. Pungo River Marl Formation (Helvetian), Aurora, Beaufort County, North Carolina. × 1.

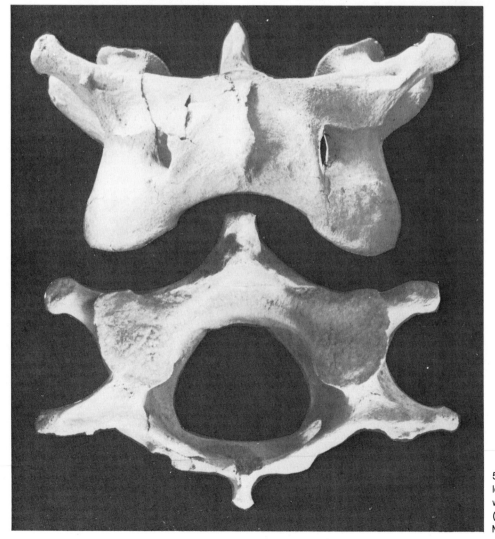

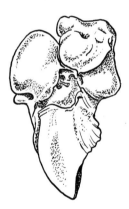

51-11. Miocene. Periotic "earbone" of an unidentified porpoise. Note that the "stapes" (small sticklike protuberance in the center of the bone mass) is intact. Pungo River Marl Formation (Helvetian), Aurora, Beaufort County, North Carolina. × 1.5.

51-10. Miocene. *Kentriodon pernix* Kellogg. Another atlas vertebra in two views. Pungo River Marl Formation (Helvetian), Aurora, Beaufort County, North Carolina. × 1.

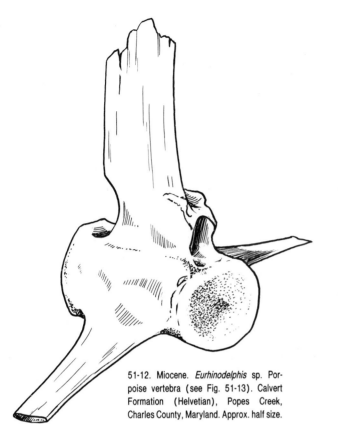

51-12. Miocene. *Eurhinodelphis* sp. Porpoise vertebra (see Fig. 51-13). Calvert Formation (Helvetian), Popes Creek, Charles County, Maryland. Approx. half size.

51-13. Miocene. *Eurhinodelphis* sp. Porpoise vertebrae. Calvert Formation (Helvetian), Popes Creek, Charles County, Maryland. Approx. half size.

51-14. Miocene. A dolphin tooth. (porpoise family!). Note the distinctive root flare. Actually, the tooth is connected to the jawbone, as opposed to the generalized porpoise jawbones which have a socket for each tooth. Bone Valley Sediments, Phosphate pits, Brewster, Polk County, Florida. ×1.

51-15. Miocene. *Kentriodon pernix* Kellogg. A fragment of the dentary (lower jaw) of a porpoise, with a series of teeth in their respective "sockets". Pungo River Marl Formation (Helvetian), Aurora, Beaufort County, North Carolina. ×1.5.

51-16. Miocene. Digit and calcaneum bones from a pinniped, possibly a small seal. Pungo River Marl Formation (Helvetian), Aurora, Beaufort County, North Carolina. ✕ 2.

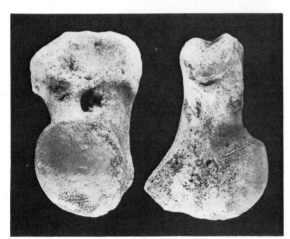

51-17. Miocene. Two views of a limb bone of a large pinniped (seal). Pungo River Marl Formation (Helvetian), Aurora, Beaufort County, North Carolina. Approx. half size.

51-18. Miocene. Two views of a digit bone of a pinniped (seal). Pungo River Marl Formation (Helvetian), Aurora, Beaufort County, North Carolina. ✕ 1.

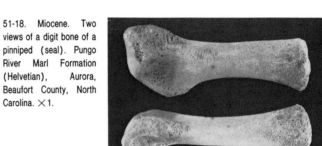

51-19. Miocene. Limb bone of a pinniped (species indeterminate). Pungo River Marl Formation (Helvetian), Aurora, Beaufort County, North Carolina. Slightly less than actual size.

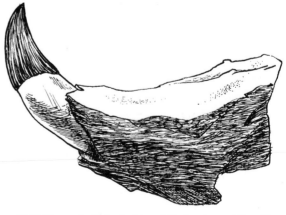

51-20. Miocene. Seal incisor tooth in partial jawbone fragment. Pungo River Marl Formation (Helvetian), Aurora, Beaufort County, North Carolina. ✕ 1.

51-21. Modern seals and sea lions sunning themselves on rocks along the Pacific coastline of the United States of America.

crichtiidae (Rhachianectidae): ?PLEISTOCENE: *Eschrichtius*. Family **Balaenopteridae**: MIOCENE: *Balaenoptera, Idiocetus,* and *Mesoteras;* PLIOCENE: *Balaenoptera, Burtinopsis, Idiocetus, Megaptera, Megapteropsis,* and *Palaeocetus;* PLEISTOCENE TO RECENT: *Balaenoptera*. Family **Balaenidae**: MIOCENE: *Morenocetus;* PLIOCENE: *Balaenula* (see Fig. 51-8), and *Protobalaena;* PLEISTOCENE: *Balaena* and *Eubalaena.*

The suborder Pinnipedia in the order Carnivora has three families: **Otariidae, Odobenidae,** and **Phocidae.** In the family **Otariidae** ("eared seals"), *Allodesmus* from the Miocene is a typical representative. In the family **Odobenidae** (walruses, *Odobenus* (see Fig. 51-22) from the Pliocene is typical. Finally, the family **Phocidae** ("earless seals") has as a typical representative in the fossil record, *Phoca* from the Miocene. *Phoca* is also a living genus.

51-22. Pliocene. *Odobenus antwerpiensis.* A walrus tusk and partial jaw fragment. Dredgings in the North Sea (30 miles out) off Westkapelle, Zeeland, Holland. Slightly smaller than actual size.

52 MAMMALIA (Rodents)

The order Rodentia (rodents) comprises a varied group of generally small mammals which have large incisors used for gnawing. They include primitive forms, i.e., *Paramys* and *Ischyromys* (see Fig. 52-1); squirrels; various South American rodents (including chinchillas); mice and rats; and beavers. The rabbits and their relatives have superficially similar dentitions, but are a separate mammalian order (Lagomorpha).

The order Rodentia has three basic suborders: Sciuromorpha (Protrogomorpha), Caviomorpha, and Myomopha. However, some well-known forms are not yet assigned to suborders. There are three superfamilies in the suborder Sciuromorpha: Ischyromyoidea, Aplodontoidea, and Sciuroidea. There are five families comprising the superfamily Ischyromyoidea: **Paramyidae, Sciuravidae, Cylindrodontidae, Protoptychidae,** and **Ischyromyidae.**

Typical genera for the above families: **Paramyidae:** PALEOCENE: *Paramys;* EOCENE: *Leptotomus;* OLIGOCENE: *Pelycomys.* **Sciuravidae:** EOCENE: *Sciuravus.* **Cylindrodontidae:** EOCENE: *Presbemys;* OLIGOCENE: *Cylindrodon.* **Protoptychidae:** EOCENE: *Protoptychus.* **Ischyromyidae:** OLIGOCENE: *Ischyromys* (see Fig. 52-1).

There are two families in the superfamily Aplodontoidea: **Aplodontidae** and **Mylagaulidae.** Typical genera for the families: **Aplodontidae:** OLIGOCENE: *Allomys;* MIOCENE: *Sciurodon;* PLIOCENE: *Tardontia;* PLEISTOCENE: *Aplodontia.* **Mylagaulidae:** MIOCENE: *Mylagaulodon;* PLIOCENE: *Mylagaulus.*

There is only one family in the superfamily Sciuroidea: **Sciuridae.** Typical genera: OLIGOCENE: *Sciurus;* MIOCENE: *Eutamias;* PLIOCENE: *Cynomys;* PLEISTOCENE: *Burosor.* (*Sciurus, Eutamias,* and *Cynomys* are all typical recent genera as well.)

In the suborder Caviomorpha, there are four superfamilies: Octodontoidea, Chinchilloidea, Cavoidea, and Erethizontoidea. There are five families comprising the superfamily Octodontoidea: **Octodontidae, Echimyidae, Ctenomyidae, Abrocomidae,** and **Capromyidae.**

Typical genera for the above families: **Octodontidae:** OLIGOCENE/MIOCENE: *Acaremys;* PLIOCENE: *Phthoramys,* PLEISTOCENE: *Plataeomys.* **Echimyidae:** OLIGOCENE: *Deseadomys;* MIOCENE: *Spaniomys;* PLIOCENE: *Eumysops;* PLEISTOCENE: *Echimys.* **Ctenomyidae:** PLIOCENE: *Ctenomys;* PLEISTOCENE: *Megactenomys.* **Abrocomidae:** PLIOCENE: *Abrocoma.* **Capromyidae:** PLIOCENE: *Paramyocastor;* PLEISTOCENE: *Capromys.*

In the superfamily Chinchilloidea, there are four families: **Chinchillidae, Dasyproctidae, Dinomyidae,** and **Elasmodontomyidae.**

Typical genera for the above families: **Chinchillidae:** OLIGOCENE: *Perimys;* MIOCENE: *Prolagostomus;* PLIOCENE: *Euphilus;* PLEISTOCENE: *Lagidium.* **Dasyproctidae:** OLIGOCENE: *Cephalomys;* MIOCENE: *Neoreomys;* PLIOCENE: *Olenopsis;* PLEISTOCENE: *Dasyprocta.* **Dinomyidae:** MIOCENE: *Dinomys;* PLIOCENE: *Gyriabrus;* PLEISTOCENE: *Clidomys.* **Elasmodontomyidae:** PLEISTOCENE: *Elasmodontomys.*

There are three families in the superfamily Cavoidea: **Eocardiidae, Caviidae,** and **Hydrochoeridae.**

Typical genera for the above families: **Eocardiidae:** OLIGOCENE: *Asteromys;* MIOCENE: *Eocardia.* **Caviidae:** MIOCENE: *Prodolichotis;* PLIOCENE: *Allocavia;* PLEISTOCENE: *Galea.* **Hydrochoeridae:** PLIOCENE: *Neoanchimys;* PLEISTOCENE: *Hydrochoerus.*

In the superfamily Erethizontoidea, there is only one family: **Erethizontidae,** represented by *Erethizon* from the Pleistocene. *Erethizon* is also a well-known modern genus as well.

There are five superfamilies in the suborder Myomorpha: **Muroidea, Diplodoidea, Geomyoidea, Gliroidea,** and **Spalacoidea.** See Fig. 52-2 for an example of the teeth for a Pleistocene "cotton rat." There are only two families represented in the superfamily Muroidea: **Cricetidea** and **Muridae.** In the superfamily Dipodoidea, there are two families: **Dipodidae** and **Zapodidae.** The superfamily Geomyoidea has three families: **Eomyidae, Geomyidae,** and **Heteromyidae;** while only two families each represent the superfamilies Gliroidea and Spalacoidea. they are, respectively: **Gliridae (Myoxidae)** and **Seleviniidae,** and **Spalacidae** and **Rhizomyidae.**

52-1. Oligocene. *Ischyromys typus* Leidy. Two skulls of a sciuromorph rodent, from the White River Badlands. Scenic Member, Brulé Formation, Pennington County, South Dakota. Courtesy: South Dakota School of Mines and Technology, Rapid City.

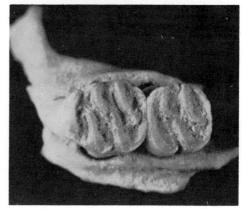

52-2. Pleistocene. *Sigmodon hispidus* Say and Ord. Two cheek teeth in a jaw section of a myomorph (Cotton-rat). Bell County, Texas. ×8.

52-3. Miocene. *Palaeocastor* sp. Beaver skeletons in a mount showing them making their homes in the "devil's corkscrews" (burrows). Nebraska. Courtesy: University of Nebraska State Museum, Lincoln.

The superfamily Castoroidea is not assigned to a suborder. It consists of a well-known group, the beavers. Its two basic families are **Castoridae** and **Eutypomyidae.** *Castoroides ohioensis* (see Figs. 52-4 through 52-7) is a good example of a huge rodent from the Pleistocene, while a smaller ancestor, *Palaeocastor* (see Fig. 52-3) is a good example from the Miocene epoch.

The order Lagomorpha (rabbits and hares) will not be discussed here as there are no representatives in this book.

The families of rodents which are not represented in this book are: **Pseudosciuridae, Theridomyidae, Phiomyidae, Thryonomyidae, Petromuridae, Hystricidae, Ctenodactylidae, Anomaluridae, Pedetidae.** Also unrepresented are the lagomorph families: **Eurymylidae, Ochotonidae, and Leporidae.**

52-4. Pleistocene. *Castoroides ohioensis* Foster. A metallic cast of the giant beaver. On exhibit at the New York State Museum in Albany, Courtesy: New York State Museum, Albany.

52-5. Pleistocene. *Castoroides ohioensis* Foster. Three views of a cheek tooth of a giant beaver. Big Brook, Marlboro Township, Monmouth County, New Jersey. Slightly less than actual size.

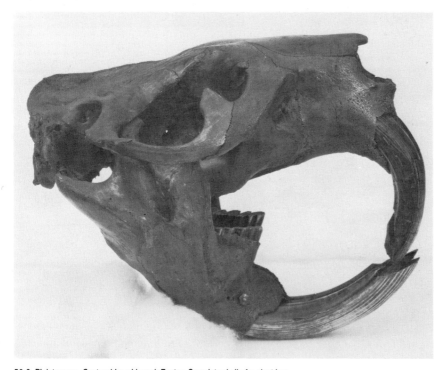

52-6. Pleistocene. *Castoroides ohioensis* Foster. Complete skull of a giant beaver. Fairmount, Indiana. Courtesy: Field Museum of Natural History, Chicago.

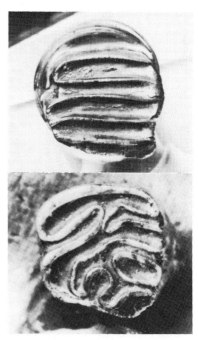

52-7. Occlusal view of two beaver teeth. Top specimen: *Castoroides ohioensis* Foster, the giant beaver from the Pleistocene of New Jersey. Bottom specimen: *Castor* sp. A modern beaver tooth from New Jersey. To show patterns of the tooth enameloid. Both specimens from Big Brook, Marlboro Township, Monmouth County, New Jersey. Top specimen ×2, bottom specimen ×3.

ADDENDA

The following material was left out of Chapter 5—Echinoidea. This faunal listing completes the information in that chapter.

Echinoid faunal listing (genera): ORDOVICIAN: *Aulechinus* and *Eothuria*. SILURIAN: *Palaeodiscus*. MISSISSIPPIAN: *Archaeocidaris (Echinocrinus)*, *Hyattechinus*, *Lepidesthes*, *Lovenechinus (Oligoporus)*, and *Melonechinus*. PERMIAN: *Archaeocidaris (Echinocrinus)* and *Meekechinus*. TRIASSIC: *Tiarechinus* and *Triadocidaris*. JURASSIC: *Clypeus*, *Diademopsis*, *Galeropygus*, *Hemicidaris*, *Holectypus*, *Palaeopedina*, *Pedinothuria*, *Pelanechinus*, *Plesiocidaris*, *Pseudocidaris*, and *Pygaster*. LOWER TO MIDDLE CRETACEOUS: *Coenholectypus (Caenholectypus)*, *Cyphosoma* (see Figs. 5-5 and 6), *Douvillaster (Epiaster)*, *Dumblea*, *Hemiaster*, *Heteraster (Enallaster)* (see Fig. 5-7), *Holaster*, *Holectypus*, *Macraster* (see Fig. 5-8), *Phymosoma*, *Salenia*, and *Washitaster*. UPPER CRETACEOUS: *Archiacia*, *Fibularia*, *Hardouinia* (see Figs. 5-13 and 14), *Hemiaster*, *Hyposalenia*, *Linthia*, *Micraster*, *Pygurus*, *Stereocidaris* (see Fig. 5-15a), and *Tylocidaris* (see Fig. 5-15b). PALEOCENE: *Linthia*. EOCENE: *Clypeaster*, *Eoscutella*, *Eupatagus* (see Fig. 5-16), *Fibularia*, *Laganum*, *Macropneustes*, *Oligopygus*, *Periarchus*, *Phyllacanthus*, *Protoscutella*, and *Schizaster*. OLIGOCENE: *Cassidulus* and *Eupatagus*. MIOCENE: *Arbacia*, *Astrodapsis (Astrodaspis)* (see Fig. 5-19), *Echinarachnius*, *Echinocardium*, *Encope* (see Fig. 5-17), *Mellita*, *Psammechinus*, and *Scutella (Albertella)* (see Figs. 5-18 and 5-21). PLIOCENE: *Astrodapsis (Astrodaspis)*, *Clypeaster*, *Dendraster* (see Fig. 5-22), *Encope*, and *Rotula*.

Additional text from Chapter 24 (Sharks, Skates, Sawfishes, Rays, Chimaeras, and Iniopterygians):

Note: The dorsal fin spines or clasper appendages attributed to *Physonemus* (see Figs. 24-10 through 24-12) are considered to be ichthyodorulites Incertae Sedis. But, Romer, 1966, places them with the family **Edestidae** in the suborder (superfamily) Hybodontoidea.

On the collecting of fossilized shark remains—particularly in the form of isolated teeth: The fossilized teeth of sharks are very popular with fossil collectors. Along with trilobites, ammonites, and fish skeletons, shark's teeth are prized as collectibles.

The individual teeth in most families (genera) of fossil sharks are identifiable as to type. Knowing the species name of the particular tooth or teeth makes the specimen or specimens more valuable to the collector.

Fossilized shark remains (as well as the remains of skates, rays, sawfishes, and chimaeroids) are known from most marine (and some freshwater) deposits as far back as the Late Devonian, when the first "shark" appeared in the area now known as Cleveland, Ohio.

These skeletal remains of *Cladoselache* (see Figs. 24-1 through 24-3) were entombed in the muds (now hardened shales and concretions) of an ancient inland sea deposit given the name of Cleveland Shales.

In general, isolated shark's teeth are very difficult to find intact in shales, concretions, and mudstones. The limestones of the Carboniferous (Mississippian and Pennsylvanian) are loaded with such shark's teeth, but it is extremely difficult to get them out of the matrix without breaking off lateral (side) denticles or cusps. Acid-etching of the rock is necessary to try and loosen the teeth from the sand particles comprising the limestone matrix. Later, in the Late Mesozoic and throughout the Cenozoic, isolated teeth of sharks and their relatives are easiest to find in loosely packed sands and phosphates. These Late Cretaceous and Tertiary teeth are by far the easiest to collect and are most popular with all collectors.

ACKNOWLEDGMENTS

The following scientists have contributed much to the preparation of the material presented in this book—not only by supplying photographs or loan specimens for photography, but by offering advice, and, most importantly, by reviewing the body copy (text) and pictorial copy (picture captions), thus giving much more accuracy to the material presented in this work:

Dr. Donald Baird, Director, Museum of Natural History, Princeton University, Princeton, New Jersey

Dr. David Bardack, Associate Dean, University of Illinois at Chicago Circle, Chicago, Illinois

Mr. Samuel J. Ciurca, Jr., Rochester, New York

Mr. Richard D. Hamell, Technical Assistant, Department of Geosciences, Monroe Community College, Rochester, New York

Dr. Richard L. Leary, Curator of Geology, Illinois State Museum, Springfield, Illinois

Mr. David C. Parris, Associate Curator, New Jersey State Museum, Trenton, New Jersey

Mr. Charles R. Pellegrino, Ph.D. candidate, Zoology Department, Victoria University, Wellington, New Zealand

Dr. Patty C. Rice, Mt. Clemens, Michigan

Dr. Eugene S. Richardson, Jr., Curator of Invertebrates, Field Museum of Natural History, Chicago, Illinois

Dr. Dwayne D. Stone, Professor of Geology, Marietta College, Marietta, Ohio

Dr. Rainer Zangerl, Rockville, Indiana (Curator emeritus, Field Museum of Natural History, Chicago, Illinois)

The following collectors and students of natural history were most helpful to the author with advice, recommendations, loan of specimens for photography, actual photography itself, and, most importantly, with their encouragement of the author to produce the best work for the benefit of his fellow natural science enthusiasts throughout the world.

Jerrine Anthony, White Plains, New York
Bret S. Beall, Columbia, Missouri
Kenneth Brancato, Brooklyn, New York
Lawrence Brancato, Brooklyn, New York

Frank Bukowski, Philadelphia, Pennsylvania
Philip Cambridge, Cambridge, England
Harold Denison, Marion, Indiana
Stephen Englebright, Stony Brook, New York
Gary A. Enters, Hackensack, New Jersey
Wolfgang Freess, Leipzig, GDR (East Germany)
Robert Garcet, Eben-Emael, Belgium
Calvin George, Napierville, Illinois
Allen Graffham, Ardmore, Oklahoma
Richard E. Grant, Farmers Branch, Texas
Richard Hamilton, Rockaway, New York
Paul Harris, Mountain Home, Arkansas
Coenraad V. Hartman, Middleburg, The Netherlands
Mr. and Mrs. Stan Hyne, Ann Arbor, Michigan
Donald J. Kenney, Andover, Massachusetts
Robert J. Koestler, Hasbrouck Heights, New Jersey
Neil Lafon, Albuquerque, New Mexico
Al Lang, Cranford, New Jersey
Tom Laronge, Vancouver, Washington
Peter L. Larson, Hill City, South Dakota
Raymond Lasmanis, Victoria, British Columbia, Canada
Thomas Lehman, Albuquerque, New Mexico
Walter Leitz, Joliet, Illinois
John Lino, Jr., Rochelle Park, New Jersey
F. van Nieulande, The Netherlands
William Novak, Garfield, New Jersey
Barry Richards, Victoria, British Columbia, Canada
Baron van Tuyll van Serooskerke, Walchern, The Netherlands
James and Sandra Shaw, Bloomfield, New Jersey
Kirby Siber, Aathal, Switzerland
Richard Steiner, Worland, Wyoming
Harrell L. Strimple, Iowa City, Iowa
Prof. Dr. Wilhelm Stürmer, Erlangen, West Germany
Eric Vanderhoeft, Bruxelles, Belgium
Mr. Visser, The Netherlands
Vincent J. Vozza, Jr., Glendale, Arizona
Julius Weber, Mamaroneck, New York
George Weir, Irvington, New Jersey
Warren D. White, Omaha, Nebraska

The following organizations and institutions have contributed specimens, photographs, and artwork for use in this book:

The Academy of Natural Sciences, Philadelphia, Pennsylvania

Black Hills Institute of Geological Research, Hill City, South Dakota

Brooklyn Childrens Museum, Brooklyn, New York

Des Moines Register and Tribune, Des Moines, Iowa

Field Museum of Natural History, Chicago, Illinois

Fossielen Museum, Walchern, Zeeland, The Netherlands

General Foods Corporation, White Plains, New York

Geological Enterprises, Ardmore, Oklahoma

Greybull Museum, Greybull, Wyoming

Illinois State Museum, Springfield, Illinois

Los Angeles County Museum, Los Angeles, California

Marietta College, Marietta, Ohio

Monroe Community College, Rochester, New York

Musée du Silex, Eben-Emael, Belgium

National Museum of New Zealand, Wellington, New Zealand

New Jersey State Museum, Trenton, New Jersey

New York State Museum, Albany, New York

New Zealand Geological Survey, Lower Hutt, New Zealand

The Geology Department, Princeton University, Princeton, New Jersey

Museum of Natural History, Princeton University, Princeton, New Jersey

Richard Rush Studio, Chicago, Illinois

Museum of Natural History, Rutgers University, New Brunswick, New Jersey

Natur und Museum, Senckenberg, Frankfurt am Main, W. Germany

Siber and Siber Ltd., Aathal, Switzerland

South African Museum, Cape Town, Republic of South Africa

Museum of Geology, South Dakota School of Mines, Rapid City, South Dakota

Southern Methodist University, Dallas, Texas

United States National Museum (Smithsonian Institute), Washington, D.C.

Veb Gustav Fischer Verlag, Jena, GDR (East Germany)

Geology Museum, Victoria University, Wellington, New Zealand

Museum of Natural History, University of Kansas, Lawrence, Kansas

University of Colorado, Boulder, Colorado

University of Utah, Vernal, Utah

Roger Williams Mint, Attleboro, Massachusetts

Additionally, the following have been of assistance to the author:

Bruce M. Bell, New York State Museum, Albany, New York

Dr. David S. Berman, Division of Vertebrate Fossils, Carnegie Museum, Pittsburgh, Pennsylvania

Dr. Philip R. Bjork, Director, Museum of Geology, South Dakota School of Mines, Rapid City, South Dakota

Michael K. Braun, Cape Province, Republic of South Africa

Dr. Gilbert J. Brenner, State University College, New Paltz, New York

J. Clay Bruner, Geology Department, Field Museum of Natural History, Chicago, Illinois

Larry Carney, Jacksonville, Florida

Kenneth Carpenter, Museum, University of Colorado, Boulder, Colorado

John Chorn, Department of Geology, University of Kansas, Lawrence, Kansas

M. A. Cluver, Assistant Director, South African Museum, Cape Town, Republic of South Africa

Dr. Donald W. Fisher, State Paleontologist, New York State Museum, Albany, New York

Thomas X. Grasso, Chairman, Department of Geosciences, Monroe Community College, Rochester, New York

Dr. James H. Gunnerson, Museum Director, University of Nebraska State Museum, Lincoln, Nebraska

Dean Hannotte, New York, New York

Eugene F. Hartstein, Wilmington, Delaware

Dr. Yoshikazu Hasegawa, Geology Department, Yokohama National University, Yokohama, Japan

Mitchell E. Hope, Wauchulla, Florida

Dr. Farish A. Jenkins, Jr., Professor of Biology and Curator of Vertebrate Paleontology, Museum of Comparative Zoology (Agassiz Museum), Harvard University, Cambridge, Massachusetts

Ian W. Keyes, Technical Assistant, New Zealand Geological Survey, Lower Hutt, New Zealand

David F. Kilmartin, President, Roger Williams Mint, Attleboro, Mass.

Edward Lauginiger, Wilmington, Delaware

James Leonard, Millburn, New Jersey

Dr. Arno Hermann Müller, Freiberg, GDR (East Germany)

Nancy Paine, Public Relations Director, Brooklyn Childrens Museum, Brooklyn, New York

Mrs. Mirl Panzenberger, Cape Town, Republic of South Africa

Louis J. Pinto, Department of Geosciences, Monroe Community College, Rochester, New York

Robert W. Purdy, Museum Specialist, National Museum of Natural History, Washington, D.C.

Mike Quinn, Public Relations Director, Academy of Natural Sciences, Philadelphia, Pennsylvania

Mrs. Alfred Sherwood Romer, Cambridge, Massachusetts

Martin Roos, Illinois State Museum, Springfield, Illinois

Raymond T. Rye, II, Museum Specialist, National Museum of Natural History, Washington, D.C.

Dr. R. William Selden, Curator, The Geology Museum, Rutgers University, New Brunswick, New Jersey

Dr. Bob H. Slaughter, Department of Geology, Southern Methodist University, Dallas, Texas

Steven and Trini Stelz, Leonia, New Jersey

Ron Testa, Photography Department, Field Museum of Natural History, Chicago, Illinois

Thomas A. Wells, Department of Geosciences, Monroe Community College, Rochester, New York

Len Willis, General Foods Corporation, White Plains, New York

Dr. Bruce J. Welton, Assistant Curator, Division of Vertebrate Paleontology, Los Angeles County Museum, Los Angeles, California

Helmut Zimmermann, Freiberg, GDR (East Germany)

Special appreciation to:

Mr. Edward Cowal, Coral Gables, Florida

Mrs. Agnes Halejian, Teaneck, New Jersey
James and Cathy Leggett, Bayonne, New Jersey

The author appreciates the assistance of the following people at the Van Nostrand Reinhold Publishing Company in New York City:

Larry Hager, Senior Editor, Professional Books Division
Alberta Gordon, Managing Editor
Eric Fallick, Manuscript Editor
Karen Kirshner, Advertising Copywriter
Laurie Ecker, Production Manager
Randi Book, Production Manager
Diane Lomonaco, Production Assistant

ILLUSTRATION CREDITS

Photographers' and artists' names and the caption numbers of their illustrations:

BAIRD, DONALD. 16-4, 16-10, 16-11, 16-12, 16-13, 25-35, 26-11, 26-19, 27-4, 27-12, and 40-13.

BENNETT, BRUCE. 26-12B.

BERMAN, DAVID S. 25-44, 26-2, and 26-3.

BRANCATO, KENNETH W. 2-6, 6-3, 10-23, 10-24, 10-28B, 10-35, 12-71, 13-2, 15-6, 24-10, 24-35, 24-37, 24-57, 24-73, 24-117, 24-118, 25-13, 25-71B, 25-101B, 25-145, 25-154B, 27-5, 30-22, and 38-2.

BRANCATO, LAWRENCE. 14-5B.

BRENNER, GILBERT. 20-45, 20-46, and 20-47.

CARPENTER, KENNETH. 32-1, 32-2, 32-3, 32-4, 33-3, 33-4, 33-5, 33-6, 33-7, 33-12, 33-13, 34-2, 35-1, 35-3, 35-4, 35-8, 35-9, 35-12, 35-13, 35-18, 35-19, 36-3, 37-1, 37-10, 37-11, 37-12, and 41-3.

CASE, GERARD R. 4-50A, 6-1, 6-5, 7-2, 7-3, 7-4, 7-6, 9-32, 9-44, 10-19B, 11-32, 14-6, 15-35, 24-1, 24-6, 24-14A, 24-19, 24-69, 24-100B, 24-102A, 24-103, 24-108B, 24-111A, 24-120, 24-121, 24-122, 24-123, 24-124, 24-125, 24-126, 24-127, 24-128A,B,&C, 24-129, 24-130, 24-131A,B,&C, 24-132, 24-133B, 24-142, 24-143, 24-144, 24-147, 24-149, 24-150, 24-151, 24-152, 24-153, 24-154, 24-155, 24-156, 24-158, 24-159, 24-160, 25-29, 25-36, 25-37, 25-78, 25-90, 25-101A, 25-154A, 25-156, 26-10, 29-3, 30-12, 35-20B, 39-9, 39-10, 39-11, 41-2, 43-2, 44-11, 49-1, 49-2, 49-27B, 51-2, 51-3B, 51-11, 51-12, 51-14, and 51-20.

CAVENDER, TED. 25-43.

CIURCA, SAMUEL J., Jr. 3-30, 16-1, 18-5, 18-13, 18-23, and 18-24.

DENISON, HAROLD. 4-4, 4-6, 4-9, 4-22, 4-31, 4-35, 4-36, 4-39B, 4-43, 4-44B, 4-57, 4-64, 5-15, 12-12, 12-22, 12-38, 12-39, 13-7, 13-13B, 13-35, 13-63, 13-72, 13-73, 15-25A, 16-25, 18-16, 18-22, 20-8, 20-44, 23-5, 23-6, 24-89, 24-95, 24-115A, 25-2, 25-41, 25-45, 25-94, 25-95, 25-98, 25-112, 25-113, 25-121, 25-132, 25-136, 25-141, 26-6, 26-24, 26-26, 28-9, 30-24, 34-5, 39-6, 39-7A, 42-4, 48-12, 48-15, and 49-20.

ENGLEBRIGHT, STEPHEN. 9-1.

ENTERS, GARY A. 13-13A, 13-38, 13-86, and 25-147.

FISHER, DONALD W. 20-2.

FORMICOLA, JUDITH L. 5-19, 10-4, 13-33

FREESS, WOLFGANG. 4-8A, 4-58, 13-9, 15-24B, 20-66, 20-67, 24-99, 25-65, 25-67, 25-110, 25-111, 25-135, 25-146, 25-162, and 26-27.

FUJIHIRA, TOD. 2-5, 3-1, 3-3, 3-8, 3-9, 4-26, 4-28, 4-41, 4-42, 4-56, 5-2, 5-6, 5-7, 5-21, 8-2, 9-3, 9-8, 9-15, 9-19, 9-40, 10-8, 10-10, 10-14, 10-15, 10-16, 10-22, 10-29, 10-36, 10-39, 10-46, 10-47, 10-48, 10-51, 11-8, 11-29, 11-30, 11-33, 11-38, 12-10, 12-11, 12-13, 12-17, 12-18, 12-21, 12-37, 12-42, 13-1, 13-3, 13-15, 13-16, 13-18, 13-19B, 13-21, 13-22, 13-23, 13-26, 13-30, 13-37, 13-42, 13-43, 13-52, 13-75, 14-2, 14-5A, 15-4, 15-5, 15-17, 15-19, 15-21, 15-22, 15-31, 15-33, 15-39, 15-52, 16-3, 17-2, 17-9, 18-7, 18-9, 18-12, 18-15, 20-1, 20-9, 20-13, 20-16, 20-19, 20-23, 20-55, 20-56, 20-57, 20-58, 20-59, 20-60, 23-1, 23-2, 23-7, 23-8, 23-10B, 24-4, 24-7, 24-8, 24-9, 24-11, 24-12, 24-13, 24-14B, 24-15, 24-16, 24-17, 24-20, 24-21, 24-22, 24-23, 24-24, 24-26, 24-31, 24-32, 24-33, 24-34, 24-36, 24-39, 24-40, 24-41, 24-43, 24-45, 24-46, 24-50, 24-55, 24-56, 24-59, 24-61, 24-67, 24-68, 24-96, 24-102B,C,&D, 24-105, 24-106, 24-107, 24-110, 24-113, 24-133A, 24-136, 24-137, 24-139, 24-141, 24-145, 24-146, 25-5, 25-19, 25-20, 25-22, 25-28, 25-31, 25-34, 25-40, 25-42, 25-47, 25-50, 25-51, 25-56, 25-68, 25-71A, 25-73, 25-80, 25-82, 25-87, 25-88, 25-89, 25-93, 25-118, 25-120, 25-124, 25-127, 25-128, 25-142, 25-143, 25-151, 25-155, 25-160, 25-161, 25-163, 25-164, 25-165, 26-12A, 26-28, 28-12, 28-17, 28-18, 29-2, 30-7, 30-8, 30-9, 30-10, 30-13, 31-20, 31-21, 33-2, 34-7, 35-10, 38-4, 38-5, 39-8, 39-12, 44-12, 46-7, 46-14, 46-19, 46-21, 48-3B, 48-5, 48-7, 48-9, 48-13, 49-13, 49-15, 49-16, 49-17A, 49-21, 49-24, 49-27A&C, 50-1, 51-4, 51-7, 51-9, 51-13, and 51-16.

GARCET, ROBERT. 25-79, 28-7, 30-16, and 30-18.

GASKILL, PAMELA. 27-2 and 27-3.

GRANT, RICHARD E. 2-3, 3-4, 3-11, 3-17, 3-18, 3-20, 3-23, 3-28, 4-5, 4-48, 5-8, 6-6, 6-7, 8-3, 9-11, 9-18, 9-22, 9-26, 9-28, 9-33, 9-41, 9-42, 9-43, 10-3, 10-17, 10-20, 10-25, 10-26, 10-27, 10-28A, 10-30, 10-37, 10-38, 10-40, 10-41, 10-43, 10-49, 10-50, 10-52, 11-11, 11-12, 11-13, 11-14, 11-15, 11-17, 11-18, 11-19, 11-28, 11-31, 11-34, 11-35, 11-36, 11-37, 12-31, 12-33, 12-

15-10, 15-23, 15-26, 15-28, 15-37, 16-24, 16-53, 16-54, 16-55, 16-58, 17-8, 18-8, 18-14, 20-3, 20-4, 20-6, 20-12, 20-15, 20-17, 20-22, 20-24, 20-25, 20-40, 20-48, 20-49, 20-52, 25-9, 25-52, 25-58, 25-69, 25-81, 25-102B, 25-103, 25-106, 25-122, 25-123, 25-133, 25-134, 30-14, 34-9, 44-18, and 51-10.

WEBER, JULIUS. 1-4, 3-21, 3-22, 3-29, 4-7, 4-8B, 4-10, 4-18, 4-21, 4-25, 4-29, 4-32, 4-37, 4-38, 4-40, 4-50B, 4-52, 4-54, 4-55, 4-60, 4-61, 4-62, 4-67, 4-70, 5-10, 5-18, 5-22, 6-2, 9-12, 9-20, 9-38, 9-39, 10-34, 11-3, 11-5, 11-16, 11-22, 11-23, 11-24, 11-25, 12-2, 12-26, 12-58, 13-5, 13-11, 13-12, 13-19A, 13-20, 13-24, 13-25, 13-39, 13-44, 13-45, 13-48, 13-55, 13-56, 13-58, 13-59, 13-62, 13-70, 13-78, 13-80, 13-81, 13-82, 13-84, 13-85, 14-1, 14-3, 15-18, 15-27, 15-49, 16-19, 16-30, 16-31, 16-36, 16-47, 17-4, 18-21, 20-14, 20-18, 20-38, 20-42, 20-53, 20-61, 20-62, 20-72, 21-1, 21-2, 24-3, 24-5, 24-44, 24-49, 24-51, 24-52, 24-54A, 24-60, 24-79, 24-102E, 24-104, 24-108A, 24-109, 24-111B, 24-116, 25-8, 25-11, 25-14, 25-15, 25-16, 25-17, 25-18, 25-24, 25-27, 25-30, 25-32, 25-33, 25-48, 25-49, 25-57, 25-70, 25-72, 25-77, 25-85, 25-86, 25-108, 25-158, 25-159, 28-5, 30-6, 30-11, 31-13, 38-1A, 48-3A, 48-4, 49-17B, 50-2, and 52-2.

WEGNER, DEAN. 4-49, 13-87, 13-88, 13-89, 24-38, 24-48, and 24-54B.

WEIBE, MAIDI. 37-5, 40-2, 40-5, and 40-7.

WOUTERS, GEORGES. 31-17 and 31-18.

ZENNARO, PIERRE. 31-16.

ZIDEK, JIRKA. 24-53, 24-62, 24-63, 24-64, 24-65, and 24-66.

ZIMMERMANN, HELMUT. 4-59.

PALEONTOLOGICAL PUBLICATIONS

The following books and articles were used for reference in compiling some of the text. No recommendations are necessarily implied by this presentation, the author merely wishes to include these titles for those readers who may be interested in furthering their knowledge of paleontology.

Archer, J. 1976. *From whales to dinosaurs: the story of Roy Chapman Andrews.* St. Martins Press.

Ascher, J. C. and Y. G. Valy. *Fossils of all ages.* Grosset and Dunlap.

Ashby, W. L. 1979. *Fossils of Calvert Cliffs.* Calvert Marine Museum.

A short guide to the Natural History Museum. 1973. British Museum (Natural History), London.

Baird, D. 1962. "A Haplolepid fish fauna in the Early Pennsylvanian of Nova Scotia." *Paleontology.* April, Vol. 5, Part 1, pp. 1–29.

Baird, D. 1978. *Studies on Carboniferous freshwater fishes.* American Museum Novitates.

Baird, D. and G. R. Case. 1966. "Rare marine reptiles from the Cretaceous of New Jersey." *Journal of Paleontology.* Vol. 40, No. 5, pp. 1211–1215.

Bardack, D. and R. Zangerl. 1968. "First fossil lamprey: a record from the Pennsylvanian of Illinois." *Science.* Dec., Vol. 162, pp. 1265–1267, 13.

Bebout, D. G. 1963. *Desmoinesian fusulinids of Missouri.* Missouri Geological Survey.

Beerbower, J. R. 1971. *Field guide to fossils.* Houghton-Mifflin Co.

Berman, D. S. 1967. *Orientation of bradyodont dentition.* Journal of Paleontology.

Berry, W. B. N. and I. R. Satterfield. 1972. *Late Silurian graptolites from the Bainbridge Formation in southeastern Missouri.* Journal of Paleontology.

Beveridge, T. R. 1955. *An introduction to the geological history of Missouri.* Missouri Geological Survey.

Black, R. M. 1970. *The elements of Paleontology.* Cambridge University Press.

Bolt, J. 1976. "Pterosaur." *Field Museum Bulletin,* May, Vol. 47, No. 5.

Boreske, J. R., Jr. 1968. *Study of the association of some Pennsylvanian age invertebrates.* Earth Science.

Boreske, J. R., Jr., G. R. Case, and W. J. Hlavin. 1970. "Mutual association of marine and freshwater sharks from the Conemaugh Series of Athens County, Ohio. Abstract." *Ohio Academy of Sciences, Columbus meeting: April 17, 1970.*

Brain, C. K. and L. H. Brain. *How life arose in South Africa.* Cape Town Star.

Bridges, W. 1970. *The New York Aquarium book of the water world.* New York Zoological Society.

British Paleozoic fossils. 1969. British Museum (Natural History), London.

British Cenozoic fossils. 1971. British Museum (Natural History), London.

British Mesozoic fossils. 1972. British Museum (Natural History), London.

Brodkorb, P. 1955. *The avifauna of the Bone Valley Formation.* Florida Geological Survey.

Brouwer, A. 1967. *General Palaeontology.* University of Chicago Press.

Brues, C. T. 1939. *Fossil phoridae in Baltic amber.* Museum of Comparative Zoology (Harvard Press).

Brues, C. T., A. L. Melander, and F. M. Carpenter. 1954. *Classification of insects.* Museum of Comparative Zoology (Harvard Press).

Burton, J. 1976. *Fossils.* Grosset and Dunlap.

Cappetta, H. and G. R. Case. 1975. "Contribution a l'étude des sélaciens du Groupe Monmouth (Campanien-Maestrichtien) du New Jersey." *Palaeontographica-Stuttgart.* Abt. A., Bd. 151, pp. 1–46.

Cappetta, H. and G. R. Case. 1975. "Sélaciens nouveaux du Crétacé du Texas." *Géobios* (Lyon, France). No. 8, f. 4, pp. 303–307.

Carpenter, F. M. 1926. *Fossil insects from the Lower Permian of Kansas.* Museum of Comparative Zoology (Harvard Press).

Carpenter, F. M. 1969. *Adaptations among Paleozoic insects.* Proceedings of the North American Paleontological Convention.

Case, G. R. 1964. Shark tooth hunting along the Calvert Cliffs of Maryland. *Rocks and Minerals.* Vol. 39, No. 9–10, p. 467.

Case, G. R. 1965. "Fossil hunting in central New Jersey." *Rocks and Minerals.* Vol. 40, No. 3, pp. 214–215.

Case, G. R. 1965. Hunting fossils in Florida's phosphate pits. *Rocks and Minerals.* Vol. 40, No. 4, pp. 294–295.

Case, G. R. 1965. "An occurrence of the sawfish, *Onchopristis dunklei* in the Upper Cretaceous of Minnesota." *Journal of the Minnesota Academy of Science.* Vol. 32, No. 3, p. 183.

Case, G. R. 1966. "Localities for collecting teeth of the Miocene giant white shark." *Earth Science*. Vol. 19, No. 2, pp. 65–68.

Case, G. R. and W. A. Owens. 1966. "The mound-builders of the St. Johns River Valley, central Florida." *Chesopien*. Vol. 4, No. 1, pp. 6–12.

Case, G. R. 1966. "Cretaceous fossils of New Jersey." *Earth Science*. Vol. 19, No. 5, pp. 200–202.

Case, G. R. 1967. "Collecting fossil vertebrates in Florida." *Earth Science*. Vol. 20, No. 1, pp. 10–12.

Case, G. R. 1967. "Eocene fossils of the Aquia Formation of Virginia." *Earth Science*. Vol. 20, No. 5, pp. 211–214.

Case, G. R. 1967. *Fossil shark and fish remains of North America*. pp. 1–20.

Case, G. R. 1968. *Fossils illustrated*. pp. 1–32.

Case, G. R. 1968. "The fossil fishes of Granton quarry (New Jersey)." *Rocks and Minerals*. Vol. 43, No. 3, pp. 169–172.

Case, G. R. 1968. "Inland shark occurrence." *Bulletin of the American Littoral Society,* Vol. 5, No. 1, pp. 20–21 and 37.

Case, G. R. 1968. "A cladodont shark from the Pennsylvanian of Nebraska." *Earth Science*. Vol. 21, No. 6, p. 266.

Case, G. R. and H. Densmore. 1970. "Report of a beached bottlenosed whale on the shores of Cobequid Bay, Nova Scotia, Canada." *Bulletin of the American Littoral Society*. Vol. 6, No. 3, pp. 18–20.

Case, G. R. 1970. "New species of fossil catfish found." *Earth Science*. Vol. 23, No. 6, p. 285.

Case, G. R. 1970. "A fossil shark in Nebraska." *Bulletin of the American Littoral Society*. Vol. 6, No. 4, pp. 29–30.

Case, G. R. 1970. "The occurrence of *Petrodus* and other fossil shark remains in the Pennsylvanian of Iowa." *Annals of Iowa*. Vol. 40, No. 6, 3rd series, pp. 445–449.

Case, G. R. 1971. *Fossils illustrated* (2nd edition). pp. 1–32.

Case, G. R. and C. Bigbie. 1972. "Gene Wilson: obituary." *Bulletin of the Society of Vertebrate Paleontology*. No. 94, pp. 81–82.

Case, G. R. and M. K. Braun. 1972. Capture of a chimaera in False Bay, South Africa. Bulletin of the American Littoral Society. Vol. 7, no. 3. pp. 28–30 and 48.

Case, G. R. 1972. *Handbook of fossil collecting*. pp. 1–64.145 figs.

Case, G. R. 1973. *Fossil sharks: a pictorial review*. pp. 1–64. 257 figs.

Case, G. R. 1973. "L'arte di collezionare fossili." *Nota Gruppo Min. Firenze, Italia*. Ann. 1, No. 2, pp. 11–12.

Case, G. R. and J. Herman. 1973. "Une épine dorsale du chiméroïde *Edaphodon* cfr. *bucklandi* (Agassiz) dans l'Yprésien du Maroc." *Bulletin Societe belge Geol., Paleont. et Hydrologie*. T. 82, f. 3, pp. 445–449. Brussels, Belgium.

Case, G. R. 1974. "Fossil collecting in Morocco, North Africa." *Earth Science*. Vol. 27, No. 1, pp. 14–22.

Case, G. R. 1974. "Denti fossili di Squalo." *Nota Gruppo Mineralogie Firenze, Italia*. Ann. 2, no. 2. pp. 14–15.

Case, G. R. 1975. "Shark's teeth." *Outdoors in Georgia*. Vol. IV, No. 3, pp. 4–9.

Case, G. R. 1975. "L'evoluzione del Pescecane —Part I: Il periodo Carbonifero (Devonian/Mississippiano/Pennsylvaniano)." *Nota Gruppo Min. Fiorenze, Italia*. Ann. 2, No. 3, pp. 20–22.

Case, G. R. 1975. "L'evoluzione del Pescecane—Part II: Riguarda importanti forme di Pesci simili al pescecane del Carbonifero: ed una discussione sui nuovi pesci: iniopterygian (un nuovo ordine di Chondrichthyans)." *Nota Gruppo Fiorentino Minerali, Italia*. Ann. 2, No. 4, pp. 18–22.

Case, G. R. 1976. "L'evoluzione del Pescecane—Part III: Permiano/Triassico/Giurassico Gli inizi dei giorni nostri." *Nota Gruppo Fiorentino Minerali d'Italia*. Ann. 3, No. 1, pp. 19–23.

Case, G. R. 1976. "L'evoluzione del Pescecane—Part IV: Continuazione." *Nota Gruppo Minerali di Fiorenze, Italia*. Ann. 3, No. 3, pp. 16–19.

Case, G. R. 1976. "L'eyoluzione del Pescecane—Part V: I moderni pescecani dei nostri Oceani." *Nota Gruppo Fior. Min. d'Italia*. Ann. 3, No. 4, pp. 15–19.

Case, G. R. 1977. "Mosasaurs in the catacombs, a visit to Temple Eben-Ezer, Belgium." *Bulletin of the Society of Vertebrate Paleontology*. No. 110, June, p. 45.

Case, G. R. 1978. "*Ischyodus bifurcatus*, a new species of Chimeroid fish from the Upper Cretaceous of New Jersey." *Géobios* (Lyon, France). Feb., No. 11, f. 1, pp. 21–29.

Case, G. R. 1978. "A new selachian fauna from the Judith River Formation (Campanian) of Montana." *Paleontographica* (Stuttgart, West Germany). Abt. A, Bd. 160, May, pp. 176–205.

Case, G. R. 1978. "Edgard Casier: an obituary." *Bulletin of the Society of Vertebrate Paleontology*. No. 113, June, pp. 49–50.

Case, G. R. 1979. "Philately: fossils on stamps." *TV News (Hudson County, New Jersey)*. March 17–23, p. 16.

Case, G. R. 1979. "Collecting shark's teeth at Big Brook, Monmouth County, New Jersey." *Bulletin of the Bergen County Mineralogical and Paleontological Society,* Vol. 13, No. 5, May, pp. 12–14.

Case, G. R. 1979. "Additional fish records from the Judith River Formation (Campanian) of Montana." *Géobios* (Lyon, France). No. 12, f. 2, April, pp. 223–233.

Case, G. R. 1979. "Cretaceous selachians from the Peedee Formation (Late Maestrichtian) of Duplin County, North Carolina." *Brimleyana*. No. 2, Nov., pp. 77–89.

Case, G. R., V. R. Koestler, and R. J. Koestler. 1980. "The shortchanged shark." *Bulletin of the American Littoral Society (Underwater Naturalist)*. Vol. 12, No. 2, pp. 10–14.

Case, G.R. 1980. "A selachian fauna from the Trent Formation, Lower Miocene (Aquitanian) of Eastern North Carolina." *Palaeontographica* (Stuttgart, West Germany), Abt. A, Bd. 171, Nov., pp. 75–103.

Caster, K. E. "Two siphonophores from the Paleozoic." *Palaeontographica Americana*.

Cavender, T. M. and G. R. Case. 1973. "A new long-snouted chondrosotean fish from the Upper Pennsylvanian of New Mexico." *Abstr. Soc. Vert. Paleo.*, Dallas, Texas, Nov.

Charig, A. 1979. *A new look at the dinosaurs*. Mayflower Books.

Colbert, E. H. 1965. *The age of reptiles*. The Norton Library.

Compagno, L. J. V. 1973. "Interrelationships of living elasmobranchs." In: *Interrelationships of fishes*. Zoological Journal-Linnean Society, London.

Condit, C. 1957. *Fossils of Illinois*. Illinois State Museum.

Copeland, C. W., Jr. 1963. *Curious creatures in Alabama rocks*. Alabama Geological Survey.

Cox, B. 1970. *Prehistoric animals*. Bantam Books.

Croneis, C. and H. L. Geis. 1940. "Microscopic pelmatozoa: Part 1, ontogeny of the Blastoidea." *Journal of Paleontology*.

Crossing the bridge of time. 1979. The Texaco Star.

Davis, R. A. and H. A. Semken, Jr. 1975. "Fossils of uncertain affinity from the Upper Devonian of Iowa." *Science.*

Dimes, F. G. 1979. *Fossil collecting.* E. P. Publishing Ltd., West Yorkshire, England.

Downs, T. 1968. *Fossil vertebrates of Southern California.* University of California Press.

Duluk, C. E. 1965. "Fossil fauna of the Silica Formation (Ohio and Michigan)." *Earth Science.*

Dunbar, C. O. and K. M. Waage. 1969. *Historical Geology.* John Wiley and Sons, Inc.

Eastman, C. R. 1898. "Dentition of Devonian Ptyctodontidae." *American Naturalist.*

Eastman, C. R. 1903. "Carboniferous fishes from the central western States." *Bulletin of the Museum of Comparative Zoology (Harvard).*

Eastman, C. R. 1904. "Descriptions of Bolca fishes." *Bulletin of the Museum of Comparative Zoology (Harvard).*

Edwards, W. N. 1967. *The early history of Paleontology.* British Museum (Natural History), London.

Fenton, C. L. and M. A. Fenton. 1932. "A new species of *Cliona* from the Cretaceous of New Jersey." *American Midland Naturalist.*

Fenton, C. L. and M. A. Fenton. 1958. *The fossil book.* Doubleday and Company.

Fenton, C. L. 1962. "New Jersey's geologic past." *New Jersey State Museum Bulletin.*

Foster, M. W. 1979. "Soft-bodied coelenterates in the Pennsylvanian of Illinois." In: *Mazon Creek fossils.* Academic Press, New York.

Fritsch, A. 1890. *Fauna der Gaskohle und der Kalksteine der Permformation Bohems.* Prague.

Fritz, W. H. 1969. "Geological setting of the Burgess shale." *Proceedings of the North American Paleontological Convention, Chicago.*

Gillespie, W. H., J. A. Clendening, and H. J. Pfefferkorn. 1978. *Plant fossils of West Virginia.* West Virginia Geological Survey.

Glut, D. F. 1972. *The dinosaur dictionary.* Citadel Press.

Grande, L. 1980. "Paleontology of the Green River Formation, with a review of the fish fauna." *Geological Survey of Wyoming.* Bull. #63.

Haddow, J. G. 1891. *Amber—all about it.* Cope's Tobacco Plant, Liverpool, England.

Hager, M. W. 1970. *Fossils of Wyoming.* Wyoming Geological Survey.

Halstead, B. 1975. *The world of dinosaurs.* Derrydale Books.

Hamilton, W. R., A. R. Wooley, and A. C. Bishop. 1977. *Minerals, rocks, and fossils.* Larousse and Co.

Hardt, H. 1954. *Der Bernstein (Amber).* A. Ziemsen Verlag, Lutherstadt, Germany.

Harksen, J. C. and J. R. Macdonald. 1969. *Type sections for the Chadron and Brulé Formations.* South Dakota Geological Survey.

Hartman, W. D. and T. F. Goreau. 1970. "Jamaican coralline sponges: their morphology, ecology and fossil relatives." *Symposium. Zoological Society of London.*

Hauk, J. K. 1969. *Badlands: its life and landscape.* National Parks Service, Washington, D.C.

Hogberg, R. K., R. E. Sloan, and S. Tufford. 1965. *Guide to fossil collecting in Minnesota.* University of Minnesota.

Hoskins, D. M. 1969. *Fossil collecting in Pennsylvania.* Pennsylvania Geological Survey.

Howard, R. W. 1975. *The dawnseekers.* Harcourt Brace Javonovich.

Ice age mammals and the emergence of man. Elephant Press. (An exhibit at the National Museum of Natural History-Smithsonian Institution, Washington, D.C.)

Jeppsson, L. 1979. *Conodonts.* Sveriges Geologiska Undersökning, Stockholm.

Jeppsson, L. 1979. *Conodont element function.* Lethaia, Oslo, Norway.

Johannesen, R. 1974. "Once in a lifetime." *Earth Science.*

Johnson, H. and J. E. Storer. 1974. *A guide to Alberta vertebrate fossils from the age of dinosaurs.* Provincial Museum of Alberta, Canada.

Johnson, R. G. and E. S. Richardson, Jr. 1969. "The morphology and affinities of *Tullimonstrum.*" *Fieldiana/Geology.*

Jordan, H. 1967. *Trilobiten.* A. Ziemsen Verlag, Lutherstadt, Germany.

Keefer, W. R. 1972. *The geologic story of Yellowstone National Park.* United States Geological Survey, Denver.

Kjellesvig-Waering, E. N. 1969. "Scorpionida: the holotype of *Mazonia woodiana* Meek and Worthen." *Fieldiana/Geology.*

Koken, E. 1907. *Ueber* Hybodus. Gustav Fischer Verlag, Jena, Germany.

Kuhn, O. 1961. *Die tierwelt der Bundenbacher schiefer.* A. Ziemsen Verlag, Lutherstadt, Germany.

Kuhn, O. 1966. *Die tierwelt des Solnhofener Schiefers.* A. Ziemsen Verlag, Lutherstadt, Germany.

Kurtén, B. 1968. *The age of dinosaurs.* McGraw Hill Co.

Lambert, D. 1978. *Dinosaurs.* Crown Publishers.

Lane, N. G. 1969. "A crinoid from the Pennsylvanian Essex fauna of Illinois." *Fieldiana/Geology.*

LaRocque, A. and M. F. Marple. 1955. *Ohio fossils.* Ohio Geological Survey, Columbus, Ohio.

Lasmanis, R. 1971. "Upper Devonian fossil locality." *Earth Science.*

Lasmanis, R. and G. R. Case. 1973. "Fossil collecting at Glenns Ferry, Idaho." *Earth Science,* Vol. 26, No. 6, pp. 299–303.

Lieber, W. 1961. "Mineral and fossil areas in Germany." *Rocks and Minerals.*

Life before Man. 1972. Time-Life Books, Chicago.

Lillegraven, J. A., Z. Kielan-Jaworowska, and W. A. Clemens. 1979. *Mesozoic mammals: the first two-thirds of mammalian history.* University of California Press.

Lundberg, J. G. and G. R. Case. 1970. "A new catfish from the Eocene Green River Formation, Wyoming." *Journal of Paleontology,* Vol. 44, No. 3, pp. 451–457.

Macfall, R. P. and J. C. Wollin. 1972. *Fossils for amateurs.* Van Nostrand Reinhold Co., New York.

McGrew, P. O. and M. Casilliano. 1975. *The geologic history of Fossil Butte National Monument and Fossil Basin.* National Parks Service.

McKerrow, W. S. (editor). *The ecology of fossils.* M. I. T. Press.

Man, J. 1978. *The day of the dinosaur.* Bison Books Ltd.

Marks, E. 1972. "Fossils of Pit 14, Grundy County, Illinois."

Earth Science.

Matthews, W. H., III. 1962. *Fossils: an introduction to prehistoric life.* Barnes and Noble.

Mayr, E. 1978. "Evolution." *Scientific American.*

Mehl, M. G. 1962. *Missouri's ice age animals.* Missouri Geological Survey.

Mehl, M. G. 1966. *The Grundel Mastodon.* Missouri Geological Survey.

Moodie, R. L. 1933. *A popular guide to the nature and the environment of the fossil vertebrates of New York.* University of the State of New York.

Moody, R. *The fossil world.* Chartwell Books.

Moody, R. 1977. *A natural history of dinosaurs.* Chartwell Books.

Moore, R. C., C. G. Lalicker, and A. G. Fischer. 1952. *Invertebrate Fossils.* McGraw-Hill Book Co.

Motz, L. 1979. *Rediscovery of the Earth.* Van Nostrand Reinhold Co., New York.

Moy-Thomas, J. A. 1939. *Palaeozoic fishes.* W. B. Saunders Co.

Müller, A. H. and H. Zimmermann. 1962. *Aus Jahrmillionen: tiere der vorzeit.* Veb Gustav Fischer Verlag, Jena, East Germany (GDR).

Murphy, D. 1977. *Lewis and Clark—voyage of discovery.* KC Publications.

Murray, M. 1967. *Hunting for fossils.* Collier Books.

Nitecki M. H. (editor). 1979. *Mazon Creek fossils.* Academic Press, New York.

Nitecki, M. H. and E. S. Richardson, Jr. 1972. "A new hydrozoan from the Pennsylvanian of Illinois." *Fieldiana/Geology.*

Obruschev, D. 1952–53. "*Helicoprion.*" *Academy of Sciences, U.S.S.R.*

O'Harra, C. C. 1920. *The White River Badlands.* South Dakota School of Mines, Rapid City, South Dakota.

Olsen, S. J. 1959. *Fossil mammals of Florida.* Florida Geological Survey, Tallahassee, Florida.

Osborn, J. W. and A. W. Crompton. 1973. "The evolution of mammals from reptile dentitions." *Breviora.*

Ostrom, J. H. 1978. "A new look at dinosaurs." *National Geographic.*

Pajaud, D. *Marvellous world of fossils.* Abbey Library, London.

Palmer, E. L. 1971. *Fossils.* Palaeontological Research Institute, Ithaca, New York.

Palmer, K. and D. Brann. 1966. *Illustrations of fossils in the Ithaca area.* Paleontological Research Institute, Ithaca, New York.

Parris, D. C. and G. R. Case. 1980. "*Castoroides* from New Jersey: possible association with artifacts reexamined." *Bulletin of the Archaeological Society of New Jersey.* Vol. 30, Bull. 36, pp. 22–24.

Pellegrino, C. R. 1978. "Life in an Upper Cretaceous sea." *Earth Science.*

Perry, T. G. 1959. *Fossils: prehistoric animals in hoosier rocks.* Indiana Geological Survey.

Planet earth. 1975. Doubleday and Company, Garden City, N.Y.

Rader, E. K. 1964. *Guide to fossil collecting in Virginia.* Virginia Geological Survey.

Raymond, P. E. 1914. "Notes on the ontogeny of *Paradoxides* with the description of a new species from Braintree, Massachusetts." *Bulletin of the Museum of Comparative Zoology (Harvard).*

Raymond, P. E. 1914. "Notes on the ontogeny of *Isotelus gigas* Dekay." *Bulletin of the Museum of Comparative Zoology (Harvard).*

Raymond, P. E. 1916. *New and old Silurian trilobites.* Museum of Comparative Zoology (Harvard Press).

Raymond, P. E. 1920. *Some new Ordovician trilobites.* Museum of Comparative Zoology (Harvard Press).

Raymond, P. E. 1925. *Some trilobites of the Lower Middle Ordovician of Eastern North America.* Museum of Comparative Zoology (Harvard).

Rhodes, F. H. T., H. S. Zim, and P. R. Schaeffer. 1962. *Fossils: a guide to prehistoric life.* Golden Books.

Rice, P. C. 1980. *Amber, the golden gem of the ages.* Van Nostrand Reinhold Co., New York.

Richards, H. G. 1953. *Record of the rocks.* Ronald Press.

Richards, H. G. 1958. *The Cretaceous fossils of New Jersey. Part 1.* New Jersey Geological Survey.

Richards, H. G. 1962. *The Cretaceous fossils of New Jersey. Part 2.* New Jersey Geological Survey.

Richardson, E. S., Jr., and R. G. Johnson. 1969. "The Mazon Creek fossils." *Proceedings of the North American Paleontological Convention.*

Riek, E. F. "Lower Cretaceous fleas." *Nature.*

Romer, A. S. 1966. *Vertebrate Paleontology* (3rd edition). University of Chicago Press.

Romer, A. S. 1968. *Notes and comments on vertebrate paleontology.* University of Chicago Press.

Rose, J. N. 1967. *Fossils and rocks of eastern Iowa.* Iowa Geological Survey.

Rothe, H. W. 1969. *Les fossiles.* Librairie Payot, Lausanne, Switzerland.

Sass, D. B. and B. N. Rock. 1975. "The genus *Plumalina* Hall, 1858 (Coelenterata) reexamined." *Bulletin of American Paleontology.*

Schaeffer, B. 1967. "Comments on elasmobranch evolution." In: *Sharks, skates and rays.* Johns Hopkins Press, Baltimore.

Shaver, R. H. 1959. *Adventures with fossils.* Indiana Geological Survey.

Schimer, H. W. and R. R. Shrock. 1944. *Index fossils of North America.* M. I. T. Press.

Stearn, C. W. 1972. "The relationship of the stromatoporoids to the sclerosponges." *Lethaia,* Oslo, Norway.

Stevens, G. R., N. de B. Hornibrook, and W. F. Harris. 1966. *New Zealand fossils.* New Zealand Geological Survey, Lower Hutt, New Zealand.

Stokes, W. L. 1978. "Transported fossil biota of the Green River Formation, Utah." *Palaeogeography, Palaeoclimatology, and Palaeoecology.* Amsterdam, the Netherlands.

Stone, J. L. 1963. *The oldest forest and the Naples tree.* New York State Museum and Science Service, Albany, New York.

Stucker, G. F. 1966. "Mountain of stone fishes." *National Parks Magazine,* Washington, D.C.

Stürmer, W. and J. Bergström. 1973. "New discoveries on trilobites by X-rays. "*Paläont. Z.* (Stuttgart, West Germany).

Stürmer, W. and J. Bergström. 1976. "The arthropods *Mime-*

taster and *Vachonisia* from the Devonian Hunsrück Shale." *Paläont. Z.* (Stuttgart, West Germany).

Stürmer, W. and J. Bergström. 1978. "The arthropod *Cheloniellon* from the Devonian Hunsrück shale." *Paläont. Z.* (Stuttgart, W. Germany).

Sullivan, W. 1974. *Continents in motion.* McGraw-Hill. Co. *The encyclopedia of prehistoric life.* 1979. McGraw-Hill.

Thompson, T. L. 1967. *Conodont zonation of Lower Osagean rocks (Lower Mississippian) of southwestern Missouri.* Missouri Geological Survey.

Thompson, T. L. 1972. *Conodont biostratigraphy of Chesterian strata in southwestern Missouri.* Missouri Geological Survey.

Tidwell, W. D. 1975. *Common fossil plants of western North America.* Brigham Young University Press.

Unklesbay, A. G. 1973. *The common fossils of Missouri.* University of Missouri Press.

Vokes, H. E. 1957. *Fossils of Maryland.* Maryland Geological Survey.

Wanner, H. E. 1926. "Some additional faunal remains from the Triassic of York County, Pennsylvania." *Proceedings of the Philadelphia Academy of Sciences.*

Weinman, P. L. 1966. *An introduction to invertebrate fossils of New York.* University of the State of New York.

West, S. 1979. "Dinosaur head hunt." *Science News.* November 3, Vol. 116.

Whittington, H. B. 1969. "The Burgess Shale: history of research and preservation of fossils." *Proceedings of the North American Paleontological Convention.*

Williams, M. and K. Elbaum. 1973. "Bendix-Almgreen's recent investigations on *Helicoprion,* etc." *Journal of Insignificant Research.*

Williams, R. B. 1975. *Ancient life found in Kansas rocks.* Kansas Geological Survey.

Witzke, B. 1972. "The Badlands and their fossils." *Earth Science.*

Wolfe, P. E. 1977. *The geology and landscapes of New Jersey.* Crane Russak Co.

Wollin, J. C. 1965. "The Mazon Creek fossils." *Earth Science.*

Yolton, J. S. 1965. *Fossils of New Jersey.* Geological Survey of New Jersey.

Zangerl, R. S. and E. S. Richardson, Jr. 1963. "The paleoecological history of two Pennsylvanian black shales. *Fieldiana/Geology Memoir.*

Zangerl, R. and G. R. Case. 1973. "Iniopterygia, a new order of chondrichthyan fishes from the Pennsylvanian of North America." *Fieldiana/Geology Memoirs.*

Zangerl, R. and G. R. Case. 1976. "*Cobelodus aculeatus* (Cope), an anacanthous shark from the Pennsylvanian black shales of North America." *Palaeontographica-Stuttgart.* Abt. A, Bd. 154, pp. 107–157.

Zangerl, R. 1969. "*Bandringa rayi,* a new ctenacanthoid shark from the Pennsylvanian Essex fauna of Illinois." *Fieldiana/Geology.*

Zangerl, R. 1979. "New chondrichthyes from the Mazon Creek fauna. In: *Mazon Creek fossils.* Academic Press, New York.

Zangerl, R. 1981. "Paleozoic elasmobranchs." *Handbook of Palaeoichthyology.* Gustav Fischer Verlag, Stuttgart, West Germany.

Zittel, K. A., von. 1932. *Textbook of Paleontology. Vol. II: Vertebrates.* Macmillan and Co.

INDEX

The following is an alphabetical listing of generic names (and some synonyms) of the fossils illustrated in this volume.

Along with these generic names, there is a listing and crossreferencing of the general subject material contained within the book.

The listings are by chapter figure numbers and not by pagination.